SPRINGER STUDY EDITION

Springer
Berlin
Heidelberg
New York
Barcelona
Budapest
Hong Kong
London
Milan
Paris
Santa Clara
Singapore
Tokyo

Hans Lüth

Surfaces and Interfaces of Solid Materials

Third Edition

With 359 Figures and 12 Tables

Professor Dr. Hans Lüth
Institut für Schicht- und Ionentechnik
Forschungszentrum Jülich GmbH
D-52425 Jülich, Germany
and
Rheinisch-Westfälische Technische Hochschule
D-52062 Aachen, Germany

Library of Congress Cataloging-in-Publication Data.

Lüth, H. (Hans)
Surfaces and interfaces of solid materials / Hans Lüth. – 3rd ed., updated print. p. cm.
"Springer study edition."
Includes bibliographical references and index.
ISBN 3-540-58576-1 (softcover: alk. paper)
1. Surfaces (Physics) 2. Solids–Surfaces. I. Title.
QC173.4.S94L88 1996
530.4'17–dc20 96-38355

Third Edition 1995
Updated Printing 1997

ISBN 3-540-58576-1 3rd Edition Springer-Verlag Berlin Heidelberg New York

ISBN 3-540-56840-9 2nd Edition Springer-Verlag Berlin Heidelberg New York

Typesetting: PS™ Technical Word Processor

SPIN: 10543123 54/3144-5 4 3 2 1 0 - Printed on acid-free paper

Preface

This edition has been revised for still better readibility, but no main changes within the main text were made since it covers the material comprehensively to my feeling. In response to several suggestions from students and colleagues typograpical errors were eliminated and some improvements made for reasons of clarity. Furthermore, some chapters were updated, and more recent references included. The ability to controllably modify solid-state surfaces on a nanometer scale by means of scanning-probe techniques such as scanning-tunneling and atomic-force microscopy has had an increasing impact on fundamental research, also with respect to future nanoelectronics. This research field is a further example that surface and interface science forms the basis for many noval exciting developments in solid-state phasics. Correspondingly a short addition about this topic was made to Panel VI.

The book is increasingly used in connection with university courses and examinations. Therefore, I have revised and supplemented the problem sections at the end of each chapter. Students will certainly benefit from solving these problems.

For the pleasant collaboration during the final production I thank Gertrud Dimler of Springer Verlag; many thanks are due to Helmut Lotsch for his continuous support of this text and his involment in this present edition.

Aachen and Jülich
October 1996

Hans Lüth

Preface to the Second Edition

Surface and interface physics has in recent decades become an ever more important subdiscipline within the physics of condensed matter. Many phenomena and experimental techniques, for example the quantum Hall effect and photoemission spectroscopy for investigating electronic band structures, which clearly belong to the general field of solid-state physics, cannot be treated without a profound knowledge of surface and interface effects. This is also true in view of the present general development in solid-state research, where the quantum physics of nanostructures is becoming increasingly relevant. This also holds for more applied fields such as microelectronics, catalysis and corrosion research. The more one strives to obtain an atomic-scale understanding, and the greater the interest in microstructures, the more surface and interface physics becomes an essential prerequisite.

In spite of this situation, there are only a very few books on the market which treat the subject in a comprehensive way, even though surface and interface physics has now been taught for a number of years at many universities around the world. In my own teaching and research activities I always have the same experience: when new students start their diploma or PhD work in my group I can recommend to them a number of good review articles or advanced monographs, but a real introductory and comprehensive textbook to usher them into this fascinating field of modern research has been lacking.

I therefore wrote this book for my students to provide them with a text from which they can learn the basic models, together with fundamental experimental techniques and the relationship to applied fields such as microanalysis, catalysis and microelectronics.

This textbook on the physics of surfaces and interfaces covers both experimental and theoretical aspects of the subject. Particular attention is paid to practical considerations in a series of self-contained panels which describe UHV technology, electron optics, surface spectroscopy and electrical and optical interface characterisation techniques. The main text provides a clear and comprehensive description of surface and interface preparation methods, structural, vibrational and electronic properties, and adsorption and layer growth. Because of their essential role in modern microelectronics, special emphasis is placed on the electronic properties of semiconductor interfaces and heterostructures. Emphasizing semiconductor microelectronics as one of the major applications of interface physics is furthermore justified by the fact that here the gap between application and basic research is small, in contrast, for example, with catalysis or corrosion and surface-protection research.

The book is based on lectures given at the Rheinisch-Westfälische Technische Hochschule (RWTH) Aachen and on student seminars organized with my colleagues Pieter Balk, Hans Bonzel, Harald Ibach, Jürgen Kirchner, Claus-Dieter Kohl and Bruno Lengeler. I am grateful to these colleagues and to a number of students participating in these seminars for their contributions and for the nice atmosphere during these courses. Other valuable suggestions were made by some of my former doctoral students, in particular by Arno Förster, Monika Mattern-Klosson, Richard Matz, Bernd Schäfer, Thomas Schäpers, Andreas Spitzer and Andreas Tulke. For her critical reading of the manuscript, as well as for many valuable contributions, I want to thank Angela Rizzi.

The English text was significantly improved by Angela Lahee from Springer Verlag. For this help, and also for some scientific hints, I would like to thank her. For the pleasant collaboration during the final production of the book I thank Ilona Kaiser. The book would not have been finished without the permanent support of Helmut Lotsch; many thanks to him as well.

Last, but not least, I want to thank my family who missed me frequently, but nevertheless supported me patiently and continuously during the time in which I wrote the book.

Aachen and Jülich *Hans Lüth*
October 1992

Contents

1. Surface and Interface Physics: Its Definition and Importance

A solid interface is defined as a small number of atomic layers that separate two solids in intimate contact with one another, where the properties differ significantly from those of the bulk material it separates. A metal film deposited on a semiconductor crystal, for example, is thus separated by the semiconductor-metal interface from the bulk of the semiconductor.

The surface of a solid is a particularly simple type of interface, at which the solid is in contact with the surrounding world, i.e., the atmosphere or, in the ideal case, the vacuum. The development of modern interface physics is thus basically determined by the theoretical concepts and the experimental tools being developed in the field of surface physics, i.e., the physics of the simple solid-vacuum interface. Surface physics itself has meanwhile become an important branch of microscopic solid-state physics, even though its historical roots lie both in classical bulk solid-state physics and physical chemistry, in particular the study of surface reactions and heterogeneous catalysis.

Solid-state physics is conceptually an atomic physics of the condensed state of matter. According to the strength of chemical bonding, the relevant energy scale is that between zero and a couple of electron volts. The main goal consists of deriving an atomistic description of the macroscopic properties of a solid, such as elasticity, specific heat, electrical conductance, optical response or magnetism. The characteristic difference from atomic physics stems from the necessity to describe a vast number of atoms, an assembly of about 10^{23} atoms being contained in 1 cm^3 of condensed matter; or the 10^8 atoms that lie along a line of 1 cm in a solid. In order to make such a large number of atoms accessible to a theoretical description, new concepts had to be developed in bulk solid-state physics. The translational symmetry of an ideal crystalline solid leads to the existence of phonon dispersion branches or the electronic band structure and the effective mass of an electron. Because of the large number of atoms involved, and because of the difference between the macroscopic and the atomic length scale, most theoretical models in classic solid-state theory are based on the assumption of an infinitely extended solid. Thus, in these models, the properties of the relatively small number of atoms forming the surface of the macroscopic solid are neglected. This simplifies the mathematical description considerably. The infinite translational symmetry of the idealized crystalline solid allows the application of a number of symmetry operations, which makes a handy mathematical treatment possible. This description of the solid in terms of an infinitely extended object, which neglects the properties of the few different atomic layers at the surface, is a good approximation for deriving macroscopic properties that depend on the total number of atoms contained in this solid. Furthermore, this description holds for all kinds of spectroscopic experiments, where the probes (X rays, neutrons, fast electrons, etc.) penetrate deep into the solid material and where the effect of the relatively few surface atoms ($\approx 10^{15}\,cm^{-2}$) can be neglected.

The approach of classical solid-state physics in terms of an infinitely extended solid becomes highly questionable and incorrect, however, when probes are used which "strongly" interact with solid matter and thus penetrate only a couple of Ångstroms into the solid, such as low-energy electrons, atomic and molecular beams, etc. Here the properties of surface atoms, being different from those of bulk atoms, become important. The same is true for spectroscopies where the particles detected outside the surface originate from excitation processes close to the surface. In photoemission experiments, for example, electrons from occupied electronic states in the solid are excited by X rays or UV light; they escape into the vacuum through the surface and are analysed and detected by an electron spectrometer. Due to the very limited penetration depth of these photoelectrons ($5 \div 80$ Å depending on their energy)[1] the effect of the topmost atomic layers below the surface cannot be neglected. The photoelectron spectra carry information specific to these topmost atomic layers. Characteristic properties of the surface enter the theoretical description of a photoemission experiment (Panel XI: Chap.6). Even when bulk electronic states are studied, the analysis of the data is done within the framework of models developed in surface physics. Furthermore, in order to get information about intrinsic properties of the particular solid, the experiment has to be performed under Ultra-High Vacuum (UHV) conditions on a freshly prepared clean sample surface. Because of the surface sensitivity, the slightest contamination on the surface would modify the results.

The concepts of surface and interface physics are important in solid-state physics not only in connection with special experimental tools, but also for certain physical systems. A thin solid film deposited on a substrate is bounded by a solid-solid interface and by its surface (film-vacuum interface). The properties of such a thin film are thus basically determined by the properties of its two interfaces. Thin-film physics cannot be reduced to the concepts of bulk solid-state physics, but instead the models of interface physics have to be applied. Similarly, the physics of small atomic clusters,

---

[1] The symbol $\div$ is used throughout the text as a shorthand for "from - to" or "between".

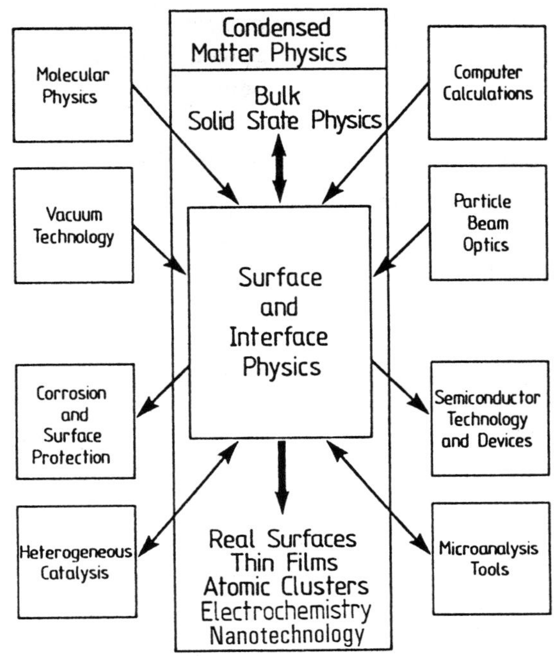

Fig.1.1. Interrelation of surface and interface physics as a subdiscipline of condensed-matter physics with other research fields

which often possess more surface than "bulk" atoms, must take into account the results from surface physics.

Surface and interface physics, as a well-defined sub-discipline of general condensed-matter physics, is thus interrelated in a complex way with a number of other research fields (Fig.1.1). This is particularly true if one considers the input from other domains of physics and chemistry and the output into important fields of application such as semiconductor electronics and the development of new experimental equipment and methods. The scheme in Fig.1.1 emphasises the way in which surface and interface physics is embedded in the general field of condensed-matter physics, as well as the strong impact of the models of bulk solid-state physics (phonon dispersion, electronic bands, transport mechanisms, etc.) on the concepts of interface physics.

On the other hand, within general solid-state physics, interface physics provides a deeper understanding of the particular problems related to the real surfaces of a solid and to thin films, dealing with both their physical properties and their growth mechanisms. The physics of small atomic clusters also benefits from surface physics, as does the wide field of electrochemistry, where the reaction of solid surfaces with an ambient electrolyte is the central topic. Furthermore, the new branch of nanotechnology, i.e.

engineering on a nanometer scale (Panel VI: Chap.3), which has emerged as a consequence of the application of scanning tunneling microscopy and related techniques, uses concepts that have largely been developed in surface sciences.

Modern surface and interface physics would not have been possible without the use of results from research fields other than bulk solid-state physics. From the experimental viewpoint, the preparation of well-defined, clean surfaces, on which surface studies are usually performed, became possible only after the development of UHV techniques. Vacuum physics and technology had a strong impact on surface physics. Surface sensitive spectroscopies use particles (low-energy electrons, atoms, molecules, etc.) because of their "strong" interaction with matter, and thus the development of particle beam optics, spectrometers and detectors is intimately related to the advent of modern surface physics. Since adsorption processes on solid surfaces are a central topic in surface physics, not only the properties of the solid substrate but also the physics of the adsorbing molecule is an ingredient in the understanding of the complex adsorption process. The physics and chemistry of molecules also plays an essential role in many questions in surface physics. Last, but not least, modern surface and interface physics would never have reached the present level of theoretical understanding without the possibility of large and complex computer calculations. Many calculations are much more extensive and tedious than in classical bulk solid-state physics since, even for a crystalline solid, a surface or interface breaks the translational symmetry and thus considerably increases the number of equations to be treated (loss of symmetry).

From the viewpoint of applications, surface and interface physics can be considered as the basic science for a number of engineering branches and advanced technologies. A better understanding of corrosion processes, and thus also the development of surface protection methods, can only be expected on the basis of surface studies. Modern semiconductor device technology would be quite unthinkable without research on semiconductor surfaces and interfaces. With an increasing trend towards greater miniaturization (large-scale integration) surfaces and interfaces become an increasingly important factor in the functioning of a device. Furthermore, the preparation techniques for complex multilayer device structures – Molecular Beam Epitaxy (MBE), metal organic MBE (Chap.2) – are largely derived from surface-science techniques. In this field, surface science research has led to the development of new technologies for semiconductor-layer preparation.

An interdependence between surface physics and applied catalysis research can also be observed. Surface science has contributed much to a deeper atomistic understanding of important adsorption and reaction mechanisms of molecules on catalytically active surfaces, even though practical heterogeneous catalysis occurs under temperature and pressure conditions totally different from those on a clean solid surface in a UHV vessel. On the other hand, the large amount of knowledge derived from classical catalysis studies under less well-defined conditions has also influenced surface science research on well-defined model systems. A similar interdependence exists between surface physics and the general field of applied microanalysis. The demand for extremely surface-sensitive probes in surface and interface physics has had an enormous impact on the development and improvement of new particle spectroscopies. Auger Electron Spectroscopy (AES), Secondary Ion Mass Spectroscopy (SIMS) and High-Resolution Electron Energy Loss Spectroscopy (HREELS) are good examples. These techniques were developed within the field of surface and interface physics [1.1]. Meanwhile they have become standard techniques in many other fields of practical research, where microanalysis is required.

Surface and interface physics thus has an enormous impact on other fields of research and technology. Together with the wide variety of experimental techniques being used in this field, and with the input from various other branches of chemistry and physics, it is a truly interdisciplinary field of physical research.

Characteristic for this branch of physics is the intimate relation between experimental and theoretical research, and the application of a wide variety of differing experimental techniques having their origin sometimes in completely different fields. Correspondingly, this text follows a concept, where the general theoretical framework of surface and interface physics, as it appears at present, is treated in parallel with the major experimental methods described in so-called **panels**. In spite of the diversity of the experimental methods and approaches applied so far in this field, there is one basic technique which seems to be common to all modern surface and interface experiments: UHV equipment is required to establish clean conditions for the preparation of a well-defined solid surface or the performance of in situ studies on a freshly prepared interface. If one enters a laboratory for surface or interface studies, large UHV vessels with corresponding pumping stations are always to be found. Similarly, the importance of particle-beam optics and analytical tools, in particular for low-energy electrons, derives from the necessity to have surface sensitive probes available to establish the crystallographic perfection and cleanliness of a freshly prepared surface.

Panel I
Ultrahigh Vacuum (UHV) Technology

From the experimental point of view, the development of modern surface and interface physics is intimately related to the advent of UltraHigh Vacuum (UHV) techniques. The preparation of well-defined surfaces with negligible contamination requires ambient pressures lower than 10^{-10} Torr (= 10^{-10}mbar or approximately 10^{-8}Pa) (Sect.2.1). Typical modern UHV equipment consists of a stainless-steel vessel, the UHV chamber, in which the surface studies or processes (epitaxy, sputtering, evaporation, etc.) are performed, the pumping station including several different pumps, and pressure gauges covering different pressure ranges. In many cases a mass spectrometer (usually a Quadrupole Mass Spectrometer, QMS, Panel IV: Chap.2) is also attached to the main vessel in order to monitor the residual gas. Figure I.1 shows a schematic view over the whole set-up. A combina-

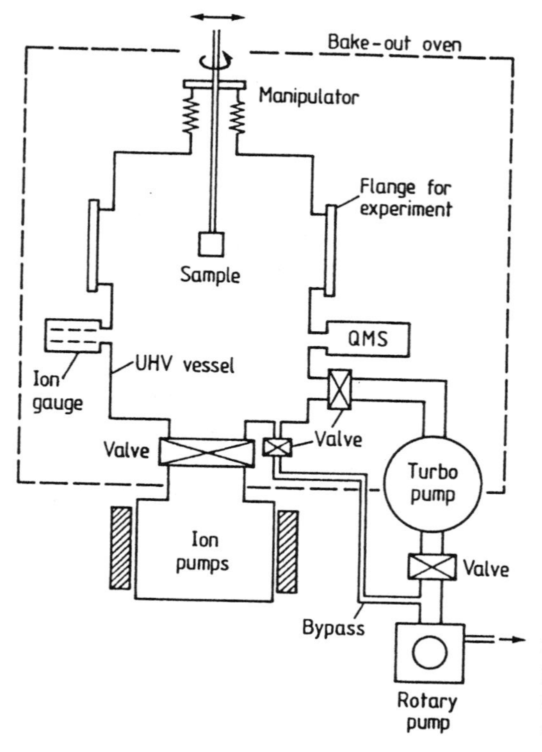

Fig.I.1. Schematic view of an Ultrahigh High Vacuum (UHV) system: stainless steel UHV vessel pumped by different pumps; the rotary backing pump can be connected to the main chamber in order to establish an initial vacuum before starting the ion pumps. Quadrupole mass spectrometer (QMS) and ion gauge are used for monitoring the residual gas. All parts enclosed by the dashed line (bake-out oven) must be baked in order to achieve UHV conditions

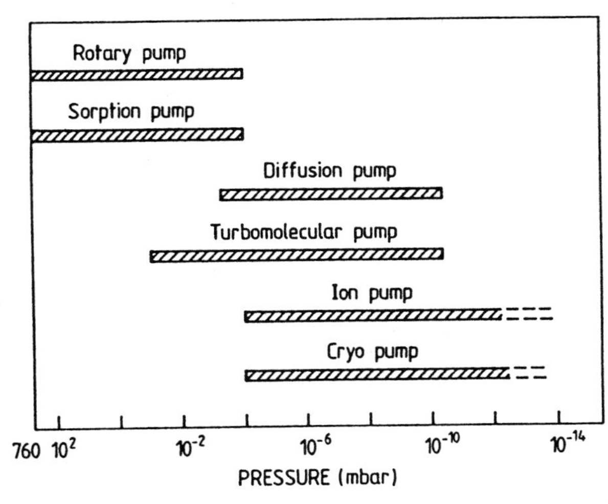

Fig.I.2. Pressure ranges in which different types of pumps can be employed

tion of different pumps is necessary in order to obtain background pressures in the main UHV chamber on the order of 10^{-10} Torr, since each pump can only operate over a limited pressure range. The UHV range (lower than 10^{-9} Torr) is covered by diffusion and turbomolecular pumps, and also by ion and cryopumps (Fig.I.2). Starting pressures for diffusion, ion, and cryopumps are in the $10^{-2} \div 10^{-4}$ Torr range, i.e. rotary or sorption pumps are needed to establish such a pressure in the main vessel (e.g., using a bypass line as in Fig.I.1). A turbomolecular pump can be started at atmospheric pressure in the UHV chamber and can operate down into the UHV regime, but a rotary pump is then needed as a backing pump (Fig.I.1). Valves are used to separate the different pumps from one another and from the UHV chamber, since a pump that has reached its operating pressure, e.g. 10^{-3} Torr for a rotary pump, acts as a leak for other pumps operating down to lower pressures.

An important step in achieving UHV conditions in the main vessel is the bake-out process. When the inner walls of the UHV chamber are exposed to air, they become covered with a water film (H_2O sticks well due to its high dipole moment). On pumping down the chamber, these H_2O molecules would slowly desorb and, despite the high pumping power, 10^{-8} Torr would be the lowest pressure obtainable. In order to get rid of this water film the whole equipment has to be baked in vacuum for about 10 h at a temperature of $150 \div 180\,°C$. When a pressure of about 10^{-7} Torr is reached in the chamber the bake-out oven (dashed line in Fig.I.1) is switched on. After switching off the bake-out equipment, again at $\approx 10^{-7}$ Torr the pressure falls down into the UHV regime.

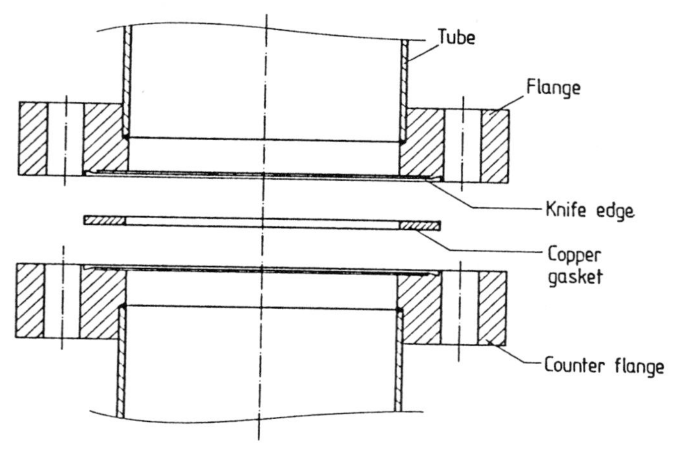

Fig.I.3. Cross section through a stainless steel Conflat flange which is used in UHV equipment for sealing

After this rough overview of the whole system, the main parts of the equipment will now be described in somewhat more detail. The different parts of a UHV system are joined together by standard flange systems. Apart from minor modifications the so-called **conflat flange** (in different standard sizes: miniconflat, 2 2/3", 4", 6", 8", etc.) is used by all UHV suppliers (Fig.I.3). Sealing is achieved by a copper gasket which, to avoid leaks, should only be used once. This conflat-flange system is necessary for all bakeable parts of the equipment. Backing pumps, bypass lines, and other components not under UHV, are usually connected by rubber or viton fittings.

In order to establish an initial vacuum ($10^{-2} \div 10^{-3}$ Torr) prior to starting a UHV pump, sorption pumps or rotary pumps are used. This procedure is known as **roughing out the system** and such pumps often go by the corresponding name of **roughing** pumps.

A **sorption pump** contains pulverized material (e.g., zeolite) with a large active surface area, the so-called **molecular sieve**, which acts as an adsorbant for the gas to be pumped. The maximum sorption activity, i.e. the full pumping speed, is reached at low temperature. The sorption pump is thus activated by cooling its walls with liquid nitrogen. From time to time regeneration of the sorbant material is necessary by means of heating under vacuum. Since the sorption process will saturate sooner or later, the sorption pump cannot be used continuously.

In combination with turbomolecular pumps, one thus uses **rotary pumps** to obtain the necessary backing pressure (Fig.I.4). The rotary pump functions on the basis of changing gas volumes produced by the rotation of an eccentric rotor, which has two blades in a diametrical slot. During the

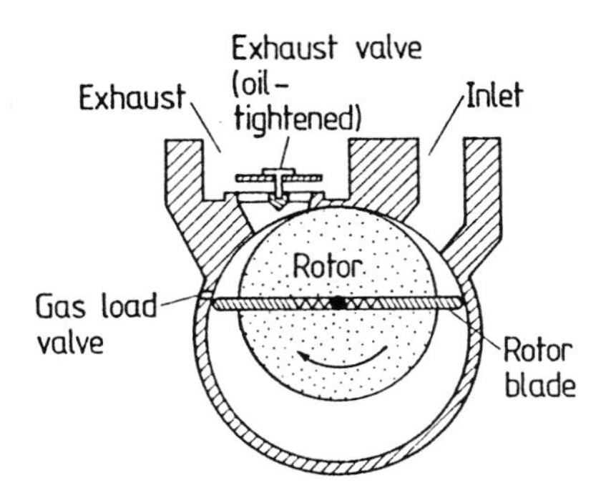

Fig.I.4. Schematic cross section through a rotary roughing pump. During the gas inlet phase the inlet volume expands. Further rotation of the eccentric rotor causes compression of this volume until the outlet phase is reached

gas inlet phase, the open volume near the inlet expands, until, after further rotation, this volume is separated from the inlet. Then, during the compression phase, the gas is compressed and forced through the exhaust valve (oil tightened). In order to avoid the condensation of vapour contained in the pumped air, most pumps are supplied with a gas load valve, through which a certain amount of air, the gas load is added to the compressed gas. Sealing between rotary blades and inner pump walls is performed by an oil film.

The pumps that are regularly used in the UHV regime are the turbomolecular pump, the diffusion pump, the ion pump and the cryopump.

The principle of the **turbomolecular pump** (or turbopump) rests on the action of a high-speed rotor (15000 ÷30000 rpm) which "shuffles" gas molecules from the UHV side to the backing side, where they are pumped away by a rotary pump (Fig.I.5). The rotor, which turns through, and is interleaved with, the so-called **stator**, has "shuffling" blades, which are inclined with respect to the rotation axis (as are the inversely inclined stator blades). This means that the probability of a molecule penetrating the rotor from the backing side to the UHV side is much lower than that of a molecule moving in the reverse direction. This becomes clear if one considers the possible paths of molecules moving through the assembly of rotor blades (Fig.I.5b). A molecule hitting the rotor blade at point A (least favorable case) can, in principle, pass from the UHV side to the backing side, if it impinges at an angle of at most β_1 and leaves within δ_1. For a molecule to pass through the rotor in the opposite direction, it must impinge within an angular range β_2 and leave within δ_2 in the least favorable case in which it arrives at point B. The probabilities of these two paths can be estimated from the ratios of angles δ_1/β_1 and δ_2/β_2. Since δ_2/β_2 is considerably smaller than δ_1/β_1, the path from the UHV side outwards is favored and pumping action occurs. This purely geometric pumping effect is strongly enhanced by the high blade velocity. Because of the blade inclination, molecules hitting the blade gain a high velocity component away from the UHV region. Compression is further enhanced by the presence of the stator

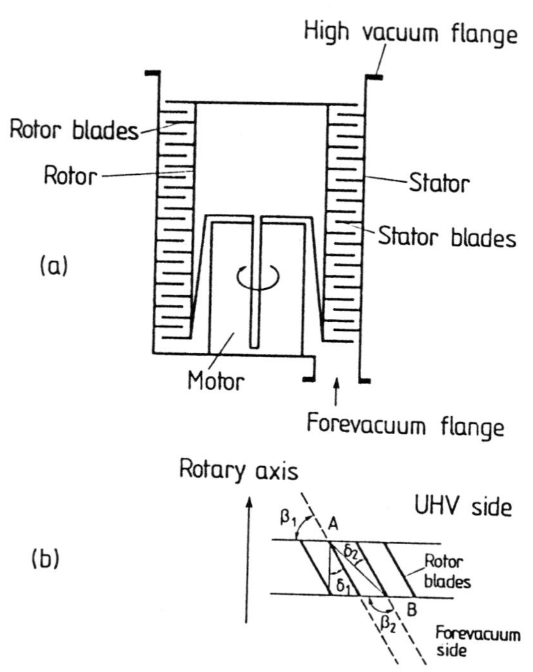

Fig. I.5. Schematic representation of a turbomolecular pump (**a**) general arrangement of rotor and stator. Rotor and stator blades (not shown in detail) are inclined with respect to one another. (**b**) Qualitative view of the arrangement of the rotor blades with respect to the axis of rotation. The possible paths of molecules from the UHV side to the backing side and vice versa are geometrically determined by the angles β_1, δ_1, and β_2, δ_2, respectively

blades with their reverse inclination. A molecule moving in the "right" direction always finds its way open into the backing line.

Since the pumping action of a turbomolecular pump relies on impact processes between the pumped molecules and the rotor blades, the compression ratio between backing and UHV sides depends on the molecular mass of the gases and on the rotor velocity (Fig. I.6). A disadvantage of turbomolecular pumps is thus their low pumping speed for light gases, in particular for H_2 (Fig. I.6a). An important advantage is the purely mechanical interaction of the gas molecules with the pump; no undesirable chemical reactions occur. Turbopumps are employed mainly when relatively large quantities of gas have to be pumped out, e.g., during evaporation or epitaxy (MOMBE).

Ion-getter pumps, which have no rotating parts, are very convenient as standby pumps for maintaining UHV conditions for an extended period (Fig. I.7). Modern ion-getter pumps are designed as multicell pumps (Fig. I.7a), in which the pumping speed is enhanced by simple repetition of the action of a single pump element (Fig. I.7b). Within each element an electri-

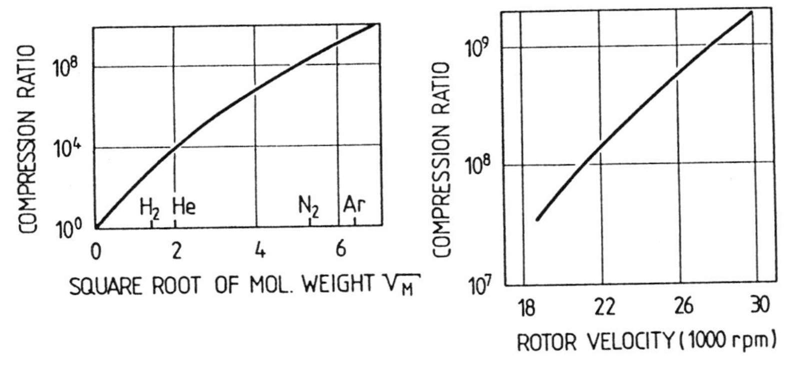

Fig. I.6. Compression ratio of a turbomolecular pump as a function of molecular weight M of the molecules pumped (*left*) and of the rotor velocity (*right*). (After Leybold Heraeus GmbH)

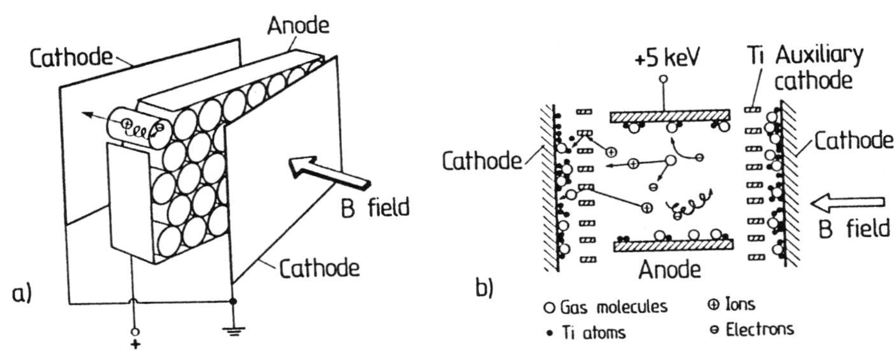

Fig. I.7a,b. Schematic view of an ion-getter pump: (**a**) The basic multicell arrangement. Each cell consists essentially of a tube-like anode. The cells are sandwiched between two common cathode plates of Ti, possibly together with auxiliary cathodes of Ti. (**b**) Detailed representation of the processes occurring within a single cell. Residual gas molecules are hit by electrons spiralling around the magnetic field B and are ionized. The ions are accelerated to the cathode and/or auxiliary cathode; they are trapped on the active cathode surface or they sputter Ti atoms from the auxiliary cathode, which in turn help to trap further residual gas ions

cal discharge is produced between the anode and the cathode at a potential of several thousand volts and in a magnetic field of a few thousand Gauss (produced by permanent magnets outside the pump). Since the magnetic field causes the electrons to follow a helical path, the length of their path is greatly increased. A high efficiency of ion formation down to pressures of 10^{-12} Torr and less is assured by this long path length. The ions so-formed are accelerated to the Ti cathode, where they are either captured or chemisorbed. Due to their high energies they penetrate into the cathode material and sputter Ti atoms, which settle on the surfaces of the anode where they also trap gas atoms. To enhance the pumping speeds, auxiliary cathodes are used (triode pump, Fig.I.7b). One problem with ion pumps is caused by Ar, which is usually the determining factor for the pumping speed (the atmosphere contains 1% Ar). This problem can be tackled to some extent by using auxiliary cathodes. Sputter ion pumps are available with a wide range of pumping speeds, between 1 ℓ/s and 5000 ℓ/s. The pressure range covered is 10^{-4} to less than 10^{-12} Torr; thus a backing pump is needed to start an ion-getter pump. Ion-getter pumps should not be used in studies of adsorption processes and surface chemistry with larger molecules, since cracking of the background gas molecules might occur, thereby inducing additional unwanted reactions.

In those cases, **vapor pumps** are a convenient alternative. The general term *vapor pump* includes both ejector pumps and diffusion pumps. In both types of pump, a vapor stream is produced by a heater at the base of the pump (Fig.I.8b). The vapor, oil or mercury, travels up a column (or a combination of several columns) and reaches an umbrella-like deflector placed at the top. There the vapor molecules collide with the gas molecules entering through the intake part. When the mean-free path of the gas molecules is greater than the throat width, the interaction between gas and vapor is based on diffusion, which is responsible for dragging the gas molecules towards the backing region. Thus diffusion induces the pressure gradient between the UHV and backing sides. When the mean free path of the gas molecules at the intake is less than the clearance, the pump acts as an ejector pump. The gas is entrained by viscous drag and turbulent mixing, and is carried down the pump chamber and through an orifice near the backing side. In some modern types of vapor pumps, combinations of the diffusion and ejector principles are used; these pumps are called **vapor booster pumps**. Diffusion pumps suffer from two drawbacks that limit their final pressure. Back-streaming and back-migration of molecules of the working fluid give rise to particle migration in the wrong direction. The vapor pressure of the working fluid is thus important for the finally obtainable pressure. The same is true for molecules of the pumped gas which can also back-diffuse to the high vacuum side. Both effects can be reduced by using baffles and cold traps, which obviously lower the net pumping speed but are necessary to reach UHV conditions. The baffles contain liquid-nitrogen cooled blades, on which the back-streaming species condense (Fig.I.8a). The consistency of the working fluid is thus very important for the performance of vapor pumps. Mercury, which was in exclusive use in former times, has now been largely displaced by high-quality ultrahigh vacuum oils, which enable pressures in the 10^{-10} Torr range to be reached when cooling traps are used.

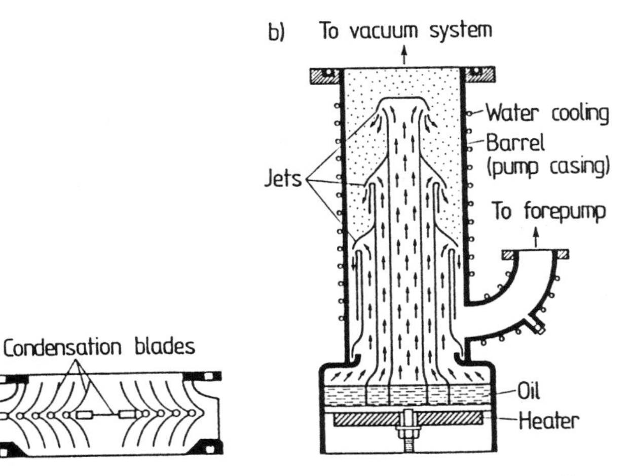

Fig. I.8. Simplified representation of a vapor (diffusion) pump (**b**) together with a baffle or cold trap (**a**) on the high vacuum side. Baffle (**a**) and pump (**b**) are arranged one on top of the other in a pumping station

Because of their extremely high pumping speeds **cryopumps** are gaining popularity for large UHV systems. Cryogenic pumping is based on the fact that if a surface within a vacuum system is cooled, vapor (gas molecules) tends to condense upon it, thus reducing the ambient pressure. A typical cryopump is sketched in Fig. I.9. The main part is a metallic helix which serves as the condenser surface. It is mounted in a chamber that is directly flanged to the UHV vessel to be evacuated. The coolant, usually liquid helium, is supplied from a dewar to the helix through a vacuum-insulated feed tube. It is made to flow through the coil by means of a gas pump at the

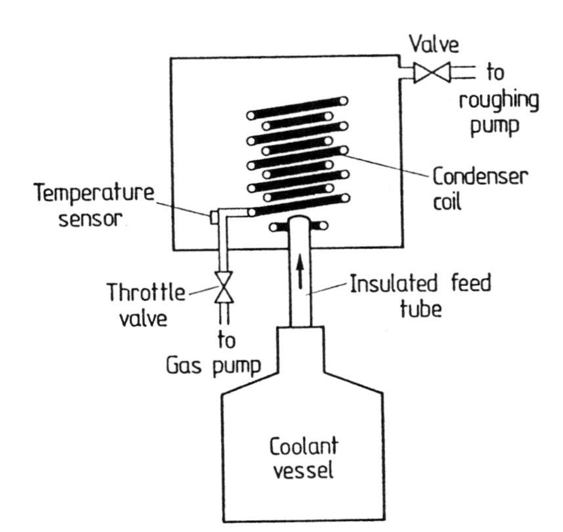

Fig. I.9. Schematic diagram of a cryopump. (After Leybold Heraeus GmbH)

outlet end of the helix. The coolant boils as it passes through the coil, hence cooling the tube. A throttle valve in the gas exhaust line controls the flux and thus the cooling rate. A temperature sensor fixed to the coil automatically controls the valve setting. Closed-loop systems are also in use; here, the pump coil is directly connected to a helium liquifier and a compressor. The helium gas from the exhaust is fed back into the liquifier.

A second type of cryopump is the so-called **bath pump**, whose coolant is contained in a tank which must be refilled from time to time. The ultimate pressure p_{min} of such a cryopump for a given gas is determined by the vapor pressure P_v at the temperature T_v of the condenser surface. According to Fig.I.10 most gases, except He and H_2, are effectively pumped using liquid He as a coolant (4.2K) such that pressures below 10^{-10} Torr are easily obtained. Extremely high pumping speeds of between 10^4 and 10^6 ℓ/s are achieved. Cryopumps cannot be used at pressures above 10^{-4} Torr, partly because of the large quantities of coolant that would be required, and partly because thick layers of deposited solid coolant would seriously reduce the pump efficiency.

The most important aspect of UHV technology is of course the generation of UHV conditions. However, a further vital requirement is the ability to measure and constantly monitor the pressure. In common with pumps, pressure gauges can also operate only over limited pressure ranges. The entire regime from atmospheric pressure down to 10^{-10} Torr is actually covered by two main types of manometer. In the higher-pressure regime, above 10^{-4} Torr, **diaphragm gauges** are used. The pressure is measured as a volume change with respect to a fixed gas volume by the deflection of a (metal) diaphragm or bellow. The reading is amplified optically or electrically, e.g. by a capacitance measurement (capacitance gauge). A further type of gauge that can operate in the high-pressure range is the molecular **viscosity gauge**. Since the viscosity of a gas is a direct function of its pres-

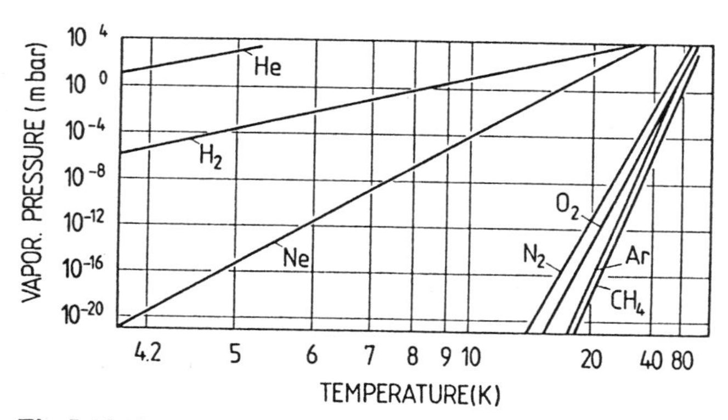

Fig.I.10. Saturation vapor pressures of various coolant materials as a function of temperature

sure, the measurement of the decay of a macroscopic motion induced by molecular drag can be used to determine the pressure. Even pressures as low as 10^{-10} Torr can be measured by spinning-ball manometers. In this equipment a magnetically suspended metal ball rotates at high speed and its deceleration due to gas friction is measured, also magnetically.

Very commonly used in vacuum systems are **thermal conductivity (heat loss) gauges**, which can be used from about 10^{-3} up to about 100 Torr. These manometers rely on the pressure dependence of the thermal conductivity of a gas as the basis for the pressure measurement. The essential construction consists of a filament (Pt or W) in a metal or glass tube attached to the vacuum system. The filament is heated directly by an electric current. The temperature of the filament then depends on the rate of supply of electrical energy, the heat loss due to conduction through the surrounding gas, the heat loss due to radiation and the heat loss by conduction through the support leads. The losses due to radiation via the support leads can be minimized by suitable construction. If the rate of supply of electrical energy is constant, then the temperature of the wire and thus also its resistance depend primarily on the loss of energy due to the thermal conductivity of the gas. In the so-called **Pirani gauge**, the temperature variations of the filament with pressure are measured in terms of the change in its resistance. This resistance measurement is usually performed with a Wheatstone bridge, in which one leg of the bridge is the filament of the gauge tube. The measured resistance versus pressure dependence is, of course, nonlinear. Calibration against other absolute manometers is necessary.

The most important device for measuring pressures lower than 10^{-4} Torr, i.e., including the UHV range ($<10^{-10}$ Torr), is the **ionization gauge**. Residual gas atoms exposed to an electron beam of sufficient kinetic energy (12.6eV for H_2O and O_2; 15÷15.6eV for N_2, H_2, Ar; 24.6eV for He) are subject to ionization. The ionization rate and thus the ion current produced are a direct function of the gas pressure. Hot-cathode ionization gauges as in Fig.I.11 consist essentially of an electrically heated cathode filament (+40V), an anode grid (+200V) and an ion collector. The thermally emitted electrons are accelerated by the anode potential and ionize gas atoms or molecules on their path to the anode. The electron current I^- is measured at the anode, whereas the ion current, which is directly related to the ambient pressure, is recorded as the collector current I^+. In operation as a pressure gauge the electron emission current I^- is usually kept fixed, such that only I^+ needs be recorded in order to determine the pressure. Since the ionization cross section is specific to a particular gas, a calibration against absolute standards is necessary and correction factors for each type of residual gas molecule have to be taken into account. Commercially available instruments are usually equipped with a pressure scale appropriate for N_2. Modern, so-called **Bayard-Alpert gauges** are con-

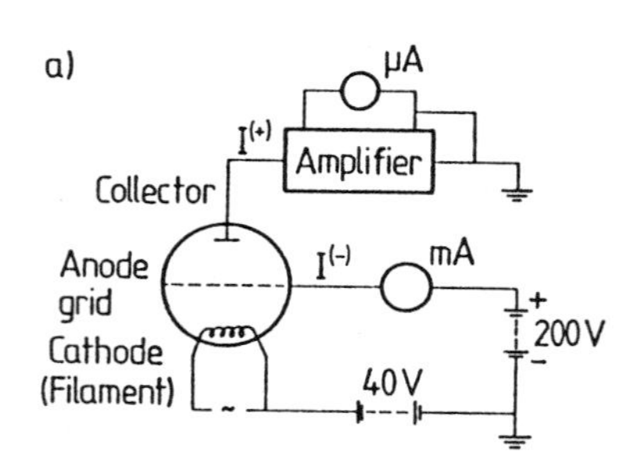

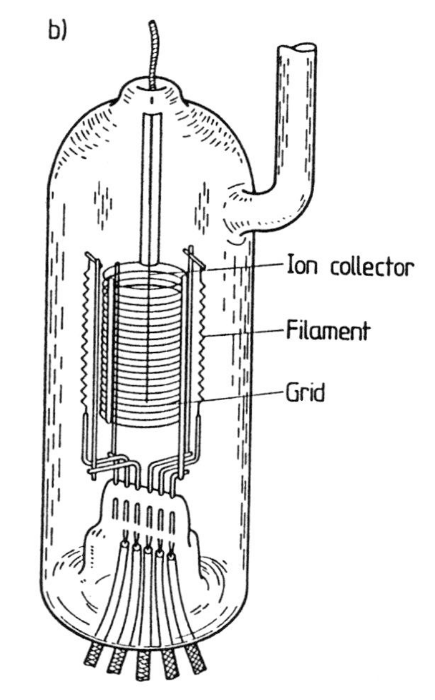

Fig.I.11a,b. Ionization gauge for pressure monitoring between 10^{-4} and 10^{-10} Torr. (**a**) Electric circuit for measuring the electron emission current I^- and the ion (collector) current I^+. (**b**) Typical construction of a modern Bayard-Alpert type ionization gauge. Cathode filament, anode grid and ion collector are contained in a glass (pyrex) tube which is attached to the UHV chamber. The electrode arrangement can also be put directly into the UHV chamber

structed as in Fig.I.11b, with several filaments (as spares), a cylindrical grid structure and a fine wire as ion collector in the center of the tube. Only this particular construction with a very thin ion collector enables pressure measurements below 10^{-8} Torr (UHV range). Spatially more extended ion collectors give rise to considerable production of soft X-rays by the electrons. These X-rays possess sufficient energy to cause photoemission of electrons from the anode. Electrically, the emission of an electron by the anode is equivalent to the capture of a positive ion, leading to a net excess current and thus to a lower limit for the detectable ion currents.

Having described the general set-up and the major components of a UHV system, some basic relations will be given, which can be used to calculate the performance and parameters of vacuum systems.

When a constant pressure p has been established in a UHV vessel, the number of molecules desorbing from the walls of the vessel must exactly balance the amount of gas being pumped away by the pumps. This is expressed by the so-called **pumping equation**, which relates the pressure change dp/dt to the desorption rate v and the pumping speed $\tilde{S}$ [ℓ/s]. Since, for an ideal gas, volume and pressure are related by

$$pV = NkT \ , \tag{I.1}$$

the requirement of particle conservation yields

$$vA_v = \frac{V_v}{kT}\left(\frac{dp}{dt} + \frac{\tilde{S}p}{V_v}\right) , \tag{I.2}$$

where A_v and V_v are the inner surface area and the volume of the vacuum chamber, respectively. Stationary conditions are characterized in (I.2) by $dp/dt = 0$, whereas the pump-down behavior is found by solving (I.2) for dp/dt. Pumping speeds in vacuum systems are always limited by the finite conductance of the tubes through which the gas is pumped. The conductance C is defined as in Ohm's law for electricity by

$$I_{mol} = C\Delta p/RT \ , \tag{I.3}$$

where I_{mol} is the molecular current, Δp the pressure difference along the tube and R the universal gas constant; C has the units ℓ/s as has the pumping speed $\tilde{S}$. In analogy with the electrical case (Kirchhoff's laws), two pipes in parallel have a conductance

$$C_p = C_1 + C_2 \ , \tag{I.4}$$

whereas two pipes in series must be described by a series conductance C_s satisfying

$$1/C_s = 1/C_1 + 1/C_2 \ . \tag{I.5}$$

Given that a certain pump is connected to a tube with a conductance C_p, the effective pumping speed $\tilde{S}_{eff}$ of the pump is

$$1/\tilde{S}_{eff} = 1/\tilde{S}_p + 1/C_p \tag{I.6}$$

where $\tilde{S}_p$ is the pumping speed of the isolated pump without tube connection. The conductance of a pipe depends on the flow conditions, i.e. on the ratio between geometrical dimensions and mean free path of the molecules.

For viscous flow ($pd > 10\,\mathrm{mbar \cdot mm}$) the conductance of a tube of circular cross section with diameter d and length L is obtained as

$$C\,[\ell/s] = 137\,\frac{d^4\,[cm^4]}{L\,[cm]}\,p\,[mbar]\,. \tag{I.7}$$

In the low-pressure regime of molecular flow ($pd < 0.1\,\mathrm{mbar \cdot mm}$) one has

$$C\,[\ell/s] = 12\,\frac{d^3\,[cm^3]}{L\,[cm]}\,. \tag{I.8}$$

Further Reading

Diels K., R. Jaeckel: *Leybold Vacuum Handbook* (Pergamon, London 1966)
Leybold broschure: *Vacuum Technology – its Foundations, Formulae and Tables*, 9th edn., Cat. no. 19990 (1987)
O'Hanlon J.F.: *A Users' Guide to Vacuum Technology* (Wiley, New York 1989)
Roth J.P.: *Vacuum Technology* (North-Holland, Amsterdam 1982)
Wutz M., H. Adam, W. Walcher: *Theorie und Praxis der Vakuumtechnik*, 4.Aufl. (Vieweg, Braunschweig 1988)

Panel II
Basics of Particle Optics and Spectroscopy

Electrons and other charged particles such as ions are the most frequently used probes in surface scattering experiments (Chap.4). The underlying reason is that these particles, in contrast to photons, do not penetrate deep into the solid. After scattering, they thus carry information about the topmost atomic layers of a solid. On the other hand, the fact that they are charged allows the construction of imaging and energy-dispersive equipment, e.g. monochromators for electrons, as used for photons in conventional optics. The basic law for the refraction (deflection) of an electron beam in an electric potential is analogous to Snell's law in optics. According to Fig.II.1, an electron beam incident at an angle α on a plate capacitor (consisting of two metallic grids) with applied voltage U is deflected. Due to the electric field $\mathcal{E}$ (normal to the capacitor plates) only the normal component of the velocity is changed from $v_1(\perp)$ to $v_2(\perp)$, the parallel component is unchanged, i.e.

$$\sin\alpha = \frac{v(\|)}{v_1}, \quad \sin\beta = \frac{v(\|)}{v_2}, \tag{II.1a}$$

$$\frac{\sin\alpha}{\sin\beta} = \frac{v_2}{v_1} = \frac{n_2}{n_1}. \tag{II.1b}$$

This refraction law is analogous to the optical law if one identifies the velocity ratio with the ratio of refractive indices n_2/n_1. Assuming that the incident beam with velocity v_1 is produced by an accelerating voltage U_0, and that energy is conserved within the capacitor, i.e.,

$$\frac{m}{2}v_2^2 = \frac{m}{2}v_1^2 + eU, \tag{II.2}$$

one obtains the law of refraction

$$\frac{\sin\alpha}{\sin\beta} = \frac{v_2}{v_1} = \frac{n_2}{n_1} = \sqrt{1 + U/U_0}. \tag{II.3}$$

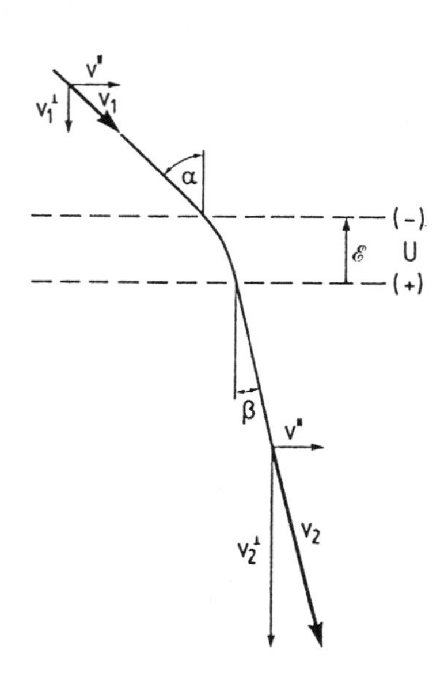

Fig.II.1. Classical electron trajectory through a parallel-plate capacitor. The electric field between the two transparent electrodes (*dashed lines*) changes the electron velocity component $v_1(\perp)$ into $v_2(\perp)$ but leaves the component $v(\|)$ unchanged

Inversion of the bias U_0 deflects the electron beam away from the normal. Describing the capacitor grids as equipotential surfaces for the field $\mathcal{E}$, the general description is that the electron beam is refracted towards or away from the normal to the equipotential lines depending on the gradient of the potential. In principle, (II.3) is also sufficient to construct, step by step, the trajectory of electrons moving in an inhomogenous electric field $\mathcal{E}(\mathbf{r})$. That is, of course, only true in the limit of classical particle motion, where interference effects due to the wave nature of the particle can be neglected (Sect.4.9).

A simple but instructive model for an *electron lens* might thus appear as in Fig.II.2, in complete analogy to an optical lens. The metallic grid itself is not important, but rather the curvature of the non-material equipotential surfaces. Electron lenses can therefore be constructed in a simpler

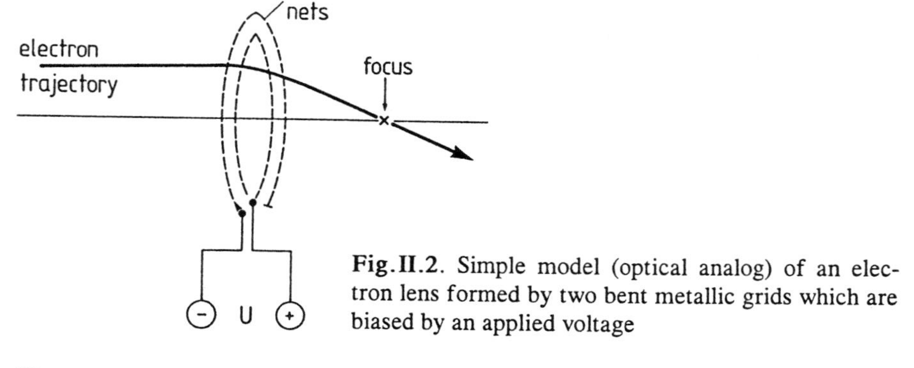

Fig.II.2. Simple model (optical analog) of an electron lens formed by two bent metallic grids which are biased by an applied voltage

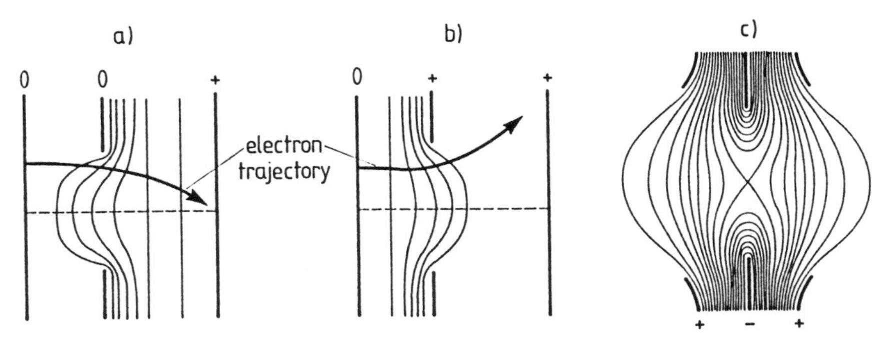

Fig. II.3. Three examples of electron lenses formed by metallic apertures (**a**) focussing arrangement (**b**) defocussing arrangement (**c**) symmetrical single lens with focussing property. In each case characteristic equipotential lines are shown

fashion, using metallic apertures which are themselves sufficient to cause curvature of the equipotential lines in their vicinity. The examples in Fig. II.3a,b act as focussing and defocussing lenses because of their characteristic potential contours. The single lens in Fig. II.3c consists of three apertures arranged symmetrically in a region of constant ambient potential U_0. Although the field distribution in this lens is completely symmetric about the central plane with a saddle point of the potential in the center, the lens is always either focussing or, for extreme negative potentials at the center aperture, acts as an electron mirror. When the potential of the middle electrode is lower than that of the two outer electrodes, the speed of the electron decreases as it approaches the saddle point of the potential. The electron remains longer in this spatial range and the central region of the potential distribution has a more significant effect on the movement than do the outer parts. The central part of the potential, however, has a focussing effect, as one can see qualitatively by comparison with Fig. II.3a,b.

On the other hand, when the inner electrode is positive with respect to the outer electrodes, the electron velocity is lower in the outer regions of the lens, the declination to the central axis is dominant, and thus, in this case too, the lens has a focussing action.

For calculating the focal length f of an electrostatic lens we use the optical analog (Fig. II.4). A simple focussing lens with two different radii of curvature embedded in a homogeneous medium of refraction index n_0 has an inverse focal length of

$$\frac{1}{f} = \frac{n - n_0}{n_0}\left(\frac{1}{|r_1|} + \frac{1}{|r_2|}\right) = \frac{\Delta n}{n_0}\left(\frac{1}{r_1} - \frac{1}{r_2}\right) . \tag{II.4}$$

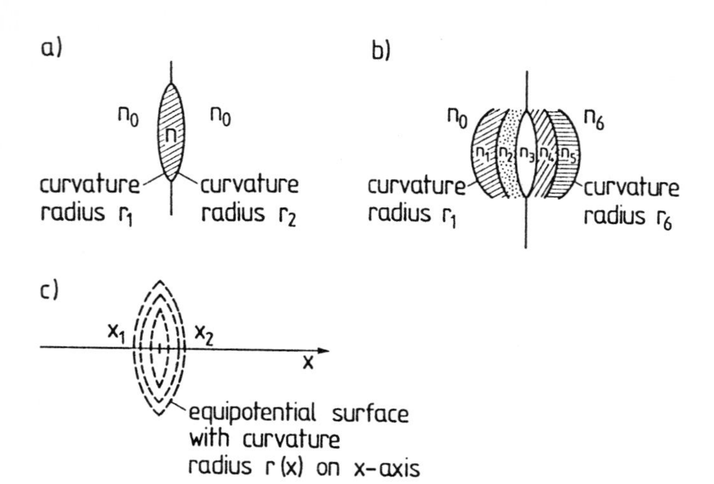

Fig.II.4. Comparison of the action of an electron lens formed by several different curved equipotential surfaces (**c**) with the optical analog, a single optical lens (refractive index n embedded in a medium with refractive index n_0) (**a**), and a multilayer lens consisting of differently curved layers with different refractive indices n_1 to n_5 embedded in two semi-infinite halfspaces with refractive indices n_0 and n_6 (**b**)

The formula for a layered lens system, as shown in Fig.II.4b, is obtained by simple generalization as

$$\frac{1}{f} = \frac{1}{n_0} \sum_{v=1}^{5} \frac{\Delta n_v}{r_v}, \quad v = 1, 2, 3, 4, 5 . \tag{II.5}$$

The formulae are, of course, only correct for trajectories close to the axis. The focal length for an electron lens with equipotential lines as in Fig.II.4c is obtained by analogy as

$$\frac{1}{f} = \frac{1}{n_2} \int_{n_1}^{n_2} \frac{dn}{r(x)} = \frac{1}{n_2} \int_{x_1}^{x_2} \frac{1}{r(x)} \frac{dn}{dx} dx , \tag{II.6}$$

where r(x) and n(x) are the radius of curvature and the "electron refractive index" according to (II.3) at a point x on the central axis. The field distribution extends from x_1 to x_2 on the axis. The refractive index for electrons

(II.3) depends on the square root of the potential U(x) on the axis and the electron velocity v(x) as

$$n(x) = \frac{v(x)}{v_1} = \text{const}\,\frac{[U(x)]^{1/2}}{v_1}\,. \tag{II.7}$$

For electron trajectories close to the axis one therefore obtains an inverse focal length of

$$\frac{1}{f} = \frac{v_1}{\text{const}(U_2)^{1/2}} \int_{x_1}^{x_2} \frac{1}{r(x)} \frac{\frac{1}{2}\text{const}\,U'(x)/v_1}{[U(x)]^{1/2}} dx$$

$$= \frac{1}{2(U_2)^{1/2}} \int_{x_1}^{x_2} \frac{1}{r(x)} \frac{U'(x)}{[U(x)]^{1/2}} dx\,. \tag{II.8}$$

Apart from boundary conditions (boundary potentials U_2, U_1) the focal length f results as a line integral over an expression containing $[U(x)]^{1/2}$ and $U'(x)$, the first derivative of the potential.

Since the charge/mass ratio e/m does not enter the focussing conditions, not only electrons but also positive particles such as protons, He^+ ions, etc. are focussed at the same point with the same applied potentials, provided they enter the system with the same geometry and the same primary kinetic energy.

This is not the case for magnetic lenses, which are used mainly to focus high energy particles. For electrons, the focussing effect of a magnetic field is easily seen for the example of a long solenoid with a nearly homogeneous magnetic field in the interior (Fig. II.5)

An electron entering such a solenoid (with velocity **v**) at an angle φ with respect to the **B** field, is forced into a helical trajectory around the field lines. This motion is described by a superposition of two velocity components $v_{\parallel} = v\cos\varphi$ and $v_{\perp} = v\sin\varphi$, parallel and normal to the **B** field. Parallel to **B** there is an unaccelerated motion with constant velocity $v_{\parallel}$; normal to **B** the particle moves on a circle with angular frequency

$$\omega = \frac{2\pi}{\tau} = \frac{e}{m}B\,, \tag{II.9}$$

i.e., the time $\tau = 2\pi m/eB$ after which the electron recrosses the same field line is not dependent on the inclination angle φ. All particles entering the

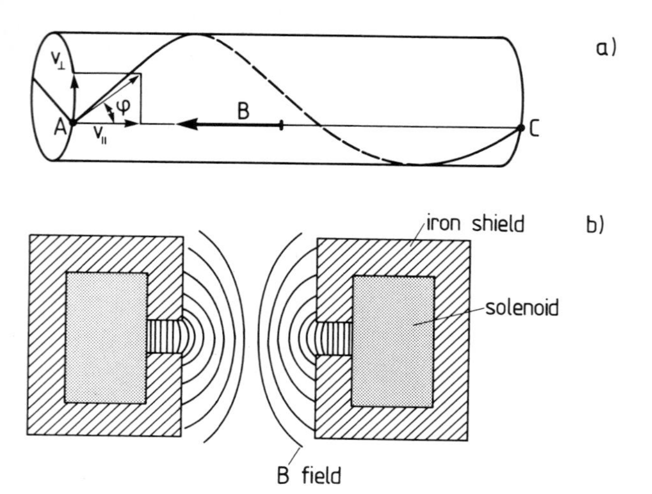

Fig. II.5. Magnetic lens for electrons (**a**) Schematic explanation of the focussing action. An electron entering a "long" solenoid adopts a helical trajectory around the magnetic field **B** such that after a certain time, τ, it recrosses the same **B**-field line through which it entered (points A and C) (**b**) Practical form of a magnetic lens consisting of a short ion-shielded solenoid

solenoid at point A at different angles reach the point C after the same time τ. Particles originating from A are thus focussed at C. The distance AC is given by the parallel velocity $v_{\|}$ and the time τ as

$$AC = v_{\|}\tau = \frac{2\pi m v \cos\varphi}{eB} . \tag{II.10}$$

For a magnetic lens the focussing condition is thus dependent on e/m, i.e. on the charge and mass of the particles. Furthermore the image is tilted in relation to the object due to the helical motion of the imaging particles. Practical forms of the magnetic lens are sometimes constructed by means of short iron-shielded solenoids with compact, concentrated field distributions in the interior (Fig. II.5b)

Also important in surface physics, in addition to the construction of imaging electron optics in electron microscopes, scanning probes, etc. (Panel V: Chap. 3), is the availability of dispersive instruments for energy analysis of particle beams. The main principles of an electrostatic electron energy analyzer are discussed in the following.

This type of analyzer has as its main components two cylinder sectors as electrodes and is thus called a **cylindrical analyzer** (Fig. II.6). A well-defined pass energy E_0 for electrons on a central circular path between the electrodes is defined by the balance between centrifugal force and the elec-

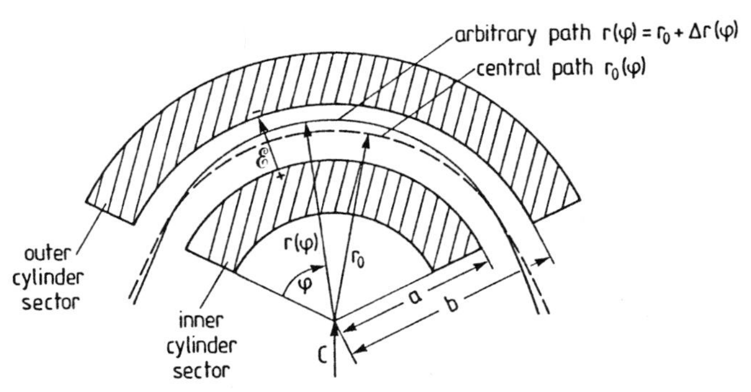

Fig. II.6. Electrostatic cylinder sector energy analyzer (schematic). The electric field between the two cylinder sectors (*shaded*) exactly balances the centrifugal force for an electron on the central path r_0. An arbitrary electron path around the central path is described by the deviation $\Delta r(\varphi)$ from the central path

trostatic force of the field $\mathcal{E}$ due to the voltage U_p applied across the electrodes. The field is a logarithmic radial field:

$$\mathcal{E} = -\frac{U_p}{r\ln(b/a)} \tag{II.11}$$

with a and b as the inner and outer radii of the region between the cylindrical sectors (Fig. II.6); r is the radius vector to an arbitrary point of the field, whereas r_0 and v_0 are the radius vector and tangential velocity on the central trajectory, i.e. they satisfy

$$\frac{mv_0{}^2}{r_0} = -e\mathcal{E}_0 = \frac{eU_p}{r_0\ln(b/a)}\,. \tag{II.12}$$

The pass energy is thus obtained as

$$E_0 = \frac{1}{2}mv_0{}^2 = \frac{1}{2}\frac{eU_p}{\ln(b/a)}\,. \tag{II.13}$$

Under certain conditions such a cylindrical analyzer has additional focussing properties, which considerably enhance its transmittance and are thus advantageous for investigations with low beam intensities. This can be seen by considering an electron path inclined at a small angle α to the central path at the entrance (Fig. II.6). This path is described by the dynamic equation.

$$m \frac{d^2 r}{dt^2} = m \frac{v^2}{r} - e\mathcal{E} = mr\omega^2 - e\mathcal{E} , \tag{II.14}$$

where ω is the angular velocity around the center C. For small deviations from the central path r_0 one has

$$r \simeq r_0 + \Delta r , \tag{II.15a}$$

$$\mathcal{E} \simeq \mathcal{E}_0 r_0 / r = \mathcal{E}_0 (1 - \Delta r / r_0) . \tag{II.15b}$$

The angular momentum around C must be conserved, i.e., neglecting small quantities we have

$$\omega r^2 \simeq \omega_0 r_0{}^2 . \tag{II.16}$$

From (II.14-16) one obtains the approximate dynamic equation

$$\frac{d^2 r}{dt^2} = \frac{\omega_0{}^2 r_0{}^4}{r^3} - \frac{e\mathcal{E}}{m} , \tag{II.17}$$

or

$$\frac{d^2 (\Delta r)}{dt^2} = \omega_0{}^2 \frac{r_0{}^4}{(r_0 + \Delta r)^3} - \frac{e}{m} \mathcal{E}_0 \left[1 - \frac{\Delta r}{r_0}\right] . \tag{II.18}$$

Using (II.12) and neglecting small quantities in second order yields

$$\frac{d^2 (\Delta r)}{dt^2} + 2\omega_0{}^2 \Delta r = 0 . \tag{II.19}$$

The solution for the deviation Δr from the central path is thus obtained as

$$\Delta r = \text{const} \times \sin(\sqrt{2}\, \omega_0 t + \delta) , \tag{II.20}$$

i.e. the deviation oscillates with a period $(2)^{1/2} \omega_0$. An electron entering the analyzer on the central path ($\delta = 0$) crosses that path again after a rotation angle $\phi = \omega_0 t$ (around C), which is given by

$$\sqrt{2}\omega_0 t = \sqrt{2}\phi = \pi , \tag{II.21a}$$

or

$$\phi = 127° \; 17' . \tag{II.21b}$$

This is independent of α, i.e. focussing occurs for cylindrical sectors with an angle of 127° 17′. In reality the electric field $\mathcal{E}$ is perturbed at the entrance and at the exit of the analyzer. A field correction is performed by so-called **Herzog apertures** which define the entrance and exit slits. This modifies the condition (II.21) and leads to a sector angle for focussing of 118.6°. For judging the performance of such analyzers the energy resolution $\Delta E/E$ is an important quantity. From an approximate, more general solution for the electron trajectories one obtains

$$\frac{\Delta E}{E} = \frac{x_1 + x_2}{r_0} + \frac{4}{3}\alpha^2 + \beta^2 , \qquad \text{(II.22)}$$

where x_1 and x_2 are the width and length of the rectangular entrance (and exit) slits, and α and β are the maximum angular deviations of the electron trajectories at the entrance, in the plane and normal to the plane of Fig. II.6. Present-day instruments employed in electron scattering experiments (HREELS, Panel IX: Chap. 4) achieve a resolution $\Delta E/E$ of $10^{-3} \div 10^{-4}$. Decreasing the slit width to improve the resolution is only possible within certain limits, since it simultaneously reduces the transmitted current. This deterioration is mainly due to space-charge effects. Electrons moving parallel to one another through the analyzer interact with each other via their Coulomb repulsion and their mutually induced magnetic field (due to the current). For high electron densities these space-charge effects distort the electron trajectories and limit the resolution. A semi-quantitative estimate of the effect can be made using the classical formula for space-charge-limited currents in radio tubes:

$$j \propto U^{3/2} . \qquad \text{(II.23)}$$

Using (II.22) one therefore obtains for the current density at the entrance

$$j_i \propto E_0^{3/2} \propto (\Delta E)^{3/2} . \qquad \text{(II.24)}$$

The final current density at the exit, then follows as

$$j_f \propto j_i \Delta E \propto (\Delta E)^{5/2} . \qquad \text{(II.25)}$$

This dependence, which is confirmed well by experiment, causes a strong reduction of the transmitted current with narrower entrance slits.

Electron analyzers can be used in two modes: Scanning the pass voltage U_p by an external ramp varies the passing energy E_0 (II.13). Because of (II.22) the ratio $\Delta E/E$ remains the same, i.e. when the pass energy is var-

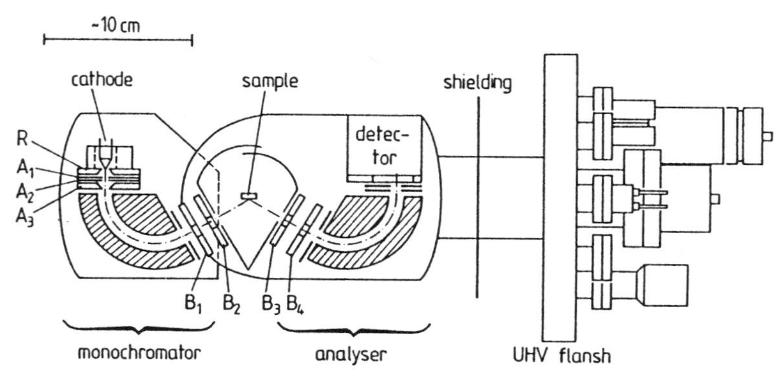

Fig.II.7. Schematic plot of a high-resolution electron energy loss spectrometer consisting of a cathode system (filament with lens system), a monochromator (cylindrical sectors), a similar analyzer and a detector. The monochromator can be rotated around an axis through the sample surface. The whole set up is mounted on a UHV flange

ied, the resolution ΔE also changes continuously across the spectrum (constant $\Delta E/E$ mode). In order to achieve a constant resolution over the entire spectrum, the pass energy of the analyzer and thus also the resolution ΔE can be held constant; but the electron spectrum being measured, must then be "shifted" through the fixed analyzer window ΔE by variation of an acceleration or deceleration voltage in front of the analyzer (constant ΔE mode).

A complete electron spectrometer (Fig.II.7) such as those used for High-Resolution Electron Energy Loss Spectroscopy (HREELS, Sect.4.6) consists of at least two analyzers with focussing apertures (lenses) at the entrances and exits. A cathode arrangement with a lens system produces an electron beam with an energetic halfwidth of $0.3 \div 0.5$ eV (Maxwell distribution of hot electrons). The first electron monochromator (the same as the analyzer) is fixed at a constant pass energy and selects electrons in an energy window of width $1 \div 10$ meV from the broad Maxwell distribution. This primary beam is focussed onto the crystal surface by a lens system. The voltage applied between this lens and the sample determines the primary energy. The backscattered electrons are focussed by a second lens system onto the entrance slit of the second analyzer which is usually used in the constant ΔE mode, i.e. with a fixed pass energy and a variable acceleration voltage between sample and analyzer. A lens system behind the exit slit focusses the analyzer beam onto a detector, a Faraday cup or, more often, a channeltron electron multiplier.

In addition to the cylindrical electron analyzer discussed so far, a number of other electrostatic analyzers are in use. Similar in its principles is the so-called **hemispherical analyzer** (Fig.II.8) in which the electric field balancing the centrifugal force of the electrons on their trajectory is formed between two metallic hemispheres. Entrance and exit apertures are

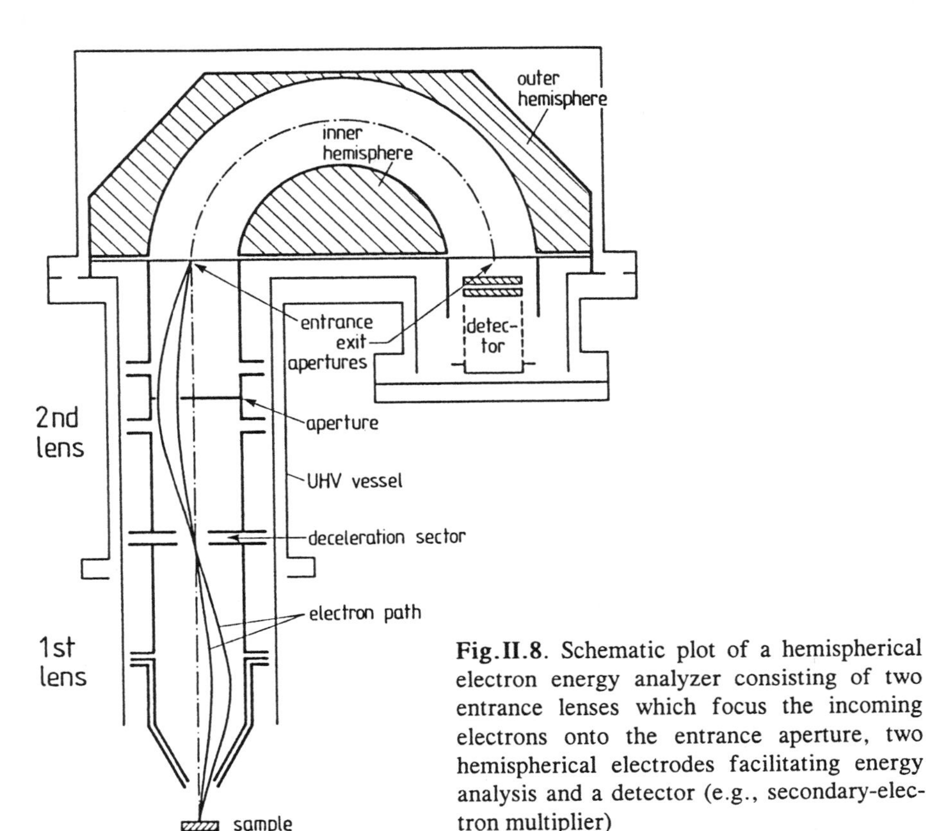

Fig. II.8. Schematic plot of a hemispherical electron energy analyzer consisting of two entrance lenses which focus the incoming electrons onto the entrance aperture, two hemispherical electrodes facilitating energy analysis and a detector (e.g., secondary-electron multiplier)

circular holes, which also produce a circular image, in contrast to the rectangular image of the cylindrical analyzer. In common with the latter, a focussing condition exists for electrons that have been deflected through an angle of 180°. Thus the hemispherical shape of the electrodes is an essential requirement for focussing. The energy resolution is calculated as

$$\frac{\Delta E}{E} = \frac{x_1 + x_2}{2\bar{r}} + \alpha^2 , \quad \bar{r} = \frac{a + b}{2} , \tag{II.26}$$

where x_1 and x_2 are the radii of the entrance and exit apertures, and a and b are the radii of the inner and outer spheres; α is the maximum angular deviation of the electron trajectories at the entrance with respect to the center line (normal to entrance aperture). Lens systems such as that shown in Fig. II.8 are also used in combination with this type of analyzer.

A widely used analyzer, in particular for Auger Electron Spectroscopy (AES), is the so-called **Cylindrical Mirror Analyzer** (CMA). Here the

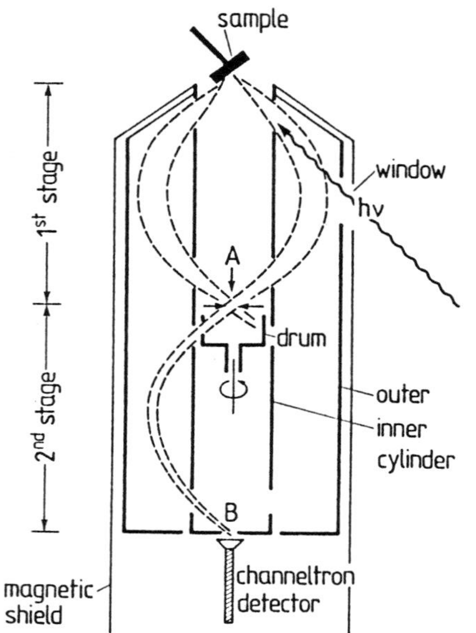

Fig. II.9. Double stage Cylindrical Mirror Analyzer (CMA) consisting of two analyzer units. For photoemission experiments the exciting light beam enters through a window and hits the sample surface. The emitted electrons enter the analyzer (outer and inner cylindrical mirror) within a certain cone. Measurements with angular resolution can be performed by means of a rotatable drum in front of the entrance to the second stage. A window in the drum selects electrons from one particular direction only. The second stage images point A into point B at the channeltron detector

electrons entering from an entrance point into a certain cone are focussed by two concentric, cylindrical electrodes onto an image point, where the detector (e.g., a channeltron) is positioned. One also encounters double-stage CMAs with two successive analyzer units (Fig. II.9). The electric field determining the pass energy is a radial field between the two concentric electrodes. Focussing occurs for electrons entering near a cone with an apex angle $\phi = 42° \; 18.5'$ (Fig. II.9). This acceptance cone and the total acceptance aperture (circle) around the cone are determined by appropriate windows in the cylindrical electrodes. The pass energy E_0 and pass voltage U_p between the two cylinders are related by

$$E_0 = \frac{eU_p}{0.77 \ln(b/a)} . \tag{II.27}$$

The energy resolution of this CMA is determined both by the angular deviation α of the incoming electron trajectory from the well-defined acceptance cone and by the axial shifts x_1 and x_2 of the actual electron trajectories with respect to the ideal entrance and image points on the cylinder axis (Fig. II.9) An approximate numerical calculation yields:

$$\frac{\Delta l}{l_0} = \frac{x_1 + x_2}{l_0} = \frac{\Delta E}{E}(1 + 1.84\alpha) - 2.85\alpha^2 . \tag{II.28}$$

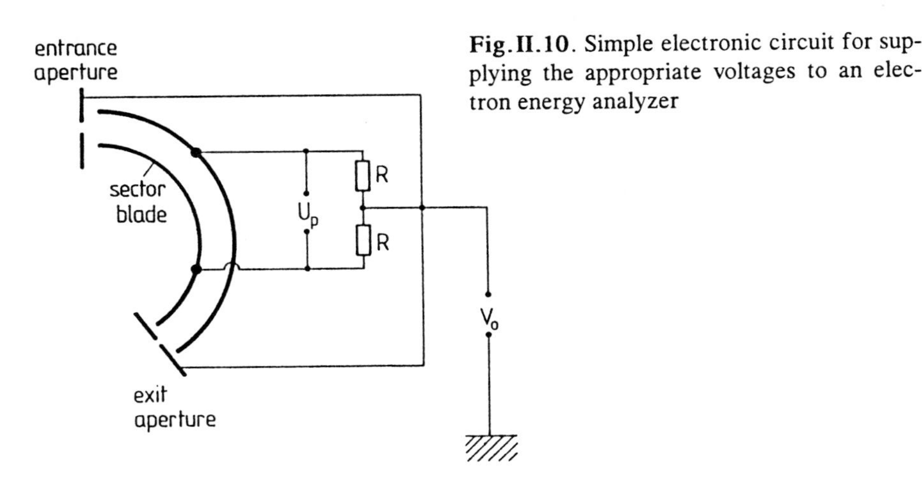

Fig. II.10. Simple electronic circuit for supplying the appropriate voltages to an electron energy analyzer

Finally, we mention briefly the electronic circuitry common to the control units of all these electrostatic analyzers. It is always necessary to scan the potential V_0 at the entrance and exit slits, which is identical to the potential of the central path within the analyzer. Simultaneously the potentials on the two main electrodes (cylinder sectors, hemispheres, etc.) must remain symmetrically disposed with respect to the potential on the central path when the pass voltage U_p is varied. This is achieved by means of a circuit such as that shown schematically in Fig. II.10.

Besides the types of analyzer described above, retarding field analyzers are also in use. In principle, these are high-pass filters in which a variation of the retarding voltage determines a variable cut-off kinetic energy for the electrons. These retarding field analyzers can simultaneously be used as optical LEED display units in electron diffraction experiments. A detailed description is given in Panel VIII: Chap. 4.

Further Reading

Grivet P.: *Electron Optics* (Pergamon, Oxford 1965)
Ibach H.: *Electron Energy Loss Spectrometers*, Springer Ser. Opt. Sci., Vol.63 (Springer, Berlin, Heidelberg 1990)

Problems

Problem 1.1. A spherical UHV chamber with a volume of 0.5 m^3 is pumped through a circular tube with a diameter of 20 cm and a length of 50 cm. The used sputter ion pump has a pumping speed of 1000 ℓ/s. The end pressure reached after baking out the vessel is $7 \cdot 10^{-11}$ Torr. What is the gas desorption rate from the walls of the UHV chamber at steady state?

Problem 1.2. In the low-pressure regime of molecular flow the molecules do not hit each other, they are only reflected from the walls of the tube. The molecular flow through a tube of diameter d and length L might thus be described roughly by a diffusion process in which the mean-free path is approximated by the tube diameter and the gradient of the particle density along the tube by $dn/dx = \Delta n/L$. Using the standard formulae for current density, diffusion coefficient, etc., derive the conductance equation (I.8) for molecular flow: $C \propto d^3/L$.

Problem 1.3. An electron beam is accelerated by a voltage of 500 V and penetrates an arrangement of two parallel metallic grids under an angle of 45° toward the grid plane. Between the grids a bias of 500 V is applied (minus at first, plus at second grid). Under what angle, with respect to the grid normal, is the electron beam detected behind the grid arrangement. What is the kinetic energy of the electrons behind the grids?

Problem 1.4. An electron penetrates a short region of varying potential U(x). Calculate the curvature radius of its trajectory from the acting centripetal force.

2. Preparation of Well-Defined Surfaces and Interfaces

As is generally true in physics, in the field of surface and interface studies one wants to investigate model systems which are simple in the sense that they can be characterized mathematically by a few definite parameters that are determined from experiments. Only for such systems can one hope to find a theoretical description which allows one to predict new properties. The understanding of such simple model systems is a condition for a deeper insight into more complex and more realistic ones.

The real surface of a solid under atmospheric pressure is far removed from the ideal system desirable in interface physics. A fresh, clean surface is normally very reactive towards the particles, atoms and molecules, impinging on it. Thus, real surfaces exposed to the atmosphere are very complex and not well defined systems. All kinds of adsorbed particles – from the strongly chemisorbed to the weakly physisorbed – form an adlayer on the topmost atomic layers of the solid. This contamination adlayer, whose chemical composition and geometrical structure are not well defined, hinders the controlled adsorption of a single and pure, selected species. A better understanding of adsorption first requires the preparation of a clean, uncontaminated surface before a well defined adsorbate of known consistency and quantity is brought into contact with the surface. Similarly, interfaces between two different crystalline materials can only be produced by epitaxy in a well-controlled way, when the topmost atomic layer of the substrate material is clean and crystallographically ordered.

2.1 Why is Ultrahigh Vacuum Used?

As "clean" surfaces, one might also think of the electrode surfaces in an electrochemical cell, or of a semiconductor surface at an elevated temperature in a flux reactor where Vapor Phase Epitaxy (VPE) is performed at standard pressure conditions. The possibility of good epitaxy in the latter case shows that contaminants may play a minor role in the presence of the chemical reactions involved. Both systems mentioned, however, are fairly complex and difficult to characterize, as every electrochemist or semiconductor technologist admits. The simplest interface one can think of is that between a crystalline surface and vacuum. We shall see in Chap.3 that even such an interface sometimes poses severe problems when one is seeking a theoretical description. But when the vacuum is sufficiently high, one can at least neglect the influence of the gas phase or of adsorbed contaminants. The surface, of course, has to be prepared as a fresh, clean surface within such a vacuum. Most of the remainder of this chapter is concerned with several methods for preparing clean, "virgin-like" crystalline surfaces in vacuum. Here we want to make a simple estimate of how good the vacuum must be, to ensure stable, well-defined conditions for experiments on such a freshly prepared surface.

The ambient pressure p determines how many particles of the residual gas impinge on a surface area of 1 cm^2 per second (impinging rate: $\dot{z}$ $[cm^{-2} s^{-1}]$) through the relation

$$p = 2m\langle v\rangle \dot{z} \, . \tag{2.1}$$

Here m is the mass of the gas atoms or molecules, and $\langle v\rangle$ is their average thermal velocity with

$$m\langle v\rangle^2/2 \simeq m\langle v^2\rangle/2 = 3kT/2 \tag{2.2}$$

where T is the temperature in Kelvin, and k is Boltzmann's constant. Thus, for the relation between pressure and impinging rate we obtain

$$p \simeq 6kT\dot{z}/\langle v\rangle \, . \tag{2.3}$$

Assuming the capacity to accomodate a surface monolayer of $3 \cdot 10^{14}$ particles, an average molecular weight of 28 and a temperature T of 300 K one obtains

$$\dot{z} = 3 \cdot 10^{14} \frac{1}{s \cdot cm^2} \simeq 5 \cdot 10^{-6} \frac{p}{Torr} \, . \tag{2.4}$$

This means that at a pressure of approximately 10^{-6} Torr (standard high vacuum conditions) the number of molecules necessary for building up a monolayer of adsorbate strikes the surface every second. The actual coverage also depends on the sticking coefficient S, which is the probability that an impinging atom or molecule remains adsorbed. For many systems, in particular for clean metal surfaces, S is close to unity, i.e. nearly every impinging atom or molecule sticks to the surface. In order to keep a surface fairly clean over a period of the order of an hour (time for performing an experiment on a "clean" surface) it is therefore necessary to have a vacuum with a residual gas pressure lower than 10^{-10} Torr ($\approx 10^{-8}$ Pa). These are UHV conditions which are necessary for experiments on "clean", well-defined solid/vacuum interfaces, where contamination effects can be neglected.

Based on (2.4) surface physicists have introduced a convenient exposure (pressure times time) or dosage unit: 1 Langmuir (1L) is the dosage corresponding to exposure of the surface for 1 s to a gas pressure of 10^{-6} Torr (or, e.g., for 100s to 10^{-8} Torr). For a surface with a sticking coefficient of unity this exposure of 1 L causes an adsorbate coverage of approximately one monolayer. Exposure values given in Langmuir, therefore, convey a direct "feeling" for the maximum amount of adsorbate coverage.

2.2 Cleavage in UHV

Fresh, clean and well-defined surfaces of brittle materials can be prepared in UHV by cleavage. This technique is derived from the classical method of preparing an alkali halide surface by cleaving the crystal with a razor blade. In an UHV system this method can also be applied by transferring mechanical pressure to the razor blade via a feedthrough bellow. Every cleavage setup in UHV is based on the application of mechanical pressure and, therefore, needs mechanical feedthroughs, or magnetic or electromagnetic devices that can be controlled magnetically or electrically from the outside of the chamber. A commonly used method is the so-called **double-wedge technique**, where two stainless steel wedges are pressed into two notches cut into opposite sides of the crystal (Fig.2.1). Depending on the crystal material, the mechanical pressure can be applied continuously with a pressure screw although, in other cases, good flat cleaves are only obtained when a pressure shock is transferred to the wedges by a hammer. The main disadvantage of the double-wedge technique is the fact that only one single surface is obtained from one sample, i.e. after one run of experiments the UHV system has to be opened and a new sample must be mounted. A modification of the double-wedge technique is often applied in cases where multiple cleavages from one single crystal bar are desirable to enable more than one set of experiments within a single UHV cycle. A long crystal bar is prepared with equidistant notches on one side, into which a wedge is pressed, the sample being supported from the other side by a flat support (Fig.2.2). After termination of the first set of experiments, a new, freshly cleaved surface is prepared by shifting the bar forwards to bring the next notch under the wedge.

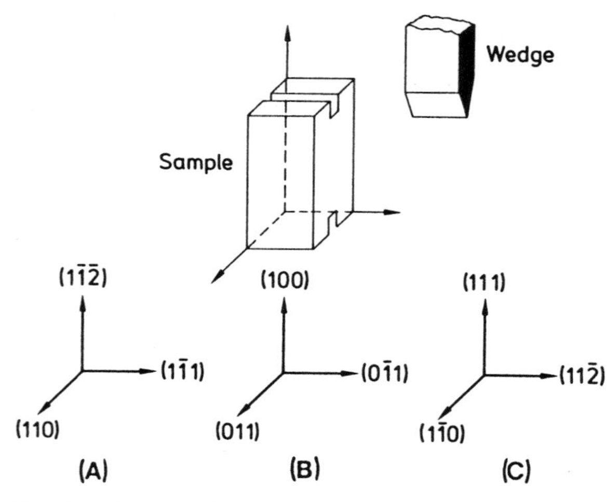

Fig. 2.1. Semiconductor sample prepared for cleavage by the double-wedge technique. For Si and Ge the cleavage plane is (111); for III-V semiconductors it is (110). (A-C) are three possible crystal orientations for cleavage along (110), the most favorable one is orientation (B)

If there is no need to produce a single-crystal surface with distinct orientation, cleaved surfaces can be obtained much more simply by crushing thin slabs of the material with a magnetically operated hammer in UHV. Large total areas, consisting of many small parts with differing orientations, are thus obtained and can be used for adsorption studies.

Cleavage in UHV is a simple and straightforward way to prepare a fresh, clean surface. Such surfaces are in general stoichiometric (important for compound materials) but they usually contain defects such as steps which expose edge atoms that are in a different surrounding compared with those in the flat areas. Another important limitation applies to cleavage. Only brittle materials like alkali halides (NaCl, KCl, etc.), oxides (ZnO, TiO_2, SnO_2, etc.) and semiconductors (Ge, Si, GaAs, etc.) can be studied in this way. Furthermore, cleavage is only possible along certain crystallographic directions which are determined by the geometry and nature of the chemical bond. The number of covalent bonds being cut, or the compensa-

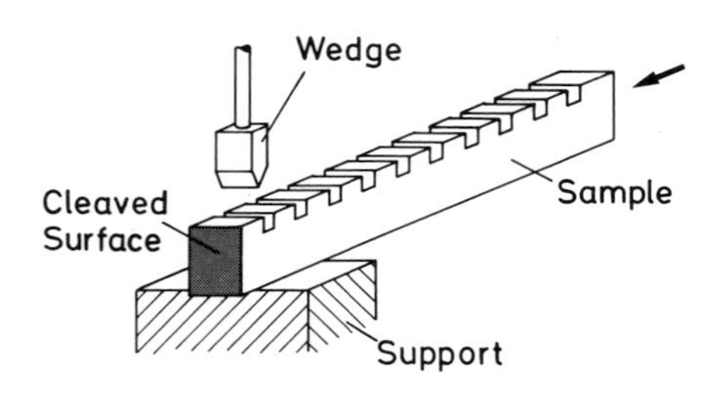

Fig. 2.2. Scheme of a multiple cleavage set-up. The semiconductor sample shaped as a rod with notches cut into the upper side can be shifted forwards to bring the next notch under the cleavage wedge

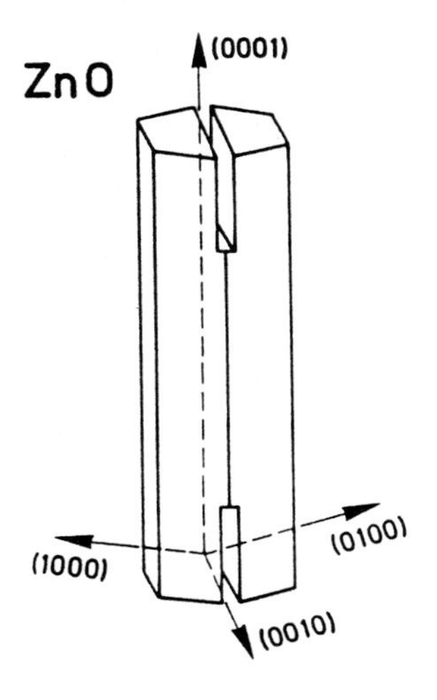

Fig.2.3. ZnO sample (hexagonal wurtzite lattice) prepared for cleavage along the non-polar prism face

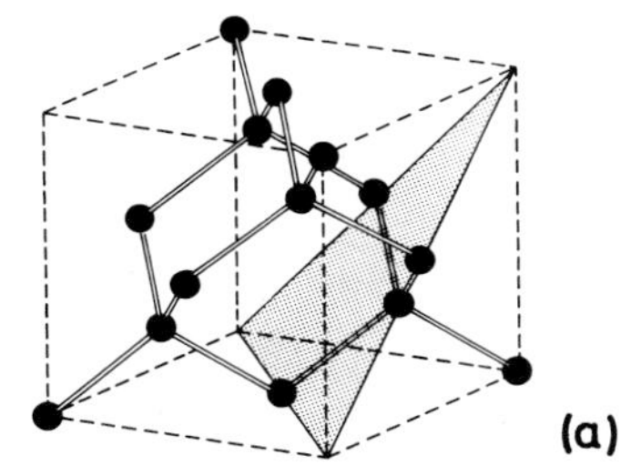

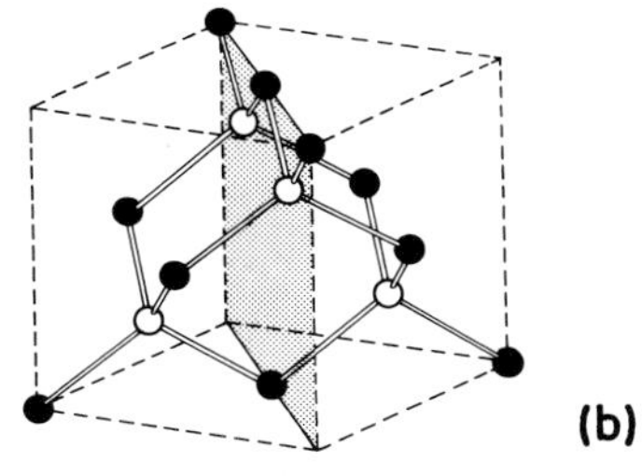

Fig.2.4. **(a)** (111) cleavage plane (*hatched*) in the diamond lattice of Si or Ge, and **(b)** (110) cleavage plane (*hatched*) in the zinc blende lattice of III-V semiconductors

tion of electric fields within the cleavage plane in the case of ionic crystals, are determining factors. Cubic alkali-halide crystals cleave along the {100} faces, which are nonpolar; i.e. they contain equal numbers of both types of ions such that the fields between the opposite charges can be compensated within the surface. The same argument explains why crystals with the wurzite structure such as ZnO cleave well along the nonpolar (prism) faces $\{10\bar{1}0\}$ (Fig.2.3). Cleavages along the polar (0001) and $(000\bar{1})$ surfaces are also possible, but the cleavage quality is much poorer and a high density of steps and other defects is usually found on these polar cleaves. Elemental semiconductors like Ge and Si can only be cleaved along {111} (Fig.2.4a). A detailed atomistic calculation which explains the occurrence of only this cleavage plane on the basis of the sp^3 chemical bond is lacking so far. For III-V compound semiconductors such as GaAs, InP or GaP, with a considerable amount of ionic bonding character, only the {110} faces are cleavage planes. These are nonpolar faces with an equal amount of positive and negative ionic charge (Fig.2.4b). The other low-index faces like {100} and {111} are polar, and cannot be produced by cleavage.

If one wants to produce a cleaved (110) surface of a III-V compound there are several possible ways to cut the sample in order to apply the double-wedge technique (Fig.2.1). Experience has shown that within the three different orientations (A,B,C) orientation B is the most favorable. For the orientations A and C one obtains a higher number of miscleavages where the crystal breaks or cleaves along undesired directions. The reason is that six different planes of the type {110} exist, and these are differently oriented with respect to the desired cleavage direction. Since the wedges induce stress not only along the desired (110) plane, there is a certain probability that cleavage occurs along more than one direction. This probability for cleavage along one of these undesired directions depends on the angle between the particular {110} face and the desired one. Let γ be the angle between normal directions by which a particular {110} face is tilted relative to the plane which is oriented normal to the slits and parallel to the longest dimension (cleavage direction) of the H-shaped sample (Fig.2.1). The smaller the value of γ, the higher is the probability for cleavage along this particular direction. For orientation A in Fig.2.1 one has to calculate the components of the different surface normals (110), (101), (011), $(1\bar{1}0)$, $(10\bar{1})$ along the $[1\bar{1}1]$ direction. Table 2.1 lists the calculated γ for the set of possible {110} planes in each orientation A, B and C. For orientation A γ vanishes three times, i.e. beside the desired cleavage plane (110) there is a high probability for cleavage along the (011) and $(10\bar{1})$ faces, too. For orientation C beside the desired cleavage plane ($\gamma = 0$) there are two other planes (101) and (011) with small γ values of $-16.78°$ and therefore high cleavage probability. Orientation B exposes only one desired cleavage plane with $\gamma = 0$, the other planes exhibit γ values of $\pm 30°$; the $(01\bar{1})$ plane with $\gamma = 0$ is perpendicular to the desired (011) plane and need therefore not be considered. Orientation B is therefore the most favorable one for cleavage of III-V compound semiconductors. Similar considerations can also be applied to other

Table 2.1. Angle γ by which the normals to the different {110} faces in a zinc blende lattice are tilted against a plane normal to the desired (110) cleavage plane. Three different crystal orientations, A, B, C are possible for cleavage along {110} (Fig.2.1). An asterisk denotes the desired cleavage plane for the particular crystal orientation

γ [°]	(110)	(101)	(011)	$(1\bar{1}0)$	$(10\bar{1})$	$(01\bar{1})$
Orient. A	0*	54.74	0	54.74	0	−54.74
Orient. B	−30	30	0*	30	−30	0
Orient. C	35.26	−16.78	−16.78	0*	60	60

materials which cleave along (111) like Ge or Si. But it should be emphasized that other factors, e.g. the dimensions of the sample, may also play a considerable role in obtaining cleaves of good quality.

2.3 Ion Bombardment and Annealing

While cleavage as a surface preparation method is restricted to certain materials and to certain crystallographic surface planes, there are essentially no such limitations to the cleaning method by ion bombardment and annealing. Contaminants, and usually the topmost atomic layers of the crystal as well, are sputtered off by bombardment with noble gas ions (Ar, Ne, etc.) and subsequent annealing is necessary to remove embedded and adsorbed noble gas atoms and to recover the surface crystallography. The whole procedure may be performed several times; each time a bombardment cycle is followed by annealing; between each step the surface has to be controlled by Auger Electron Spectroscopy (AES) (Panel III: Chap.2) for cleanliness and by Low-Energy Electron Diffraction (LEED) (Panel VIII: Chap.4) for crystallographic order. The procedure is stopped when the checking methods, AES and LEED, show satisfactory results concerning contamination and the degree of crystallographic order (sharp LEED pattern). One should keep in mind that the sensitivity of AES for contamination is usually not better than 10^{-3} monolayers and that a good LEED pattern with sharp Bragg spots and low background intensity does not give much information about long-range order extending over distances longer than the coherence length of the electrons (typically 100 Å) [2.1].

In detail, the cleaning procedure starts by admitting a noble gas, preferably Ar, into the UHV chamber (or into the ion gun) up to a pressure between 10^{-4} and 10^{-3} Torr. Then the crystal surface is bombarded by an ion current which is produced by a noble-gas ion gun positioned in front of the crystal surface. A conventional version of such a gun is schematically illustrated in Fig.2.5. Noble-gas ions are produced by electron impact with gas atoms. The ions are then accelerated by a voltage of a few kilovolts towards the sample surface. The ion current and the duration of the bombardment depend on the kind of material and on the thickness of the layer to be removed, i.e. on the degree of contamination. Typical values, e.g. for cleaning of a 1 cm^2 polished Cu surface in UHV are a current of 5 μA for a period of half an hour.

The temperature necessary in the subsequent annealing step is also very much dependent on the material; semiconductor, oxide or metal. For Cu surfaces, temperatures in the region of 500 °C are typical, whereas for Pt and Si up to 1200 °C are applied. The annealing is often done by Ohmic

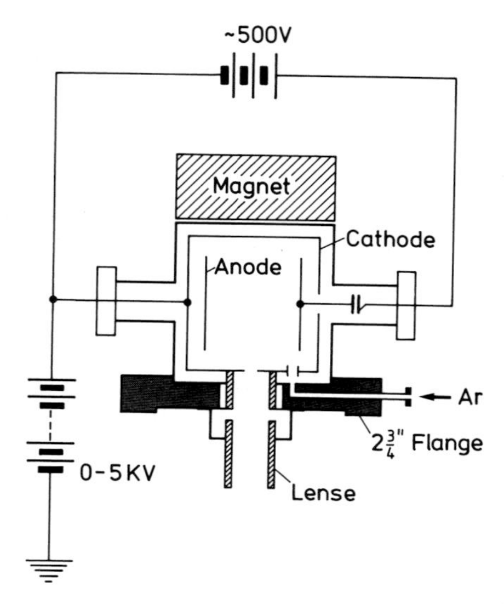

Fig. 2.5. Cold-cathode Ar ion sputter gun. A discharge is burning between cathode and anode ($U \simeq 500\,V$) where the Ar ions are produced by electron impact. The magnet increases the path of the electrons and the lens is used for focussing of the ion current

heating, but also very convenient is electron bombardment from behind the crystal. For this purpose a heated filament at a distance of about 2 cm from the rear side of the crystal is necessary and the crystal is biased at about +2000 V relative to the filament. It is possible that during this heat treatment further impurities from the bulk diffuse to the surface, so that the ion bombardment/annealing cycle has to be repeated several times. Sometimes it is favorable to keep the sample at elevated temperatures during the ion bombardment. It is of particular importance to minimize the partial pressures of residual gases (mainly CO) during the sputtering process since these molecules may become adsorbed and implanted into the lattice after ionisation in the ion gun. Once a surface has been cleaned in UHV by ion bombardment and annealing, subsequent cleaning procedures for later experiments are usually much easier; sometimes only flashing is sufficient.

While noble-gas ion sputtering with subsequent annealing is the most versatile cleaning technique for metal surfaces and elemental semiconductors like Si and Ge, the method has severe disadvantages when applied to composite materials like compound semiconductors, oxides or alloys. Differing sputtering rates for the various components generally cause a high degree of non-stoichiometry on the oxide or semiconductor surface. The composition of alloys can be strongly changed near the surface after such a treatment. Since AES and other analysis techniques are sensitive to at best about 10^{-3} of a monolayer, the non-stoichiometry or the change in composition cannot usually be controlled. For compound semiconductors, therefore, cleavage in UHV is much preferred as a preparation technique. It should be mentioned that in the particular case of Si and Ge(111), cleavage in UHV and ion bombardment and annealing lead to different types of atomic surface structure. Cleaved Si and Ge(111) surfaces exhibit a (2×1) superstructure in LEED, whereas the annealed Si(111) and Ge(111) surfaces exhibit a (7×7) and (2×8) LEED pattern, respectively. Good clean Si(111)-(7×7) surfaces can also be obtained by simply flashing a Si(111) wafer with an epitaxial overlayer up to 1200°C; ion bombardment in this case of an epitaxially grown Si(111) surface is mostly not necessary.

2.4 Evaporation and Molecular Beam Epitaxy (MBE)

One of the classical methods to prepare a fresh, clean surface in UHV is evaporation and condensation of thin films. Polycrystalline metal films of Pt, Pd, Ni, Au, Cu, Al, etc. can easily be prepared in this way. Pt, Pd and Ni can be evaporated from a suitable electrically heated filament; Au and Cu are usually sublimated from a tungsten crucible. Materials with a high melting temperature are most conveniently evaporated by electron bombardment. Depending on the melting point and on wetting properties of the melt, a variety of evaporation methods and devices are used [2.2, 3]. A compilation for a number of materials is given in Table 2.2.

The usual way to control the thickness of a sublimated film is by means of a quartz balance which is mounted close to the sample. In order to avoid contamination levels in the film which exceed the AES detection limit, the background pressure in the UHV system should not be higher than 10^{-9} Torr during evaporation. This can be achieved by sufficient outgasing procedures in advance. Clean films of metals or elemental semiconductors prepared in this way are usually polycrystalline or amorphous, i.e. surface studies which require a specific crystallographic surface orientation are not possible. But a lot of adsorption studies have been carried out on such films in the past. The deposition of metal films on clean semiconductor surfaces has furthermore attracted considerable interest because of their importance in the field of semiconductor-device technology (Schottky barriers, Chap. 8).

Depending on the substrate surface and on the evaporation conditions, sublimated films can also be monocrystalline. In this case they are epitaxial films and the preparation method is then called Molecular Beam Epitaxy (MBE) [2.4-6]. When the monocrystalline film grows on a substrate different from that of the film material the process is called **heteroepitaxy**; when substrate and film are of the same material, it is called **homoepitaxy**. Surfaces of such epitaxial films grown under UHV conditions are ideal for surface studies, they are clean, monocrystalline, their stoichiometry can mostly be controlled by the growth process and there are only few limitations

Table 2.2. Evaporation source table. For a number of materials the melting points (M.P.) are given as well as different possible means (including material) for evaporation [2.2] (W: Tungsten, Ta: Tantalum, Mo: Molybdenum, Ao: Alumina, Bo: Beryllium oxide, Q: Quartz, P: Porcelain, C: Carbon, Fe: Iron, Bn: Boron nitride)

Evaporation material	MP [°C]	Basket	Filament	Boat	Crucible	Remarks and techniques
Aluminum (Al)	660	W	W	W	Bn, C	Wets and alloys readily with tungsten, stranded wire superior to single strand. Aluminum 99.9% ÷pure is necessary for good optical films. Sputters very slowly.
Alumina (Al_2O_3)	2050	W		W		Recommend RF sputtering or reactive sputtering
Antimony (Sb)	630		Mo,Ta	Mo, Ta	Ao,C	Evaporates readily from all sources. Use external heater with crucibles. Toxic.
Arsenic (As)	613		(Sub-limes)		Ao, C	Sublimes, use external heaters with crucible. Toxic.
Barium (Ba)	717	W	W	W,Ta Mo		Readily evaporates from refractory metals without alloying. Chips must be scraped clean of oxides. Sputters moderately.
Beryllium (Be)	1284	W	W	Ta,W Mo	Bo,C	Wets refractory metals. Heat Bo crucible with electron bombardment. Sputters moderately. Very toxic.
Bismuth (Bi)	271	W	W	Ta, W,Mo	Ao,C	Heat crucibles with external heater. Vapors are toxic
Boron (B)	2000			C		Carbon resistance heated boat or strips.
Cadmium (Cd)	321	W	W (Sub-limes)	W,Ta Mo	Ao,P Q	Heat crucible with external basket heaters. Evaporate Cd rapidly to insure condensation on substrate. Sputters moderately. May contaminate vacuum system.
Calcium (Ca)	810	W			Ao	Use external heaters with crucible.
Carbon (C)	3650					Carbon rods pointed and pressed together to make high resistance contact. Use RF Sputtering technique.
Chromium (Cr)	1900	W	W (Sub-limes)	W		Electrodeposit Cr on filaments. Use chips in tapered baskets and dimpled boats. Sputters well.

Evaporation material	M.P [°C]	Basket	Filament	Boat	Crucible	Remarks and techniques
Cobalt (Co)	1478			W	Ao,Bo	Cobalt alloys readily with refractory metals, therefore evaporant should not exceed 30 % of weight of source. Use embedded heater with crucible (H type). Sputters well.
Copper (Cu)	1083	W	W	W,Ta Mo	Ao	Cu evaporates well from all refractory sources. Sputters well, Ao Coated Boat.
Gallium (Ga)	30				Ao,Q,Bo	Reacts with metals and will attack crucibles above 1000°C. Heat crucible with external heater. Ao coated boat
Germanium (Ge)	959	W	W	W,Ta Mo	Ao,C	Ge Wets W, Ta and Mo. Densified graphite resistance source recommended. Sputters fair.
Germanium Oxide GeO_2)	1115	W	W	W,Ta Mo		Evaporates similar to SiO. Use covered boat or box sources.
Gold (Au)	1063	W	W	W,Mo	Ao	Gold wets W and Mo. Filament should be wrapped tightly with fine wire. Sputters well. Ao coated boat.
Indium (In)	157	W		W,Mo		Use Mo for boats.
Ind. oxide (In_2O_3)				Pt		Use platinum sources, Reactively sputter.
Iron (Fe)	1535	W	W	W	Ao,Bo,C	Fe must not exceed 30 % of the refractory metal source. Use embedded W heater with crucibles. Sputters well.
Lead (Pb)	328	W	W	W,Mo	Ao,Fe	Lead does not wet refractory metals well. Use external heaters with crucibles. Sputters well. Toxic.
Lithium (Li)	179				Fe,Q	Use external heater for crucible Attacks Q.
Magnesium (Mg)	651	W	W (sub-limes)	W,Mo Ta	C	Mg sublimes from refractory sources. Produces high reflective coating. Sputters very slow.
Manganese (Mh)	1244	W	W	W,Mo Ta	Ao	Wets refractory metals well. Use either internal or external crucible heaters.
Molybdenum (Mb)	2622					Sputters well. Evaporation source material.

Evaporation material	MP [°C]	Basket	Filament	Boat	Crucible	Remarks and techniques
Nickel (Ni)	1455		W	W	Ao,Bo	Electroplated coating on heavy tungsten filament not to exceed 9% of the filament mass. Use dimple type boat. Sputters very well.
Nichrome (Ni/Cr)	1360			W	Ao	Reacts with refractory source. Use 20% or less. Fractionates. Sputters well, deposits as the alloy.
Palladium (Pd)	1555		W	W	Ao,Bo	Evaporates rapidly. Alloys with refractory metals. Sputters well.
Platinum (Pt)	1774	W	W			Alloys with refractory sources and must be evaporated rapidly. Sputters well.
Selenium (Se)	217			Mo, Ta 304 stain-less steel	Ao, Mo Ta, Q	Wets all sources and may contaminate vacuum system. Toxic.
Silicon (Si)	1420				Bo,C	Use only small amounts with crucible source. Use external heater. Difficult to evaporate free of SiO. Use RF sputtering technique
Silicon Dioxide (SiO_2)	1800					Use RF or reactive Sputtering technique.
Silver (Ag)	961		Ta,Mo	Ta,Mo		Wing Ag wire tightly on filament. Sputters well. Ao Coated Boats.
Tantalum (Ta)	2996					Sputters well. Evaporation source material. Getters O_2.
Tellurium (Te)	452	W		W	Ao,Q	Wets without alloying refractory metals. Heat crucible externally. May contaminate vacuum system. Toxic.
Thallium (T1)	304	Ta		Ta	Ao,Q	Wets without alloying Ta. Use external heater with crucible. Toxic
Thorium	1827	W		W		Wets tungsten readily.
Tin (Sn)	232	Mo,Ta	Mo	Mo, Ta	Ao,C	Wets Mo readily. Sputters well.

Evaporation material	MP [°C]	Basket	Filament	Boat	Crucible	Remarks and techniques
Titanium (Ti)	1727	W,Ta	W,Ta	W	C	Sputters well. Reacts with W and deposits contain traces.
Tungsten (W)	3382					Evaporation source material. Sputters well.
Uranium (U)	1132	W				Forms oxided deposits Sputters well.
Vanadium (V)	1697	W			W,Mo	Alloys slightly with W, wets and does not alloy with Mo. Sputters.
Zinc (Zn)	419	W	(Sub-limes)	W, Ta	Ao, P	Wets all refractory metals. Use external heater with crucibles. May contaminate vacuum system. Sputters well.
Zirconium (Zr)	2127	W	W	W		Zr wets and alloys slightly with W, film contains traces of W. Requires good vacuum to avoid oxidation. Sputters well.

concerning the crystallographic type of the surface. Molecular beam epitaxy is therefore the most versatile technique for preparing clean and well-defined surfaces and interfaces of elemental semiconductors like Si, Ge of compound materials like III-V semiconductors (GaAs, InP, InSb, etc.) and of II-VI compounds such as CdTe, Pbs, etc. Furthermore, there is a strong technological interest in the preparation of epitaxial films of these semiconducting materials.

MBE allows a controlled growth of films with sharp doping profiles and different chemical composition changing over a spatial depth of several Ångstroms. Multilayer structures with alternating doping (n-intrinsic-p-intrinsic = n-i-p-i) [2.7] or alternating bandgap (GaAs-GaAlAs-GaAs-...) can be grown; a whole new field of semiconductor material "tailoring" was made possible by the development of MBE [2.4-8].

It is therefore useful to present some more details about MBE. This technique is discussed with reference to III-V compound semiconductors because of their technological importance and since they also serve as an example for other systems. Figure 2.6 shows the scheme of a typical UHV chamber with facilities for MBE. Knudsen-type crucibles are used as effusion cells for the evaporation. For most purposes these cells are tubular crucibles, open at one end and made from pyrolytic BN (boron nitride) or high-purity graphite. The crucibles are mounted within spiral Ta heater windings which are themselves enclosed within Ta-foil radiation shields. A require-

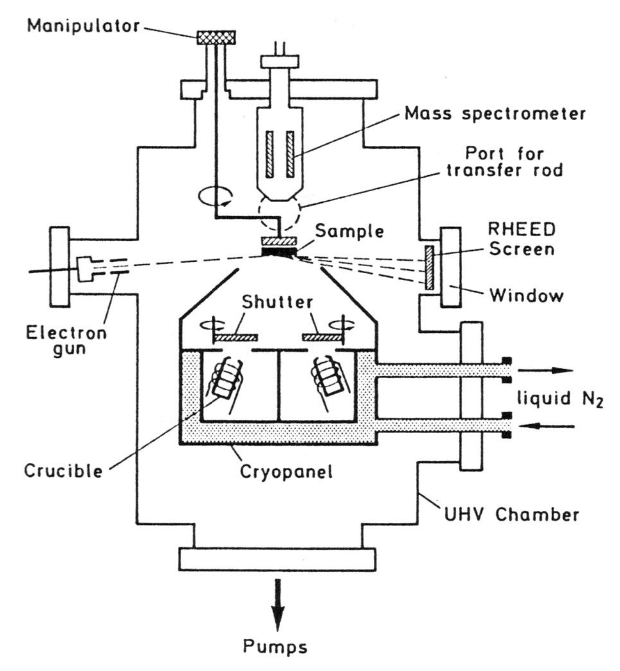

Fig.2.6. Scheme of a simple UHV growth chamber for molecular beam epitaxy (MBE). The surrounding of the evaporation crucibles is cooled by liquid N_2. Facilities for RHEED and mass spectroscopy as well as for transferring the sample into a second UHV chamber are also shown

ment for the source oven and the whole unit is a very low production of impurities in the molecular beam. The oven set-up is therefore surrounded by a liquid-nitrogen cooled cryo-panel on which shutters are mounted which can close and open one or the other effusion cell (mostly automatically controlled). Also the space between sources and the sample is shielded by a cooling shield at liquid nitrogen. Often a mass spectrometer is mounted at a position close to the sample, in order to control and adjust the beam fluxes from the sources. The sample (in the present case a III-V wafer) can be heated to temperatures of at least 700 °C. Homoepitaxial GaAs growth requires substrate temperatures between 500° and 600 °C. In order to obtain an extremely homogeneous temperature profile over the growing surface, the wafer is sometimes "glued" by means of liquid In or Ga to a Mo sample holder, which is electrically heated.

A very useful technique to control the crystallographic structure of the epitaxial surface is RHEED (Panel VIII: Chap.4). Because of the long geometric distances between RHEED electron gun, sample surface and phosphorus screen there is no problem in incorporating this technique into an

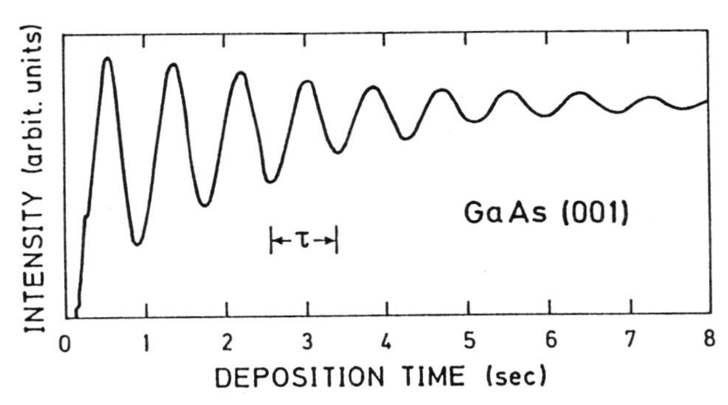

Fig.2.7. RHEED oscillations measured during MBE growth of GaAs(001). The intensity of the particular RHEED spot is measured as a function of deposition time. The oscillation period τ indicates the completion of a monatomic layer

MBE chamber as an in situ method to control the crystallographic structure of the growing surface during the growth process itself.

An excess of Ga or of As on a growing GaAs surface is usually connected with the appearance of particular non-integral-order spots between the regular Bragg spots in the diffraction pattern in RHEED (superstructure, Chap.3). Furthermore, during the growth process itself, the intensity of a single Bragg spot on the RHEED screen can be monitored by an optical device (Fig.2.7). Within a certain range of growth conditions it shows oscillations with a regular period (**RHEED oscillations**) [2.9]. The interpretation of these oscillations is based on the growth mechanism. When a full atomic layer is completed during growth, the 2D periodicity of the topmost atomic layer is nearly ideal and the diffraction on this regular periodic array of atoms causes a certain maximum spot intensity (Sect. 4.3). Further growth leads to irregularly distributed atoms or little islands on top of this complete atomic layer before the next full atomic layer is deposited. In between two complete layers a certain degree of disorder is present on the growing surface. Like in optical diffraction this disorder causes an increase in background intensity, which corresponds to a decrease in intensity of the sharp Bragg spots. The situation of an incomplete topmost atomic layer is characterized by a decreased Bragg spot intensity. Maxima in Fig.2.7 thus indicate completion of growing layers. RHEED oscillation curves as in Fig. 2.7, therefore, allow a layer-by-layer control of the growing crystal surface. By counting the number of maxima one can monitor the number of deposited atomic layers and the thickness of the grown crystal on an atomic scale.

Let us discuss in a little more detail the homoepitaxial growth of GaAs overlayers on a GaAs substrate [2.10], to see the characteristic problems of MBE. Ideally the beam source should be a Knudsen cell containing vapor and condensed phase, e.g. of Ga and As, of Al or of S, Sn, Te, etc. as doping materials, at equilibrium. In this case the flux F at the substrate can be calculated from the equilibrium vapor pressure P(T) in the cell at temperature T. In equilibrium, one assumes that the particle current density leaving the liquid or solid phase by evaporation equals the flux of particles impinging on the liquid or solid surface at a pressure P(T) from the gas phase side. By means of (2.1-3) one obtains

$$F = \frac{P(T)\,a}{\pi L^2 \sqrt{2\pi m k T}} \frac{1}{[s \cdot cm^2]}, \tag{2.5}$$

where a is the area of the cell aperture, L the distance to the substrate, and m the mass of the effusing species. In practice, a true Knudsen source is not very convenient because a wide aperture is needed to provide a useful rate of mass transfer to the growing surface.

For source materials to generate the molecular beams, either pure elements (Ga, As, Al, etc.) or suitable compounds are useful sources of group-V-element molecular beams since they provide stable, well-determined beam fluxes until nearly all of the group-V element is exhausted. Convenient growth rates in MBE are $1 \div 10$ monolayers/s, i.e. $1 \div 10$ Å/s or $0.1 \div 1$ μm/h. This corresponds to an arrival rate F at the substrate of $10^{15} \div 10^{16}$ molecules/$(s \cdot cm^2)$. With typical geometrical factors $L \simeq 5$ cm and $a = 0.5$ cm^2 the equilibrium vapor pressure in the Knudsen cell is obtained according to (2.5) to be in the range $10^{-2} \div 10^{-3}$ Torr. The temperatures needed to establish a pressure of 10^{-2} Torr in the cell can be evaluated from the vapor pressure plots in Fig.2.8. For Ga and As these temperatures T_s are very different (Table 2.3); nevertheless, it is possible to grow GaAs using only a single source with polycrystalline GaAs material. There is an important underlying reason why stoichiometric growth of GaAs is possible even with non-stoichiometric fluxes of Ga and As: The growth rate is limited by the Ga arrival rate. At a growth temperature between 500 and 600°C (substrate temperature) As sticks to the surface in measurable quantities only if Ga is present in excess. Without Ga being present the sticking coefficient for As is negligibly small. Sophisticated procedures to establish a stoichometric flux of the two elements in MBE, are therefore not necessary to enable GaAs epitaxy. Furthermore, growth is always possible under slight As excess flux conditions. The growth is usually started by heating up the substrate to growth conditions, i.e. to at least 500°C in an As beam and then the Ga beam is switched on. This procedure prevents the formation of a high density of As vacancies during annealing to 500°C.

Epitaxial growth on the substrate surface is, of course, only possible, if this surface is free of contamination. Before mounting the wafer into the UHV chamber, the mechanically polished surface is usually etched in 5%

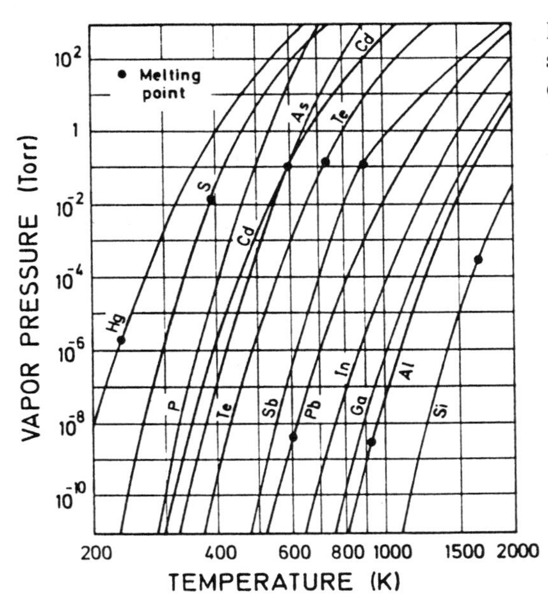

Fig. 2.8. Equilibrium vapor pressure of important compound semiconductor source materials

Table 2.3. Melting temperature T_m for selected materials, and crucible source (Knudsen type) temperature T_s necessary to establish an equilibrium vapor pressure $P(T_s)$ of 10^{-2} Torr. This pressure is convenient to achieve reasonable evaporation rates in MBE

Material	Melting temp. T_m [°C]	Source temp. T_S [°C] for $P(T_S) = 10^{-2}$ Torr
Al	660	1220
Cu	1084	1260
Ge	940	1400
Si	1410	1350
Ga	30	1130
As	613	300

Br-methanol and rinsed in methanol and water. After baking out the system for at least 8 h at 100 ÷ 200 °C, epitaxial growth is possible after heating the sample in the As beam to growth conditions, i.e. to about 500 ÷ 600 °C. Sometimes ion bombardment and annealing cycles are also applied for cleaning before starting the growth procedure.

It is interesting to mention the differences between GaAs layers which have been grown from one single source with polycrystalline GaAs, and those grown using separate sources for Ga and As. In the latter case As evaporates as As_4 and the epitaxial layers usually exhibit a p-type background doping (due to C) in the range of some 10^{14} cm^{-3}. If one single GaAs

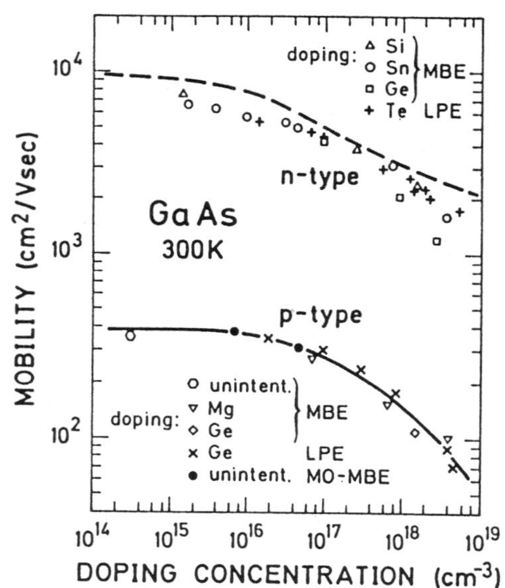

Fig.2.9. Room temperature electron and hole mobilities of GaAs epitaxial layers which have been grown by different groups. Methods: MBE molecular beam epitaxy; MOMBE metal-organic molecular beam epitaxy using triethyl gallium (TEGa) and AsH_3 as sources. p-type doping in MOMBE by carbon is performed by using admixtures of TMGa. The n-type doping in MOMBE is due to Si by using predissociated SiH_4

source is used, As arrives predominantly as As_2 and the films are usually n-type doped in the $5 \cdot 10^{15}$ cm^{-3} range due to contaminants (mostly Si) in the GaAs source material. For applications in semiconductor device technology, intentional doping is performed by additional Knudsen-type sources for Sn, S, Te, Si (n-type) or Mn, Mg, Be (p-type). The quality of an epitaxially grown layer can easily be seen from the free-carrier mobility at a particular carrier concentration or doping level. Results for room-temperature mobilities of GaAs measured by Hall effect are depicted in Fig.2.9.

If epitaxially grown surfaces are to be used as fresh, clean surfaces in surface studies, it is convenient to grow the layers in a separate growth chamber (Figs.2.6 and 10) and to transfer them under UHV conditions in situ by a transfer mechanism (magnetically or mechanically operated) into a second UHV chamber where the investigations are made.

Many aspects of the GaAs homoepitaxy considered here as an example, can also be extended to other systems. Of high practical importance is also the homoepitaxy of Si on Si wafers in a UHV system [2.11]. Because of the high melting temperature of Si, an electron gun evaporator is used here as the Si source. Si is evaporated from a crucible by bombardment with an electron beam.

Good examples of heteroepitaxy are alloys of III-V compounds which are epitaxially grown in MBE systems both for scientific and for commercial reasons [2.12]. The main technical applications of these materials are in the field of very fast devices (because of their high electronic mobility) and in optoelectronics. For the latter field, it is important that many of the III-V

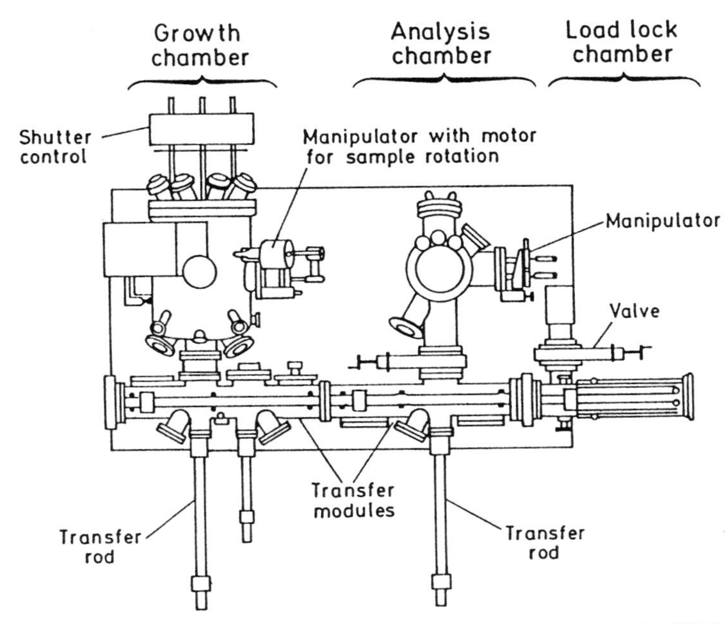

Fig. 2.10. Schematic top view over a combination of growth UHV chamber for MBE (or MOMBE) with analysis and load-lock chamber. All UHV units are separately pumped (ion, cryo and turbo pumps are not shown). The sample, usually wafers, can be moved through the various transfer modules (mechanically or magnetically operated) and transferred into the corresponding chambers by the transfer rods

compounds are direct-gap semiconductors where the conduction-band minimum and valence-band maximum are at the same **k** vector in reciprocal space. Electronic transitions between states with equal initial and final **k** vector can couple strongly to electromagnetic fields.

On the other hand, by alloying a third component, e.g. P into GaAs, it is possible to change the direct semiconductor GaAs step by step through different $GaAs_{1-x}P_x$ compositions into the indirect semiconductor GaP. Figure 2.11 shows how the bulk electronic band structure of the ternary compound $GaAs_{1-x}P_x$ depends on the mole fraction x of P [2.13]. At $x = 0.45$ the compound switches from a direct to an indirect semiconductor, the conduction band minimum at the symmetry point X of the Brillouin zone drops to a lower energy than the Γ-point ($\mathbf{k} = 0$) minimum.

It is evident that MBE is an ideal experimental technique to grow such ternary or even quaternary compounds in a well defined manner – just by controlling the flux rate of the different compound substituents being evaporated from different crucibles. In contrast to other methods of epitaxy (liquid-phase or conventional, normal-pressure vapor-phase epitaxy) MBE allows a fast change (within fractions of a second) of the composition. For

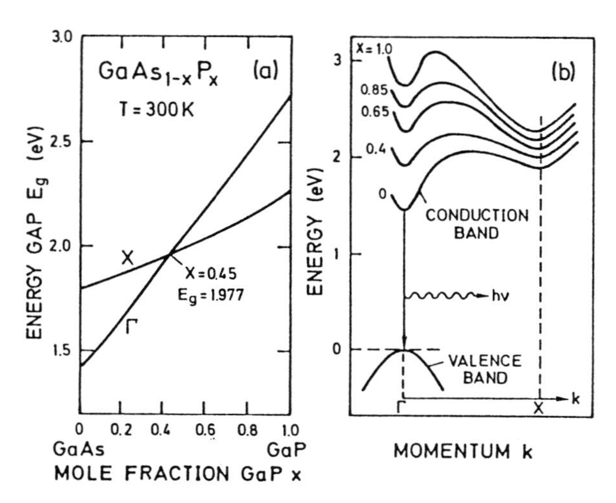

Fig.2.11. (**a**) Compositional dependence of the direct and indirect energy bandgap for $GaAs_{1-x}P_x$ at 300 K, and (**b**) schematic energy-versus wavevector dependence E(k) of valence and conduction band for various alloy compositions x [2.13]

growth rates of μm/h (i.e., Å/s) one can switch over from one component to another during the growth time for a single monolayer. Thus atomically sharp growth of profiles can be generated by MBE. This facilitates the fabrication of extremely well-defined interfaces between two semiconductors (Chap. 8).

As one would expect for simple geometric reasons, two materials can grow epitaxially on each other with a high-quality interface, and particularly as thicker films, if the crystallographic mismatch is low. The diagram (Fig. 2.12) of the lattice constants of major elemental and compound semiconductors helps in choosing appropriate materials for certain applications. Because of very similar lattice constants, good epitaxial films with high-quality interfaces can be grown within the GaAs/AlAs system. In MBE, GaAs can be grown on AlAs, or vice versa; and more interesting ternary compounds of arbitrary composition $Al_xGa_{1-x}As$ can be grown on GaAs or on AlAs wafers. According to Fig. 2.12, where the corresponding band-gap energies and the type of gap (full line: direct; broken line: indirect) are also indicated, $Al_xGa_{1-x}As$ switches from a direct- into an indirect-gap material at about $x \simeq 0.5$. With changing composition x, the band-gap energy ranges from about 1.4 eV on the GaAs side to 2.15 eV for AlAs (indirect gap). According to Fig. 2.12 other III-V alloys suitable for good heteroepitaxial growth in MBE are AlP/GaP and AlSb/GaSb. But II-VI materials can also be combined with III-V semiconductors, e.g., InP/CdS or InSb/PbTe/CdTe.

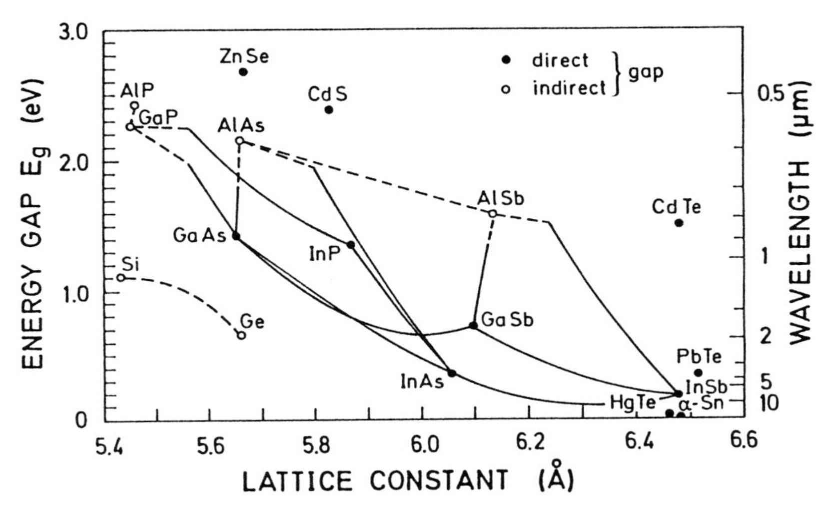

Fig.2.12. Energy gap [eV] and corresponding optical wavelengths [µm] versus lattice constant at 300 K for important semiconductors. The connection lines describe the behavior of corresponding alloys. Direct gap materials are plotted as solid lines, whereas indirect gap semiconductors are shown by broken lines

Interesting heteroepitaxy is also possible for elemental semiconductors on III-V compounds, and vice versa. A lot of experimental work has been done, e.g., for Ge on GaAs and Si on GaP (Fig.2.12). On GaAs(110) surfaces cleaved in UHV Ge grows epitaxially at substrate temperatures above 300°C. Below this temperature the deposited Ge layer is polycrystalline.

Another interesting case is that of Sn on InSb (Fig.2.12). The tetrahedrally (sp^3) bonded α-Sn modification (a zero-gap semiconductor) is stable as bulk material only below 287 K; above this temperature the stable phase of Sn is its metallic β-modification. If however Sn is deposited by MBE onto an ion bombarded and annealed InSb(100) surface at 300 K, Sn grows as a semiconductor in its α-modification. The films are stable up to about 150°C and for higher temperatures a change into β-Sn is observed [2.14]. On a clean-cleaved InSb(110) surface a polycrystalline, tetrahedrally bonded Sn species grows at 300 K. As one would expect from Fig.2.12, α-Sn can also be grown by MBE on CdTe or PbTe as substrates.

Heteroepitaxy in MBE is, of course, also possible for metals. Interesting systems for future applications in semiconductor device technology might be related to epitaxy of metals on semiconductors, and vice versa. Epitaxially grown cobalt silicides on Si (Sect.8.5) have been used to fabricate a fast metal-base transistor [2.15]. Since GaAs and Fe differ in their lattice constant by a factor of about two, it is possible to grow crystalline Fe films on both GaAs(100) and on GaAs(110) at a substrate temperature of 180°C [2.16]. Even though the Fe/GaAs interface is not sharp and well defined because of considerable interdiffusion and interface reaction, the iron films exhibit crystallographic structure seen in electron diffraction (LEED, RHEED, Panel VIII: Chap.4).

2.5 Epitaxy by Means of Chemical Reactions

Under standard pressure conditions, epitaxy of III-V compound semiconductors by means of chemical reactions is well established as an important method to produce epilayers for semiconductor device technology. In a cold-wall flux reactor GaAs can be grown on a GaAs substrate at $500 \div 600\,^{\circ}C$ from gas streams of arsine (AsH_3) and trimethyl gallium [TMGa = $Ga(CH_3)_3$] using H_2, N_2 etc. as a carrier gas. This technique belongs to a more general class of processes, which are called **Metal-Organic Chemical Vapor Deposition** (MOCVD). This technique of growing GaAs from arsine and TMGa or TEGa [$Ga(C_2H_5)_3$] can also be applied under UHV conditions [2.17]. The method which is, of course, also applicable to many other III-V systems (GaInAs, InP, etc.), then combines the advantages of a UHV technique like MBE with those of a continuously running preparation method, useful in applied device technology. In the present context it is relevant that this method, which is called **Metal-Organic MBE** (MOMBE), or Chemical Beam Epitaxy (CBE), can be used in situ to prepare clean, well-defined surfaces of compound semiconductors [2.18, 19]. The epitaxial layers can be transferred from the growth UHV chamber into a separate chamber, where surface or interface studies can be performed under UHV conditions. Figure 2.13a shows schematically a UHV growth chamber in which GaAs can be grown epitaxially by means of MOMBE. As is the case in MBE, the sample and the sources of AsH_3 and TMGa or TEGa are surrounded by a liquid nitrogen cooled cryo-panel to minimize the contamination level during growth. The mounting for the GaAs wafer used as substrate are the same as in MBE; the wafer might be glued by Ga or In to a Mo support which is heated during growth to temperatures between 500° and $600\,^{\circ}C$. In more recent equipment the wafers are only mechanically clamped to the holder and heated by radiation from the rear. The sources of AsH_3 and the Metal-Organic (MO) compound, both in the gas phase, consist in the simplest case of UHV leak valves followed by the inlet capillaries, which form the molecular source beams. While the inlet capillary for the metal-organic compound has to be heated only slightly above room temperature (in order to avoid condensation due to the adjacent cryoshield) the gaseous group-V starting materials (AsH_3, PH_3, etc.) with high thermal stability have to be precracked in the injection capillary. The simplest experi-

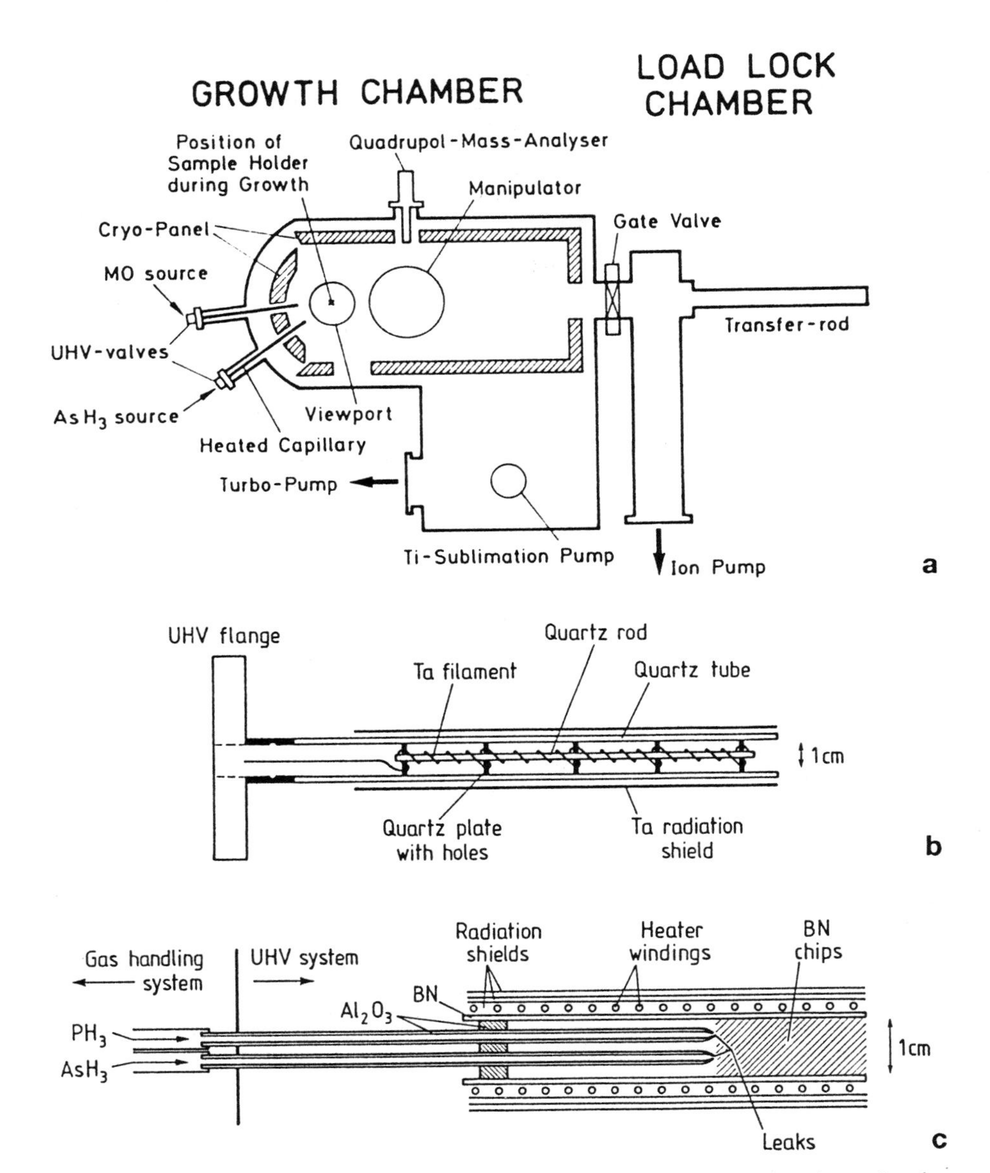

Fig.2.13a-c. Scheme of the experimental equipment for metal-organic molecular beam epitaxy (MOMBE): (**a**) Side view of the growth chamber with load-lock unit. (**b**) Low-pressure inlet capillary for precracking the hydrides (AsH_3, PH_3, etc.) by a heated Ta filament [2.17, 18]. (**c**) High-pressure source for simultaneous inlet of PH_3 and AsH_3. The precracking is performed by gas phase collisions in the heated Al_2O_3 capillaries [2.20]

mental approach consists in a so-called **low-pressure cracking capillary**, where the hydrides AsH_3, PH_3, etc. are thermally decomposed by means of a heated metal filament (e.g., Ta or W at $\approx$1000K) (Fig.2.13b). Typically the AsH_3 gas is injected into the UHV system via a controllable leak valve, which allows a reproducible adjustment of the beam pressure, and thus of the flux to within about 0.2%. During its passage through the quartz capillary along the heated filament, the gas beam changes from laminar (hydrodynamic) flow (10$\div$300Pa) to molecular flow conditions ($\approx 10^{-3}$ Pa). In such a set-up, up to 90% decomposition of AsH_3 is achieved.

A different principle operates in the high-pressure effusion source (Fig.2.13c) [2.20]. AsH_3 and PH_3 at a pressure between 0.2 and 2 atm are injected through alumina tubes with fixed, small leaks into the UHV growth chamber. These tubes are mounted within the UHV system in an electri-

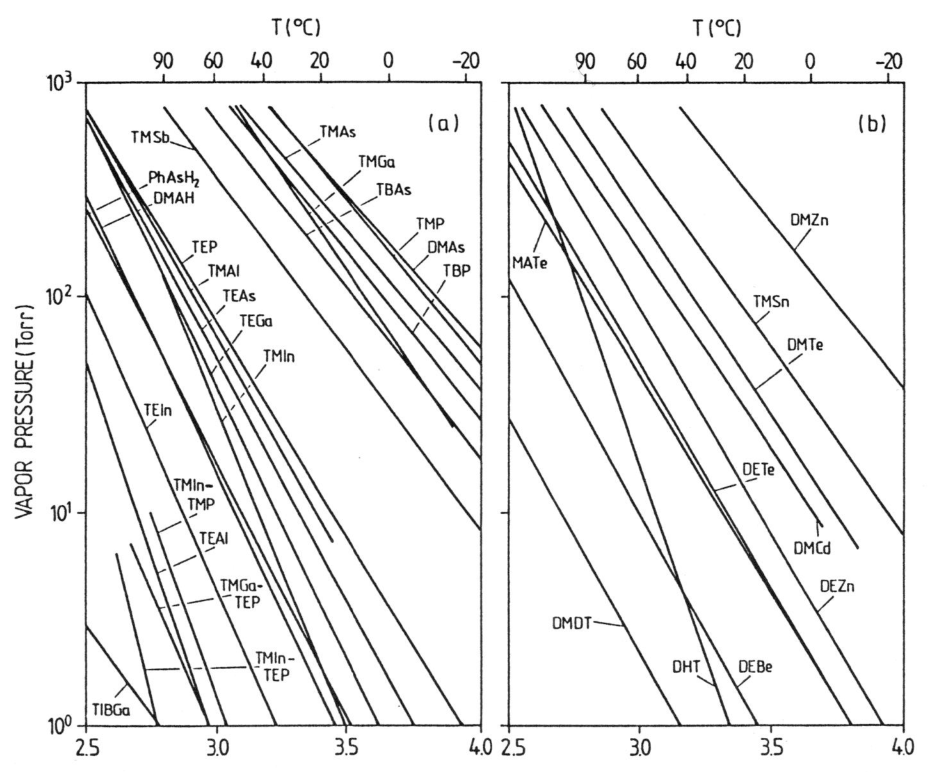

Fig.2.14. Temperature dependence of vapor pressures for the common group III and group V (**a**) as well as for the group II and group VI (**b**) organometallic source materials. The group II and group VI materials can be used for p- and n-type doping of III-V semiconductor layers. The explanation of the symbols is as follows: TIBGa: Triisobutyl gallium, TMAl: Trimethyl aluminun, TEAs: Triethyl arsenic, $PhAsH_2$: Phenylarsine, etc. [2.21]

cally heated oven, where the hydrides are thermally decomposed by gas phase collisions at temperatures between 900° and 1000°C. On their path through the leak inlet, the transition from hydrodynamic to molecular flow occurs. The leak is essentially a free jet. Behind this leak the decomposition products are injected into a heated low-pressure zone, where the pressure is in the millitorr range or lower, since this region is directly pumped by the system vacuum. In contrast to the simpler low-pressure sources, the flux in this high-pressure cell is controlled by pressure variation in the alumina tube with its constant leak. In designing the supply of the metal-organic compounds one has to take account of the fact that, in contrast to TMGa and TEGa, a number of interesting materials such as TEIn (triethyl-indium) and TEAl (triethyl-aluminum) for the growth of InAs and AlAs, have an extremely low vapor pressure at room temperature (Fig.2.14) [2.21]. To achieve reasonable growth rates in the μm/h range the storage vessel and the gas lines up to the UHV inlet valve thus have to be heated moderately to increase the MO vapor pressure.

In contrast to standard MBE, where the UHV system is usually pumped by ion pumps, MOMBE is performed by means of cryopumps in combination with turbomolecular or diffusion pumps (Fig.2.13a). But in common with MBE, the substrate surface has to be cleaned by chemical etching and final rinsing in water before mounting it into the growth chamber. Starting from a background pressure of about 10^{-10} Torr after bake-out, the growth process begins with the heating of the substrate in the cracked AsH_3 molecular beam to $500 \div 600$°C (growth temperature); then the metal-organic molecular beam is switched on. During growth the total pressure in the chamber rises up to about 10^{-6} Torr, with hydrogen and hydrocarbons as decomposition products being the major components within the residual gas. For the growth of GaAs the metal-organic components TMGa [$Ga(CH_3)_3$] and TEGa [$Ga(C_2H_5)_3$] are used. In contrast to AsH_3 they are not predissociated but react on the surface to form Ga which becomes incorporated into the growing crystal.

With both TMGa and TEGa, carbon resulting from the decomposition of the metal-organic is also incorporated into the growing GaAs film as an acceptor (C on an As site). The III/V ratio, of course, can be used as a parameter to control the p-type doping level. The higher the MO beam pressure at constant AsH_3 flux, the higher the resulting p-doping level. There is, however, a considerable difference in the minimum attainable doping levels. The use of TMGa always gives rise to hole concentrations between 10^{19} and 10^{21} cm^{-3} at 300 K, whereas the less stable TEGa must be used to obtain doping levels between 10^{14} and 10^{16} cm^{-3}. Using well-defined mixtures of the two metal-organics, the intermediate doping regime can also be covered and p-doped GaAs layers with hole concentrations between 10^{14} and 10^{21} cm^{-3} can be grown in a controlled way (Fig.2.15) [2.22].

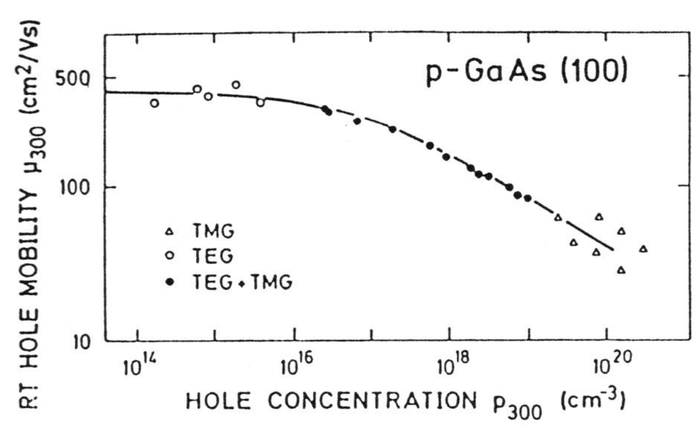

Fig.2.15. Room-temperature Hall mobilities μ_{300} of MOMBE-grown, intentionally p-doped (carbon) GaAs layers [orientation (100)] by the use of TMGa only (Δ), TEGa only (o), and mixtures of both alkyls (●). Solid line: best mobilities from literature [2.22]

The room-temperature Hall mobility, i.e. the electrical (crystallographic) quality of these layers is comparable with that of layers obtained by other techniques (MOCVD, MBE) (Figs.2.9, 15)

The n-type doping of III-V compound layers can also be performed by means of gas line sources. Using SiH_4 (5% in H_2 carrier gas) for Si doping of GaAs, electron concentrations between 10^{15} and 10^{19} cm^{-3} can be achieved at room temperature. For doping levels above 10^{16} cm^{-3}, the SiH_4 has to be predissociated in a similar way to the AsH_3 using the same type of inlet capillary. The electron mobilities obtained for GaAs are also comparable to characteristic values for MBE grown layers (Fig.2.9).

It has been shown in general that both the electrical, and the crystallographic and morphological quality of MOMBE grown layers of GaAs, GaInAs, etc. is as good as that of layers produced by MBE. Concerning the density of the surface defects that inevitably occur during the growth process, the MOMBE process is superior to MBE. Larger, so-called **oval defects** having typical diameters of about 10 μm are absent on MOMBE-grown surfaces [2.23].

From the experience with III-V epitaxy from gaseous, in particular MO sources, it seems likely that a number of other semiconductor and metal overlayers might also be deposited in a UHV system by means of gaseous source materials. Beside metal-organics such as TMGa, TEGa, TEIn, etc., carbonyls such as $Ni(CO)_5$, $Fe(CO)_5$, etc., might also be interesting [2.24].

Another interesting aspect emerges when one compares MBE and MOMBE. As compared with MBE the surface chemical reactions leading to film growth in MOMBE are far more complex (dissociation of TEG, etc.). This allows a defined control of the reactions and thus of the growth process by external parameters such as light irradiation or electron bombardment. By scanning a focused light beam of appropriate photon energy over the growing surface, growth can be laterally enhanced and suppressed in other areas which are not hit by the light beam. Two-dimensional structures can thus be produced during the growth process itself.

Techniques such as MBE and MOMBE are best applied to surface research as in situ methods, i.e. the freshly grown surface is transferred under UHV conditions from the growth chamber into another so-called **analysis chamber**, which is coupled by a UHV valve and a transfer unit to the growth chamber (Fig.2.10). In the particular case of GaAs or related compound surfaces prepared by MBE or MOMBE, however, an interesting method exists which allows preparation and analysis of the clean surfaces in separate UHV systems. The freshly grown surface is passivated in the growth chamber by the deposition of an amorphous As layer after completion of the growth process [2.25]. For this purpose the epitaxial film is cooled down after growth in the arsenic beam (MBE) or in the predissociated AsH_3 beam (MOMBE). The surface is thus "capped" by an amorphous As film which protects it against contamination when brought to atmosphere.

After loading the sample into a separate UHV chamber for surface research, mild annealing up to about 300 °C causes desorption of the As film and the MBE or MOMBE grown surface is exposed in a clean and crystallographically well-preserved state. The desorption temperatures of around 300 °C are below the limit where decomposition of the surface or the formation of defects (arsenic vacancies, etc.) occurs. Thus, freshly prepared III-V compound semiconductor surfaces can even be transferred between different laboratories.

In conclusion, MBE and MOMBE open new possibilities for preparing clean and well-defined surfaces and interfaces in a UHV environment, without many of the restrictions concerning the specific crystallographic type of surface or interface. In this field, the demands of research on interface physics are simultaneously beneficial in the interests of technology, where these techniques allow the production of sophisticated semiconductor layer structures, devices and circuits [2.26].

Panel III
Auger Electron Spectroscopy (AES)

Auger Electron Spectroscopy (AES) is a standard analysis technique in surface and interface physics [III.1-3]. It is used predominantly to check the cleanliness of a freshly prepared surface under UHV conditions. Other important fields of application include studies of film growth and surface-chemical composition (elemental analysis) as well as depth profiling of the concentration of particular chemical elements. The last of these applications involves alternate sputtering and AES stages.

AES is an electron core-level spectroscopy, in which the excitation process is induced by a primary electron beam from an electron gun. The Auger process results in secondary electrons of relatively sharply-defined energy, which are energy analysed and detected by a standard electron analyser (Panel II: Chap. 1). Cylindrical Mirror Analyzers (CMA) are widely used in this connection. As with all other electron spectroscopies, AES is surface sensitive because of the limited escape depth of electrons. According to Fig. 4.1 observation of Auger electrons with a kinetic energy around 1000 eV means an observation depth of about 15 Å. Typical probing depths in AES are in the range 10 ÷ 30 Å.

The principle of the Auger process is explained in Fig. III.1. The primary electron produces an initial hole by ionization of a core level (K or L shell). Both primary electron and core electron then leave the atom with an ill-defined energy; the escaping primary electron has lost its "memory" due to the complexity of the scattering process. The electronic structure of the ionized atom rearranges such that the deep initial hole in the core level is filled by an electron originating from an energetically higher-lying shell. This transition may be accompanied by the emission of a characteristic X-ray photon, or alternatively the deexcitation process might be a radiationless Auger transition, in which the energy gained by the electron that "falls" into the deeper atomic level is transferred to another electron of the same or a different shell. This latter electron is then emitted with a characteristic Auger energy, thereby leaving the atom in a double-ionized state [two holes in different (or the same) core levels]. The characteristic Auger energy is close to the characteristic X-ray photon energy but, due to many-body interactions, it is not identical. In comparison to the X-ray emission process the final state of the atom now has one more hole and is thus more highly ionized.

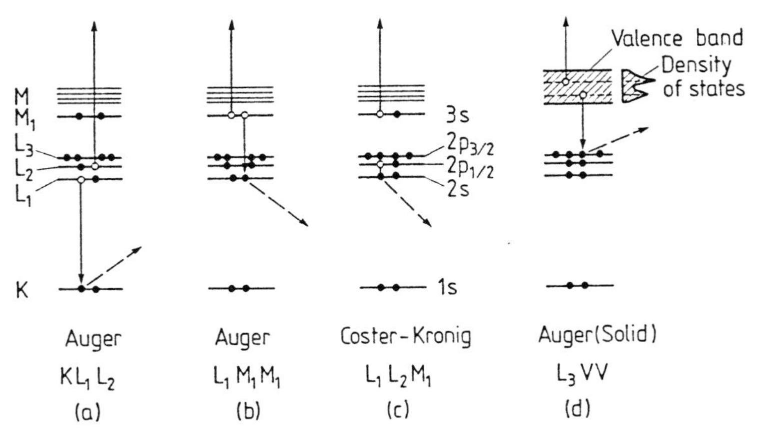

Fig. III. 1a-d. Explanation of the Auger process on the basis of atomic-level schemes. A primary electron produces an initial hole in a core level and the escaping electron is indicated by a broken arrow; another electron is deexcited from a higher shell, core levels in (**a**, **b**, **c**) and the valence band of a solid in (**d**). The deexcitation energy is then transferred to a third electron, which leaves the system as an Auger electron

Since the emitted Auger electron carries a well-defined kinetic energy that is directly related to differences in core-level energies, measurement of this energy can be used to identify the particular atom. Chemical element analysis is possible in the same way as with characteristic X-ray emission, but in AES with much higher surface sensitivity.

The nomenclature of Auger transitions reflects the core levels involved (Fig. III.1). When the primary hole is produced in the K shell, the Auger process is initiated by an outer electron from the L shell, e.g. the L_1 level as in Fig. III.1. This electron falls into the initial K vacancy giving up its transition energy to another electron from the L shell, e.g. the L_2 shell; such an Auger process is termed a **KL_1L_2 process** (Fig. III.1a). Another possibility is shown in Fig. III.1b. In this case the two final holes are both in the M_1 shell. Since the initial hole was in the L_1 shell, this is known as an $L_1M_1M_1$ process. If the initial hole is filled by an electron from the same shell (Fig. III.1c), the process is called a **Coster-Kronig transition** (e.g., $L_1L_2M_1$).

When the Auger process occurs in an atom that is bound in a solid, electronic bands may be involved in the transition, in addition to sharply-defined core levels. The process shown in Fig. III.1d involves the formation of a primary hole in the L_3 shell and deexcitation via an electron from the valence band (V), which transfers its transition energy to another valence-band electron. This process is correspondingly called an **L_3VV process**. The strongest intensity is observed for processes in which the two final holes are produced in regions of a high valence-band density of states.

To illustrate the calculation of the characteristic energy of an Auger transition, we consider as an example the KL_1L_2 process of Fig.III.1a. In a simple, one-electron picture the kinetic energy of the outgoing Auger electron would be given by a difference between the corresponding core-level energies: $E_{kin} = E_K - E_{L_1} - E_{L_2}$. These energies can be obtained from X-ray photoemission spectroscopy (XPS, Sect.6.3). By using experimental XPS data, one already takes into account many-electron relaxation effects (Sect. 6.3). However, the Auger process differs from photoemission by the formation of an additional core hole. A further correction term ΔE is therefore used to describe the many-electron effects related to the corresponding rearrangement of the other electrons. The KL_1L_2 process thus yields an Auger electron with the energy

$$E^Z_{KL_1L_2} = E^Z_K - E^Z_{L_1} - E^Z_{L_2} - \Delta E(L_1L_2) , \tag{III.1}$$

where Z is the atomic number of the element concerned. The correction term $\Delta E(L_1, L_2)$ is small; it involves an increase in binding energy of the L_2 electron when the L_1 electron is removed, and of the L_1 electron when an L_2 electron is removed. The detailed calculation of the correction term is, of course, difficult, but there is a reasonable empirical formula, which relates the higher ionization states of atom Z to the core-level energies of the atom with the atomic number $Z+1$. The average increase in binding energy due to a missing electron in the L_1 shell is thus approximately expressed by $(E^{Z+1}_{L_1} - E^Z_{L_1})/2$ and the correction term follows as

$$\Delta E(L_1L_2) = \frac{1}{2}\left[E^{Z+1}_{L_2} - E^Z_{L_2} + E^{Z+1}_{L_1} - E^Z_{L_1}\right] . \tag{III.2}$$

As an example we consider the KL_1L_2 process for an Fe atom ($Z = 26$). The experimentally observed Auger energy is $E^{Fe}_{KL_1L_2} = 5480$ eV. For an approximate calculation of this energy we use the core level energies determined from XPS: $E^{Fe}_K = 7114$ eV, $E^{Fe}_{L_1} = 846$ eV, $E^{Fe}_{L_2} = 723$ eV. Furthermore, for the correction term (III.2) the corresponding experimental binding energies for Co ($Z = 27$) are taken as $E^{Co}_{L_2} = 794$ eV, $E^{Co}_{L_1} = 926$ eV. As an approximate value one thus obtains $E^{Fe}_{KL_1L_2} \simeq 5470$ eV, a value which deviates only by 10 eV from the Auger energy actually observed.

The principal Auger electron energies of the elements are given versus atomic number in Fig.III.2. Three main branches, the KLL, LMM and the MNN processes can be distinguished. The stronger transitions are indicated by heavier points. The various transitions within a single group KLL, LMM

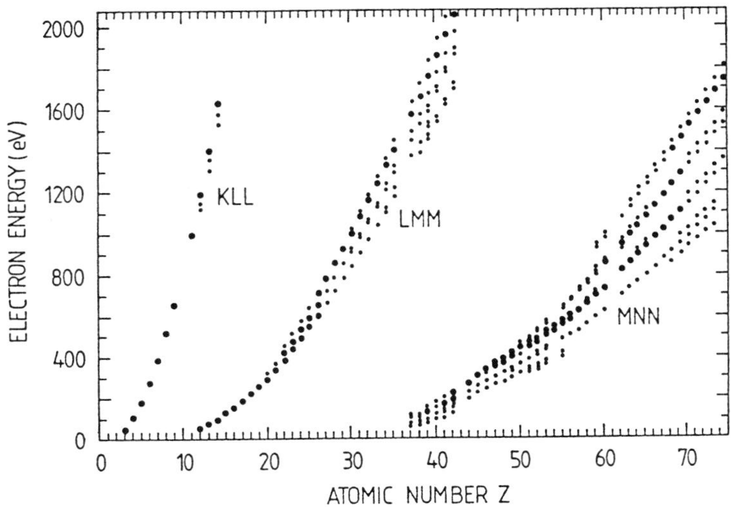

Fig.III.2. Principal Auger electron energies as a function of the atomic number Z. The strongest transitions of each element are indicated by bold points [III.2]

or MNN result from different spin orientations in the final state of the atom. One realizes that atoms with less than 3 electrons cannot undergo Auger transitions. The strong Z dependence of the binding energies and of the Auger energies (Fig.III.2) is important for the application of AES as a chemical analysis technique.

An Auger transition is a complex process involving several different steps, but the crucial interaction is between the electron filling the initial core hole and the electron taking up the corresponding energy to be emitted from the atom as an Auger electron. This energy transfer from one electron to the other is facilitated mainly by the Coulomb interaction. Auger transition probabilities can thus be approximately calculated using Coulomb interaction potentials of the type $e^2/|\mathbf{r}_1-\mathbf{r}_2|$ for two electrons at $\mathbf{r}_1$ and $\mathbf{r}_2$. The transition probability for a KLL process (Fig.III.1a) is thus obtained as

$$w_{KLL} \propto \left| \psi_{1s}^*(\mathbf{r}_1) e^{-i\mathbf{k}\cdot\mathbf{r}_2} \frac{e^2}{|\mathbf{r}_1 - \mathbf{r}_2|} \psi_{2s}(\mathbf{r}_1) \psi_{2p}(\mathbf{r}_2) \right|^2 . \qquad \text{(III.3)}$$

The two-electron initial state $\psi_{2s}(\mathbf{r}_1)\psi_{2p}(\mathbf{r}_2)$ is described by the two single-electron wave functions 2s and 2p. The final state $\psi_{1s}(\mathbf{r}_1)\exp(i\mathbf{k}\cdot\mathbf{r}_2)$ contains electron *1* in its 1s state and electron *2* escapes as a free electron with the wave vector **k** (plane wave state). More complex many-electron effects are not contained in this simple description. A detailed calculation yields, as its main result, that the Auger-transition probability is roughly independent of Z, in contrast to the strong Z dependence of radiative transitions. Furthermore Auger processes do not obey the dipole selection rules that govern optical transitions. The transition probability is determined essentially by the Coulomb interaction and not by a dipole matrix element. For example, the prominent KL_1L_1 Auger transition is forbidden optically, since it does not satisfy $\Delta\ell = \pm 1$ and $\Delta j = \pm 1, 0$.

The standard equipment for AES consists of an electron gun, which produces the primary electron beam with a typical energy of 2000 to 5000 eV. The most commonly used energy analysers for Auger electrons are hemispherical or Cylindrical Mirror Analysers (CMA). The electron gun is sometimes integrated into the CMA on its central axis (Fig.III.3). This is particularly useful for depth profiling, where AES is combined with ion sputtering. Because of the small Auger signals AES is usually carried out in the derivative mode to suppress the large background of true secondary electrons. The differentiation is performed by superimposing a small alternating voltage $v = v_0 \sin\omega t$ on the outer cylinder voltage V and synchronously detecting the in-phase signal from the electron multiplier with a lock-in amplifier. In this mode the detector current

$$I(V + v_0 \sin\omega t) \simeq I_0 + \frac{dI}{dv} v_0 \sin\omega t + \dots \qquad \text{(III.4)}$$

contains the first derivative dI/dv as the prefactor of the phase-sensitively detected AC signal with the angular frequency ω.

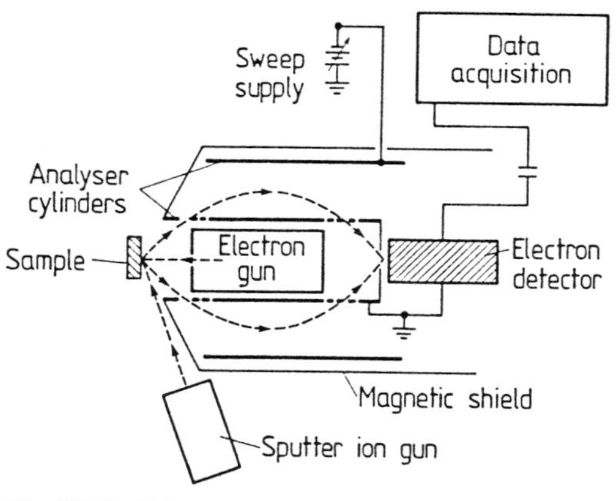

Fig.III.3. Schematic plot of a standard experimental set-up for Auger Electron Spectroscopy (AES). The primary electron beam is generated by an electron gun which is integrated on the central axis of a Cylindrical Mirror Analyser (CMA). An additional sputter ion gun provides the possibility of depth analysis

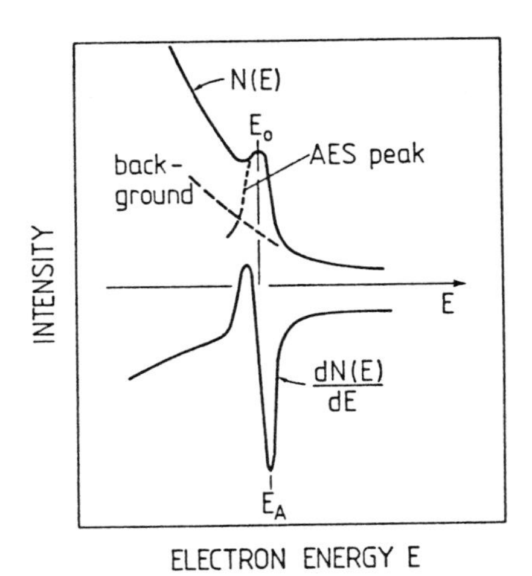

Fig.III.4. Qualitative comparison of a non-differentiated Auger spectrum N(E) with its differentiated counterpart dN(E)/dE (*lower plot*). The AES peak with a maximum at E_0 generates a "resonance"-like structure in dN/dE, whose most negative excursion at E_A corresponds to the steepest slope of N(E)

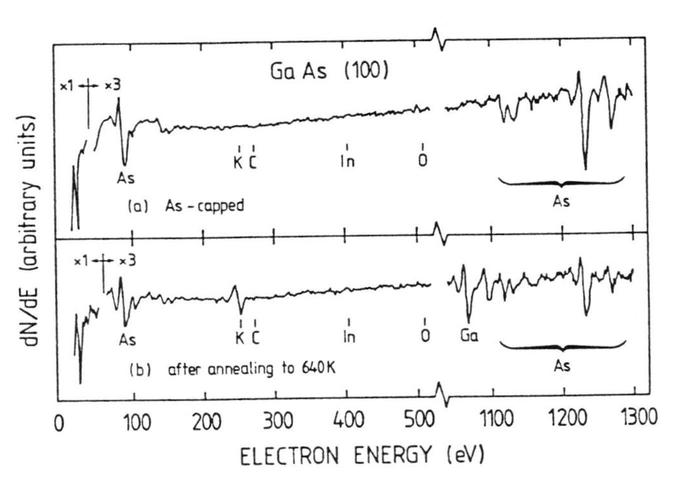

Fig.III.5a,b. Differentiated Auger electron spectra dN(E)/dE measured with a primary electron energy of 2000 eV on a GaAs(100) surface prepared by Metal-Organic Molecular Beam Epitaxy (MOMBE). (**a**) After the epitaxy process the surface was covered in the MOMBE system by an amorphous arsenic film and transferred through air into the analysis chamber for the AES analysis. This spectrum corresponds to the As-capped surface. (**b**) After mild annealing to about 300°C the arsenic film is desorbed and the characteristic spectrum of the GaAs surface appears, with very slight contamination due to K [III.4]

On the basis of this detection mode, Auger line energies are usually given in reference works as the position of the minimum of the derivative spectrum dN/dE (Fig.III.4). This energy, of course, does not coincide with the maximum of the Auger peak in the non-differentiated spectrum. As an example of the application of AES, Fig.III.5 shows differentiated dN/dE spectra measured for a nearly clean GaAs surface (*b*) and for the same surface covered with an amorphous As film (*a*). The GaAs(100) surface had been grown in a Metal-Organic Molecular Beam Epitaxy (MOMBE) system (Sect.2.5). In order to transfer the clean, freshly grown surface through atmosphere into another UHV system, after growth the surface was covered with a passivating As layer. This As layer is removed in the second UHV system by mild annealing to about 350°C and a well-defined, clean, and well-ordered GaAs surface appears. The Auger spectrum of the As-covered surface (Fig.III.5a) shows a number of high-energy As peaks between 1100 and 1300 eV due to LMM processes (Fig.III.2). In addition, low-energy As lines are seen at energies below 100 eV. The flat curve between 200 and 500 eV reveals that none of the common contaminants K, C, In or O are present in the As overlayer, at least not within the detection limit of AES. After desorption of the passivating As film (Fig.III.5b) new peaks appear slightly below 1100 eV. These are Ga LMM peaks stemming from the topmost Ga atoms of the "clean" GaAs surface. Around 250 eV a small K Auger signal (LMM transition) indicates a slight K contamination of the GaAs(100) surface. From a differentiated spectrum such as that of Fig.III.5 it is difficult to accurately evaluate the Auger line shape and detect minor shifts due to a different chemical surrounding. In order to use AES to investigate these details, non-differentiated spectra must be measured.

The sensitivity of AES to small amounts of surface contamination is usually no better than 1% of a monolayer. Surfaces that appear clean in AES might well show contamination when examined with photoemission (UPS, XPS) or Electron Energy Loss Spectroscopy (EELS, HREELS, Panel IX: Chap.4). A further severe disadvantage of AES relates to its applicability to semiconductor surfaces. Due to the high energy and current density of the primary electron beam, defects are produced at a relatively high density, particularly on compound semiconductors such as GaAs or ZnO. These defects can cause dramatic changes in the electronic properties of the surface: electronic surface states and thus space-charge layers (Sect.6.7) are formed. Thus AES can usefully be combined with optical or electronic studies of a semiconductor surface only in a final analysis step.

References

III.1 B.K. Agarwal: *X-Ray Spectroscopy*, 2nd edn., Springer Ser. Opt. Sci., Vol.15 (Springer, Berlin, Heidelberg 1991)

III.2 L.E. Davis, N.C. McDonald, P.W. Palmberg, G.E. Riach, R.E. Weber: *Handbook of Auger Electron Spectroscopy* (Physical Electronics Industries, Eden Prairie, MN 1976)

III.3 J.C. Fuggle, J.E. Inglesfield: *Unoccupied Electronic States*, Topics Appl. Phys., Vol.69 (Springer, Berlin, Heidelberg 1992)

III.4 B.J. Schäfer, A. Förster, M. Londschien, A. Tulke, K. Werner, M. Kamp, H. Heinecke, M. Weyers, H. Lüth, P. Balk: Surf. Sci. **204**, 485 (1988)

Panel IV
Secondary Ion Mass Spectroscopy (SIMS)

In Secondary Ion Mass Spectroscopy (SIMS) [IV.1,2] a primary ion beam consisting, e.g., of Ar^+ ions with a typical energy between 1 and 10 keV is incident on a surface. Due to the transferred impact energy neutral atoms, molecules and ions – so-called **secondary ions** – are emitted from the surface; they are analysed and detected by a mass spectrometer. The measured mass spectrum then yields information about the chemical composition of the surface. This type of static SIMS is used in surface physics to study the composition of the topmost atomic layer, including the nature and properties of adsorbed layers. In a second type, the dynamic SIMS, higher primary-beam currents are used. Thus, a much higher rate of emission of secondary ions results, and the sputtering process removes considerable quantities of material. During this process one monitors the secondary-ion mass spectra, which yield information about the chemical elements contained in the removed material. This kind of measurement allows a layer by layer analysis of the substrate, i.e. a depth profiling of the chemical composition. This method is extremely useful in studies of thin films.

The main components of a typical static SIMS set-up are illustrated in Fig.IV.1. The whole apparatus (Fig.IV.1a) operates under UHV conditions and is connected via a flange to a UHV analysis chamber containing sample-handling systems, the mass spectrometer (usually a quadrupole mass spectrometer) and often other equipment for LEED (Panel VIII: Chap.4), AES (Panel III: Chap.2), etc. The ion source is usually a discharge source located in the ionization chamber. The primary ions are accelerated to energies of $1 \div 10$ keV and are focussed prior to the passing through a magnetic mass separator (magnetic sector field normal to ion beam). Having acquired a well-defined ion mass and energy, the ion beam is then directed through an aperture (beam centering plates) onto the sample. The primary beam should have a homogeneous current density over its cross section (typically $\approx 0.1\,cm^2$), and this should be controllable over a wide range between 10^{-4} and 10^{-10} A/cm^2, in order to be able to change the sputtering conditions. The secondary-ion beam emitted from the sample surface is accelerated by an applied voltage before entering the quadrupole mass filter. After mass separation an electrostatic mirror reflects the secondary ions into the detector, usually an electron multiplier. The detector is positioned out of "direct sight" of the beam, in order to avoid the measurement of neutral

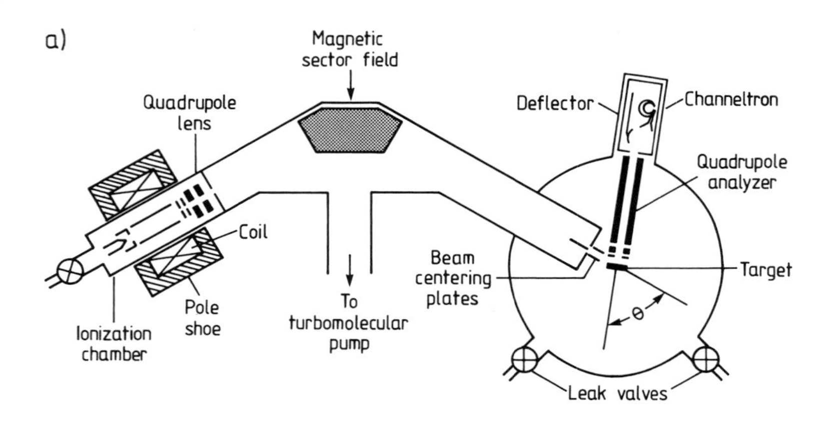

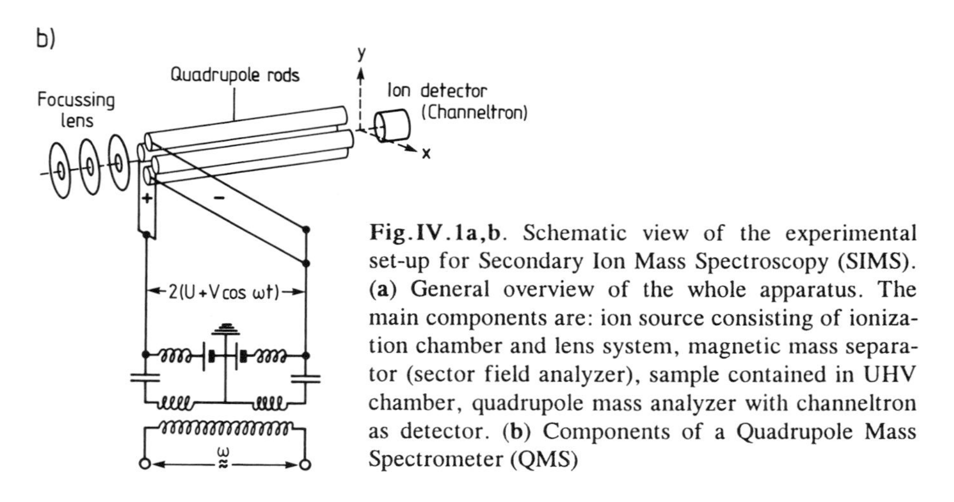

Fig.IV.1a,b. Schematic view of the experimental set-up for Secondary Ion Mass Spectroscopy (SIMS). (**a**) General overview of the whole apparatus. The main components are: ion source consisting of ionization chamber and lens system, magnetic mass separator (sector field analyzer), sample contained in UHV chamber, quadrupole mass analyzer with channeltron as detector. (**b**) Components of a Quadrupole Mass Spectrometer (QMS)

particles and photons, etc., which would give rise to a considerable background.

An important part of the equipment is the Quadrupole Mass Spectrometer (QMS), which contains as basic components four quadrupole rods (Fig. IV.1b). The rods are pairwise biased with a superposition of a DC voltage U and an AC component $V\cos\omega t$ ($\omega/2\pi$ being typically 1MHz). The electric potential near the z-axis of the analyzer is thus

$$\phi(x,y,t) = \frac{1}{{r_0}^2}(U + V\cos\omega t)(x^2 - y^2)\,, \qquad \text{(IV.1)}$$

where r_0 is the radius of the rod (typically 5mm). For the mass analysis U and V are scanned. V is typically 1 kV at maximum, whereas the DC component U is chosen to be about V/6 for optimum performance. With the electric field $\mathcal{E} = -\nabla\phi$, (IV.1) yields the following dynamic equations for a positive ion (charge q, mass m) entering the rod system along z through an aperture:

$$\begin{bmatrix} \ddot{x} \\ \ddot{y} \\ \ddot{z} \end{bmatrix} = \frac{2q}{m r_0^2}(U + V\cos\omega t)\begin{bmatrix} -x \\ y \\ 0 \end{bmatrix}. \qquad \text{(IV.2)}$$

In the z direction the ion is not accelerated. The solution of (IV.2) for the x and y coordinates leads to differential equations of the Mathieu type. But the following qualitative argument is sufficient to show that the dynamics described by (IV.2) causes mass separation. Because of their inertia, heavy ions cannot follow the high-frequency field. For high enough masses (IV.2) can be approximated by

$$\begin{bmatrix} \ddot{x} \\ \ddot{y} \end{bmatrix} \simeq \frac{2qU}{m r_0^2}\begin{bmatrix} -x \\ y \end{bmatrix}, \qquad \text{(IV.3)}$$

i.e., a harmonic oscillation with frequency $\Omega = (2qU/mr_0^2)^{1/2}$ in the x-direction, whereas in the y-direction the motion is unbounded with $y(t) \propto \exp(\Omega t)$. For rods of length 10 ÷ 20 cm the ion hits the electrodes and is discharged. Low masses, however, can follow the high-frequency field, which dominates ($V > U$) the DC bias. In the x-y plane the light ions perform an oscillatory motion with the frequency of the AC bias. Their amplitude increases, and discharging at the rods might result. In the y-direction the DC field component has a defocussing effect; for small y values at least, the resulting force is directed outwards. With increasing amplitude, i.e. increasing distance from the central axis, the effect of the AC field becomes stronger and causes the ion to move back to the axis. The result is a stable oscillation around the central axis, a motion which allows the ion to pass through the filter. To summarize, heavy ions can carry out stable oscillations in the x-direction, whereas in the y-direction light ions can pass the filter under certain conditions. For a narrow mass range there is overlap between the two regimes, ions can pass both in x- and y-directions, and can thus pass the whole rod arrangement. When the ratio of AC/DC components (V/U) is about 6, this critical pass regime contains only one mass. For a mass scan, therefore, the voltages V and U are varied simultaneously, maintaining a constant ratio. In this way the masses are successively "shifted" through the "window", where stable oscillations between the rods are possible.

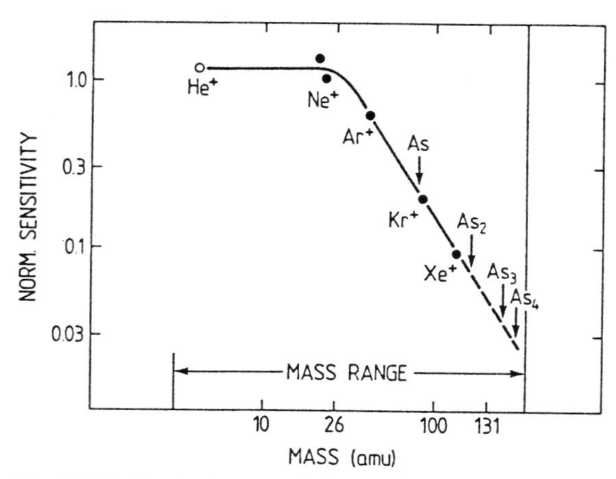

Fig.IV.2. Typical transmission ratio of a quadrupole mass filter as a function of mass number (amu) in the range $4 \div 350$. Important arsenic masses As_x are indicated by arrows [IV.3]

When a mass spectrum needs to be analyzed quantitatively, one must keep in mind that the transmittance of a QMS is dependent on mass (Fig. IV.2). Calibration of the signal intensity versus mass must therefore be performed by means of well-defined inert gas mixtures. It is worth mentioning that, for mass spectroscopic measurements other than those involved in normal SIMS, the ionization chamber in front of the quadrupole rods causes partial cracking of the incoming molecules. The dissociation products are detected in well-defined ratios, the so-called **cracking pattern**, which have to be known for a detailed analysis of mass spectroscopic data.

Besides the physics related to mass analysis in the QMS, the sputtering process due to the action of the primary ion beam is also of importance in SIMS [IV.4]. The energy transfer from the incoming primary ion to a substrate atom near the surface occurs via a cascade of two-body collisions (Fig.IV.3). This collision cascade is more or less destructive for the sample. Lattice defects are produced, primary ions are implanted in the topmost atomic layers and, finally, surface substrate (or adsorbate) atoms are removed as neutrals or as secondary ions which are detected in the QMS. Removal of a surface atom in the sputtering process requires that the elastically transferred energy exceeds the binding energy. Correspondingly, three regimes of sputtering by elastic collisions can be distinguished (Fig.IV.4). In the **single-knock-on regime**, atoms recoiling from the ion-target collision receive sufficient energy to be sputtered out of the sample, but not enough to generate recoil cascades (Fig.IV.4a). This regime applies mainly to primary ion energies below 1 keV. In the **linear-cascade regime** (primary energy $1\,\text{keV} \div 1\,\text{MeV}$) the recoil atoms themselves carry enough energy to

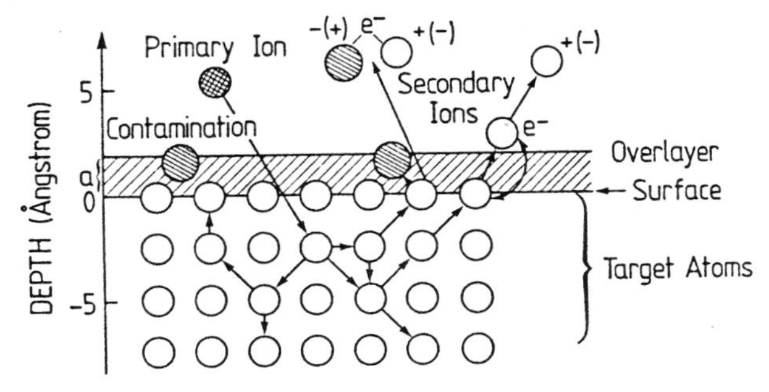

Fig. IV.3. Schematic representation of the sputtering process. The cascade of single particle collisions involves the impact of a primary ion, formation of defects, implantation of ions, and removal of a surface atom (substrate or adsorbate) as a neutral or a secondary ion

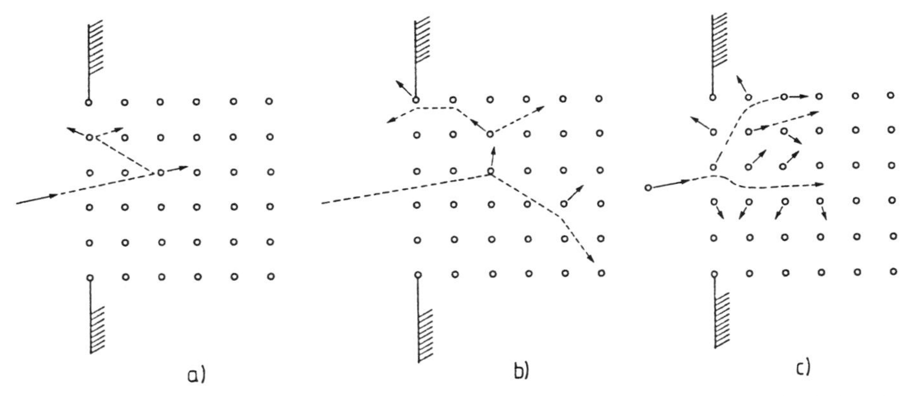

Fig. IV.4a-c. Three regimes of sputtering by elastic collisions. **(a)** The single knock-on regime; recoil atoms from ion-target collisions receive sufficiently high energy to be sputtered. **(b)** The linear cascade regime; recoil atoms from ion-target collisions receive sufficiently high energy to generate recoil cascades. **(c)** The spike regime; the density of recoil atoms is so high that the majority of atoms within a certain volume are in motion

produce further recoils. A cascade is generated, but the density of recoil atoms is low enough that knock-on collisions dominate and collisions between moving atoms are infrequent (Fig. IV.4b). For heavy primary ions with high energy the so-called **spike regime** is attained, where the density of recoil atoms is so high that within a certain volume (spike volume), the majority of atoms are in motion.

The basic parameter for a quantitative description of the sputtering process is the sputter yield Y (number of sputtered surface particles per incident particle). With a primary ion current density of $j_{PI} = e\nu$ (ν being the primary ion flux density), the number of sputtered particles during time dt is given by

$$- dN = NY\nu A dt \tag{IV.4}$$

where A is the surface area hit by the beam and N the density of surface atoms. For an adsorbate layer which is sputtered off, the coverage $\theta(t)$ is given by the ratio $N(t)/N_{max}$. Equation (IV.4) has the solution

$$N(t) = N_{max} \exp\left[\frac{-Yj_{PI}}{eN_{max}} t\right], \tag{IV.5}$$

i.e., the sputtering efficiency can be described by the average lifetime

$$\bar{\tau} = \frac{eN_{max}}{Yj_{PI}} \tag{IV.6}$$

for a target atom in the surface. The important parameter, the primary ion-current density j_{PI}, controls the sputtering rate as expected. The degree of surface destruction induced by the method depends sensitively on the magnitude of the primary ion current density.

For current densities of 10^{-4} and 10^{-9} A/cm^2 the average lifetimes are in the order of 0.3 s and 9 h, respectively; i.e. about 3 and $3 \cdot 10^{-5}$ monolayers of atoms are removed per second, respectively. This estimate clearly highlights the very different experimental conditions involved in static (less destructive) and dynamic SIMS.

Another factor determining the signal strength in SIMS is the degree of ionization. The secondary particles (Fig.IV.3) are mostly emitted as neutrals. For example, for clean metal surfaces, less than 5% of the secondaries are ionized. During the emission process the secondary particle is in permanent interaction with the surface and possibly with other particles whose chemical bonds are likewise being disrupted (Fig.IV.3). This complex interaction is responsible for excitation, ionization or deexcitation and neutralization of the emitted particle. To a good approximation, one can assume that the formation of the collision cascade and its propagation have no effect on the excitation and ionization state of the individual secondary particle. Although model descriptions in terms of Auger processes (Panel III: Chap.2) and auto-ionization processes of highly excited atoms and molecules do exist for the surface excitation processes, practical SIMS analysis is based mainly on empirical data for sputter yields and ionization cross sections (Table IV.1) [IV.5, 6].

Table IV.1. Empirical values for absolute positive ion yields S^+ of clean metals and their respective oxides. The experimental values have been determined using Ar^+ as primary ions [IV.5]. The degree of ionization $\alpha = S^+/S$ was calulated assuming that the sputter yield is the same for metals and oxides [IV.6]

Element	S^+_{clean}	α	S^+_{oxide}	α	S^+_{oxide}/S^+_{clean}	$\bar{Y}$
Mg	$8.5\cdot10^{-3}$	$4\cdot10^{-3}$	$1.6\cdot10^{-1}$	$8\cdot10^{-2}$	20	2.1
Al	$2\cdot10^{-2}$	$1\cdot10^{-2}$	2	1	100	2
V	$1.3\cdot10^{-3}$	$7\cdot10^{-4}$	1.2	$6\cdot10^{-1}$	1000	1.9
Cr	$5\cdot10^{-3}$	$3\cdot10^{-3}$	1.2	$6\cdot10^{-1}$	200	1.8
Fe	$1\cdot10^{-3}$	$5\cdot10^{-4}$	$3.8\cdot10^{-1}$	$2\cdot10^{-1}$	380	2
Ni	$3\cdot10^{-4}$	$2\cdot10^{-4}$	$2\cdot10^{-2}$	$1\cdot10^{-2}$	70	1.7
Cu	$1.3\cdot10^{-4}$	$7\cdot10^{-5}$	$4.5\cdot10^{-3}$	$2\cdot10^{-3}$	30	2.4
Sr	$2\cdot10^{-4}$	$1\cdot10^{-4}$	$1.3\cdot10^{-1}$	$7\cdot10^{-2}$	700	1.3

Panel IV

In **static SIMS** primary current densities in the 10^{-9} to 10^{-10} A/cm² range are used, and the sputtering rate is correspondingly extremely low, on the order of 10^{-4} to 10^{-5} monolayers per second. The accompanying destruction of the surface is very minor, and the method is mainly used to study the topmost atomic layer; investigations of surface composition, adsorption processes and surface chemical reactions are the major fields of application. The low primary current densities result in very low secondary ion-current densities ($<10^{-16}$ A/cm²), which require sensitive detection equipment, e.g. pulse counting. Using empirical data (Table IV.1) for the sputtering yield Y and the degree of ionization, α, even a quantitative determination of adsorbate (mass M) coverages θ_M is possible. From the expressions (IV.4-6) one obtains for low sputtering rates, a secondary ion current $I_{SI}(M)$ at mass M of

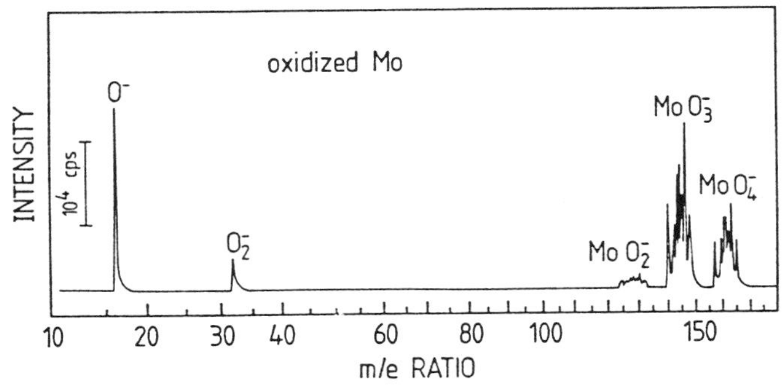

Fig.IV.5. Spectrum of negative secondary ions sputtered from a Mo surface which was cleaned by ion bombardment and subsequently oxidized in 100 L oxygen [IV.7]

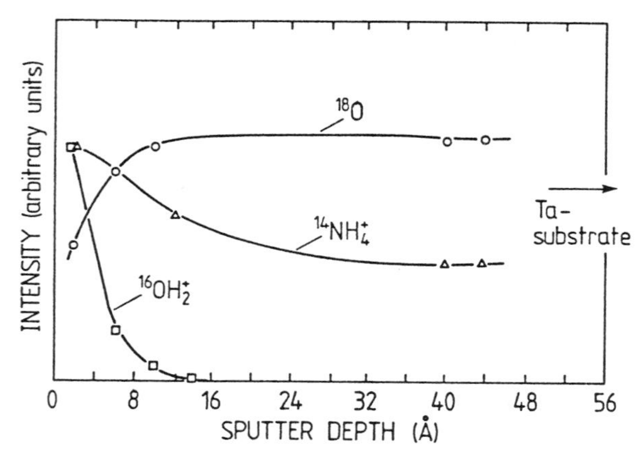

Fig.IV.6. Depth profile (ion intensity versus sputtering depth) of a Ta oxide layer obtained by anodic oxidation of a Ta substrate in an aqueous solution of ammonium citrate enriched with ^{18}O [IV.8]

$$I_{SI}(M) = I_{PI} Y \alpha \eta \theta_M \ , \tag{IV.7}$$

where I_{PI} is the primary ion current, and η the transmissivity of the QMS. Using characteristic data from Table IV.1 and a typical QMS transmissivity η of about 0.01, one can estimate that in static SIMS the detection limit for adsorbed atoms can be as low as 10^{-6} monolayers in favourable cases. The surface sensitivity of SIMS is therefore higher by several orders of magnitude than that of most electron spectroscopies (AES, UPS, XPS, EELS, etc.). As an example Fig.IV.5 shows the negative secondary ion spectrum of a clean Mo surface, which was exposed to 10^{-4} Torr·s (100 Langmuir) oxygen. The spectrum was obtained by sputtering off less than 1% of a monolayer with 3 keV Ar^+ ions at a current density of 10^{-9} A/cm^2. The observation of MoO_x^- ions indicates the presence of an oxide. A problem arising in the interpretation of SIMS data is that, for more complex systems, a certain fraction of the observed ions may originate from interactions between the primary beam and the target atoms or molecules.

In **dynamic SIMS** the high sputtering rates needed for depth profiling (several monolayers per second) are obtained by primary ion current densities of 10^{-4} to 10^{-5} A/cm^2. More than one mass signal is usually recorded versus time during the erosion process. For a constant incident current, the time elapsed can be directly related to a depth scale. High sputtering rates limit the depth resolution since mass signals from several atomic layers are mixed in the detector causing sharp profiles to be smeared out. In practice, the sputtering rate must be carefully adjusted in order to combine the optimum depth resolution with the required maximum erosion depth. Figure IV.6 exhibits a SIMS depth profile recorded with high spatial and mass reso-

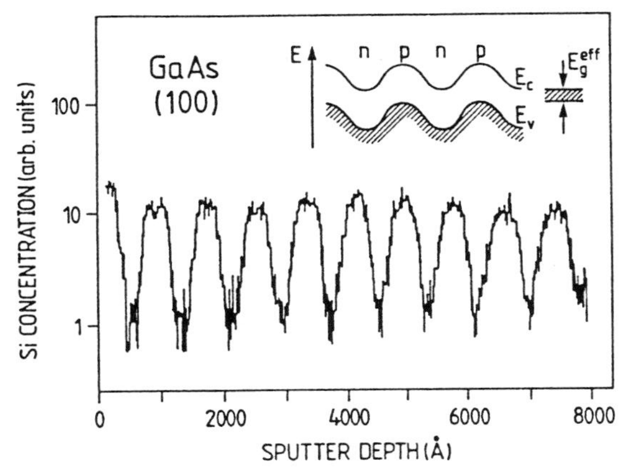

Fig. IV.7. SIMS depth profile of a GaAs(100) nipi doping superlattice with 10 periods (800Å thick). p-doping was performed with carbon (C) and n-doping with Si. Nominal concentrations are about 10^{18} cm^{-3}. Inset: qualitative band structure of a nipi superlattice [IV.9]

lution (the three mass signals belong to M = 18, but mass separation is achieved by measuring the different isotopes). One data point required a measurement time of 0.2 s, during which about two monolayers were sputtered off. In Fig. IV.7, on the other hand, a depth profile of an alternately n- and p-doped GaAs film (produced in MOMBE; Sect. 2.5) is displayed. The Si mass signal (Si is the n-type dopant) was recorded versus sputtering depth.

References

IV.1 H.W. Werner: Introduction to secondary ion mass spectroscopy (SIMS), in *Electron and Ion Spectroscopy of Solids*, ed. by L. Fiermans, J. Vennik, W. Dekeyser (Plenum, New York 1977) p.342
A. Benninghoven, F.G. Rüdenauer, H.W. Werner: *Secondary Ion Mass Spectrometry* (Wiley, New York 1987)
D.J. O'Connor, B.A. Sexton, R.St.C. Smart (eds.): *Surface Analysis Methods in Materials Science*, Springer Ser. Mater. Sci., Vol.23 (Springer, Berlin, Heidelberg 1992) Chap.5

IV.2 Oechsner H. (ed.): *Thin Film Depth Profile Analysis*, Topics Curr. Phys., Vol.37 (Springer, Berlin, Heidelberg 1984)

IV.3 J. Fischer: Entwicklung quantitiver Meßverfahren für die Massenspektrometrie in der MOMBE und Erprobung an einer Arsenquelle, Diploma Thesis, Rheinisch-Westfälische Technische Hochschule Aachen (1986)

IV.4 R. Behrisch (ed.): *Sputtering by Particle Bombardment I*, Topics Appl. Phys., Vol.47 (Springer, Berlin, Heidelberg 1981)

IV.5 A. Benninghoven: Surf. Sci. **28**, 541 (1971)

IV.6 A. Benninghoven: Surf. Sci. **35**, 427 (1973)
IV.7 A. Benninghoven: In Hochvakuumfachbericht der Balzers AG für Hochvakuumtechnik und dünne Schichten (FL 9496 Balzers) Analytik (Dezember 1971) KG2
IV.8 R. Hernandez, P. Lanusse, G. Slodzian, G. Vidal: Rech. Aerospatiale **6**, 313 (1972)
IV.9 H. Heinecke, K. Werner, M. Weyers, H. Lüth, P. Balk: J. Crystal Growth **81**, 270 (1987)
H. Lüth: Inst. Phys. Conf. Ser. **82**, 135 (1986)

Problems

Problem 2.1. Cubic alkali-halide crystals cleave along the {100} faces, III-V materials like GaAs along {110}, whereas silicon cleaves along {111}. Give arguments for this particular cleavage behavior.

Problem 2.2. Tellurium is a convenient n-doping material in the MBE growth of GaAs. What is the Te-beam flux F needed at the growing GaAs surface (growth rate $1\,\mu$m/h), in order to reach a bulk doping level of $N_D = 10^{17}$ cm^{-3}, when every impinging Te atom is assumed to be built into the growing layer. To what temperature must the Te crucible be heated when the distance between substrate and crucible opening (open area: $0.5\,\mathrm{cm}^2$) amounts to 5 cm?

Problem 2.3. What Ar pressure is needed for $4.5 \cdot 10^{20}$ Ar atoms to impinge on a circular surface with diamater of 1.5 mm at a temperature of 425 K.

Problem 2.4. In the MBE growth of GaAs the sticking coefficient of Ga on GaAs is assumed to be one. The arsenic vapor pressure is adjusted such that the growing surface is As-stabilized. By use of the equilibrium vapor-pressure curves (Fig. 2.8) estimate the Ga crucible temperature for a growth rate of $1\,\mu$m/h for GaAs. The opening of the Knudsen cell is assumed to be point-like and 25 cm away from the center of the substrate wafer. The lattice constant of GaAs amounts to 0.565 nm.

The growing GaAs layer shall be n-doped with Si. The probability for Si incorporation is one. What is the necessary Si crucible temperature to reach a doping level of 10^{18} cm^{-3}?

What is the achieved growth homogeneity on a non-rotating 5 cm (2 inch) wafer when the Knudsen cell is directed with its axis under an angle of 30° to the wafer-surface normal?

3. Morphology and Structure of Surfaces and Interfaces

To begin with, it will be useful to give a brief definition of the terms *morphology* and *structure*. The term morphology is associated with a macroscopic property of solids. The word originates from the Greek μορφή, which means form or shape, and here it will be used to refer to the macroscopic form or shape of a surface or interface. Structure, on the other hand, is associated more with a microscopic, atomistic picture and will be used to denote the detailed geometrical arrangement of atoms and their relative positions in space.

The distinction between the two terms, however, is sometimes not so clear, even in the case of a clean, well-defined surface prepared in UHV (Chap.2). What we consider as morphology, i.e. as shape, depends on the type of property being considered and on the resolution of the technique used for its observation. Furthermore, the atomistic structure may often determine, or at least have a significant influence on, the morphology of an interface. For example, details of the interatomic forces determine whether a metal deposited on a semiconductor surface grows layer by layer or whether islands are formed. It is thus necessary to consider both aspects, morphology and structure, in a little more detail. For this purpose one has to approach the problem of an interface from both macroscopic and atomistic viewpoints.

3.1 Surface Tension and Macroscopic Shape

The most general macroscopic approach to a problem in the physics of matter is that of thermodynamics. The specific features associated with a thermodynamic description of an interface are illustrated in Fig.3.1 for the example of a solid/vapor interface. Through this interface the solid maintains thermodynamic equilibrium with its vapor. The interface region separates the two homogeneous phases, but the interface itself is not a sharp, well-defined geometrical surface. All physical quantities, e.g. the density in Fig. 3.1, change more or less gradually from their solid to their vapor values (ρ_s and ρ_v in Fig.3.1). The interface region whose spatial extent t is of atomis-

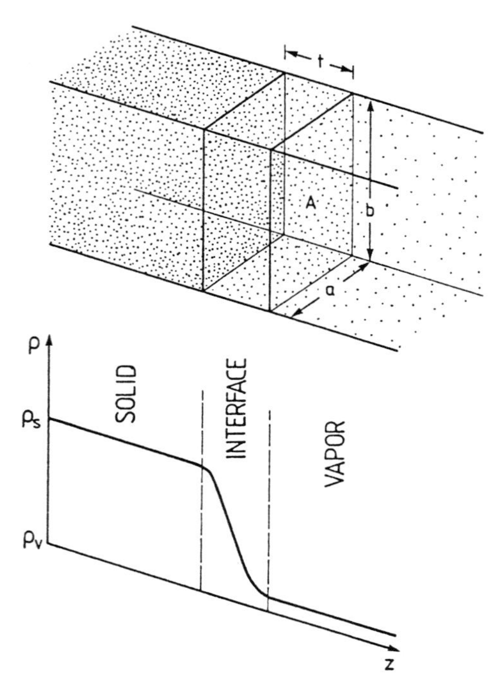

Fig.3.1. Schematic description of a solid/vapor interface (area A, thickness t). The solid is in thermodynamic equilibrium with its vapor. In the interface region the density ρ of the material changes gradually from its solid value ρ_s to its vapor value ρ_v

tic dimensions (depending on the system $2 \div 100$ Å), is thus a strongly inhomogeneous region in equilibrium with two homogeneous phases, the solid and the vapor. Since the interface layer is a material system with a well-defined volume and material content, most thermodynamic properties are defined as usual. Temperature, free energy, composition, chemical potential per particle etc., all have their usual meaning, as in the neighboring homogeneous phases. The only quantity that has to be considered more carefully, is the pressure p [3.1, 2]. In a homogeneous bulk phase the force across any unit area is equal in all directions. This is no longer true for the interface region. Within the interface, the force across any plane of unit area parallel to A (Fig.3.1) has the same value, i.e. the pressure p. It is also the same in both homogeneous phases in thermodynamic equilibrium. If, on the other hand, we choose a plane perpendicular to the interface plane, e.g. the plane with area bt, the force across this area will not be pbt, in general, but will differ from this by a certain amount. If we write the force $f_\perp$ across this plane bt normal to the interface plane as

$$f_\perp = pbt - \gamma b \;, \tag{3.1}$$

then γ defines the so-called **interfacial** or **surface tension**. We now consider a certain volume V^s of the interface region, given by the surface area A and the interface thickness t (Fig.3.1). We allow the volume to be changed into $V^s + dV^s$ by an area change to $A + dA$ and a thickness change to $t + dt$ with the material content remaining unaltered. The total work done by the different forces during this volume change is

$$- pAdt - (pt - \gamma)dA = - p(Adt + tdA) + \gamma dA = - pdV^s + \gamma dA \, . \quad (3.2)$$

For an interface region the expression (3.2) takes the place of the conventional $-pdV$ term of homogeneous bulk phases.

The variation of the free energy of a homogeneous bulk phase is

$$dF = - SdT - pdV + \sum_i \mu_i dn_i \quad (3.3)$$

where S is the entropy, T the temperature, and μ_i the chemical potential (per mole or particle) for the ith kind of particles. With (3.2) the most general variation of the free energy for a surface or interface phase is thus

$$dF^s = - S^s dT - pdV^s + \gamma dA + \sum_i \mu_i dn_i^s \, . \quad (3.4)$$

Here S^s, V^s and n_i^s denote the entropy, volume and particle number of the interface region.

To integrate (3.4) we use the homogeneity property along the interface. If we change the interface volume dV^s just by "cutting off" a certain portion at the edge, so that what remains is structured exactly like the original system (only reduced in extent), we can write

$$dV^s = - V^s d\eta \, , \quad dA = - Ad\eta \, , \quad dn_i^s = n_i d\eta \quad \text{with} \quad dT = 0 \, , \quad (3.5)$$

and η is an arbitrary parameter.

From (3.4) one thus obtains

$$dF^s = - d\eta \left[- pV^s + \gamma A + \sum_i \mu_i n_i^s \right] . \quad (3.6)$$

With the conditions (3.5) it is evident that F^s also changes by the same ratio as the quantities V^s, A and n_i^s, i.e.,

$$dF^s = - F^s d\eta \, . \quad (3.7)$$

Comparing (3.7) with (3.6) one arrives at

$$F^s = -\,pV^s + \gamma A + \sum_i \mu_i n_i^{\,s}\ . \tag{3.8}$$

Equation (3.8) is the interface analogue of the common relation

$$G = \sum_i \mu_i n_i = F + pV \tag{3.9}$$

for the Gibbs free enthalpy G of a homogeneous medium. It is clear from (3.8 and 9) that the Gibbs free enthalpy for a surface or interface results as

$$G^s = \sum_i \mu_i n_i^{\,s} = F^s + pV - \gamma A\ . \tag{3.10}$$

The variation of the surface free enthalpy dG^s is easily calculated by differentiating (3.10) and substituting (3.4), i.e.

$$dG^s = -\,S^s dT + V^s dp - A d\gamma + \sum_i \mu_i dn_i^{\,s}\ . \tag{3.11}$$

From (3.4, 8 and 10) it is clear that the surface tension γ may be regarded as an excess free energy per unit area. It is the reversible work of formation of a unit area of surface or interface at constant system volume, temperature, chemical potential, and number of components. The reversibility requirement in the definition of γ implies that the composition and atomic configuration in the interface region are those of thermodynamic equilibrium.

We now address the question of the true physical origin of this surface tension or surface energy γ. If we consider a crystalline structure with its surface next to vacuum, it costs energy to generate an additional piece of surface while keeping the crystal volume and the number of constituent atoms constant. Bonds between neighboring atoms must be broken in order to expose new atoms to the vacuum. The formation of surface defects including steps might also be involved in forming the new surface area. All these effects contribute to the excess surface free energy γ. For crystalline materials, most surface properties depend on the orientation. In particular, depending on the surface orientation (hkl) more or less bonds have to be broken to create a piece of surface (Sect.2.2); also the effect of charge compensation is completely different for polar and nonpolar surfaces of the same crystal (Sect.2.2). The surface tension of crystals $\gamma(\mathbf{n})$ is therefore strongly dependent on the orientation of the particular surface, $\mathbf{n}$. The equilibrium shape of a crystal is not necessarily that of minimum surface area, it may be a complex polyhedron. What then determines the macroscopic shape of solid matter at constant temperature with a fixed volume and chemical potential? From the requirement of minimum free energy one obtains as a necessary condition

$$\int_A \gamma(\mathbf{n})\,dA = \text{minimum}\;. \tag{3.12}$$

Thus questions concerning the morphological stability of certain surfaces and the equilibrium shape of materials involve a detailed knowledge of the surface tension $\gamma(\mathbf{n})$ and its orientational dependence (γ as a function of lattice plane (hkl) or surface normal $\mathbf{n}$).

In contrast to fluid interfaces, where the surface tension can usually be obtained quite easily by capillary, and similar experimental techniques, the determination of $\gamma(\mathbf{n})$ for the solid/vapor interfaces is extremely difficult. Accordingly, not much reliable experimental information about γ can be found in the literature at present.

In the theory for the crystal equilibrium shape and morphological stability, etc. the Wulff plot of $\gamma(\mathbf{n}) = \gamma(\text{hkl})$ plays an important role [3.3, 4]. As shown in Fig.3.2, the scalar surface tension γ is plotted in polar coordi-

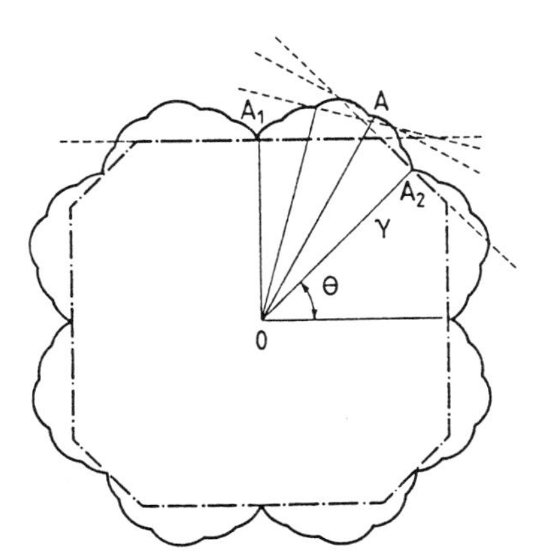

Fig.3.2. Schematic plot of the scalar surface tension γ(hkl) in polar coordinates as a function of angle Θ (describing the normal directions to the {hkl} planes). This Wulff construction [3.3, 4] yields the equilibrium shape of a solid (*dash-dotted*) as the inner envelope of the so-called Wulff planes, i.e., the normals to the radius vectors (*broken lines*)

nates versus the angle Θ between a particular fixed direction and the normals to the (hkl) planes. The length of the vector from the origin to a point on the plot represents the magnitude of γ(hkl), and the direction is that of the normal to the (hkl) plane. Because of (3.12) one can obtain the equilibrium shape of a crystal by connecting those lattice planes with minimum surface tension. The procedure of constructing the equilibrium shape by means of the γ(hkl) plot is displayed qualitatively in Fig.3.2. A set of planes, each perpendicular to a radius vector is constructed at the point where it meets the Wulff plot. The inner envelope of the planes then determines the set of surfaces that fulfills (3.12). These surfaces are assumed by the material at equilibrium. For liquids and amorphous solids, where $\gamma(\mathbf{n})$ is isotropic, both the Wulff plot and the equilibrium shape are spherical.

As a simple example for the rough calculation of $\gamma(\mathbf{n})$ let us consider a two-dimensional crystal with a so-called vicinal surface plane [3.4], i.e. a surface plane which consists of a relatively high number of areas with (01) orientation being separated by steps of atomic height (Fig.3.3a). Such a surface has an orientation angle of θ against the (01) direction. The more steps we have per unit surface area, the higher $\gamma(\mathbf{n}) = \gamma(\theta)$ will be. The step density can be written as $\tan(\theta/a)$. If now the surface tension of the low-index plane (01) is set to γ_0 and each step is assumed to make a contribution γ_1 to the total surface tension on the vicinal plane, one might express the surface tension, in general, as a series in the step density $\tan(\theta/a)$

$$\gamma(\theta) = \cos\theta\left[\gamma_0 + \gamma_1\left(\frac{\tan\theta}{a}\right) + \gamma_2\left(\frac{\tan\theta}{a}\right)^2 + \dots\right] . \tag{3.13a}$$

The quadratic and higher-order terms describe interactions between steps. Neglecting these interactions for large step distances, i.e. small θ, one gets

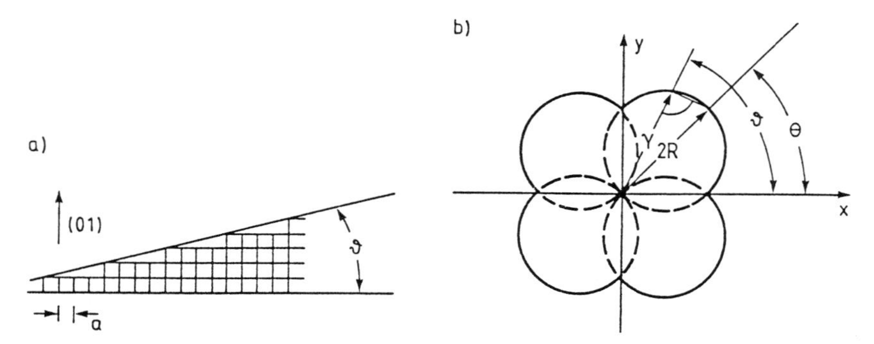

Fig.3.3. Schematic drawing of surface and simple Wulff plot (surface tension γ in polar coordinates) for a vicinal surface plane with inclination angle θ against [01] consisting of areas with [01] orientation (**a**). The Wulff plot (**b**) consists of circles passing through the origin

$$\gamma(\theta) = \cos\theta\left[\gamma_0 + \gamma_1\left(\frac{\tan\theta}{a}\right)\right] = \gamma_0\cos\theta + \frac{\gamma_1}{a}\sin\theta\,. \tag{3.13b}$$

According to the definition of γ in the Wulff plot as a radius vector from the origin, (3.13b) describes, within that polar plot, a circle (or sphere in three dimensions) passing through the origin. This can, e.g., be shown by considering such a circle (Fig.3.3b) with diameter 2R and inclination Θ with respect to the x-axis. Its mathematical description is

$$\gamma = 2R\cos(\theta-\Theta) = 2R[\cos\Theta\cos\theta + \sin\Theta\sin\theta]\,, \tag{3.14}$$

which is identical to (3.13b), and also enables a geometrical interpretation of γ_0 and γ_1/a. Taking into account the whole range of possible θ values (four quadrants in the plane) a Wulff plot as in Fig.3.3b results from (3.13b). This is, of course, a rough model in which all higher-order interactions, e.g. interactions between steps, are neglected. More realistic models yield Wulff plots which consist of a number of circles or spheres, as depicted qualitatively in Fig.3.2.

For real surfaces the surface tension has enormous importance for a number of questions related to surface inhomogeneities. If there are certain lattice planes with particularly low surface tension, a surface can lower its total free energy by exposing areas of this orientation. Thus, even though facetting produces an increase in total surface area, it may decrease the total free energy and is sometimes thermodynamically favored. Also the appearance of surface segregations might be described in terms of minimizing the total free energy of a surface by exposing patches of particularly low surface tension.

As a last remark, an intuitively reasonable, but rough estimation for the surface tension γ of a material is given as follows: the surface tension γ per atom is approximately equal to half the heat of melting per atom. This relation results from the consideration that melting involves the breaking of all the chemical bonds of an atom, whereas γ is related to breaking only about half of the bonds of a surface atom.

3.2 Relaxation, Reconstruction, and Defects

We now move on to consider in a little more detail the atomic structure of a surface. It can easily be seen that on a surface, due to the absence of neighboring atoms on one side, the interatomic forces in the uppermost lattice planes are considerably changed. The equilibrium conditions for surface

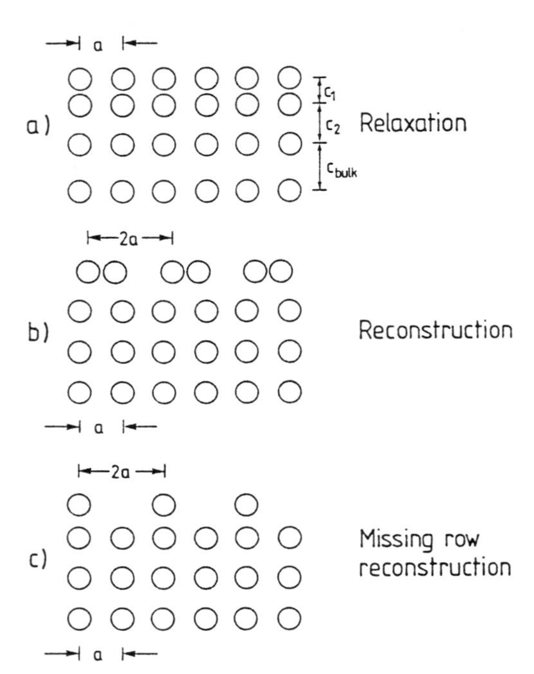

Fig. 3.4a-c. Schematic side view of the characteristic rearrangements of surface atoms of a simple cubic lattice with lattice constant a: (**a**) Relaxation of the topmost atomic layer normal to surface (different lattice spacings c); (**b**) reconstruction of the topmost atomic layer into a surface net with double periodicity distance 2a; (**c**) missing row reconstruction with missing atoms in the topmost lattice plane

atoms are modified with respect to the bulk; one therefore expects altered atomic positions and a surface atomic structure that usually does not agree with that of the bulk. Thus a surface is not merely a truncation of the bulk of a crystal. The distortion of the ideal bulk-like atom configuration due to the existence of a surface will be different for metals and for semiconductors. In metals we have a strongly delocalized electron gas and a chemical bond which is essentially not directed, whereas in tetrahedrally bonded semiconductors (Si, Ge, GaAs, InP, etc.) significant directional bonding is present. Bond breaking on one side due to the surface is expected to have a more dramatic effect on the atomic configuration at the surface of a semiconductor.

Figure 3.4 illustrates schematically some characteristic rearrangements of surface atoms. A pure compression (or possibly extension) of the topmost (or top few) interlayer separation(s) normal to the surface is called **relaxation**. In this case, the 2D lattice, i.e. the periodicity parallel to the surface within the topmost atomic layer is the same as for the bulk. More dramatic changes of the atomic configuration, as shown in Fig. 3.4b, usually related to shifts parallel to the surface, can change the periodicity parallel to the surface. The 2D unit mesh has dimensions different from those of a projected bulk unit cell. This type of atomic rearrangement is called **reconstruction**. In reconstructions the surface unit mesh must not necessarily be changed with respect to the bulk, only the atomic displacements from their bulk positions must be more complex than a pure shift normal to the surface as in Fig.3.4a. As an example we will consider below the cleaved GaAs(110) surface. Reconstructions also include surface atomic configurations in which atoms or a whole row of atoms are missing in comparison with the bulk (Fig.3.4c). In this case the surface periodicity is always different from that of the bulk.

As was already mentioned above, semiconductor surfaces with their strongly directional covalent bonding character often show quite complex reconstructions. Up to now it has been a difficult task to determine the atomic positions of these reconstructed surfaces experimentally. Usually, a variety of methods [LEED (Panel VIII: Chap.4), ARUPS (Sect.6.3), Rutherford backscattering (Sect.4.11), etc.] have to be applied to arrive at an

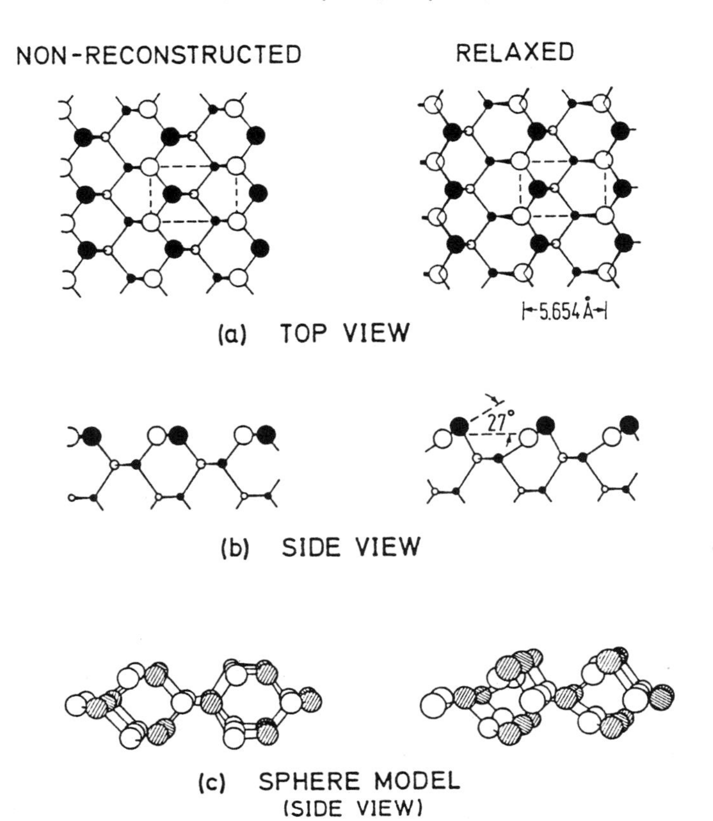

Fig.3.5a-c. Atomic positions of the GaAs(110) surface; ideal, non-reconstructed and relaxed as it appears after cleavage in UHV. (**a**) Top view; the (1×1) unit mesh is plotted as a broken line. (**b**) Side view. (**c**) Sphere model. (*Open circles* designate Ga atoms and *shaded circles* As. *Smaller circles* indicate deeper atomic layers)

unequivocal structural model. One of the well-understood examples is the clean cleaved GaAs(110) surface which exhibits in LEED the same surface unit mesh as one would expect for a truncated bulk crystal. Nevertheless, the atomic positions at the surface are quite distorted with respect to the bulk (Fig.3.5). Compared with their bulk positions the As atoms of the top-most atomic layer are raised, whereas the neighboring Ga surface atoms are pushed inwards. There is a tilt in the surface Ga-As bond of about 27° to the surface and only small changes in the Ga-As bond lengths [3.5]. In this particular case of GaAs(110) the reconstruction develops just by simple shifts of surface atoms involving tilting of covalent bonds. An even stronger perturbation of the lattice including bond breaking and reformation of new bonds is found on the Si(111) surface when prepared by cleavage in UHV. In the $[0\bar{1}1]$ direction the (2×1) LEED pattern indicates a double periodicity distance in real space. The reconstruction model which has become established for this type of surface is shown in Fig.3.6 (more details are given in Sect.6.5). The Si atoms on the surface are rearranged in such a way that neighboring dangling bonds can form zig-zag chains of π-bonds, similar to a long-chain organic molecule [3.6]. Atoms of the next deeper atomic layers also have to change their positions and participate in the reconstruction. Detailed self-consistent total energy calculations reveal that in spite of bond breaking, this so-called π-bonded chain model fulfills the requirement of minimum total energy with respect to other, may be more obvious, atomic configurations. According to the present state of knowledge, a slight buckling, i.e. outward and inward shift of neighboring Si surface atoms also has to be assumed (Fig.3.6).

One might enquire a little deeper into the reasons for relaxation and reconstruction. As the examples on semiconductor surfaces show, there is certainly a tendency to saturate free dangling bonds by forming new bonds within the surface [Si(111)-(2×1)]. This can lead to an overall decrease of the surface free energy. On the surface of a polar semiconductor like GaAs the undisturbed dangling bonds of Ga and As surface atoms carry charge and an overall decrease of surface energy is achieved by an electronic charge transfer from the Ga to the As dangling bond. The Ga dangling bond becomes more sp^2-like whereas the As bond gets more p_z character thus causing the inwards and outwards shifts of the corresponding atoms.

On metal surfaces there are usually no directional bonds, and other mechanisms might be imagined which give rise to surface relaxation and reconstruction. Figure 3.7 exhibits in a schematic way the rigid ion core positions near a metal surface. The free electrons are delocalized between the ion cores making an electrically neutral object. In order to ensure such a neutrality one can formally attribute to each core a Wigner-Seitz cell (squares in Fig.3.7a), which contains the corresponding electronic charge.

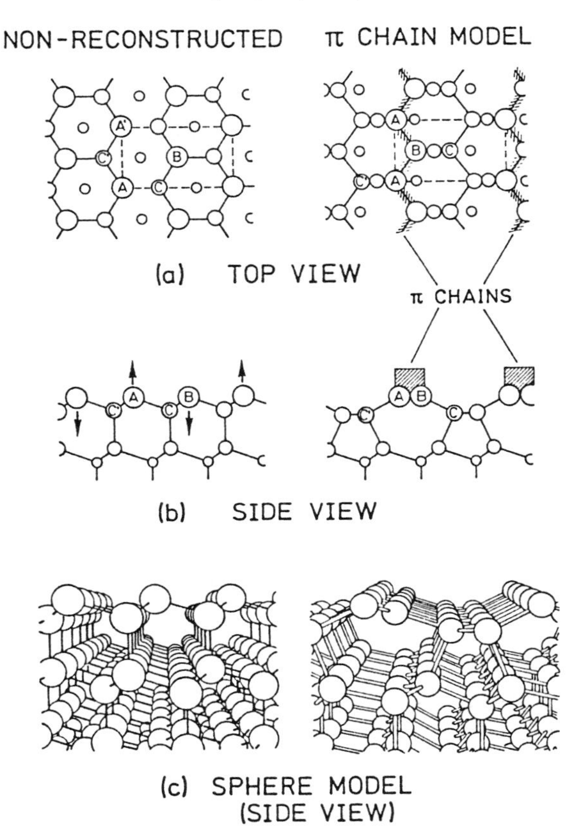

Fig. 3.6a-c. Atomic positions at the Si(111) surface; ideal, non-reconstructed and with (2×1) reconstructuion (π-bonded chain model) as occurs after cleavage in UHV. The shaded areas denote the location of the chains originating from overlap between neighboring dangling bonds. (**a**) Top view, the (2×1) unit mesh is plotted in broken line. (**b**) Side view; arrows indicate possible up- and downwards shifts of surface atoms of type A and B, which give rise to the so-called buckling reconstruction. (**c**) Sphere model. (*Smaller circles* indicate deeper atomic layers)

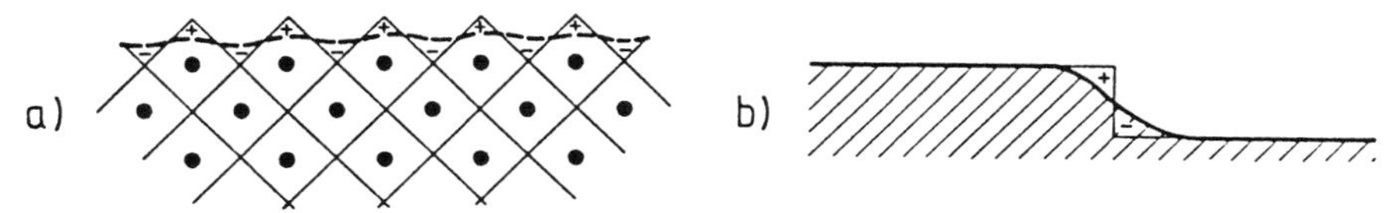

Fig. 3.7. Schematic representation of the formation of electronic surface dipoles at metal surfaces (**a**) by smearing out of the electronic charge distribution of the Wigner Seitz cells at the surface (*rectangles*), and (**b**) by smearing out of the electronic charge distribution at a step

On a surface, as depicted in Fig.3.7a, this would lead to a rapidly varying electron density at the surface, thus increasing the kinetic energy of the electrons in proportion to the square of the derivative of the wave function. As indicated in Fig.3.7a, the surface electronic charge therefore tends to smooth out and to form a surface contour (broken line) which gives rise to the formation of a surface electronic dipole contributing to the work function of that particular surface (Sect.9.3). On the other hand, due to this dipole layer, the positive ion cores in the topmost atomic layer feel a net repulsion from the charge in their Wigner Seitz cell, and an inwards displacement results. This effect is thought to be one possible source of the contraction (relaxation) of the topmost lattice plane observed on many metal surfaces (Table 3.1). Apart from this relaxation, most low-index metal surfaces do not exhibit a reconstruction, i.e. their surface unit mesh equals that of the bulk.

As is the case in the bulk, ideal surfaces with complete translational symmetry cannot exist for entropy reasons. On a real surface *defects* are always found. One can classify surface defects according to their dimensionality (Fig.3.8). The terraces represent portions of low-index planes. Zero-dimensional or point defects involve adatoms, ledge adatoms, kinks and vacancies. For a clean monatomic crystal (Si, Ge, Al, etc.) this characterization seems sufficient. Looking via an atomic scale, in particular on a surface of a compound crystal (GaAs, ZnO, etc.), one can distinguish, in more detail, between adatoms of the same kind (e.g., Ga or As on GaAs) or foreign adatoms (Si on GaAs, etc.); adatoms might be bonded on top of the uppermost atomic layer or they might be incorporated into the topmost lattice plane as interstitials. Vacancies, too, might have varying character on an atomic scale. On a GaAs surface, for example, both Ga and As vacancies

Table 3.1. Compilation of some clean metal surface relaxations. As can be seen, the topmost lattice plane distance is generally contracted by several percent of the unrelaxed value [3.4]

Surface	Spacing change of top layer [%]
Ag (110)	− 8
Al (110)	− 10
Cu (100)	0
Cu (110)	− 10
Cu (311)	− 5
Mo (100)	− 12.5

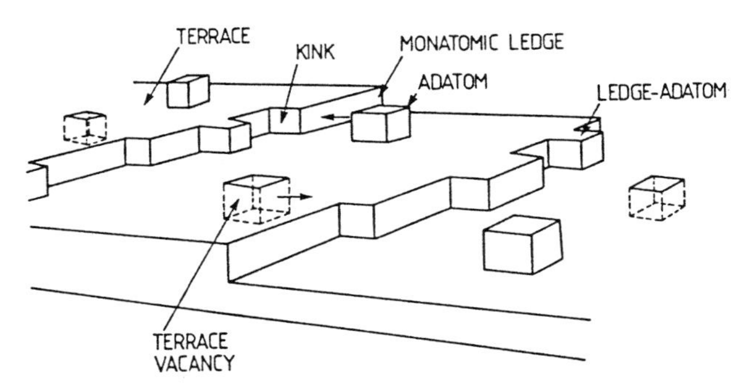

Fig. 3.8. Schematic drawing of various defects that may occur on a solid surface

may exist. Due to the different ionic charge of the missing atoms, the two types of vacancies exhibit different electronic properties (Sects. 6.2, 3). Another type of zero-dimensional defect characteristic of compound semiconductor surfaces is the so-called **anti-site defect**, where in GaAs, for example, an As atom occupies a Ga site (As_{Ga}) or vice versa (Ga_{As}). As is easily seen from the different electron orbitals and the type of chemical bonding, anti-site defects also give rise to electrically active centers (Sects. 6.2, 3).

An important one-dimensional or line defect is the step in which the ledge (Fig. 3.8) separates two terraces from each other. In many cases steps of single atomic height prevail. Depending on the orientation of the step and of the corresponding terraces, step atoms expose a different number of dangling bonds as compared with atoms in the terraces. Steps are important in the formation of vicinal surfaces (high-index surfaces), i.e. surfaces which are oriented at a small angle with respect to a low-index surface. Such vicinal surfaces are formed by small low-index terraces and a high density of regular steps (Fig. 3.3a). Steps often have interesting electronic properties. On semiconductors with strongly covalent bonds, the different dangling bond structure might modify the electronic energy levels near steps. On metal surfaces the free electron gas tends to be smoothed out at the step thus forming dipole moments due to the spatially fixed positive ion cores (Fig. 3.7b).

Other important surface defects are related to dislocations. An edge dislocation penetrating into a surface with the Burgers vector oriented parallel to the surface gives rise to a point defect. Step dislocations hitting a surface also cause point defects which are usually sources of a step line.

Because of the local variation that defects cause in all important surface quantities, such as binding energy, coordination, electronic states, etc., the defect structure of a surface plays a predominant role in processes such as crystal growth, evaporation, surface diffusion, adsorption and surface chemical reactions.

3.3 Two-Dimensional Lattices, Superstructure, and Reciprocal Space

3.3.1 Surface Lattices and Superstructures

Even though real crystalline surfaces always contain point and/or line defects, the model of a perfectly periodic two-dimensional surface is convenient and adequate for the description of well-prepared samples with large well-ordered areas and low defect density. The surface region of a crystal is, in principle, a three-dimensional entity; reconstructions usually extend into the crystal by more than one atomic layer. Space-charge layers on semiconductor surfaces can have a depth of hundreds of Ångstroms (Chap. 7). Moreover, the experimental probes in surface experiments, even slow electrons, usually have a non-negligible penetration depth. But compared to subsurface layers, the topmost atomic layer is always predominant in any surface experiment. As a result, each layer of atoms in the surface is intrinsically inequivalent to other layers, i.e. the only symmetry properties which the surface posesses are those which operate in a plane parallel to the surface. Although the surface region is three-dimensional, all symmetry properties are two-dimensional (2D). Thus surface crystallography is two-dimensional and one has to consider 2D point groups and 2D Bravais nets or lattices [3.7].

The point group operations which are compatible with 2D periodicity are the usual 1, 2, 3, 4 and 6-fold rotation axes perpendicular to the surface, and mirror planes, also normal to the surface. Inversion centers, mirror planes and rotation axes parallel to the surface are not allowed, since they refer to points outside the surface. By combining the limited number of allowed symmetry operations, one obtains 10 different *point group symmetries* denoted [3.8]

1, 2, 1m, 2mm, 3, 3m, 4, 4mm, 6, 6mm .

The numeral $\nu = 1 \ldots 6$ denotes rotations by $2\pi/\nu$ and the symbol m referrs to reflections in a mirror plane. The third m indicates that a combination of the preceding two operations produces a new mirror plane.

The operation of the 2D point groups on a 2D translational net or lattice produces the possible **2D Bravais lattices**. In contrast to three dimensions, only 5 symmetrically distinct nets are possible (Fig. 3.9). There is only one non-primitive unit mesh, the centered rectangular one. In all other cases a centered unit cell can also be described by a primitive Bravais lattice. Nevertheless, in practice one often uses, e.g., centered square meshes for the convenience of description.

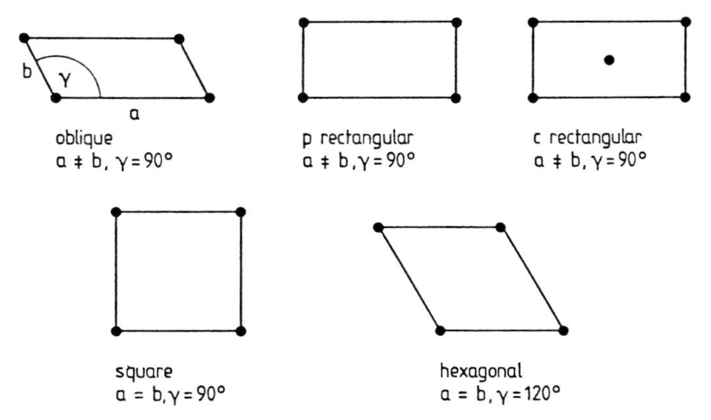

Fig. 3.9. Five possible two-dimensional (2D) Bravais lattices

As was mentioned already, surface experiments such as diffraction of low-energy electrons (Panel VIII: Chap. 4) usually probe not only the topmost atomic layer; the information obtained (e.g., in a diffraction pattern) is related to several atomic layers. A formal description of the periodic structure of the topmost atomic layers of a crystalline solid must therefore contain information about the ideal substrate as well as about the one or two topmost atomic layers which might exhibit a different periodicity due to a possible reconstruction or a well-ordered, periodic adsorbate layer (Fig. 3.10). In such a situation, where a different periodicity is present in the topmost atomic layer, a surface lattice, called a **superlattice**, is superimposed on the substrate lattice which exhibits the basic periodicity. The basic substrate lattice can be described by a set of 2D translational vectors

$$\mathbf{r_m} = \mathrm{m}\mathbf{a}_1 + \mathrm{n}\mathbf{a}_2 \tag{3.15}$$

where $\mathbf{m} = (\mathrm{m}, \mathrm{n})$ denotes a pair of integer numbers, and the $\mathbf{a}_i$'s are the two unit mesh vectors. The surface net of the topmost atomic layer may then be determined in terms of the substrate net by

$$\begin{aligned} \mathbf{b}_1 &= \mathrm{m}_{11}\mathbf{a}_1 + \mathrm{m}_{12}\mathbf{a}_2 \\ \mathbf{b}_2 &= \mathrm{m}_{21}\mathbf{a}_1 + \mathrm{m}_{22}\mathbf{a}_2 \end{aligned} \quad \text{or} \quad \begin{pmatrix} \mathbf{b}_1 \\ \mathbf{b}_2 \end{pmatrix} = \mathbf{M} \begin{pmatrix} \mathbf{a}_1 \\ \mathbf{a}_2 \end{pmatrix}, \tag{3.16}$$

where $\mathbf{M}$ is a 2×2 matrix, namely

$$\mathbf{M} = \begin{pmatrix} \mathrm{m}_{11} & \mathrm{m}_{12} \\ \mathrm{m}_{21} & \mathrm{m}_{11} \end{pmatrix}. \tag{3.17}$$

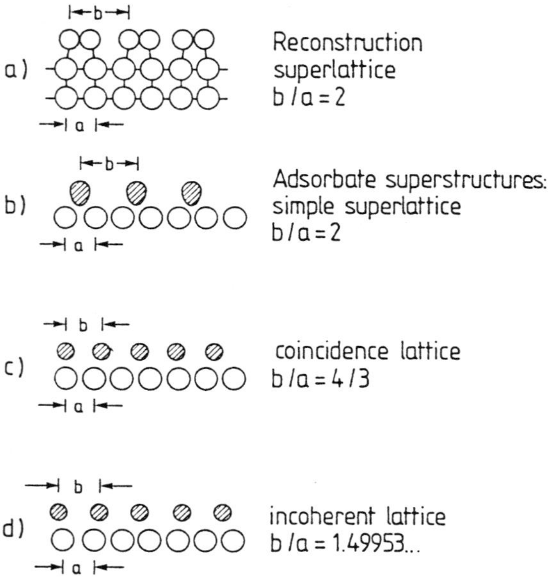

Fig.3.10a-d. Different possibilities for surface unit meshes which are different from that of the underlying bulk material: (**a**) Reconstruction of a clean surface due to lateral shift of the atoms within the topmost lattice plane. (**b** to **d**) Adsorbate reconstructions with different adsorbate atom/substrate atom lattice constant ratios b/a

For the areas B and A of the surface and of the substrate unit mesh, respectively, it follows that

$$B = |\mathbf{b}_1 \times \mathbf{b}_2| = A \det \mathbf{M} . \tag{3.18a}$$

The determinant of **M**,

$$\det \mathbf{M} = \frac{|\mathbf{b}_1 \times \mathbf{b}_2|}{|\mathbf{a}_1 \times \mathbf{a}_2|} , \tag{3.18b}$$

can be used to characterize the relation between surface and substrate lattice (Fig.3.10). When $\det \mathbf{M}$ is an integer (Fig.3.10b), the surface lattice is said to be simply related; it is called a **simple superlattice**. When $\det \mathbf{M}$ is a rational number (Fig.3.10c), the superstructure is referred to as a **coincidence lattice**.

In cases where the adsorbate/substrate interaction is much less important than interactions between the adsorbed particles itself, the substrate has no determining influence on the superstructure and the adsorbate lattice might be out of registry with the substrate net. In this case, where $\det \mathbf{M}$ is an irrational number (Fig.3.10d), the superstructure is called an **incoherent lattice** or an incommensurate structure.

According to *Wood* [3.9] there is a simple notation for superstructures in terms of the ratio of the lengths of the primitive translation vectors of the superstructure and those of the substrate unit mesh. In addition, one indicates the angle (if any) through which one mesh is rotated relative to the other. If on a certain substrate surface X{hkl} a reconstruction is given with $(\mathbf{b}_1 \parallel \mathbf{a}_1, \mathbf{b}_2 \parallel \mathbf{a}_2)$

$$\mathbf{b}_1 = p\mathbf{a}_1 \,, \quad \mathbf{b}_2 = q\mathbf{a}_2 \,, \tag{3.19}$$

then the notation is given as

$$X\{hkl\}(p \times q) \quad \text{or} \quad X\{hkl\}c(p \times q) \tag{3.20a}$$

A possible centering can be expressed by the symbol c. If in the more general case the translational vectors of substrate and of superstructure are not parallel to each other, but rather a rotation by a certain angle R° has to be taken into account, one describes this situation by

$$X\{hkl\}(p \times q) - R^\circ \,. \tag{3.21b}$$

Some examples of this notation for some adsorbate superlattices on metal surfaces M{100}, M{111} and for the clean Si{111} surface with (2×1) superstructure (after cleavage in UHV) are illustrated in Fig.3.11.

Surface scattering and diffraction experiments are best described in terms of the reciprocal lattice which will be explained in detail in Sect.3.3.2 and Chap.4).

3.3.2 2D Reciprocal Lattice

As in three-dimensional space, the translational vectors of the 2D reciprocal lattice, $\mathbf{a}_i^*$, are defined in terms of the real-space lattice vectors $\mathbf{a}_i$ by

$$\mathbf{a}_1^* = 2\pi \frac{\mathbf{a}_2 \times \hat{\mathbf{n}}}{|\mathbf{a}_1 \times \mathbf{a}_2|} \,, \quad \mathbf{a}_2^* = 2\pi \frac{\hat{\mathbf{n}} \times \mathbf{a}_1}{|\mathbf{a}_1 \times \mathbf{a}_2|} \,, \tag{3.21}$$

where $\hat{\mathbf{n}}$ is the unit vector normal to the surface, i.e.

$$\mathbf{a}_i^* \cdot \mathbf{a}_j = 2\pi\, \delta_{ij} \quad \text{and} \quad |\mathbf{a}_i^*| = 2\pi\, [a_i \sin \sphericalangle (a_i a_j)]^{-1} \,, \quad i,j = 1,2 \,. \tag{3.22}$$

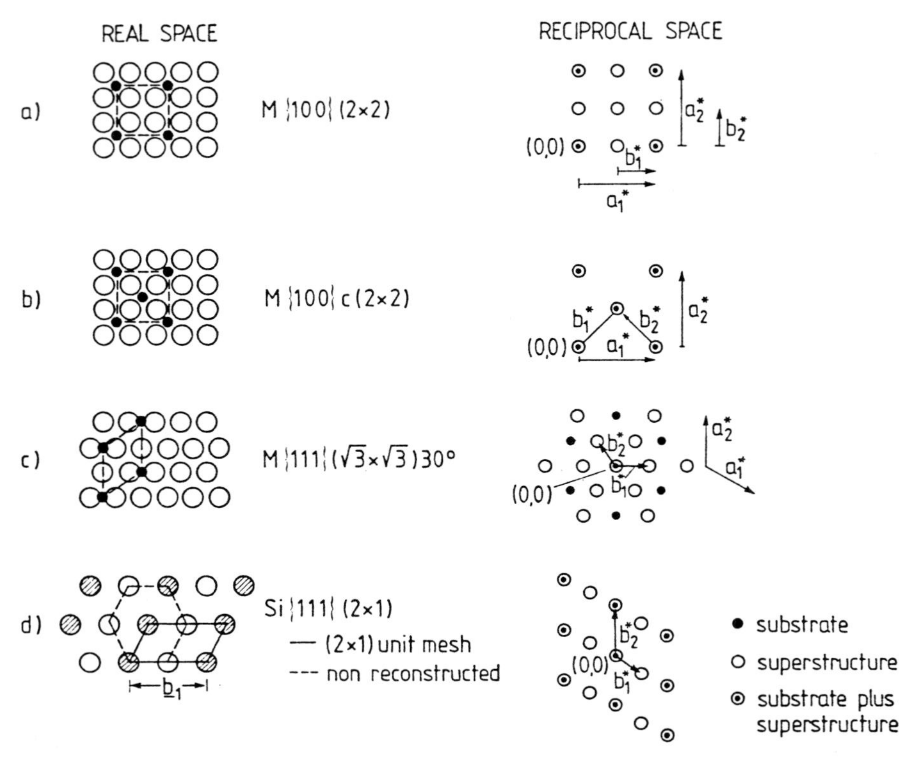

Fig.3.11a-d. Examples of different superlattices in real space and in reciprocal space: (**a** to **c**) Adsorbed atoms on several low index surfaces of a closed packed metal (M), (**d**) (2x1) unit mesh (*solid line*) of a Si(111) surface prepared by cleavage in UHV

Equations (3.21, 22) can be used to construct the reciprocal lattice geometrically to a given 2D network, as displayed in Figs.3.11.

A general translation vector in reciprocal space is given by

$$\mathbf{G}_{hk} = h\mathbf{a}_1{}^* + k\mathbf{a}_2{}^* , \tag{3.23}$$

where the integer numbers h and k are the Miller indices.

In analogy to the real-space relations (3.16-18), the reciprocal network of a superstructure ($\mathbf{b}_1{}^*, \mathbf{b}_2{}^*$) can be expressed in terms of the substrate reciprocal lattice ($\mathbf{a}_1{}^*, \mathbf{a}_2{}^*$) by

$$\begin{aligned} \mathbf{b}_1{}^* &= m_{11}{}^*\mathbf{a}_1{}^* + m_{12}{}^*\mathbf{a}_2{}^* \\ \mathbf{b}_2{}^* &= m_{21}{}^*\mathbf{a}_1{}^* + m_{22}{}^*\mathbf{a}_2{}^* \end{aligned} \quad \text{or} \quad \begin{pmatrix} \mathbf{b}_1{}^* \\ \mathbf{b}_{2*} \end{pmatrix} = \mathbf{M}^* \begin{pmatrix} \mathbf{a}_1{}^* \\ \mathbf{a}_{2*} \end{pmatrix} . \tag{3.24}$$

The matrices in (3.16, 24) are related to one another by

$$\mathbf{M}^* = (\mathbf{M}^{-1})^T , \tag{3.25a}$$

where $(\mathbf{M}^{-1})^T$ is the transpose inverse matrix with

$$m_{ii} = \frac{m_{ii}^*}{\det \mathbf{M}^*} \quad \text{and} \quad m_{ij} = \frac{-m_{ji}^*}{\det \mathbf{M}^*} . \tag{3.25b}$$

3.4 Structural Models of Solid/Solid Interfaces

The vacuum/solid interface, i.e. the surface considered so far is the simplest interface in which a crystal can participate. Many concepts of surface physics, e.g. that of defects, relaxation (Sect.3.2), surface states (Chap.6), surface collective modes such as phonons and plasmons (Chap.5) etc., can be transferred to more complex interfaces, in particular to the liquid/solid and the solid/solid interface. Important examples of solid/solid interfaces with considerable technological importance are the interfaces in semiconductor heterostructures ($GaAs/Ga_xAl_{1-x}As$, GaAs/Ge, Chap.8), the Si/SiO_2 interface in a MOS structure (Sect.7.7), the metal/semiconductor junction (Chap.8), the interfaces between optical elements and antireflecting coatings or the interface between an organic and an inorganic semiconductor (sensor applications).

In terms of the atomic structure, two main features (on different levels) have to be considered (Fig.3.12). The interface might be crystalline/crystalline (Fig.3.12a) or the crystalline substrate might be covered with a non-crystalline amorphous solid (Fig.3.12b). Polycrystalline overlayers might form, at least locally, more or less disturbed crystalline/crystalline interfaces (grain boundaries, etc.).

The second important feature is the abruptness of an interface. Solid/solid interfaces, both crystalline/crystalline and crystalline/non-crystalline, might be sharp and abrupt on an atomic scale or they might be "washed out" by interdiffusion (Fig.3.12c) and/or formation of new chemical compounds (Fig.3.12d). In such a situation more than two different phases might be in thermal equilibrium with each other. If we consider an interface between two different materials A and B at a certain temperature T_0 the "sharpness" of the interface can be described in terms of the concentrations [A] or [B]. After the materials A and B have been brought into contact at a temperature T_0, [A] might change gradually with distance after some reaction time

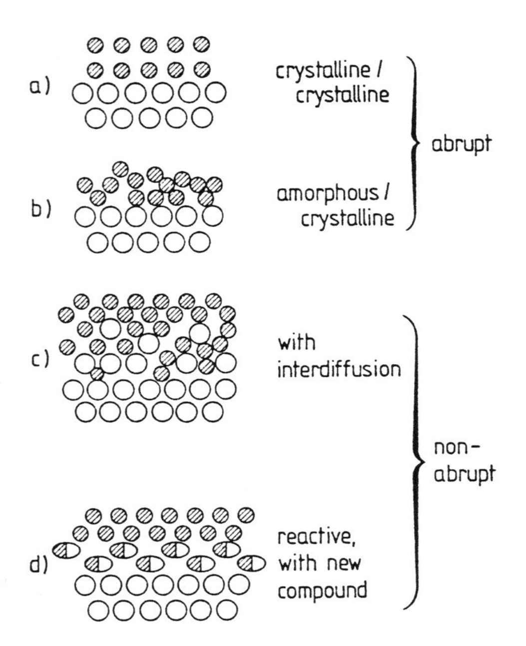

Fig.3.12a-d. Different types of solid/solid interfaces (film on substrate); substrate atoms depicted as open circles, film atoms as dark circles, reaction compounds (in d) as half-dark and half-open symbols

(Fig.3.13a, left hand) or an abrupt change in concentration indicates a more or less sharp interface at z_0 (Fig.3.13b, left hand); here [A] changes abruptly on going from phase I to phase II. The detailed shape of the concentration profiles within the two phases I and II depends, of course, on diffusion constants, reaction rates and on the reaction time during which the materials A and B are in contact at the reaction temperature T_0. Details of the solid/solid chemical reaction which forms the A/B interface are dependent both on kinetic and on thermodynamic aspects. Thermodynamic considerations can yield the limiting conditions for the formation of the particular interface, but not all thermodynamically possible phases will necessarily occur in reality because of kinetic limitations such as high activation energies for nucleation (Sect.3.5). For analyzing the various possibilities for interface formation, however, thermodynamic parameters are important. Phase diagrams can therefore give useful information about the expected properties of certain solid/solid interfaces. For example, a prerequisite for the formation of a smooth gradual interface (Fig.3.13a) at temperature T_0 is a binary phase diagram, which allows complete mixing of the two components A and B at T_0 (Fig.3.13a, right). In contrast, a phase diagram as in Fig.3.13b (right hand) allows mixing of the two components at temperature T_0 only between the concentrations [A]′ and [A]″; outside this concentration range only the phases I and II exist. Correspondingly a solid-state reaction at T_0 between A and B leads to a sharp interface (Fig.3.13b, left hand) with an abrupt change from [A]′ to [A]″ at the very interface at z_0. Of

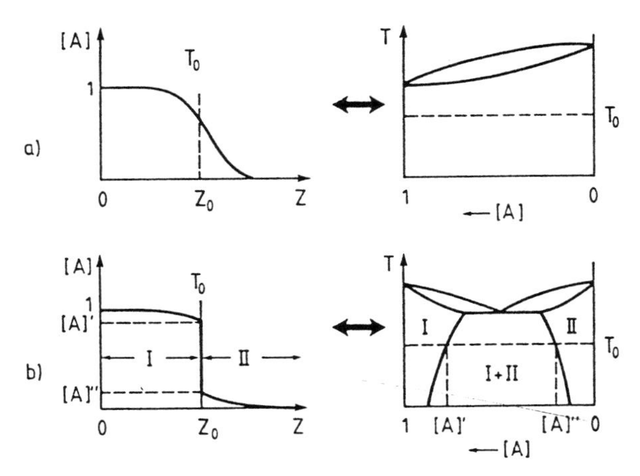

Fig. 3.13a,b. Phase diagrams (*right*) and concentration versus depth (z) plots (*left*) for two materials A and B. [A] and [B] denote the concentrations of A and B, respectively. The two materials have been brought into contact at a temperature T_0 and after a certain reaction time the interface at z_0 (*left*) might be gradual (**a**) or more or less sharp (**b**) depending on the phase diagram (*right*). In a) the phase diagram allows complete mixing of A and B whereas in b) the phase diagram (right) allows the existence of two separate phases I and II and mixing of I and II at the particular temperature T_0 only between the concentrations [A]′ and [A]″

course, such interfaces can also be related to different types of phase diagrams having a miscibility gap. More complex phase diagrams with more than two stable phases (I, II, III, etc.) at a certain temperature might be related to spatially extended solid/solid interfaces with several layered phases I, II, III, etc.

As has already been emphasized, the real atomic structure of a solid/solid interface depends in detail on the atomic properties of the two partners forming the interface. Thermodynamic considerations can only give a rough guide to the general possibilities. In this sense interfaces between two solids are more complex than vacuum/solid interfaces, and heterophase boundaries are likewise more complex than homophase interfaces. But even in the latter case of a homophase boundary, where two identical crystalline structures, but with different lattice orientation, touch each other (as in grain boundaries), different atomic models for the interface exist. According to L. Brillouin (1898) an *amorphous interlayer*, where the crystalline structure is fully disturbed (liquid-like phase) might be assumed. The idea seems consistent with the intuitive notion that atoms at an interface, as in Fig. 3.14, are in energetically less favorable positions and that a displacement from their crystallographic sites might lead to an energy reduction. This displacement disrupts the crystallographic structure close to the interface and might result in an amorphous interlayer.

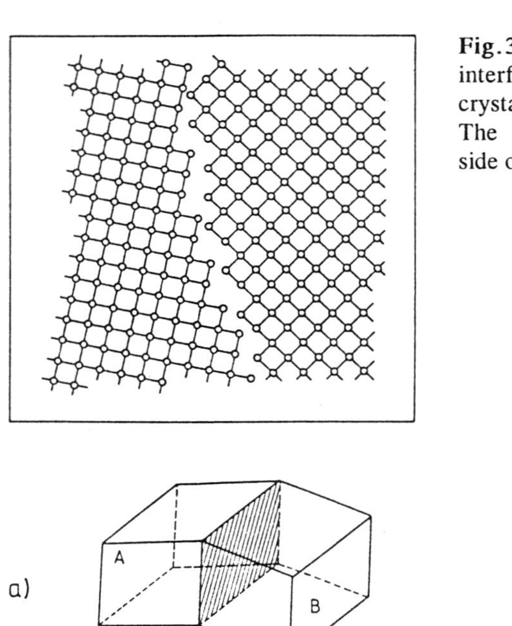

Fig.3.14. Simplest picture of a solid/solid interface between two differently oriented crystalline domains (domain boundary). The crystallographic structure on each side of the boundary is full retained

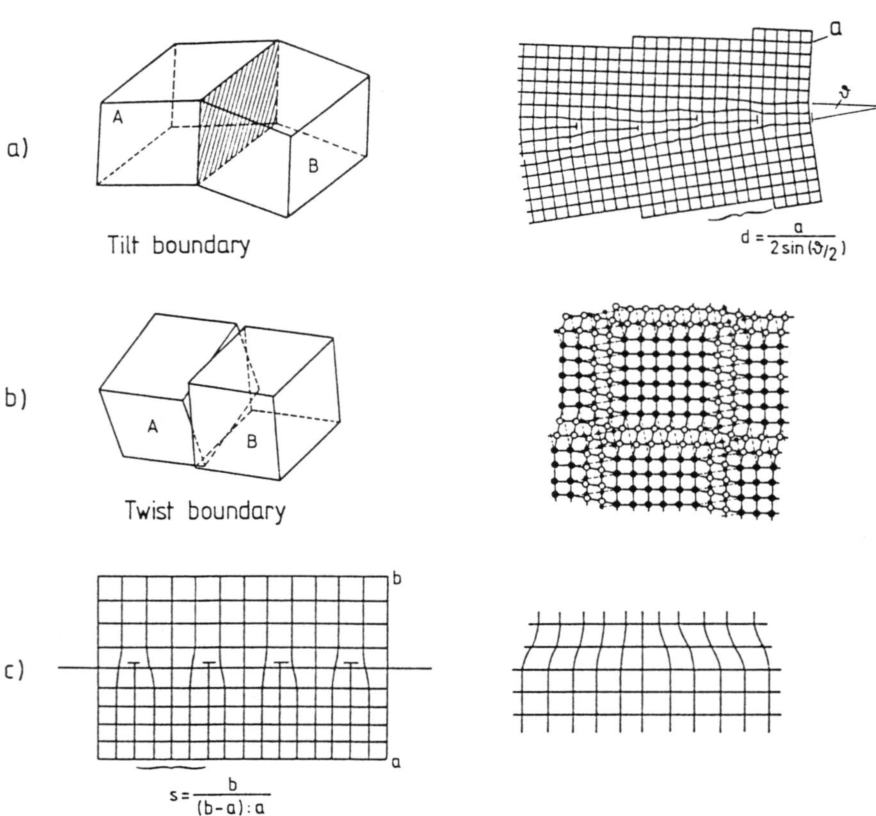

Fig.3.15a-c. Crystalline interface models (grain boundaries) based on dislocations: (**a**) Tilt boundary due to edge dislocations. (**b**) Twist boundary due to screw dislocations. (**c**) Hetero-interface between two different crystalline solids. The lattice mismatch is adjusted by edge dislocations (*left*) or by strain (*right*)

Other interface models are based on the concept of **dislocations**, which allow matching of the two half crystals. In Fig.3.15a a grain boundar-y is schematically depicted in which two tilted, but otherwise identical, crystal halves are separated by a tilt boundary. With θ as the inclination angle and a as lattice constant, the distance s between two edge dislocations is

$$s = a\,[2\sin(\theta/2)]^{-1} \simeq a/\theta \,. \tag{3.26}$$

Each dislocation contains a certain amount of elastic energy in its surrounding strain field. The energy stored in such a tilt boundary can thus be calculated by summing up the contributions of the edge dislocations. Similar considerations are valid for the twist boundary (Fig.3.15b), where screw dislocations are necessary to adjust the two twisted lattices to each other. Another interesting case is the heterophase boundary, which is always found in heteroepitaxy, where a monocrystalline overlayer is grown on a different monocrystalline substrate (see heteroepitaxy of III-V semiconductors in Sect.2.4). Usually there is a certain lattice mismatch, i.e. a difference between the lattice constants of the two materials. Only below a certain critical value for the mismatch can epitaxy occur. But even in that case the different lattice constants might adjust at the interface by the formation of edge dislocations (Fig.3.15c). With a and b as the two lattice constants, the distance beween dislocations follows as

$$s = \frac{ab}{|b - a|} \,. \tag{3.27}$$

The energy stored in such an interface between an epitaxial film and its substrate is calculated by summing up the energy contributions of the strain fields of the edge dislocations.

For thinner epitaxial films with low lattice mismatch one might image a different way of matching the various lattice constants, namely by elastic strain, i.e. a deformation of the lattice of the epitaxial film. The type of interface that is actually formed depends on the thickness of the film and on the amount of lattice mismatch. In order to predict the type of lattice matching, by strain only or by formation of edge dislocations, a calculation and comparison of the free energies involved is necessary. In particular, the interaction between dislocations has to be taken into account. If two dislocations approach each other, their strain fields overlap which might lead to a decrease in energy relative to the situation of non-overlapping strain fields (since a given dislocation already provides part of the strain required by its neighbors). Thus for higher lattice mismatch, $(b-a)/a$, the formation of dis-

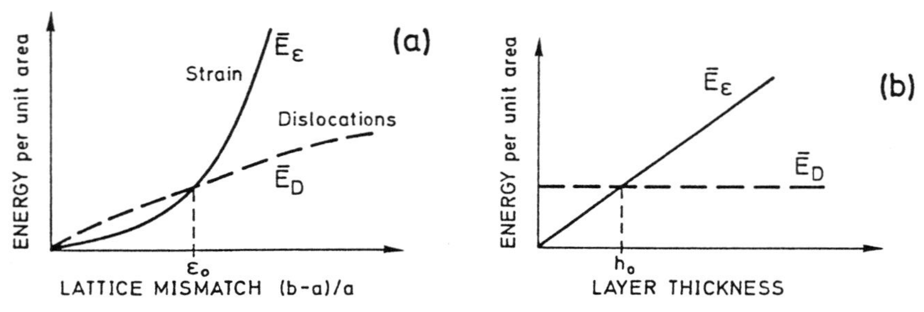

Fig.3.16a,b. Qualitative plots of lattice energy stored at a crystalline hetero-interface per unit area: (**a**) as a function of lattice mismatch; beyond a critical lattice mismatch ϵ_0 (a and b are the lattice constants of the two materials) the adjustment of the two lattices by dislocations (*broken line*) is energetically more favorable than by strain (energy $\bar{E}_D < \bar{E}_\epsilon$), (**b**) as a function of overlayer thickness; for thicknesses exceeding the critical thickness h_0 dislocations are energetically more favorable than strain (energy $\bar{E}_D < \bar{E}_\epsilon$)

locations becomes energetically more favorable than building up lattice strain in the whole layer without formation of dislocations. Qualitatively, and also by use of model calculations [3.10], one thus arrives at relationships between the corresponding energy densities $\bar{E}_\epsilon$ (with only strain), $\bar{E}_D$ (with dislocations), and the mismatch and layer thickness h, as shown in Fig.3.16. Below a critical mismatch ϵ_0 and below a critical layer thickness h_0 a purely strained epitaxial layer has a lower interface energy than do dislocations. If only dislocations are present, $\bar{E}_D$ does not change with increasing layer thickness h (Fig. 3.16b), since thicker films are not more stressed than thin ones. On the other hand, with no dislocations present, the

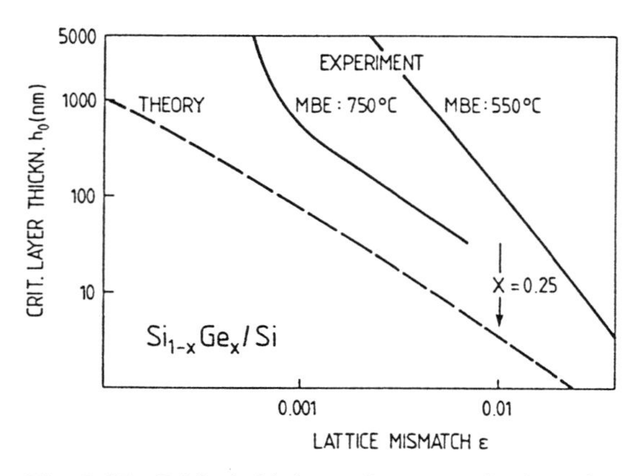

Fig.3.17. Critical thickness h_0 versus lattice mismatch (b−a)/a for $Si_{1-x}Ge_x$ overlayers on a Si substrate. The experimental data originate from overlayers deposited by MBE (Sect.2.4) at different substrate temperatures [3.10]

total stress energy $\bar{E}_\epsilon$ increases with growing layer thickness and for thicker layers the formation of dislocations eventually becomes favorable.

According to Fig.3.16 the critical layer thickness is also a function of the lattice mismatch between two materials. Theoretical values are compared with experimental ones for the system of a mixed $Si_{1-x}Ge_x$ overlayer on Si in Fig.3.17 [3.11]. Depending on the preparation method, in particular the substrate temperature, the theoretically expected curve is considerably lower than the experimental ones. This demonstrates that kinetic limitations, e.g. details of the nucleation process, are also important for the type of mismatch relaxation.

3.5 Nucleation and Growth of Thin Films

3.5.1 Modes of Film Growth

Of major importance in modern technology are solid interfaces between thin films and solid substrates, and, in particular, epitaxial films that have been grown on crystalline material by one of the methods MBE or MOMBE discussed in Sects.2.4,5. Thus it will be useful to look, in a little more detail, at the process of film growth and the underlying principles which determine the structure and morphology of a particular film and the related interface to its substrate [3.12].

The individual atomic processes which determine film growth in its initial stages are illustrated in Fig.3.18. Condensation of new material from the gas phase (molecular beam or gas phase ambient) is described by an impinging rate (number of particles per cm^2 per second)

$$r = p(2\pi MkT_0)^{-1/2} \tag{3.28}$$

where p is the vapor pressure, M the molecular weight of the particles, k Boltzmann's constant, and T_0 the source temperature. Once a particle has condensed from the vapor phase, it might immediately re-evaporate or it may diffuse along the surface. This diffusion process might lead to adsorption, particularly at special sites like edges or other defects (Sect.3.2), or the diffusing particle may re-evaporate. In all these processes, characteris-

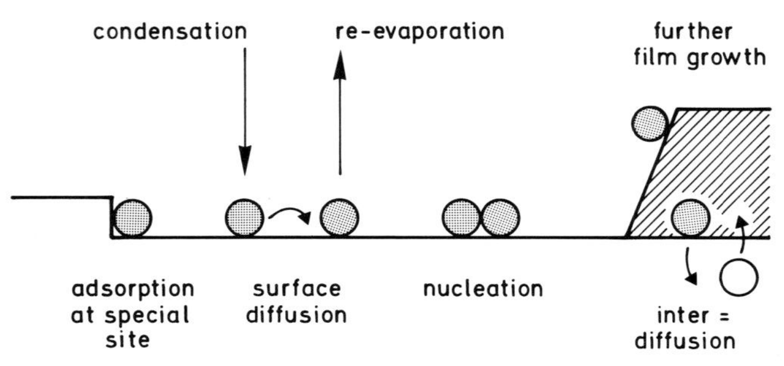

Fig.3.18. Schematic representation of atomic processes involved in film growth on a solid substrate. Film atoms shown as dark circles, substrate atom as open circle

tic activation energies have to be overcome, i.e. the number of particles being able to participate in the particular process is given by an Arrhenius-type exponential law; the desorption rate, for example, is given by (also Sect. 9.5)

$$\nu \propto \exp(E_{des}/kT) , \tag{3.29}$$

where E_{des} is the activation energy for desorption. The corresponding activation energies for adsorption or diffusion depend on the atomic details of the particular process. Their origin will be discussed in more detail in connection with the theory of adsorption in Chap.9. Besides adsorption at special defect sites and surface diffusion, nucleation of more than one adsorbed particle might occur, as might further film growth by addition of particles to an already formed island. How many particles are needed to form a new nucleus, which further grows into an island, is an interesting question, and one for which simple theoretical answers do exist. During film growth, interdiffusion is often an important process (Sect.3.4). Substrate and film atoms hereby exchange places and the film/substrate interface is smoothened.

In thermodynamic equilibrium all processes proceed in two opposite directions at equal rates, as required by the principle of "detailed balance". Thus, for example, surface processes such as condensation and re-evaporation, decay and formation of 2D clusters must obey detailed balance. Therefore, in equilibrium, there is no net growth of a film and so crystal growth must clearly be a non-equilibrium kinetic process.

The final macroscopic state of the system depends on the route taken through the various reaction paths indicated in Fig.3.18. The state which is obtained is not necessarily the most stable one, since it is kinetically determined. In general, certain parts of the overall process may be kinetically forbidden, whereas others may be in local thermodynamic equilibrium. In

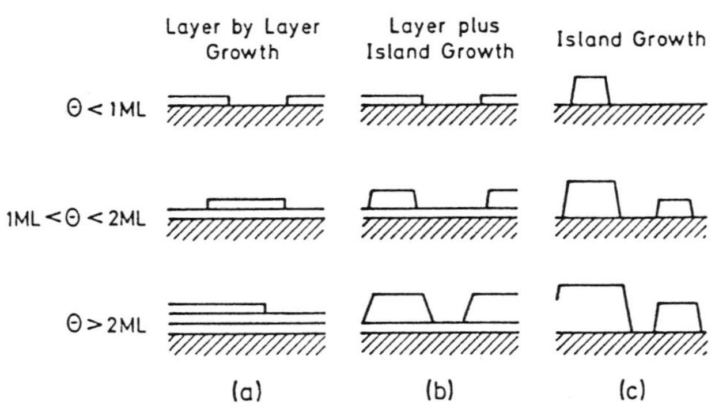

Fig.3.19a-c. Schematic representation of the three important growth modes of a film for different coverage (θ) regimes (ML means monolayer). (**a**) Layer-by-layer growth (Frank-von der Merve, FM). (**b**) Layer-plus island growth (Stranski-Krastanov, SK). (**c**) Island growth (Vollmer-Weber, VW)

this case equilibrium arguments may be applied locally even though the whole growth process is a non-equilibrium process. Due to this non-equilibrium nature of the process, a global theory of film growth requires a description in terms of rate equations (kinetic theory) for each of the processes depicted in Fig.3.18 [3.13, 14].

Instead of following this more theoretical atomistic approach we will consider the process of film growth more phenomenologically. In general, three markedly different modes of film growth can be distinguished (Fig. 3.19). In the **layer-by-layer** growth mode (or Frank-van der Merve, FM) the interaction between substrate and layer atoms is stronger than that between neighboring layer atoms. Each new layer starts to grow only when the last one has been complete. The opposite case, in which the interaction between neighboring film atoms exceeds the overlayer substrate interaction, leads to **island growth** (or Vollmer-Weber, VW). In this case an island deposit always means a multilayer conglomerate of adsorbed atoms.

The **layer-plus-island** growth mode (or Stransky-Krastanov, SK) is an interesting intermediate case. After formation of one, or sometimes several complete monolayers, island formation occurs; 3D islands grow on top of the first full layer(s). Many factors might account for this mixed growth mode: A certain lattice mismatch (Sect.3.4) between substrate and deposited film may not be able to be continued into the bulk of the epitaxial crystal. Alternatively, the symmetry or orientation of the overlayers with respect to substrate might be responsible for producing this growth mode.

A simple formal distinction between the conditions for the occurrence of the various growth modes can be made in terms of surface or interface tension γ, i.e. the characteristic free energy (per unit area) to create an additional piece of surface or interface (Sect.3.1). Since γ can also be in-

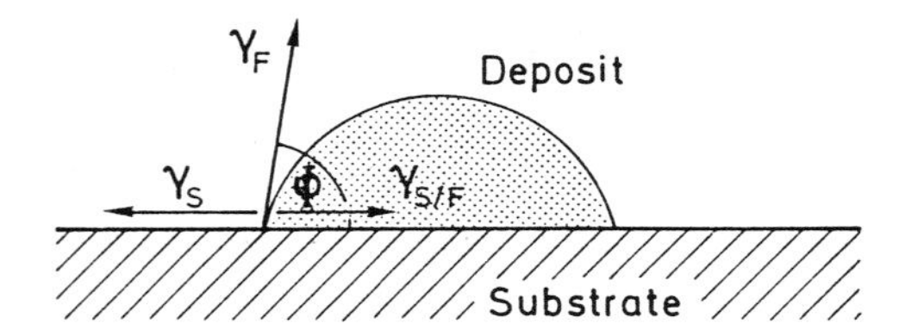

Fig.3.20. Simplified picture of an island of a deposited film; γ_S, γ_F and $\gamma_{S/F}$ are the surface tensions between substrate and vacuum, between film and vacuum and between substrate and film, respectively

terpreted as a force per unit length of boundary, force equilibrium at a point where substrate and a 3D island of the deposited film touch (Fig.3.20) requires

$$\gamma_S = \gamma_{S/F} + \gamma_F \cos\phi \ , \tag{3.30}$$

where γ_S is the surface tension of the substrate/vacuum interface, γ_F that of the film/vacuum, and $\gamma_{S/F}$ that of the substrate/film interface. Using (3.30) the two limiting growth modes, layer-by-layer (FM) and island (VW), can be distinguished by the angle ϕ, i.e.

(i) layer growth: $\phi = 0 \ , \quad \gamma_S \geq \gamma_F + \gamma_{S/F} \ ;$ (3.31a)

(ii) island growth: $\phi > 0 \ , \quad \gamma_S < \gamma_F + \gamma_{S/F} \ .$ (3.31b)

The mixed Stranski-Krastanov growth mode (layer plus island) can easily be explained in this picture by assuming a lattice mismatch between deposited film and substrate. The lattice of the film tries to adjust to the substrate lattice, but at the expense of elastic deformation energy. The transition from layer to island growth occurs when the spatial extent of the elastic strain field exceeds the range of the adhesion forces within the deposited material.

The relations (3.31) are not complete if one considers the equilibrium condition for the whole system including the gas phase above the deposited film. Since the equilibrium is determined by a minimum of the Gibbs free enthalpy G, one has to take into account a contribution $\Delta G = n\Delta\mu$ (n is particle number), which is the change in (Gibbs) free energy when a particle is transferred from the gas phase into the condensed phase of the deposited film. If this transfer occurs exactly at the equilibrium vapor pressure $p_0(T)$, then no energy is needed because of the equilibrium condition $\mu_{solid}(p_0, T) = \mu_{vapor}(p_0, T)$. If however the particle changes over from vapor to solid at a pressure p, a free enthalpy change (see any thermodynamics textbook on compression of an ideal gas)

$$\Delta G = n\Delta\mu = nkT\ln(p/p_0) \tag{3.32}$$

is involved. The ratio $\zeta = p/p_0$ is called the **degree of supersaturation**; as is easily seen from (3.32). ζ is one of the "driving forces" for the formation of a thin film deposited from an ambient vapor phase. Taking this vapor phase into account, the conditions for layer or island growth (3.31) have to be supplemented in the following way (C is a constant):

$$\text{(i) layer growth:} \quad \gamma_S \geq \gamma_F + \gamma_{S/F} + CkT\ln(p_0/p)\,, \tag{3.33a}$$

$$\text{(ii) island growth:} \quad \gamma_S < \gamma_F + \gamma_{S/F} + CkT\ln(p_0/p)\,. \tag{3.33b}$$

From (3.33) one sees that the growth mode of a certain material on a substrate is not a constant material parameter, but that the growth mode can be changed by varying the supersaturation conditions. With increasing supersaturation, layer by layer growth is favored. In vacuum deposition on solid substrates with low equilibrium vapor pressure p_0, supersaturation ζ might be high. In this case of vacuum deposition, the actual vapor pressure p is determined by the rate of impingement r, (3.28), in terms of the momentum transferred to the substrate per cm^2 per second. The supersaturation ζ might in this case, reach values of 10^{20} and more. Much lower ζ values are, of course, present in the epitaxy of III-V compounds (MBE, MOMBE, Sects.2.4, 5) [3.15], where a substrate temperature of about 500°C causes relatively high equilibrium vapor pressures, at least for the group V component.

3.5.2 "Capillary Model" of Nucleation

A simple, but intuitively very appealing theoretical approach was proposed by *Bauer* [3.16] to describe island (3D cluster) and layer-by-layer (2D cluster) growth of nuclei on an ideal, defect-free, substrate surface [3.17]. Since this approach uses only the thermodynamically defined interface (surface) tensions (Sect.3.1) γ_S, γ_F, $\gamma_{S/F}$ of the substrate, of the film material and of the interface between substrate and film, respectively, it is called the **capillary theory of nucleation**. In this approach the total free enthalpy ΔG_{3D} or ΔG_{2D} for the formation of a 3D or a 2D nucleus, i.e. an aggregation of film atoms, on a substrate is considered as a function of the volume or the number of atoms constituting this nucleus. This free enthalpy is the sum of the contribution (3.32) gained upon condensation of the vapor and an energy cost for the formation of the new surfaces and interfaces of the nucleus (surface tensions). As for every process a necessary condition for its occurrence is $\Delta G < 0$. This condition of decreasing free enthalpy then yields the limits for nucleation.

For **3D nucleation** (island growth) one obtains, with j as the number of atoms forming the nucleus,

$$\Delta G_{3D} = jkT\ln(p_0/p) + j^{2/3}X = -j\Delta\mu + j^{2/3}X\,. \tag{3.34}$$

The quantity X contains the contributions of the interface tension

$$X = \sum_k C_k \gamma_F{}^{(k)} + C_{S/F}(\gamma_{S/F} - \gamma_S)\,. \tag{3.35}$$

$C_{S/F}$ is a simple geometrical constant which relates the basis area of the nucleus $A_{S/F}$ with the number of atoms according to $A_{S/F} = C_{S/F} j^{2/3}$. C_k relates $j^{2/3}$ to a part of the surface of the nucleus (adjacent to vacuum) having the surface tension $\gamma_F{}^{(k)}$. The outer surface of the nucleus, i.e. the part exposed to the vapor phase (or vacuum), is assumed in (3.35) to be composed of several patches of different crystallographic orientation with different surface tensions $\gamma_F{}^{(k)}$.

For a simple, hemispherical nucleus (similar to that in Fig.3.20) with radius r, Eq.(3.34) would read, with Ω as atomic volume of the film material,

$$\Delta G_{3D} = -\frac{1}{2}\frac{4}{3}\pi r^3 \Omega^{-1} \Delta\mu + 2\pi r^2 \gamma_F + \pi r^2 (\gamma_{S/F} - \gamma_S)\,, \tag{3.36}$$

and $C_{S/F}$, for example, would amount to $\pi^{1/3}(3\Omega/2)^{2/3}$.

If we consider **2D nucleation**, i.e. the beginning of the growth of a layer on top of an ideally flat surface, the growth proceeds by incorporating new adsorbate atoms at the edges of the 2D cluster.

Corresponding to the surface tension there is a so-called **edge energy** γ_E per unit length, which describes the amount of energy necessary to position additional film atoms along the unit length of such an edge. The free enthalpy for 2D cluster growth, analogous to (3.34), is thus obtained as

$$\Delta G_{2D} = -j\Delta\mu + j(\gamma_F + \gamma_{S/F} - \gamma_S)\Omega^{2/3} + j^{1/2}Y\,, \tag{3.37}$$

where Ω is the atomic volume of the film material, and Y describes the effect of the adsorbate (film) atoms at the edges

$$Y = \sum_\ell C_\ell \gamma_E{}^{(\ell)}\,. \tag{3.38}$$

The sum over ℓ takes into account the fact that crystallographically different edge orientations ℓ may be associated with different edge energies $\gamma_E^{(\ell)}$. C_ℓ are geometrical factors defined as the C_k and $C_{S/F}$ according to (3.35). On the assumption of a circular, planar nucleus of radius r formed by one atomic layer and with the film lattice constant a, the general equation (3.37) simplifies to

$$\Delta G_{2D} = -\pi r^2 a \Omega^{-1} \Delta\mu + \pi r^2 a \Omega^{-1/3} (\gamma_F + \gamma_{S/F} - \gamma_S) + 2\pi r \gamma_E \, . \quad (3.39)$$

In (3.34, 37) the last terms are positive and give rise to an increase of the free enthalpy with growing size of the nuclei; they form a nucleation barrier. The first terms ($\propto \Delta\mu$) being proportional to the logarithm of the supersaturation ζ are negative; they drive the nucleation process faster with increasing supersaturation and growing nucleus size. The superposition of the two effects generates a non-monotonic dependence of ΔG on the number of atoms j within the nucleus (Fig.3.21); i.e., there exists a critical size of the nuclei both for 3D and for 2D nucleation, for which the free enthalpy ΔG becomes a maximum. If the nucleus reaches this critical size with a number of atoms j_{cr}, the cluster tends to grow rather than to decay. Below this critical cluster size, the nucleus is not stable, it tends to decay because ΔG increases with growing size.

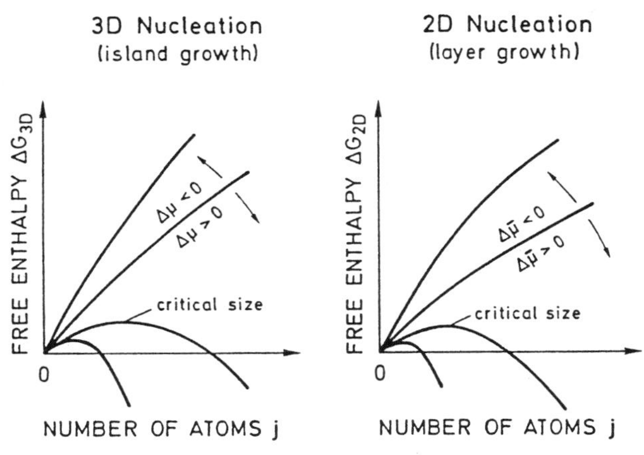

Fig.3.21. Qualitative plots of free enthalpy changes ΔG_{3D} and ΔG_{2D} for 3D (island growth) and 2D (layer growth) film growth versus number of atoms j, forming the 3D or 2D clusters, respectively. The terms $\Delta\mu$ and $\Delta\bar{\mu}$ describe the dependence on the state of supersaturation

The **critical cluster sizes**, i.e. the critical number of atoms, j_{cr}, and the corresponding ΔG values are obtained by differentiating (3.34, 37)

$$\text{3D growth:}\quad \Delta G_{3D}(j_{cr}) = \frac{4}{27}\frac{X^3}{\Delta\mu^2}, \quad j_{cr} = \left[\frac{2X}{3\Delta\mu}\right]^3 ; \tag{3.40}$$

$$\text{2D growth:}\quad \Delta G_{2D}(j_{cr}) = \frac{1}{4}\frac{Y}{\Delta\bar{\mu}}, \quad j_{cr} = \left[\frac{Y}{2\Delta\bar{\mu}}\right]^2 . \tag{3.41}$$

For layer-by-layer (2D) growth the second term in (3.37) is also negative because of the condition (3.31a); it thus augments supersaturation. In (3.41), therefore, $\Delta\bar{\mu}$ is an effective quantity

$$\Delta\bar{\mu} = \Delta\mu - (\gamma_F - \gamma_{S/F} - \gamma_S)\Omega^{2/3} , \tag{3.42}$$

which determines the critical 2D cluster size.

A comparison between (3.34 and 37) shows that, in principle, 3D cluster growth can only occur for $\Delta\mu > 0$, whereas 2D growth can also proceed for $\Delta\bar{\mu} > 0$. Because of the condition (3.31a) for layer-by-layer growth, the last term $(\gamma_F + \gamma_{S/F} - \gamma_S)$ in (3.42) is negative; 2D growth with $\Delta\bar{\mu} > 0$ can thus occur with $\Delta\mu \leq 0$, i.e. under conditions of subsaturation. In contrast, 3D growth requires supersaturation because of the requirement $\Delta\mu > 0$.

The rate J_N at which critical nuclei are formed is, of course, determined by the free enthalpies $\Delta G_{3D}(j_{cr})$ and $\Delta G_{2D}(j_{cr})$, respectively, for formation of a critical 3D or 2D nucleus (K is a constant):

$$J_N = K \exp[-\Delta G(j_{cr})/kT] . \tag{3.43}$$

The nucleation rate J_N also determines the growth rate of a film. By comparing (3.32, 40, 41, 43), one sees that certain supersaturation and temperature conditions determine the critical nucleus size or atom number j_{cr}. This, in turn, also influences $\Delta G(j_{cr})$ and through (3.43) the formation rate of nuclei. The two extreme cases are the formation of many small nuclei or of a few large ones.

With increasing supersaturation ζ the number of atoms in a critical nucleus decreases both in the 3D and 2D cases according to (3.40, 41). Increasing the supersaturation thus causes smaller aggregation sizes and higher numbers of nuclei. In vacuum deposition, which often has high supersaturation at low substrate temperatures, the critical size of a nucleus might be very small – it may contain only a few atoms. In this limit then, the applicability of the classical theory with its macroscopically determined terms γ_F, γ_S, $\gamma_{S/F}$ is questionable.

Another limiting factor for the applicability of the classical capillary theory of nucleation must also be kept in mind: Only ideal, perfect substrate surfaces are taken into account. Special defect sites (point defects, edges, dislocations, etc.), which are extremely important for nucleation, are not considered. Nevertheless, capillary theory is helpful as a guide for predicting general trends in nucleation and film growth.

3.6 Film-Growth Studies: Experimental Methods and Some Results

The standard methods for studying the modes of film growth in situ are **Auger Electron Spectroscopy (AES)** and **X-ray Photoemission Spectroscopy (XPS)** in their simplest form of application (Panel XI and Sect.6.3). Characteristic Auger transitions (usually differentiated spectra) or characteristic core-level emission lines (in XPS) of the substrate and of the adsorbate atoms are measured as a function of coverage. In both techniques the detected electrons originate from single atoms in the substrate or the adsorbate material. In AES they are released from a certain atomic level and carry an energy characteristic of a certain core-level transition (similar to characteristic X-ray lines): in XPS their energy corresponds to a characteristic core-level binding energy. But in both techniques, these electrons have to penetrate a certain amount of material, depending on the location of the emitting atom, before they can leave the crystal or the adsorbate overlayer. Thus, independent of the mode of excitation, the intensity I of the observed signal outside the film depends on the mean-free path of electrons of the particular energy and on the amount of matter which has to be penetrated. Because of the relatively "strong" interaction with matter (excitation of plasmons, etc.) the mean-free path λ of such electrons is in the order of a few Ångstroms, i.e. only electrons from the topmost atomic layers contribute to the detected AES or XPS signal. For the simplest case of layer by layer growth (FM) and assuming a simple continuum-type description the change of the AES or XPS line intensity dI is related to the change dh in film thickness by

$$dI/I = -\,dh/\lambda \,. \tag{3.44}$$

As in the case of absorption of electromagnetic radiation (then λ is the extinction length) one obtains an exponential decay of intensity due to the adsorbate thickness, i.e.,

$$I^F / I^F_\infty = 1 - \exp(-h/\lambda) = 1 - \exp(-\theta' d/\lambda) \,. \tag{3.45}$$

where I^F_∞ is the intensity measured on bulk film material, and θ' is the number of monolayers of the deposit (thickness of monolayer d). The intensity originating from substrate atoms after deposition of θ' monolayers of the film is then

$$I^S / I^S_0 = \exp(-\theta' d/\lambda) \,. \tag{3.46}$$

where I^S_0 is the intensity of the substrate material without any deposited film.

Equations (3.45, 46) do not describe the detailed functional dependence. It is evident that during the growth of a monolayer, the signal intensity changes linearly with θ'. But the envelope of the curve being composed of several linear pieces is to a good approximation, given by an exponential function (Fig. 3.22). More extended theoretical approaches, taking into account the atomistic nature of the process, are able to reveal this fine structure [3.18, 18]. It should further be emphasized that for non-normal emission, i.e. for detection of the AES or XPS electrons at an angle δ, the exponential extinction length in (3.45, 46) is not the free path λ, but must be corrected for the inclination of the surface to give $\lambda\cos\delta$.

It is evident that for island growth the intensity versus coverage dependence $I(\theta')$ will be very different from (3.45 and 46). During film growth, large areas of the substrate remain free of deposit and the substrate

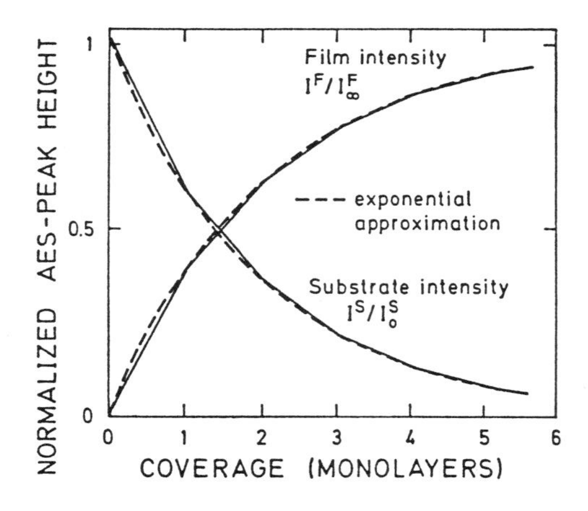

Fig. 3.22. Normalized Auger (AES) line intensities of substrate peak (I^S/I^S_0) and an adsorbate (overlayer) peak (I^F/I^F_∞) as a function of coverage (in MonoLayers, ML) for the layer-by-layer growth mode (Frank-von der Merve, FM). The solid curves are derived from a more detailed calculation [3.17]

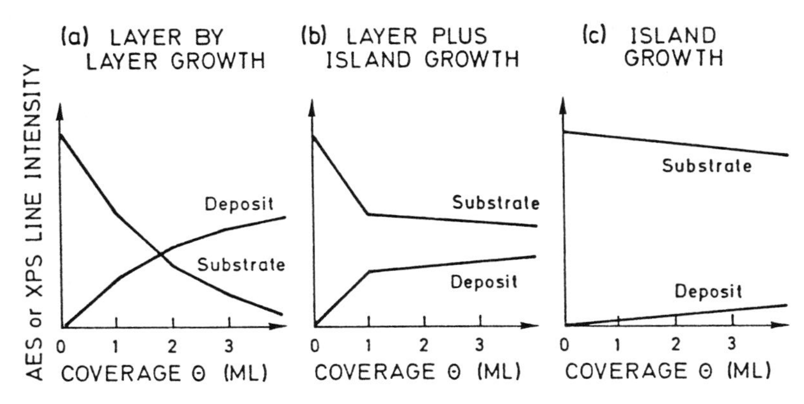

Fig.3.23. Schematic Auger (AES) line intensities from deposit and substrate versus amount of deposited material (coverage θ) for **(a)** layer growth (FM), **(b)** layer-plus-island growth (SK), and **(c)** island growth (VW)

signals are much less suppressed than in the case of layer-by-layer growth with complete overlayers. Only a slow increase and decrease of the film and of the substrate signal, respectively, is observed with the coverage θ' (Fig. 3.23c). Layer-plus-island (SK) growth is ideally characterised by a linear increase or decrease up to one or sometimes a few monolayers followed by a break point, after which the Auger or XPS amplitude increases or decreases (for film or substrate) only slowly. This regime corresponds to island formation on top of the first full monolayer(s). The gradient after the break point is dependent on the island density and shape and, of course, on the mean-free path of the particular electrons. Even though the three growth modes give qualitatively different AES and XPS patterns (Fig.3.23), a clear distinction solely on the basis of AES or XPS results may be difficult. Similar AES and XPS dependences are expected, for example, for inhomogeneous film growth with interdiffusion and precipitations of substrate material in the film. Usually information from other measurements (SEM, LEED, RHEED, etc.) is needed to analyse the experimental data. Furthermore, model calculations, assuming, e.g., simple shaped islands (half spheres, circular flat disks, etc.), can help to quantify an observed $I(\theta')$ dependence.

An example of layer-by-layer (FM) growth is provided by the system InSb(110)-Sn [3.19]. Because of the chemical and electronic similarity of α-Sn to InSb, there is a nearly perfect lattice match of the semiconducting α-Sn modification (tetrahydrally sp^3 bonded) to InSb. As is seen by electron diffraction (LEED), α-Sn grows epitaxially on InSb(100), whereas on UHV-cleaved InSb(110) surfaces, no long-range order can be detected when Sn is deposited at room temperature [3.19]. But according to Fig.3.24 the Auger intensity shows a clearly exponential behavior with coverage θ' over more than an order of magnitude, thus revealing the layer growth

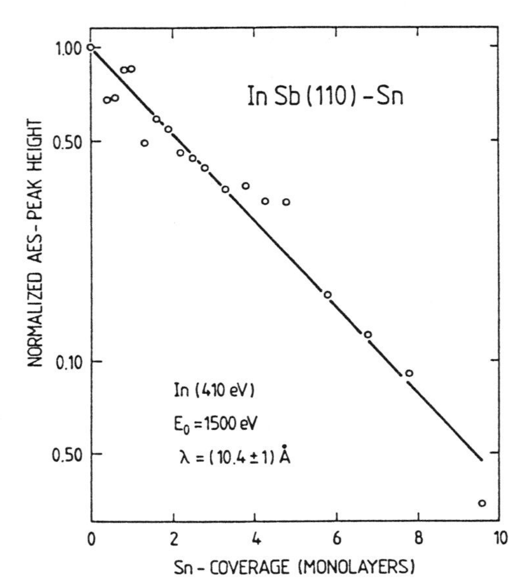

Fig.3.24. Normalized Auger peak intensity of the In (410eV) line versus coverage (in monolayers) for Sn deposited on cleaved InSb(110) surfaces. E_0 is the primary energy and λ the mean free path of 400 eV electrons in Sn [3.19]

mode (FM). As expected from the mean-free path of about 10 Å (for 400 eV electrons) derived from the slope in Fig.3.24, the AES signal intensity of the Sn film (Fig.3.25) saturates near 10 monolayers. For thicker films the intensity does not increase further since the limiting factor is then the mean free path of the electrons rather than the amount of deposit.

On UHV-cleaved GaAs(110) surfaces the growth mode of Sn deposited at room temperature is different [3.20]. Figure 3.26 reveals a break in the

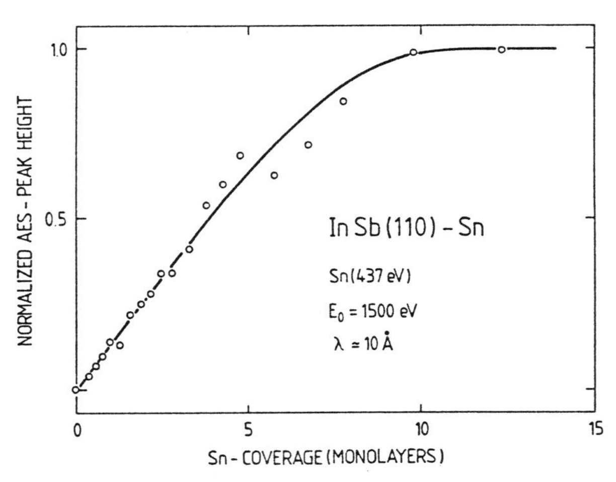

Fig.3.25. Normalized Auger peak intensity of the Sn (437eV) line versus coverage (in monolayers) for Sn deposited on cleaved InSb(110) surfaces. E_0 is the primary energy and λ the mean free path of 400 eV electrons in Sn [3.19]

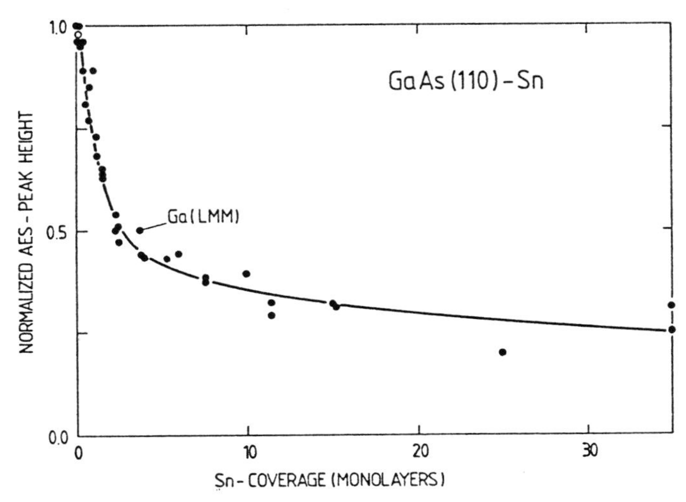

Fig. 3.26. Normalized Auger peak intensity of the Ga (LMM) line versus coverage (in monolayers) for Sn deposited on cleaved GaAs (110) surfaces [3.20]

substrate Ga emission line near a coverage of 1 to 2 monolayers, followed by a much slower decrease of the intensity. This behavior, as well as the corresponding intensity dependence of the Sn AES line (electron energy: 430 eV) is characteristic of Stranski-Krastanov (SK, layer plus island) growth. In Fig. 3.27 the full curve is obtained from a model calculation where the behavior of the AES intensity is attributed to the growth of a closed amorphous metallic β-Sn layer of about 3 Å thickness with hemispherical islands on top of this layer which grow in diameter with increasing film thickness. This model of SK growth is confirmed at higher coverages by scanning electron micrographs (Fig. 3.28), where at a nominal coverage of 38 monolayers of Sn an average island diameter of a few hundred Å is revealed.

Pure island (VW) growth is not found as often as the other two growth modes. The behavior of the characteristic Auger (or XPS) line intensities of deposit and substrate with coverage, as qualitatively shown in Fig. 3.23c, is sometimes difficult to distinguish from the case of SK growth. Different experimental techniques have to be applied to discriminate between the two modes.

A number of experimental examples for the different types of film growth are listed in Table 3.2. The compilation does not explicitly take into account the type of interface formed: abrupt, more gradual, or with the formation of new chemical compounds as in the case of transition metals on Si (layer growth), where at the interface a new crystalline compound, a Pt, Pd or Ni silicide (Sect. 8.5), is formed, depending on the substrate temperature.

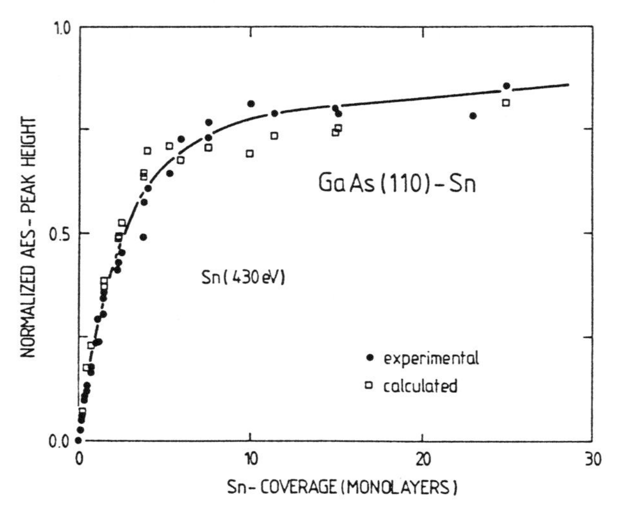

Fig.3.27. Normalized Auger peak intensity of the Sn (430eV) line versus coverage (in monolayers) for Sn deposited on cleaved GaAs(110) surfaces. As well as the experimental data points, results are also given of a theoretical model calculation (*open symbol*), where island formation on top of a closed β-Sn film of 3Å thickness is assumed [3.20]. The solid line is the best fit to the experimental data points

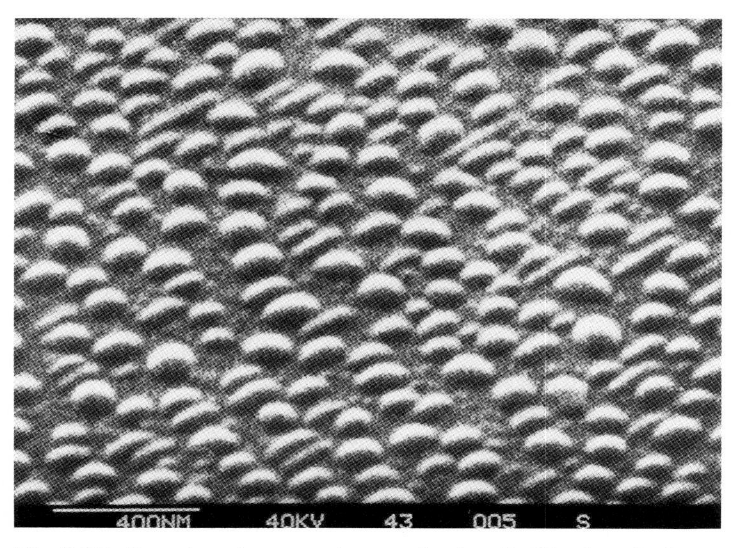

Fig.3.28. Scanning electron micrograph of an Sn film (nominal coverage 38 monolayers) deposited on a cleaved GaAs surface [3.20]

Table 3.2. Overview of some layer systems and their growth modes FM, SK, VW

Layer growth Frank-Van der Merwe (FM)	Layer plus island growth Stranski-Krastanov (SK)	Island growth Volmer-Weber (VW)
many metals on metals e.g., Pd/Au, Au/Pd, Ag/Au, Au/Ag, Pd/Ag, Pb/Ag, Pt/Au, Pt/Ag, Pt/Cu	some metals on metals, e.g., Pb/W, Au/Mo, Ag/W, rare gases on graphite	most metals on alkali halides most metals on MgO, MoS_2, graphite, glimmer
alkali halides on alkali halides	many metals on semicon- ductors, e.g.:	
III-V alloys on III-V alloys e.g., GaAlAs/GaAs, InAs/GaSb GaP/GaAsP, InGaAs/GaAs	Ag/Si, Ag/Ge Au/Si, Au/Ge Al/GaAs, Fe/GaAs, Sn/GaAs	
IV semiconductors on some III-V compounds, e.g., Ge/GaAs, Si/GaP, α-Sn/InSb	Au/GaAs, Ag/GaAs	
transition metals on Si Pt/Si, Pd/Si, Ni/Si (silicides)		

On top of this silicide, layer-by-layer growth of the pure metal then occurs. It must be further emphasized, that the type of growth indicated in Table 3.2 is usually found only within certain substrate temperature ranges. Outside this range the growth behavior may be different.

An important class of experimental techniques for investigation of thin films grown under well-defined conditions is based on diffraction. **Electron diffraction**, which includes both LEED and RHEED (Panel VIII: Chap. 4), yields information about the crystallographic order of the deposit. The information, however, is limited by the finite coherence length of the electron beam. Only ordering over a range of a few hundred Ångstroms contributes to the sharpness of the diffraction pattern. The observation of sharp LEED spots, therefore, indicates a well ordered crystal lattice only over such distances. Well developed Bragg spots do not necessarily indicate long-range crystallographic order. RHEED, in contrast to LEED, also gives information about the growth mode of the film. A flat surface, i.e. layer by layer growth means that the third Laue equation imposes no restriction; 2D elastic scattering usually gives rise to sharp diffraction stripes (Panel VIII: Chap. 4). The occurrence of 3D island growth brings the third Laue equation into play and spots rather than stripes are observed in the RHEED pattern.

Besides these diffraction techniques, which yield essentially a Fourier transform of the real-space structure, there are other techniques important in the study of film growth, that give real space images. **Scanning electron microscopy** (SEM [3.21], Panel V: Chap.3) can provide a direct image of the film morphology (Fig.3.28) down to dimensions of 10 Å in favorable cases. This lateral resolution in SEM is essentially determined by the diameter of the electron beam. SEM is usually performed ex situ, i.e. films prepared under UHV conditions have to be transferred through the atmosphere into the microscope. This might give rise to contamination-induced changes of the film structure, in particular in the low-coverage range. Only in special cases is SEM equipment available with UHV conditions and transfer units from the preparation chamber. It should also be noted that the SEM picture is produced by secondary electrons, whose emission intensity is not only affected by geometrical factors such as the type of surface, inclination to the primary beam, etc., but also by electronic properties of the surface such as work function and surface-state density, etc. Some of the intensity contrast in the image might therefore not be related to geometrical inhomogeneities but to electronically inhomogeneous areas. In particular, isolated islands of overlayer atoms should not be confused with patches of varying work function on the surface of a geometrically flat film.

Another direct imaging technique for the study of thin films is **scanning tunneling microscopy** (STM [3.22], Panel VI: Chap.3). In this, the electron tunneling current between a metal tip of atomic dimensions and the film surface is measured as a function of lateral position of the tip. Scanning the tip over the surface yields a "real-space" image of the film surface. Even though it is actually an outer electron density contour that is probed rather than a geometrical surface, this type of surface imaging resembles a real "sensing" comparable to macroscopically drawing a pencil over a rough surface. Electronically processed two-dimensional scanning images or line scans can give a clear impression of the roughness and general morphology of a film surface right down to atomic dimensions (Fig. 3.29). The spatial resolution both laterally, i.e. parallel to the surface, and vertically, i.e., normal to the surface, is in the range of Ångstroms.

Direct imaging is, of course, also performed in **transmission electron microscopy** (TEM) [3.24]. Because of the limited penetration depth of electrons through solid material, conventional TEMs with acceleration voltages below 200 keV allow the analysis of samples having a maximum thickness of about 1 μm. Using high voltage instruments (acceleration voltages up to 3MeV) thicker samples can be investigated. TEM can be used in a variety of ways to study thin films and overlayers. In the classical experiments on island growth and 3D nucleation, metal is deposited by evaporation onto alkali halide surfaces prepared by cleavage in UHV. Subsequently, the metal film consisting of more or less coalescent islands (VW growth

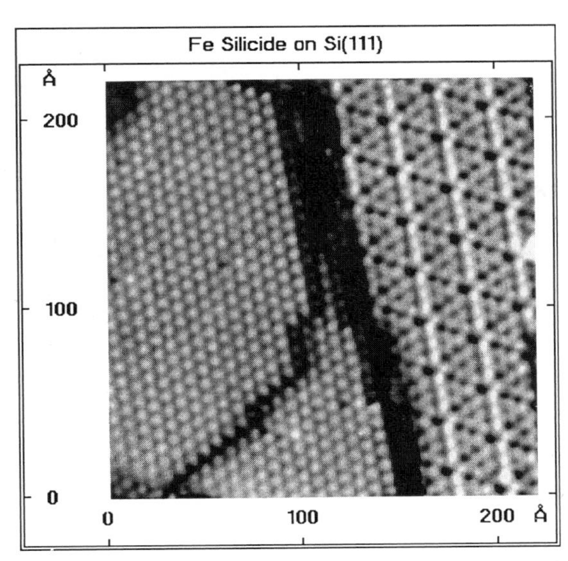

Fig.3.29. Scanning Tunneling Microscope (STM) image of a Si(111)-(7 × 7) surface (superstructure seen in the *right-hand part*) covered with a $FeSi_2$ layer (*left-hand part*). Within the $FeSi_2$ layer a step is evident. The data were taken with a tunnel current of 1 μA and a tip voltage of 1.8 V; i.e. the filled Si states are seen [3.23]

mode) is fixed by a deposited carbon film. Outside the UHV chamber the alkali halide substrate is dissolved from the carbon film by water treatment and the films with the embedded metal clusters and islands are analysed with respect to shape, distribution, and number of islands by conventional TEM techniques. In particular, nucleation rates of metals versus temperature T (or versus 1/T) have been studied for alkali-halide surfaces. For Au on KCl (100) surfaces (Fig.3.30) an exponential dependence of the nucleation rate J_N on 1/T, as predicted by (3.43), was found [3.13]. Additionally the nucleation rate depends quadratically on the adsorption rate u, the density of adsorbed atoms is $u\tau_a$ (where τ_a is the adsorption time). From plots such as Fig.3.30, the free enthalpy of cluster formation $\Delta G(j_{cr})$ (3.43) can be derived.

More refined preparation techniques are needed to study the crystallographic quality (dislocations, etc.) of thin films and the degree of perfection of heterointerfaces in TEM. The film must be imaged in a plane normal to the film/substrate interface. The sample therefore has to be cut normal to the interface, and thin slices have to be prepared by chemical etching and ion milling. Local thinning has to proceed down to dimensions of 10 to 100 nm, such that the electron beam can be transmitted. With conventional resolution, dislocations in the film can be studied and their number and density can be evaluated (Fig.3.31). For this purpose the sample is oriented in

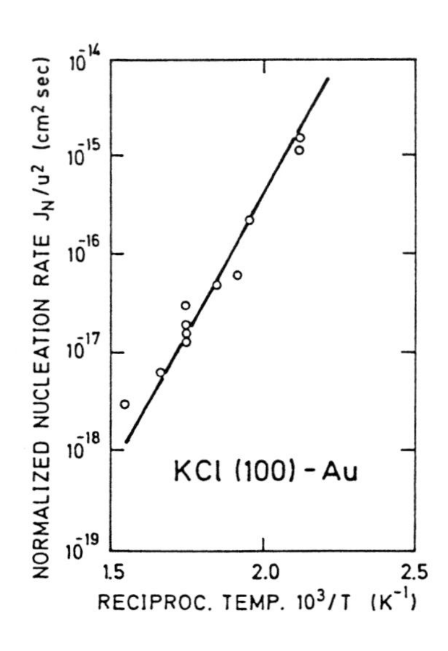

Fig.3.30. Nucleation rate J_N divided by square of adsorption rate u for Au nuclei on a KCl surface, plotted versus reciprocal substrate temperature [3.13]

the electron beam slightly off from the Bragg condition. The strain field surrounding a dislocation causes the Bragg condition to be fulfilled locally, and part of the incoming beam is diffracted out of the transmitted beam. Part of the strain field of the dislocation then appears as a bright structure in a dark field or dark in a bright field image. In high-resolution TEM even the atomic structure of an interface can be resolved (Fig.3.32). It must be

Fig.3.31. Transmission Electron Micrograph (TEM) of dislocations in an epitaxial (110)GaAs overlayer on a Si substrate [3.25]

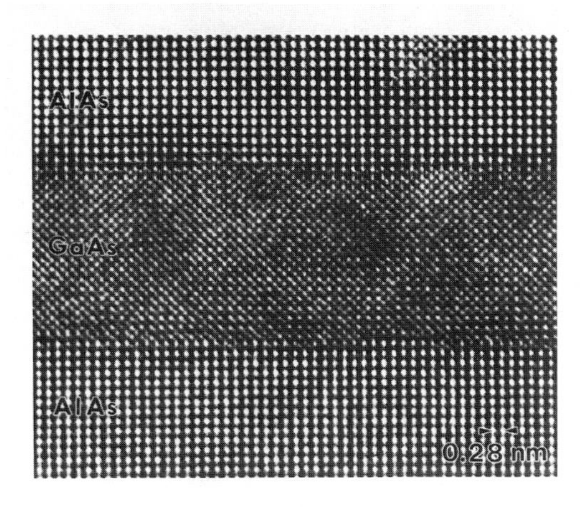

Fig. 3.32. High-resolution TEM of an AlAs/GaAs double heterostructure. The dark and bright points are correlated with single rows of atoms [3.26]

emphasized, however, that the contrast seen in such a high-resolution electron micrograph is not directly related to single atoms. Rows of atoms are imaged and an involved theoretical analysis (taking into account details of the electron scattering process) is necessary for a detailed interpretation of the dark and light spots in terms of atomic positions. Nevertheless, information about the quality of an interface, in particular about the orientation of lattice planes, etc., can be obtained by simple inspection.

Optical methods are also successfully applied to the study of thin films. The **Raman effect** [3.27, 28] (inelastic scattering of photons in the visible or UV spectral range) allows one to study the vibrational properties of thin films, which can yield interesting information about their morphology. An example is exhibited in Fig. 3.33. Sb overlayers have been deposited under

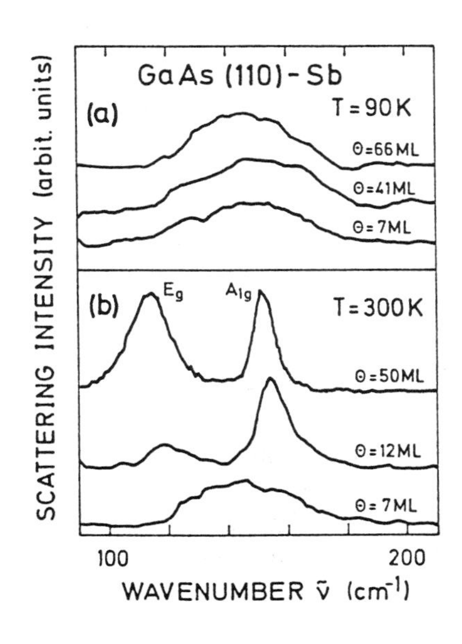

Fig. 3.33a,b. Raman spectra of Sb overlayers on lightly n-doped GaAs(110) surfaces prepared by cleavage in UHV; excitation by laser lines at 406.7 nm and 530.9 nm. The substrate temperatures T during deposition were 90 K (**a**) and 300 K (**b**). The coverage θ is given in MonoLayers (ML) on the spectra [3.28]

UHV conditions on cleaved GaAs(110) surfaces. From LEED and other techniques it is known that the first monolayer of Sb is well ordered with a (2×1) superstructure. Additional deposition of Sb on top of this first layer leads to thicker layers which do not show any crystallographic order in LEED. Characteristic phonons of Sb, as measured by the Raman effect, reveal more details about these thicker overlayers. Because of its rhombohedric (D_{3d}) symmetry, crystalline Sb has two Raman active phonon modes at the Γ point, one of type A_{1g} and the other a twofold degenerate E_g mode. The corresponding lines are seen for thicker layers ($\theta' = 50$ML, Fig.3.33b) after deposition at 300 K. Since photons transfer only a negligible wave vector **q** in the scattering process, and since for a crystalline solid with 3D translational symmetry wavevector conservation applies, only phonons with a well-defined **q** vector near Γ contribute to the relatively sharp phonon lines (FWHM $\simeq$ 10cm^{-1}). Their sharpness, together with the lack of any long range order in LEED, indicates a polycrystalline morphology of the Sb layers deposited at 300 K and with thickness exceeding about 10 ML. For thinner layers ($\theta' < 10$ML, Fig.3.33b) a broad, smeared-out structure is seen. This clearly demonstrates, that the Sb overlayers are amorphous. Due to the lack of translational symmetry in amorphous material, the wave vector is not conserved upon light scattering and phonons from all over the Brillouin zone contribute to give essentially an image of the phonon density (versus energy) rather than of single sharp phonon branches. Deposition at 90 K leads to amorphous Sb layers even in the high coverage regime as is seen from the featureless, broad spectral structure in Fig.3.33a. It is obvious that phonon-line broadening in a Raman spectrum can thus be used to get information about the degree to which wave vector conservation is violated due to finite crystallite size in a polycrystalline layer. Not only can the amorphous state be distinguished from a polycrystalline morphology, but the crystallite size can also be estimated. According to a number of experimental data, sharp phonon lines, as expected for large bulk crystals, are observed in Raman scattering for crystallites with diameters exceeding 100 to 150 Å.

Among the various optical measurements on thin films, **ellipsometry** (Panel XII: Chap.7), in particular when used as a spectroscopy, is of considerable importance. In ellipsometry, the optical reflectivity of the film/substrate system is determined by measuring the change of the polarisation state of light upon reflection [3.29]. The reflectivity measurement is thus reduced to a measurement of angles (Panel XII), which yields extremely high surface sensitivity ($<10^{-1}$ monolayers of an adsorbate) in comparison with conventional intensity measurements. The underlying reason for this surface sensitivity is the high accuracy by which the angle of polarisation can be measured in contrast to a reflection intensity measurement, which is inherently less accurate. The change of the state of polarisation upon reflection can be expressed in terms of the ratio of the two complex reflection coefficients $r_{\parallel}$ and $r_{\perp}$ for light polarized parallel and perpendicular to the plane of incidence (containing incident and reflected beam), r being the ratio of reflected and incident electric field strength. The complex quantity

$$\rho = r_{\parallel}/r_{\perp} = \tan\psi \exp(i\Delta) \tag{3.47}$$

defines the two ellipsometric angles Δ and ψ that are measured in ellipsometry (Panel XII: Chap.7). Δ and ψ completely determine the two optical constants n (refractive index) and κ (absorption coefficient) or $\mathrm{Re}\{\epsilon\}$ and $\mathrm{Im}\{\epsilon\}$ (dielectric functions) of an isotropic reflecting medium (semi-infinite half-space). For film growth and adsorption studies in interface physics one usually measures changes $\delta\Delta$ and $\delta\psi$ in the values of Δ and ψ for the clean surface and for the film (or adsorbate) covered surface. Most interesting are in situ measurements under UHV conditions. Even in the simplest *one-layer model* five parameters, the dielectric functions of substrate ($\mathrm{Re}\{\epsilon_s\}$, $\mathrm{Im}\{\epsilon_s\}$) and film ($\mathrm{Re}\{\epsilon_f\}$, $\mathrm{Im}\{\epsilon_f\}$) and the film thickness d determine $\delta\Delta$ and $\delta\psi$ [3.29]. The commonly applied analysis of experimental spectra $\delta\Delta(\hbar\omega)$ and $\delta\psi(\hbar\omega)$ is based on known substrate optical constants ($\mathrm{Re}\{\epsilon_s\}$, $\mathrm{Im}\{\epsilon_s\}$) and fitting spectra calculated with assumed film optical constants to the experimental spectra. The approximate film thickness is obtained from a measurement of the total amount of deposited film material by means of a quartz balance. The optimum fit then yields the dielectric functions $\mathrm{Re}\{\epsilon_f\}$ and $\mathrm{Im}\{\epsilon_f\}$ of the film. Ellipsometric spectroscopy, when used in film growth studies, is most frequently applied to the determination of film optical constants and thus yielding integral information about the chemical and structural nature of a film and its global electronic structure. Figure 3.34 shows, as an example, results [3.30] obtained on the same InSb(110)-Sn system, as discussed in connection with Figs.3.24 and 25. For coverages below about 500 Å ($\approx$200 monolayers of Sn) the dielectric functions (Fig. 3.34a) with their spectral structure due to electronic interband transitions are characteristic of semiconducting (gray) α-Sn, whereas for thicker Sn layers ($>$500 Å) the well-known behavior of the dielectric response of a quasi-free electron gas is observed (Fig.3.34b). Sn layers at higher coverage are thus deposited as metallic β-Sn. Simultaneous LEED studies show that neither the α-Sn nor the β-Sn modification possesses long-range order; the deposit is in both cases polycrystalline. The stabilization of the tetrahedrally bonded α-Sn species even at room temperature (usually it is stable only at low temperature) is obviously due to the isoelectronic nature of α-Sn and InSb.

In applied technology, ellipsometry using a single light wavelength is often employed to determine film thicknesses. In this case the optical constants of the substrate and of the film material must be known and only d is calculated via Fresnel's formulae from the measured $\delta\Delta$ and $\delta\psi$.

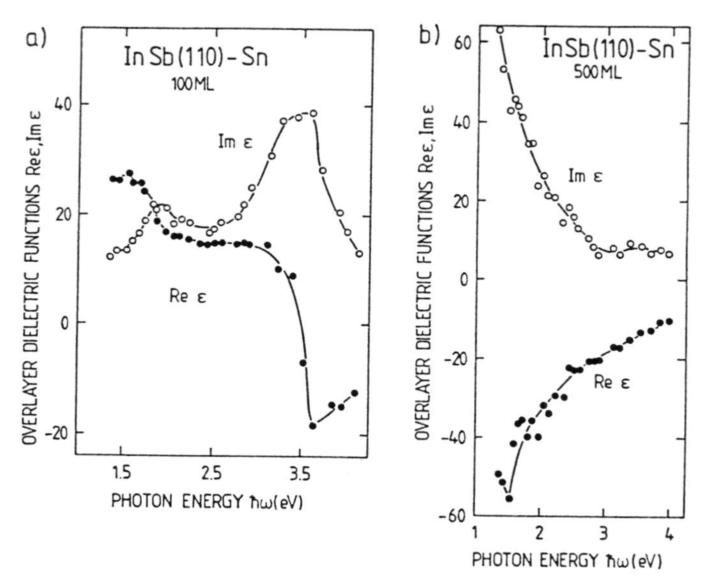

Fig.3.34a,b. Real and imaginary parts of the dielectric function ϵ_f of Sn overlayers deposited in UHV on clean cleaved InSb(110) surfaces. The ellipsometric measurements were performed in situ on overlayers with differing thickness (coverage in MonoLayers, ML) [3.30]

An important technique for studying thin films and interfaces between overlayers and substrates is **ion scattering** with medium and high-energy (Rutherford BackScattering, RBS). In Sects.4.9-11 the method and the underlying physics are described in more detail. A whole variety of information can be extracted from RBS measurements. Energy analysis of the backscattered ions (primary energy between 5keV and 5MeV) can be used as a chemical surface analysis technique, i.e., to probe the chemical nature of a particular surface layer, or a segregation on top of it (Sect.4.10, Fig.4.29). Measurement of the angular distribution of the backscattered particles and of the scattering yield as a function of angle of incidence gives detailed information about (i) the crystallographic quality of an epitaxial film and (ii) relaxations, lattice mismatch and internal strain (Sect.4.11, Figs.4.35,36). The concept on which these measurements rely is that of "shadowing and blocking" of ion beams in certain crystallographic directions (Sect.4.11).

In conclusion, the study of film growth is an expanding field, both in fundamental and in applied research. Many experimental techniques are available, on the macroscopic level as well as with atomic resolution. A complete compilation is beyond the scope of the present book; the aim of this chapter was to give the reader a first impression and an overview of this wide field of research.

Panel V
Scanning Electron Microscopy (SEM) and Microprobe Techniques

Scanning techniques (see also Panel VI: Chap.3) have the common feature that a certain physical quantity is measured with spatial resolution and recorded as a function of position on the surface. The local distribution of this quantity is obtained electronically and viewed on an optical display, usually a TV screen. Of prime importance for the characterization of microstructures on a scale down to 10 Å are Scanning Electron Microscopes (SEM) [V.1] and microprobes. In addition to simple imaging of surface topography (SEM), a local surface analysis in terms of chemical composition can also be performed by the scanning electron microprobe.

The basic principle of these techniques consists in scanning a focussed electron beam (primary energy typically $2 \div 10\,\mathrm{keV}$) over the surface under study and simultaneously detecting electrons emitted from the surface. The intensity of this emitted signal determines the brightness of the spot on a TV tube. The formation of a topographical image is due to local variations of the electron emissivity of the surface. The operation of a SEM is described in more detail in Fig. V.1. The scanned electron beam is produced in an electron microscope column. Electrons are emitted from a heated W or LaB_6 cathode (or field emission cathode) and are focussed by the Wehnelt cylinder and an anode aperture into a so-called **cross-over point**. The cross-over point is projected by a first magnetic lens onto a smaller image point, which is further reduced by a second magnetic lens onto the sample surface. This image on the sample is reduced by a factor of about 1000 with respect to the primary spot. The best SEM columns can achieve focussing on the sample surface into a spot of about 10 Å. This spot size is the essential factor determining the spatial resolution of the SEM. Magnetic lenses [V.2] are used since they are more effective for high electron energies than electrostatic lenses ($\mathbf{F} = e\mathbf{v} \times \mathbf{B}$); aberration errors are smaller, too. The x-y scan is usually performed by two magnetic coils arranged perpendicular to one another between the two magnetic lenses. The amplification of the SEM is produced simply by electronically varying the deflection angle of the scanning electron beam. A video tube is employed as optical display; its electron beam is scanned synchronously with the primary probing beam of the microscope column, i.e. both beams are controlled by the same scanning electronics and the amplification of the SEM is adjusted by a scaling factor introduced essentially by a resistance divider circuit. The in-

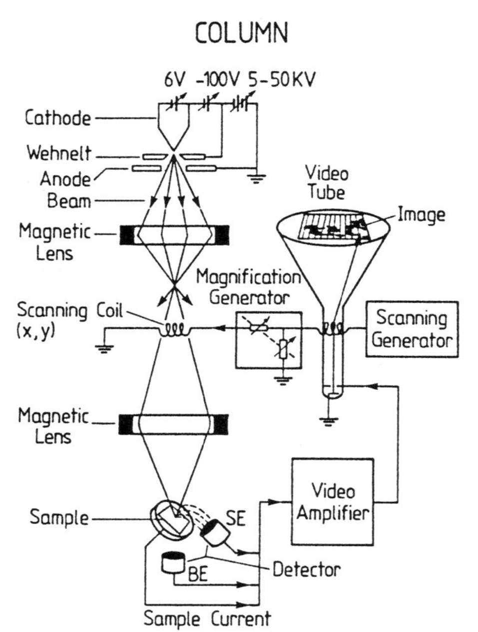

Fig. V.1. Schematic set-up of a scanning electron microscope or microprobe

tensity of the electrons emitted from the sample determines (through an amplifier) the intensity of the TV tube beam. Different detectors sensitive to various energy ranges of the emitted electrons can be used. A solid surface irradiated by an electron beam of energy E_0 emits electrons of various origin (Fig. V.2). Besides elastically backscattered electrons of energy E_0 (Region I), there is a regime of inelastically backscattered electrons (II),

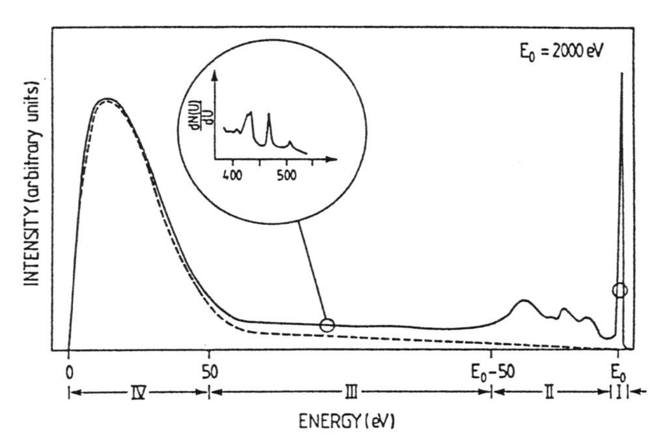

Fig. V.2. Qualitative large-scale overview of the energy distribution of electrons emitted from a surface which is irradiated by an electron beam of primary energy E_0. The true secondary-electron contribution is indicated by the dotted curve

where the spectral structure stems from energy losses due to plasmon excitation and interband transitions (Chap.4). These excitations are specific to every solid. A detector tuned to this energy range (BE in Fig.V.1) thus yields an image of the surface which is extremely sensitive to composition. A high contrast results from the material specificity of the emitted electrons. Because of the high degree of forward scattering (dielectric scattering with small wave-vector transfer $q_{\|}$, Sect.4.6) in this loss regime, strong shadowing effects are observed for the three-dimensional microstructures being imaged.

The spectrum of emitted electrons exhibits two other characteristic energy ranges, the next of these being a flat part between the above-mentioned loss regime and energies down to about 50 eV (Region III). With sufficient amplification (and detection usually in the differentiated mode dN/dE) the Auger emission lines characteristic for each chemical element can be detected in this energy range. The major part of the emission, the so-called **true secondary electrons**, form a strong and broad spectral band between 0 and 50 eV (Region IV), but extend as a weak tail up to E_0. The true secondary background originates from electrons that have undergone multiple scattering events (involving plasmons, interband transitions, etc.) on their way to the surface. Their spectral distribution and their intensity is thus not very specific to a particular material; nor is there a strong angular dependence of the true secondary emission as was the case for the inelastic losses in Region II. The detector SE in Fig.V.1 is tuned to record the low-energy true secondary electrons and thus yields a different image of the surface from that obtained by detector BE. Due to the lack of forward scattering, less shadowing is observed. The contrast due to chemical composition is poorer but instead variations in surface roughness and work function are made visible. A topographical image similar to that from the detector SE, but with inverted contrast can be obtained by recording the total electric current from the sample, i.e. the total difference between the numbers of incident and emitted electrons (essentially the true secondaries).

Although, in principle, SEM pictures are not stereoscopic, good insight into the geometry of an object or a surface can be achieved by using different detectors and different irradiation geometries. However, one must expect some difficulties in the interpretation, since not only geometrical factors (inclination of planes, etc.) but also work-function variations and other electronic factors give rise to contrast changes. As an example of the application of SEM to the study of metal overlayers on semiconductors in microelectronics (Fig.V.3) shows golds contacts (bright structures) on a GaInAs/InP heterotructure. In the upper left part of the figure the GaInAs overlayer is seen, whereas the deeper laying InP substrate is exposed in the lower-right hand [V.4]. A further example is the SEM picture of a clean GaAs(110) surface (Fig.V.4a) which has been annealed in a separate UHV

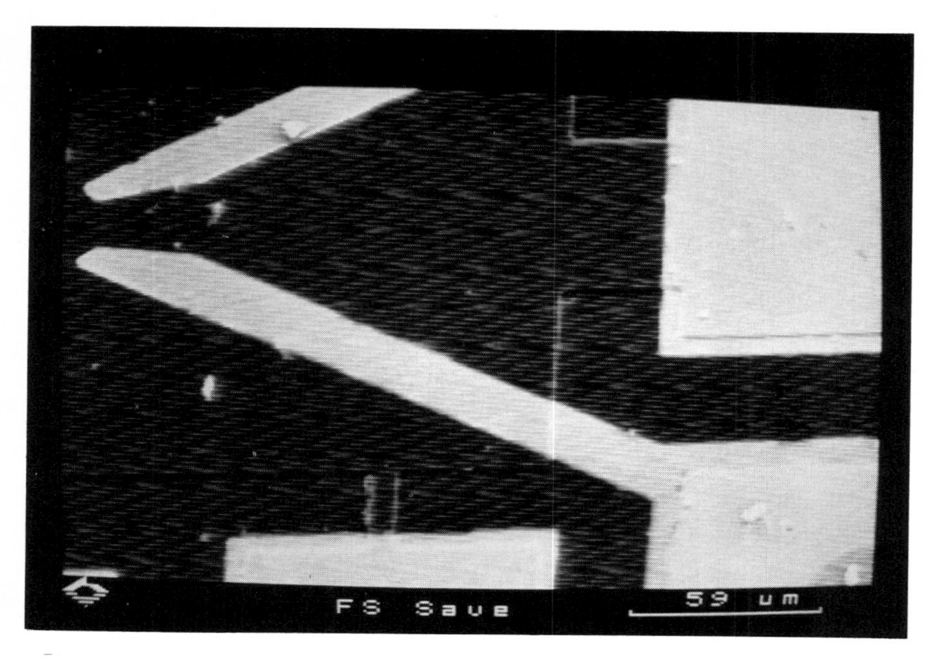

Fig. V.3. Scanning electron micrograph of gold contacts (bright structures) deposited on a GaInAs/InP heterostructure [V.4]

chamber up to temperatures around 1150 K. Segregation of material into islands is observed on the surface. The SEM pictures of Figs. V.3,4a are obtained in a standard SEM under high vacuum conditions ($\approx 10^{-5}$ Pa), whereas the surfaces were prepared under UHV conditions in a separate chamber. Between preparation and the SEM study a transfer of the sample through air is necessary. This drawback of standard equipment has now been overcome by the development of SEMs that work under UHV conditions. These more advanced instruments can be connected via load lock chambers to other UHV equipment, such that, after preparation, SEM investigations of a surface or interface can be performed in situ.

In addition to measuring the secondary electron yield as a function of spatial position as in the SEM, other quantities can also be studied. The primary electron beam of a SEM also induces Auger processes (Panel III: Chap. 2) and the emission of X-rays [V.3]. Both Auger electrons and X-ray photons can be detected as a function of position by suitable detectors, Auger electrons by an electron analyzer (e.g., CMA; Panel II: Chap. 2) and photons, e.g., by an energy dispersive semiconductor photon-sensitive detector. Both the Auger signal and the X-ray emission are specific to a given element. Recording these quantities thus yields a spatially resolved picture of the distribution of chemical elements on a surface. The apparatus which

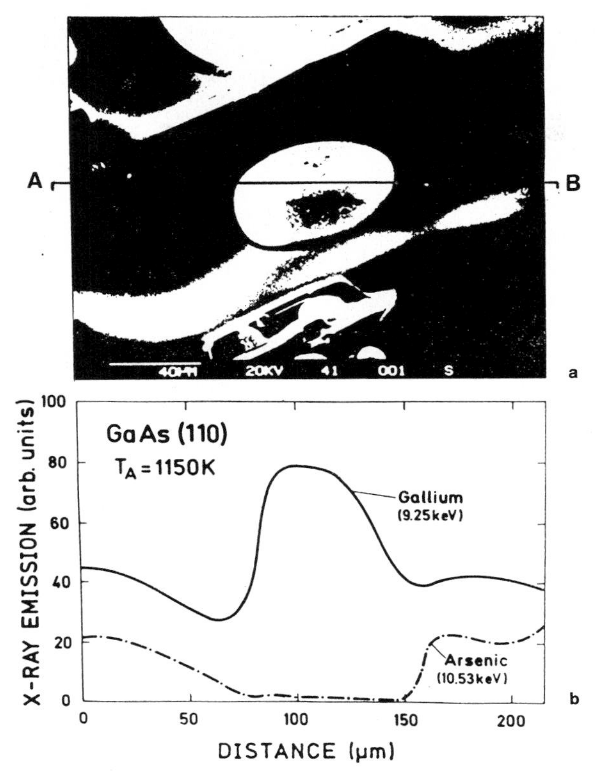

Fig. V.4. (**a**) Scanning electron micrograph of a GaAs(110) surface prepared by cleavage in UHV with subsequent annealing to 1150 K. The micrograph was taken ex situ after transfer through air. Segregation of substrate material into islands is visible. (**b**) X-ray emission signal for As (10.53keV) and Ga (9.25keV) measured along a line AB intersecting the segregation region in (**a**) [V.5]

is equipped with the appropriate instruments is sometimes called a **microprobe**. As an example Fig. V.4b shows the X-ray signals of the characteristic As line at 10.53 keV and the Ga line at 9.25 keV photon energy along a line scan intersecting the segregation in Fig. V.4a. The spatially resolved element analysis clearly exhibits the segregation of Ga islands on the GaAs surface after extended heating.

An example of a scanning Auger image is depicted in Fig. V.5. Sb has been evaporated on a UHV-cleaved GaAs(110) surface at a substrate temperature of about 300 K. After transfer through air the relative intensity of the Sb Auger line (integral over 454 and 462V lines) $(I_{Sb}-I_{back})/I_{back}$ was measured and the spatially resolved intensity distribution is displayed. The relatively homogeneous gray background indicates a quasi-uniform coverage of the GaAs surface with Sb. The dark spots are regions of lower cover-

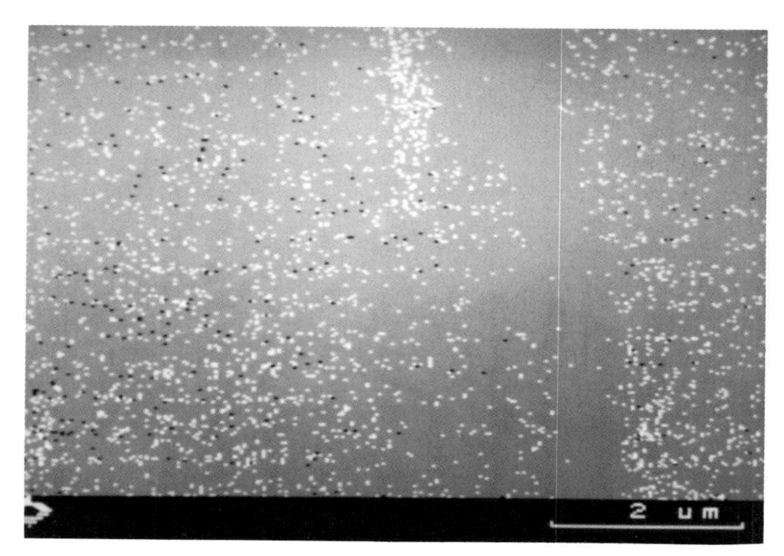

Fig.V.5. Scanning Auger electron micrograph of a GaAs(110) surface cleaved in UHV and covered with a nominal 5 monolayers of Sb. The Auger micrograph was recorded ex situ with a primary energy of 20 keV. The intensity distribution of the Sb lines (454eV, 462eV), i.e. $(I_{Sb}-I_{back})/I_{back}$ is displayed with a magnification of 10 000 [V.6]

age, i.e. they indicate deficiencies in the Sb overlayer. The bright spots (diameter ≈ 350 Å) originate from Sb islands which are formed on top of the more or less complete Sb underlayer. The AES image shows Stranski-Krastanov growth for the Sb overlayers on GaAs(110).

Although both the Auger intensity distribution and the characteristic X-ray emission pictures yield a local-element analysis of the surface, different depth information is obtained by the two probes. The absorption of X-ray photons by matter is weak, and thus the emitted photons originate from a depth range which is essentially determined by the penetration depth of the high-energy primary electrons. Depending on the material and the primary energy, the information depth for the X-ray probe is thus $0.1 \div 10$ μm. For the Auger probe, on the other hand, the information depth is given by the distance that the Auger electrons can travel without losing energy by scattering. This length is material dependent and is typically smaller by a factor of 10^2 to 10^4, as compared with the information depth of the X-ray probe. The characteristic differences between the information depth and the volume probed by the different techniques is shown schematically in Fig.V.6. Backscattered electrons and X-ray photons originate typically from a pear-shaped zone below the surface, whereas the low-energy secondary electrons and the Auger electrons carry information from the small, narrow neck of

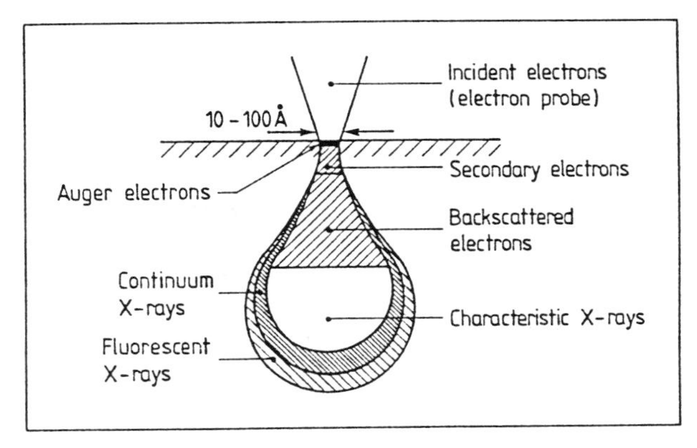

Fig. V.6. Schematic overview of the pear-shaped volume probed by different microprobe signals (electron and X-ray emission), when a primary electron beam is incident on a solid sufrace. Auger electrons originate from a depth of $5 \div 20$ Å, whereas the X-ray information depth is $0.1 \div 10$ μm with much less spatial resolution

the "pear". The pear shape of the probed region arises from elastic and inelastic scattering of the high-energy primary electrons. It is evident from the distribution in Fig. V.6 that much better spatial resolution is obtained by the Auger probe ($\geq$ 10nm as with SEM) than by using X-ray photons. On the other hand, the X-ray probe offers better depth analysis for the investigation of layer structures. For scanning Auger studies, UHV conditions are necessary because of the significant influence of surface contamination, whereas the X-ray probe can be used under conditions of poorer vacuum.

References

V.1 L. Reimer: *Scanning Electron Microscopy*, Springer Ser. Opt. Sci. Vol.45 (Springer, Berlin, Heidelberg, 1985)

V.2 P.W. Hawkes (ed.): *Magnetic Electron Lenses*, Topics Curr. Phys., Vol.18 (Springer, Berlin, Heidelberg 1982)

V.3 B.K. Agarwal: *X-Ray Spectroscopy*, 2nd edn., Springer Ser. Opt. Sci. Vol.15 (Springer, Berlin, Heidelberg, 1991)

V.4 H. Dederichs (ISI, Research Center Jülich): Priv. commun.

V.5 W. Mockwa (Phys. Inst., RWTH Aachen): Priv. commun.

V.6 H. Dederichs (ISI, Research Center Jülich): Priv. commun.

Panel VI

Panel VI
Scanning Tunneling Microscopy (STM)

The Scanning Tunneling Microscope (STM), developed by *Binnig* and *Rohrer* [VI.1], delivers pictures of a solid surface with atomic resolution. A direct real-space image of a surface is obtained by moving a tiny metal tip across the sample surface and recording the electron tunnel current between tip and sample as a function of position [VI.2]. In this sense the STM belongs to the wider class of scanning probes (Panel V: Chap. 3), in which a certain signal – in this case the tunnel current – is recorded and displayed electronically versus surface position.

Tunneling is a genuine quantum mechanical effect in which electrons from one conductor penetrate through a classically impenetrable potential barrier – in the present case the vacuum – into a second conductor [VI.3]. The phenomenon arises from the "leaking out" of the respective wave functions into the vacuum and their overlap within classically forbidden regions. This overlap is significant only over atomic-scale distances and the tunnel current I_T depends exponentially on the distance d between the two conductors, i.e. the tip and the sample surface. The now classic work of *Fowler* and *Nordheim* [VI.4] yields, as a first approximation, the expression

$$I_T \propto \frac{U}{d} \exp(-Kd\sqrt{\bar{\phi}}) \qquad \text{(VI.1)}$$

where U is the applied voltage between the two electrodes, tip and sample, $\bar{\phi}$ their average work function ($\bar{\phi} \gg eU$), and K a constant with a value of about 1.025 $\text{Å}^{-1} \cdot (\text{eV})^{-1/2}$ for a vacuum gap. I_T is easily measurable for distances d of several tens of Ångstroms and, in order to get interesting information about the surface, d must be controlled with a precision of 0.05 ≈ 0.1 Å [VI.5].

To achieve a lateral resolution that allows imaging of individual atoms, the movement of the tiny metal tip across the surface under investigation must be controlled to within 1 ÷ 2 Å. The high sensitivity of the instrument to the slightest corrugations of the surface electron density is due to the exponential dependence (VI.1) of I_T on d and $(\bar{\phi})^{1/2}$. Experimentally the stringent requirements for precise tip movement are satisfied by a two-step approach (Fig. VI.1): The sample is mounted on a piezoelectrically driven sup-

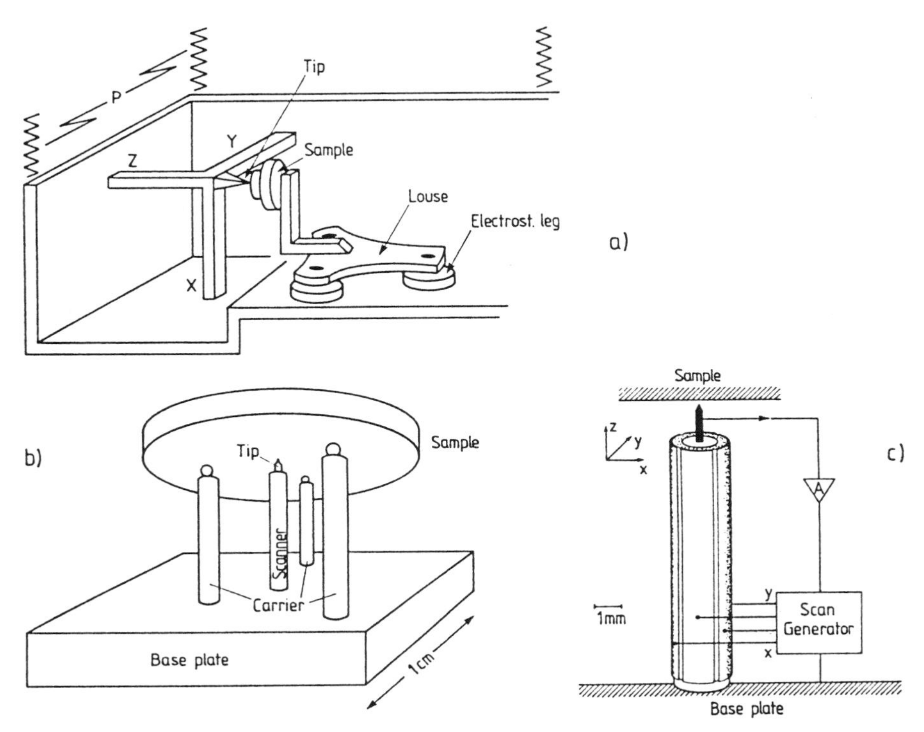

Fig. VI.1. (**a**) Schematic of the classic Scanning Tunneling Microscope (STM) [VI.1]. The metallic tip of the microscope is scanned over the surface of the sample with a piezoelectric tripod (x, y, z). The rough positioner or "louse", driven by controlled piezoelectric deformation of its main body, brings the sample within reach of the tripod. A vibration filter system P protects the instrument from external vibrations. (**b**) Schematic of the recent compact STM of *Besocke* [VI.6]. A large area sample, e.g. a semiconductor wafer, with its lower surface under study is carried by three piezoelectric actuators (carriers) which are used to produce microscopic shifts of the sample. A fourth piezoelectric rod (scanner) allows scanning of the metallic tip across the sample surface. (**c**) A function diagram of piezoelectric actuator element of part (b) equipped with tip and electronic connections for scanner operation. The carrier elements are identical in size and operation

port called a **louse**. This name derives from the fact that the stepwise movement of this support over distances of $100 \div 1000$ Å is achieved by three metallic legs carrying a piezoelectric plate. The legs act as electrostatic clamps which are successively attached to the metallic support plate by applying a voltage. When unbiased they move laterally by the action of the piezoelectric plate (voltage-induced tensions). This facility is employed for rough adjustment of the sample surface in front of the tip. The scanning of the tip across the surface is performed by means of a piezoelectric triple leg. Tip movements along three directions (x and y parallel, and z normal to the surface) with an accuracy of better than 1 Å are obtained by biasing the piezodrives by several tenths of a volt. Surface areas of typically 100×100 $Å^2$ are scanned.

Recent, more easily operated STMs have been built with particular attention to a compact arrangement of sample, piezodrives and tip. A very compact construction (Fig. VI.1b), which is relatively insensitive to thermal drifts and mechanical vibrations, uses three piezoelements as sample holder and distance control between tip and sample surface. A similarly constructed piezoelement in between carries the tip and is used as scanner. The piezoelements (Fig. VI.1c) are long rods with four separate metal electrodes. Biasing these electrodes causes bending of the rods. Actuating the x and y component of the three carriers simultaneously causes the sample to shift in the xy plane. Movements of the sample over macroscopic distances can be carried out by the simultaneous application of appropriate voltage pulses to all three carrier piezos. Adjustment and scanning of the tip are performed by suitably biasing the electrodes of the central scanner piezo.

Two major experimental difficulties had to be overcome in the development of the STM: suppression of mechanical vibrations of the whole equipment in its UHV chamber, and the preparation of a tip with atomic dimensions. Since the distance between tip and sample surface must be controlled down to an atomic radius, vibrational amplitudes must be damped down to less than an Ångstrom. In the original version of the STM, this vibrational damping was achieved by suspending the central part (Fig. VI.1a) on very soft springs and by the additional action of eddy currents induced in copper counter plates by powerful magnets. The development of more compact and simpler arrangements (Fig. VI.1b) is still continuing.

The tip is prepared from Ir or W wire (1mm in diameter), which is grounded at one end to yield a radius of curvature below 1 μm. Chemical treatment follows, and the final shape of the tip end is obtained by exposing the tip in situ to electric fields of 10^8 V/cm for about ten minutes. The detailed mechanism of tip formation (adsorption of atoms or atomic migration) is not well understood at present. The method of preparation does not usually lead to very stable tips: Optimum resolution can only be achieved for a limited time and after a while the sharpening procedure has to be repeated.

According to (VI.1) the tunnel current depends both on the distance d between tip and surface, and on the work function. Changes of I_T might therefore be due to corrugation of the surface or to a locally varying work function. The two effects can be separated by an additional measurement of the slope of the tunnel characteristic during scanning. The normal measurement of surface corrugation is based on the assumption of constant $\bar{\phi}$; then keeping I_T constant, the voltage U_z on the z piezodrive is measured as a function of the voltages U_x, U_y on the x and y piezodrives (Fig. VI.1). The

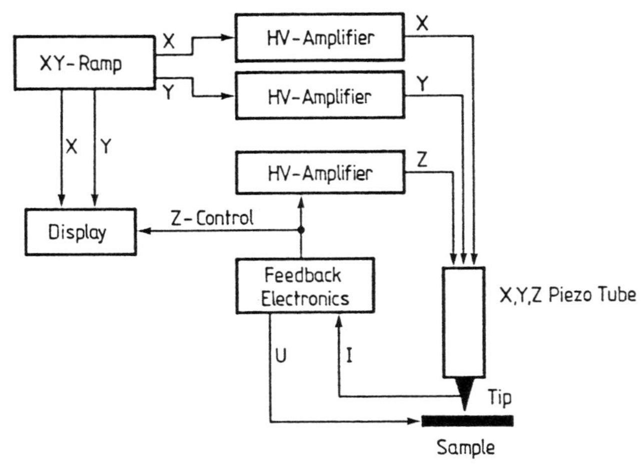

Fig. VI.2. Main components of the electronics of a STM. The tunneling current through tip and sample controls the tip (z) movement via feedback electronics and a High Voltage (HV) amplifier. A digital ramp is used for the xy scan

topography of the surface is obtained in terms of the corrugation function z(x, y). I_T usually changes by an order of magnitude for changes in the distance d on the order of an Ångstrom. By keeping the tunnel current constant, changes in the work function are compensated by corresponding changes in d. Spurious structures in the surface morphology induced by work function changes can be identified by measuring ϕ separately. This is done by recording the first derivate of I_T versus d using a modulation of the distance d, i.e. a modulation voltage U_z for the z-piezodrive ($U_z = U_z^0 + \tilde{U}_z$) and phase-sensitive detection (lock-in). According to (VI.1) the average work function $\bar{\phi}$ is easily obtained by

$$\bar{\phi} \simeq (\partial \ln I_T / \partial d)^2 \ . \qquad \text{(VI.2)}$$

By recording both the signal I_T (or equivalently the compensating voltage U_z) and the derivative (VI.2) a distinction between topographical and work function changes becomes possible. Part of a commonly used electric circuit for this measuring procedure is shown schematically in Fig. VI.2. The essential parts are the feedback electronics for the z piezodrive and a circuit (including an xy ramp) controlling the x and y piezodrives. Figure VI.3 depicts a line scan (along x) over a cleaved Si(111) surface on which Au had been deposited [VI.7]. By comparison to the direct signal I_T(d) and its derivative $[\propto (\bar{\phi})^{1/2}]$ the structures A and B are seen to be caused by work-function inhomogeneities, i.e. Au islands.

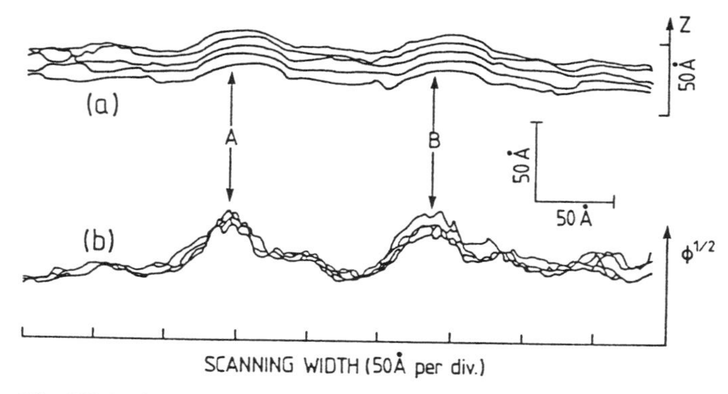

Fig. VI.3a,b. Line scans over a cleaved Si(111) surface on which Au has been deposited [VI.7]. (**a**) Direct measurement of the corrugation z(x) and (**b**) derivative measurement $\partial \ln I_T / \partial d$ according to (VI.2), which yields information about the average work-function

A classic example of the use of the STM in surface structure analysis is the study of the Si(111)-(7×7) surface (Sect. 6.5). In the STM relief of the (7×7) reconstructed surface (Fig. VI.4) several complete unit cells are discernible on the basis of their deep corner minima [VI.8]. This real-space STM image of the Si(111)-(7×7) surface allowed to discard many of the existing structural models proposed for this surface. Although precise structural details cannot be established from Fig. VI.4 alone, there were many arguments in favor of a modified adatom model (Fig. 6.37), which finally led to a structure model for the (7×7) surface of Si(111) (Sect. 6.5). The STM results demonstrated, in particular, that the two halves of the unit mesh are not completeley equivalent because of the slightly different heights of the minima and maxima (Fig. VI.4).

Since electron tunneling can occur from the metal tip to the semiconductor surface, or vice versa, depending on the direction of the bias, more

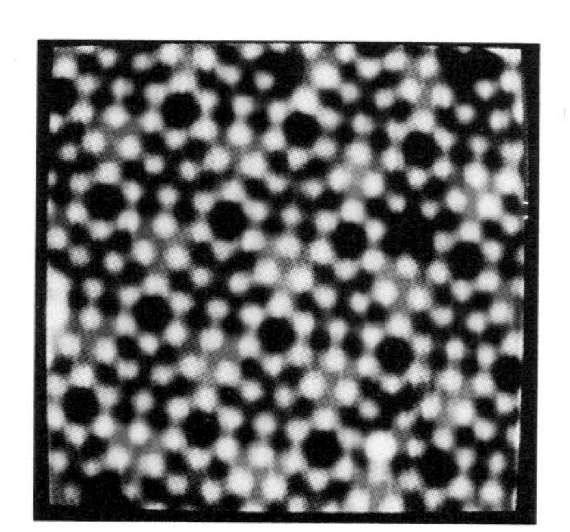

Fig. VI.4. STM relief of the (7×7) reconstructed Si(111) surface [VI.8]. The large unit mesh is discernible by the deep corner minima. The two halves of the unit mesh are not equivalent as evidenced by the different intensities of the minima and maxima

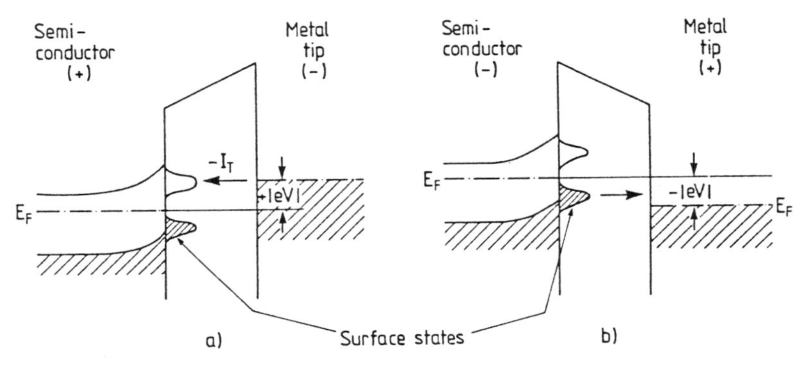

Fig. VI.5a,b. Electronic-band scheme of the semiconductor sample and the metal tip for two opposite values of bias voltage. (**a**) Tunneling of electrons from the metal tip into empty surface states in the sample, and (**b**) tunneling of electrons from occupied surface states in the sample into the metal tip

information about the electronic structure of the surface can be obtained by studying the dependence of the STM signal on the sign and magnitude of the tip-sample voltage. This is qualitatively shown for opposite bias voltages in the tunnel-band schemes in Fig. VI.5. For positive bias in Fig. VI.5a, tunneling of electrons can only occur from occupied metal states into empty surface states or conduction-band states in the semiconductor. When the metal tip is positive with respect to the semiconductor surface (Fig. VI.5b) elastic tunneling of electrons from the metal into the semiconductor is not possible, only occupied states can be reached because of the Fermi-level position. The measured tunneling current I_T therefore originates from occupied surface or valence-band states in the semiconductor. Depending on the bias direction, therefore, occupied or empty states of the surface under investigation can be probed. By measuring the dependence of the current I_T on the applied voltage one can even obtain an image of the state distribution. An example of this kind of application is the STM study of the clean, cleaved Si(111)-(2×2) surface (Sect. 6.5). A real space STM image [corrugation z(x, y)] (Fig. VI.6) depicts a linear corrugation along $(01\bar{1})$ with an amplitude of 0.54 Å and a raw spacing of about 6.9 Å [VI.9]. This spacing is not consistent with the surface atomic distances in a buckling model (Fig. VI.7a), but good agreement exists with the distance between two dangling bond chains in a π-bonded chain model (Sect. 6.5 and Fig. VI.7b). Further strong support for the π-bonded chain model is obtained from line scans of the tunnel current, i.e. the corrugation along $(2\bar{1}\bar{1})$, at opposite biases of +0.8 and −0.8 V. Although the occupied and empty surface states are probed separately in these two cases (Fig. VI.5), the maxima and minima of the corrugation occur at the same location, as is expected for the π-bonded chain model (Fig. VI.7b). The spatial phase shift that would be associated

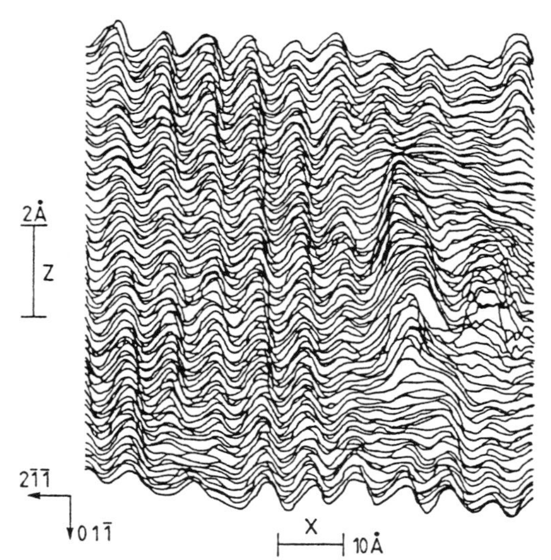

Fig. VI.6. STM relief of a cleaved Si(111) surface, measured at a sample voltage of +0.6 V. The image extends laterally over an area of 70 × 70 Å^2, with the vertical height given by the scale on the left-hand side of the figure. The (2 × 1) π-bonded chains are seen at the left-hand side, and a disordered region is seen on the right [VI.9]

with the buckling model (Fig. VI.7a) is not observed. A measurement of the differentiated tunnel characteristics dI_T/dU (lock-in technique) as a function of positive and negative bias (Fig. VI.8a) then gives a qualitative picture of the spectral distribution of the occupied and empty surface states on Si(111)-(2 × 2). Good agreement with the distribution of π and π^* states predicted by the π-bonded chain model (Figs. VI.8b and 7.13) is found. The

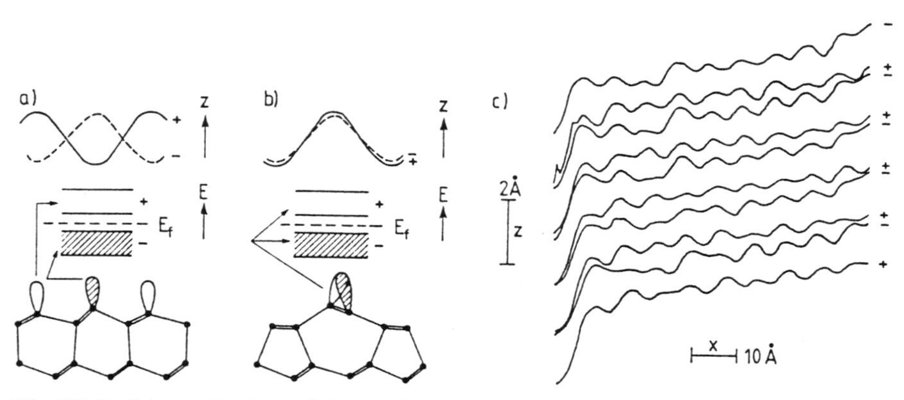

Fig. VI.7. Schematic view of the surface states on a (2×1) Si(111) surface for (**a**) the buckling model, and (**b**) the π-bonded chain model. Dangling bonds on the surface result in two bands of surface states above and below the Fermi level E_F. The corrugation z(x) of these bands is measured separately by applying positive or negative voltages. (**c**) Experimentally determined line scans of the corrugation z(x) measured with sample voltages +0.8 and −0.8 V. A monatomic step is seen at the left hand edge of the scans [VI.9]

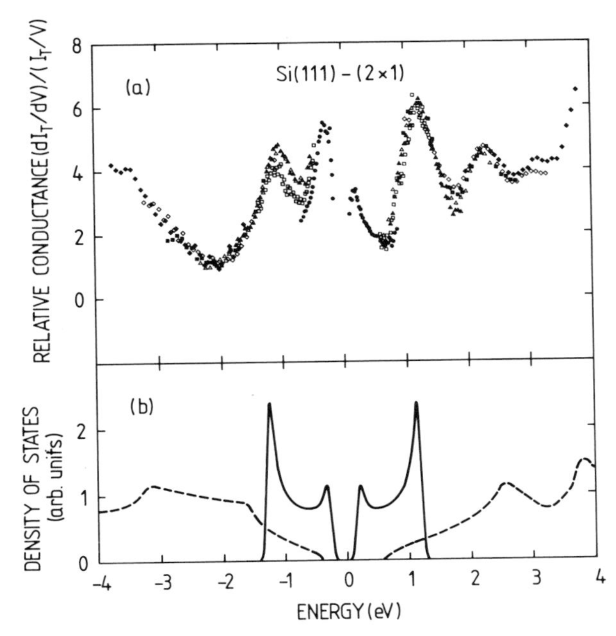

Fig. VI.8. (**a**) Ratio of differential to total tunneling conductance $(dI_T/dU)/(I_T/U)$ versus electron energy (relative to the Fermi energy E_F) measured by an STM on a cleaved Si(111)-(2×1) surface. The different symbols refer to different tip-sample separations [VI.10]. (**b**) Theoretical Density Of States (DOS) for the bulk valence and conduction bands of Si (*dashed line*) [VI.11], and the DOS from a one-dimensional tight-binding model of the π-bonded chains (*solid line*) [VI.10]

four dominant peaks are due to flat areas in the occupied bonding π- and empty antibonding π^*-state bands.

The present examples clearly demonstrate the great value of the STM in studies of the geometric and electronic structure of solid surfaces. The STM and a modification of this instrument, the Atomic Force Microscope (AFM) [VI.12, 13] has meanwhile also opened pathways to a new branch of science: **Nanotechnology**.

Nanotechnology, i.e., nanopositioning of atoms and molecules as well as controll of chemical processes on a nanometer (10^{-9}m) scale and nanoprecision machining has become possible by means of the STM and AFM. By scanning the tip of an STM while tip and substrate are in intimate contact, submicrometer-wide lines can be drawn by scratching the surface of a solid. By increasing the tip-sample interaction at a particular surface location indentations can be produced with an STM or AFM. *Van Loenen* et al. [VI.14] could produce holes of 2÷10 nm diameter in a clean Si surface by means of an STM with a tungsten tip.

Even single atoms and molecules can be pulled and pushed on a substrate surface under UHV conditions at low temperature by an STM, as was first demonstrated by *Eigler* and *Schweizer* [VI.15]. After adsorption of Xe atoms on a Ni(110) surface single Xe atoms could be moved along the surface to form special figures or letters. For this purpose the tip is first positioned above the selected atom (pictured in the microscope mode). The interaction of the tip with this Xe atom, being essentially due to van der Waals forces, is then increased by approaching the tip toward the atom (Fig. VI.9). The tip is then moved laterally under constant-current conditions while carrying the Xe atom with it. Upon reaching the desired destination, the tip is finally withdrawn by decreasing the set-value of the tunnel current, leaving the Xe atom at the desired new position. By applying the sliding procedure to other adsorbed atoms as well (Fig. VI.9), overlayer structures can be fabricated atom by atom. For another example, see [VI. 16].

A second type of moving an atom on a metal surface to a desired position is by simply pushing it by the STM tip. For this purpose the tip is lowered on top of the particular atom until strong interaction between the atom and the tip occurs. By moving the tip to the disered position the atom is dragged with it. The atom moving across the surface does not really break its chemical bond to the electron "sea" of the metal surface. Only the diffusion barrier being essentially 1/10 of the binding energy has to be surmounted by the tip adatom interaction.

A nice example of an artificial nanostructure prepared on a copper surface under UHV conditions by an STM is displayed in Fig. VI.10. *Crommie* et al [VI.17] have assembled a so-called **quantum corral** by pushing and pulling 48 Fe atoms adsorbed at 4 K on a Cu(111) surface into a ring of

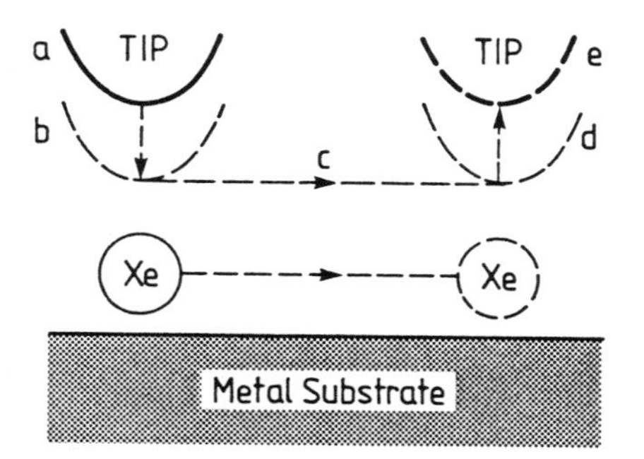

Fig. VI.9. Schematic illustration of sliding and positioning an atom across a surface by means of an STM. The tip is placed over the Xe atom (a), subsequently lowered to (b), where the atom-tip attractive force is sufficient to keep the atom beneath the tip. The tip is subsequently moved across the surface (c) to the desired destimation (d). Finally the tip is withdrawn to a position (c) where the atom-tip interaction is negligible. Thus the atom is releaved from the tip at a new location on the surface [VI.15]

Fig. VI.10. Constant-tunneling-curent STM image of a quantum corral consisting of 48 Fe atoms assembled in a ring on a Cu(111) surface at 4 K (imaging bias: 0.02 V). The ring with a diameter of 142.6 Å encloses a defect-free area of the surface. Inside of the corral a circular standing wave of electrons in sp-like surface states of the Cu(111) surface is visible. The bias parameters during the sliding process were 0.01 V and $5 \cdot 10^{-8}$ A [VI.16]

radius 71.3 Å. Beside two seperate Fe atoms the circular corral consisting of the 48Fe atoms is imaged by the STM after positioning them individually at the correct sites. Tunneling microscopy performed inside of the corral reveals a series of discrete resonances forming a kind of a ring-like standing wave. Since an STM probes electronic wave functions the standing wave within the corral must be due to electrons located at the Cu surface. As will be shown in Sect. 6.4, there are so-called *electronic surface states* at the Cu(111) surface which have free-electron character (Parabolic dispersion in Fig. 6.18). Electrons occupying these states are located at the very surface and their motion parallel to the Cu(111) surface is essentially that of free electrons. For these electrons the Fe atoms forming the circular corral are strong scattering centers, such that an electron is confined by the circular barrier. The spatial variation of the electronic density of states inside of the Fe ring can be described, even quantitatively, by the distribution of round-box eigenstates (Bessel functions) of electrons within sp-like Cu surface states near the Fermi level [VI.16].

References

VI.1 G. Binnig, H. Rohrer, Ch. Gerber, E. Weibel: Appl. Phys. Lett. **40**, 178 (1982); Phys. Rev. Lett. **50**, 120 (1983)

VI.2 H.-J. Güntherodt, R. Wiesendanger (eds.): *Scanning Tunneling Microscopy I*, 2nd edn., Springer Ser. Surf. Sci., Vol.20 (Springer, Berlin, Heidelberg 1994)
R. Wiesendanger, H.-J. Güntherodt (eds.): *Scanning Tunneling Microscopy II, III*, 2nd edn., Springer Ser. Surf. Sci., Vols.28, 29 (Springer, Berlin, Heidelberg 1995, 1996)
D.J. O'Connor, B.A. Sexton, R.St.C. Smart (eds.): *Surface Analysis Methods in Materials Science*, Springer Ser. Surf. Sci., Vol.23 (Springer, Berlin, Heidelberg 1992) Chap. 10

VI.3 H. Ibach, H. Lüth: *Solid State Physics – An Introduction to Principles of Materials Science*, 2nd edn. (Springer, Berlin, Heidelberg 1995)

VI.4 R.H. Fowler, L.W. Nordheim: Proc. Roy. Soc. A **119**, 173 (1928)

VI.5 P.K. Hansma, J. Tersoff: J. Appl. Phys. **61**, R2 (1987)

VI.6 K. Besocke: Surf. Sci. **181**, 145 (1987)

VI.7 G. Binnig, H. Rohrer, F. Salvan, Ch. Gerber, A. Baro: Surface Sci. **157**, L373 (1985)

VI.8 R. Butz (ISI, Research Center Jülich): Priv. commun.

VI.9 R.M. Feenstra, W.A. Thomson, A.P. Fein: Phys. Rev. Lett. **56**, 608 (1986)

VI.10 J.A. Stroscio, R.M. Feenstra, A.P. Fein: Phys. Rev. Lett. **57**, 2579 (1986)

VI.11 J.R. Chelikowsky, M.L. Cohen: Phys. Rev. B **10**, 5095 (1974)

VI.12 R. Wiesendanger: *Scanning Probe Microscopy and Spectroscopy* (Cambridge Univ. Press, Cambridge, UK 1994)

VI.13 C. Bai: *Scanning Tunneling Microscopy and its Applications*, Springer Ser. Surf. Sci., Vol.32 (Springer, Berlin, Heidelberg 1995)

VI.14 E.J. Van Loenen, D. Dijkkamp, A.J. Hoeven, J.M. Lenssinck, J. Dieleman: Appl. Phys. Lett. **55**, 1312 (1989)

VI.15 D.M. Eigler, E.K. Schweizer: Nature **344**, 524 (1990)

VI.16 G. Meyer, B. Neu, K.-H. Rieder: Appl. Phys. A **60**, 343 (1995)

VI.17 M.F. Crommie, C.P. Lutz, D.M. Eigler: Science **262**, 218 (1993)

Panel VII
Surface Extended X-Ray Absorption Fine Structure (SEXAFS)

Besides ion scattering, scanning tunneling microscopy (Panel VI: Chap.3) and LEED intensity analysis (Sect.4.4, Panel VIII: Chap.4), Surface Extended X-ray Absorption Fine Structure (SEXAFS) measurements have become a major source of information about surface atomic structure [VII.1-3]. In principle, the technique is an indirect surface-sensitive measurement of the X-ray absorption in the energy range up to about 10 keV. Before discussing SEXAFS it is thus useful to examine the more direct technique EXAFS, which, from the experimental point of view, is a straightforward X-ray absorption measurement [VII.4]. Since such an absorption measurement is not particularly sensitive to the topmost atomic layers, EXAFS is generally used to study the bulk atomic structure, i.e. bond length and coordination numbers of materials. Figure VII.1 presents a typical K-shell absorption spectrum of Cu [VII.5]. The insert is a schematic absorption spectrum over a wide range of photon energies. The absorption coefficient $\mu(\hbar\omega)$ for X-rays, defined by the exponential decay law of the intensity

$$I = I_0 \, e^{-\mu x} \, , \tag{VII.1}$$

decreases monotonically with photon energy except at absorption edges (L_I, L_{II}, ..., K...) where the photon energy reaches the ionization energy of a particular atomic shell; photoelectrons are produced and a steep edge, i.e. a strong enhancement of the absorption occurs. In condensed matter the absorption coefficient on the high-energy side of this excition edge exhibits a characteristic oscillatory fine structure which is clearly seen in Fig.VII.1 for the K-absorption edge of Cu. The origin of this fine structure is an interference effect of the photoelectron wave function (Fig.VII.2). The photoelectron produced by the ionization of a particular atom is scattered from neighboring atoms. The direct outgoing wave ψ_0 and the waves ψ_s backscattered from the various neighboring atoms superimpose to yield the final state of the excitation. The overlap, i.e. the interference condition of this total final-state wave function with that of the core level initial-state wave function changes with photon energy depending on the phase difference between ψ_0 and ψ_s, thus causing the observed variations in $\mu(\hbar\omega)$. The absorption coefficient μ is conveniently expressed as

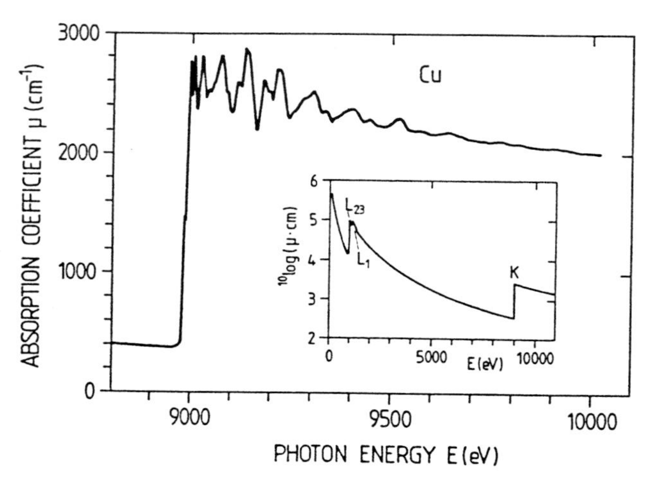

Fig. VII.1. K-shell X-ray absorption coefficient μ of Cu versus X-ray photon energy. Insert: qualitative overview of the absorption coefficient $\mu(\hbar\omega)$ for a wide range of photon energies covering two L edges and one K edge [VII.5]

$$\mu = \mu_{0K}(1+\chi) + \mu_0 , \qquad \text{(VII.2)}$$

where μ_{0K} and μ_0 are monotonic functions due to K shell excitation (μ_{0K}) and excitation of weaker bound L and M electrons (μ_0). The structural information, i.e. bond length and coordination numbers are then contained in χ. According to the experimental data (Fig. VII.1) χ is a function of photon energy $\hbar\omega$, but because of energy conservation in the photoexcitation process, i.e.

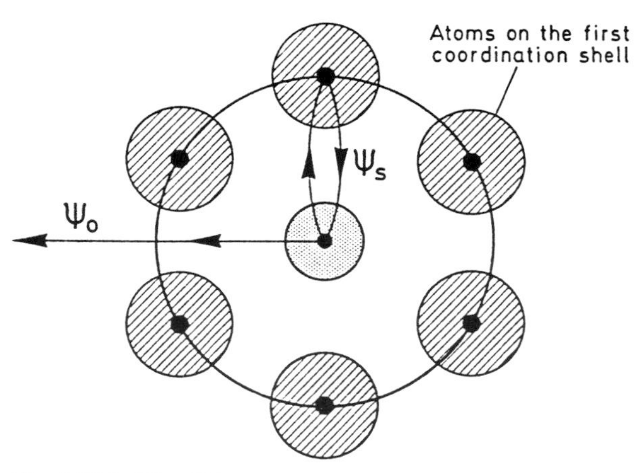

Fig. VII.2. The EXAFS mechanism (schematic). The final-state wave function is a superposition of the unscattered outgoing wave ψ_0 and waves ψ_s that are backscattered from the neighbors of the atom that has been photoexcited

$$\frac{\hbar^2 k^2}{2m} = \hbar\omega - E_B + V_0 , \qquad \text{(VII.3)}$$

one can also express χ as a function of the wave vector k. In (VII.3) E_B is the binding energy of the electron in its initial state, and V_0 the inner potential of the solid (the photoelectron is excited in this potential rather than into vacuum). V_0 is usually not well known and for the data analysis a reasonable estimate is taken. A theoretical expression for $\chi(k)$ is calculated using dipole matrix elements between the initial K-shell wave function and the superposition of ψ_0 (outgoing wave) and ψ_s (waves backscattered from neighboring atoms):

$$\chi(k) = \sum_i A_i(k) \sin[2kR_i + \rho_i(k)] . \qquad \text{(VII.4)}$$

This expression contains the structure of the photoelectron final state due to interference between ψ_0 and several other waves reflected from neighboring atoms at a distance $2R_i$ (Fig. VII.2). The distance of travel in the interference term is therefore $2R_i$. A number N_i of identical scatterers are grouped together at a distance R_i in a scattering shell. The scattering phase $\rho_i(k)$ at the ith neighbor takes into account the influence of the absorbing

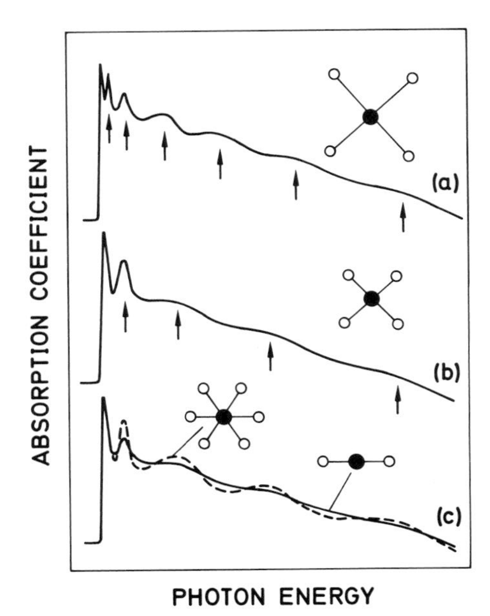

Fig. VII.3. Influence of different bonding geometries on the EXAFS oscillations (qualitative). A larger bonding length decreases the oscillation period (*a*); a greater number of scatterers increases the oscillation amplitude (*c*)

and scattering potentials on the photoelectron wave. The EXAFS amplitudes $A_i(k)$ are proportional to the backscattering amplitudes of the surrounding atoms. They allow one to distinguish between different neighbors. Furthermore thermal vibrational amplitudes and damping due to inelastic scattering of the photoelectrons are contained in $A_i(k)$.

Figure VII.3 illustrates qualitatively how the fine structure reacts to variations in the nearest neighbor distance R_i and in the coordination number. For smaller R_i the first maximum occurs at shorter wavelengths, i.e. higher photon energies. Higher coordination increases the number of scatterers and thus the oscillation amplitude.

The analysis of EXAFS data includes several steps: the smooth decreasing background μ_{0K} in the absorption coefficient (VII.2) is subtracted and the oscillating term χ amplified and recorded taking its average value as the zero line. In the simplest case R_i can be obtained according to (VII.4) simply from the energetic separation ΔE between the first maximum and the first minimum [$R_i/\text{Å} \approx (151/\Delta E)^{1/2}$; ΔE in eV]. To achieve better accuracy the measured $\chi(k)$ is Fourier-transformed according to

$$F(r) = \frac{1}{\sqrt{2\pi}} \int_{k_{min}}^{k_{max}} \chi(k)\, W(k)\, k^3\, e^{2ikr}\, d(2k) \,. \qquad \text{(VII.5)}$$

Maxima in F(r) then correspond to the major oscillation periods in $\chi(k)$ and indicate the sequence of distances R_i of shells of neighbors.

In (VII.5) the factor k^3 compensates for the rapid decrease of $\chi(k)$ at large k values. The window function W(k) is used to smooth $\chi(k)$ at the boundaries k_{min} and k_{max} in order to avoid artificial side bands in F(r).

Up to this point EXAFS is not a surface-sensitive technique; adsorbate overlayers or the topmost atomic layers of a solid give only a small, frequently negligible, contribution to the total X-ray absorption. Instead of measuring the absorption coefficient μ directly, one can also record any deexcitation product of the photoionization process. This is an indirect way to measure μ; the absorption coefficient is proportional to the probability of ionization, i.e. of producing a hole in the K (or L) shell. Every process for which the probability is proportional to the deexcitation of that hole can be used to measure the absorption coefficient. Possible deexciation channels are the emission of an X-ray photon or the emission of an Auger electron (Panel III: Chap.2). EXAFS becomes a surface-sensitive method – then called SEXAFS – when the photoemitted Auger electrons are recorded by means of an electron analyser and a detector, or without any analyser simply by means of a channeltron. The surface sensitivity arises from the limited escape depth of the Auger electrons (Fig.4.1). Because of inelastic

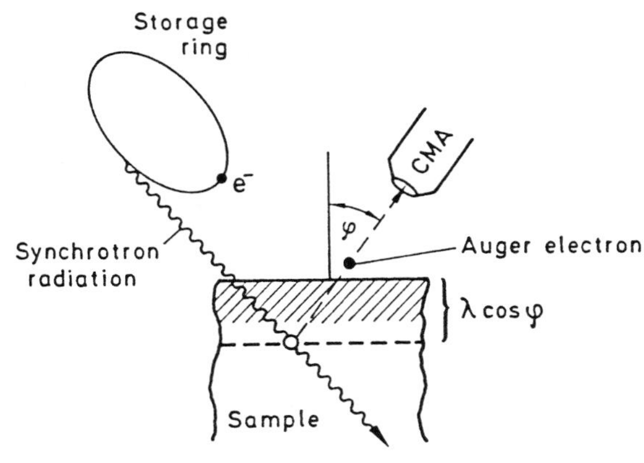

Fig. VII.4. Experimental set-up for SEXAFS. Polarized synchrotron radiation is used as the light source and the Auger yield or secondary partial yield is measured by a CMA. The information depth is essentially given by the projection of the mean-free path λ of the electrons

interactions (excitation of plasmons, etc.) the average free path of electrons varies between a few Ångstroms and about 100 Å for electron energies between 20 eV and a few keV. The surface sensitivity is particularly high if only electrons from a depth of a few Ångstroms below the surface contribute to the measured yield. As light source, a tunable high intensity X-ray source is necessary. Synchrotron radiation from storage rings is ideal. Figure VII.4 shows schematically such an experimental set-up for SEXAFS

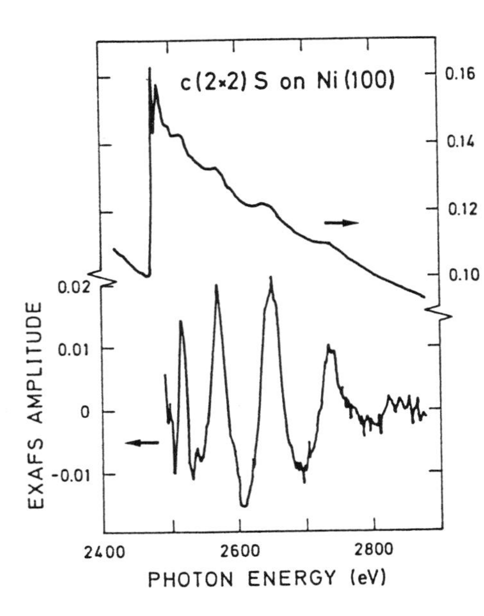

Fig. VII.5. Sulfur K-edge SEXAFS spectrum for half a monolayer of sulfur on Ni(100) [c(2× 2) LEED pattern] recorded at 45° X-ray incidence. The SEXAFS oscillations after background subtraction are shown in the lower half [VII.6]

studies at a storage ring. As in AES, a CMA (Panel II: Chap.2) is used as electron analyser: its energy window is typically set at energies between 20 and 100 eV to ensure maximum surface sensitivity.

As an experimental example we consider a SEXAFS study of sulfur (S) adsorption on Ni(100) [VII.6]. Figure VII.5 shows the corresponding Auger electron yield measured by means of a CMA. The sample was held at room temperature during adsorption and during the measurement. The energy window of the CMA was fixed at 2100 eV near the KLL-transition of S and measurements were performed with angles of incidence of the synchrotron X-ray beam of 10°, 45° and 90°. The complex Fourier transform F(r), in particular $|F(r)|$ and $Im\{F(r)\}$, of the measured SEXAFS oscillations (Fig. VII.6) clearly exhibits two main features. Peak A at 2.23 ± 0.02 Å corresponds to the S-Ni neighbor distance whereas peak *B* indicates that the second neighbor distance is 4.15 ±0.10 Å. From other measurements it is known that the bulk S-Ni distance in NiS is 2.3944 ±0.0003 Å. The S-Ni separation for the adsorbed S layer on Ni(100) is thus smaller by 0.16 ±0.02 Å.

Additional information is needed to yield an unequivocal structure model for the site geometry of the adsorbed S atoms. Taking into account the c(2×2) superstructure observed in LEED there are three different possibilities for S adsorption sites. Considering also the intensities of the Fourier bands and using information about the scattering phases, etc., the only remaining possibility is a fourfold coordinated site, where an S atom forms the top of a pyramid whose rectangular basis is formed by four surface-layer Ni atoms.

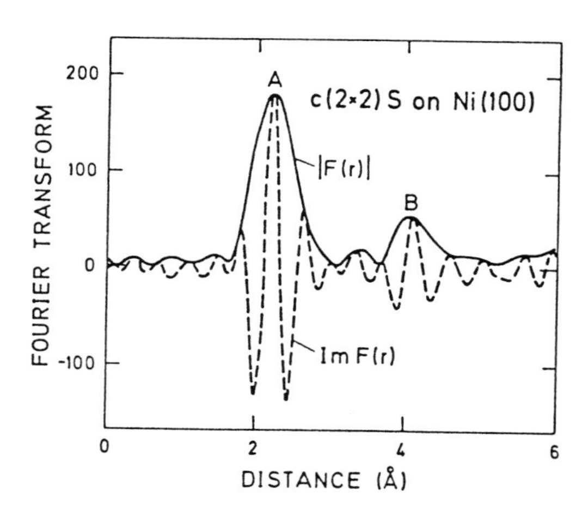

Fig. VII.6. Absolute magnitude (solid line) and imaginary part (*dashed*) of the Fourier transform of the SEXAFS signal for c(2×2) S on Ni(100) recorded at 90° X-ray incidence. Peaks *A* and *B* correspond to S-Ni nearest neighbor and second nearest neighbor distances, respectively [VII.6]

References

VII.1 P.A. Lee, P.H. Citrin, P. Eisenberger, B.M. Kincaid: Extended X-ray absorption fine structure – its strengths and limitations as a structure tool. Rev. Mod. Phys. **53**, 769 (1981)

VII.2 B. Lengeler: Adv. Solid State Phys. **29**, 53 (1989)

VII.3 D. Koningsberger, R. Prins (eds.): *Principles, Techniques and Applications of EXAFS, SEXAFS and XANES* (Wiley, New York 1988)
J. Stöhr: *NEXAFS Spectroscopy*, Springer Ser. Surf. Sci., Vol.25 (Springer, Berlin, Heidelberg 1992)

VII.4 B.K. Agarwal: *X-Ray Spectroscopy*, 2nd edn., Springer Ser. Opt. Sci., Vol.15 (Springer, Berlin, Heidelberg 1991)

VII.5 B. Lengeler (ISI, Research Center Jülich): Priv. commun. (1991)

VII.6 S. Brennan, J. Stöhr, R. Jäger: Phys. Rev. B **24**, 4871 (1981)

Problems

Problem 3.1. Prove that in two-dimensional Bravais lattices an n-fold rotation axis with n = 5 or n $\geq$ 7 does not exist. Show first that the axis can be chosen to pass through a lattice point. Then argue by "reductio ad absurdum" using the set of points into which the nearest neighbor of the fixed point is taken by the n rotations to construct a point closer to the fixed point than its nearest neighbor.

Problem 3.2. Chemisorption of group-III atoms such as B, Al, Ga and In on Si(111) surfaces [1/3 of a monolayer coverage on Si(111)-7 $\times$7] leads to the formation of a $(\sqrt{3} \times \sqrt{3})$ R30° superstructure. Construct the corresponding reciprocal lattice vectors and plot the LEED pattern.

Problem 3.3. On a crystal surface with $\langle 100 \rangle$ orientation, the deposition of a thin film in the initial growth mode leads to island formation. The crystalline islands of the deposited film material exhibit surfaces with $\langle 100 \rangle$ and $\langle 110 \rangle$ orientations. Plot different stages of the island growth under the assumption that $\langle 100 \rangle$ is a fast growing surface, whereas the $\langle 110 \rangle$ surfaces are slowly growing ones.

Problem 3.4. Upon co-deposition of arsenic and gallium on Si(111) surfaces an initial exponential decrease of the Auger intensity of the Si(LVV) line at 92 eV is observed as a function of coverage. The intensity decays from the normalized value 1 on the clean surface to 0.6 at a nominal monolayer coverage of GaAs; at higher coverages between 1 and 3 monolayers, saturation occurs and a constant intensity value of 0.55 is reached.

a) Calculate the mean free path of Si(LVV) Auger electrons in GaAs (lattice constant 5.65 Å).

b) From experiment, arsenic is known to form a saturation coverage of 0.85 monolayers on Si(111) at typical growth temperatures. Can the described experimental finding be explained in terms of layer-by-layer growth of GaAs on top of this partial arsenic layer?

4. Scattering from Surfaces

As in many branches of modern physics, scattering experiments are an important source of information in surface research. The scattering process on a surface is therefore a central topic among the various interactions of a solid. Like in bulk solid-state physics, elastic scattering can tell us something about the symmetry and the geometric arrangement of atoms near the surface, whereas inelastic scattering processes, where energy quanta are transferred to or from the topmost atomic layers of a solid, yield information about possible excitations of a surface or interface, both electronic and vibronic ones. In principle, all kinds of particles, X rays, electrons, atoms, molecules, ions, neutrons, etc. can be used as probes. The only prerequisite in surface and interface physics is the required surface sensitivity. The geometry and possible excitations of about 10^{15} surface atoms per cm^2 must be studied against the background of about 10^{23} atoms present in a bulk volume of one cm^3. In surface and interface physics the appropriate geometry for a scattering experiment is thus the reflection geometry. Furthermore, only particles that do not penetrate too deeply into the solid can be used. Neutron scattering, although it is applied in some studies, is not a very convenient technique because of the "weak" interaction with solid material. The same is true to some extent for X-ray scattering. X rays generally penetrate the whole crystal and the information carried by them about surface atoms is negligible. If used in surface analysis, X-ray scattering requires a special geometry and experimental arrangement. In this sense ideal probes for the surface are atoms, ions, molecules and low-energy electrons [4.1]. Atoms and molecules with low energy interact only with the outermost atoms of a solid, and low-energy electrons generally penetrate only a few Ångstroms into the material. The mean-free path in the solid is, of course, dependent on the energy of the electrons, as may be inferred from Fig.4.1. In particular, for low-energy electrons, the "strong" interaction with matter – i.e. with the valence electrons of the solid – leads to considerable problems in the theoretical description; in contrast to X-ray and neutron scattering multiple-scattering events must be taken into account, and thus the simple analogy to an optical diffraction experiment breaks down. In quantum-mechanical language, the Born approximation is not sufficient. A detailed treatment using the so-called **dynamic theory** (Sect.4.4) takes into ac-

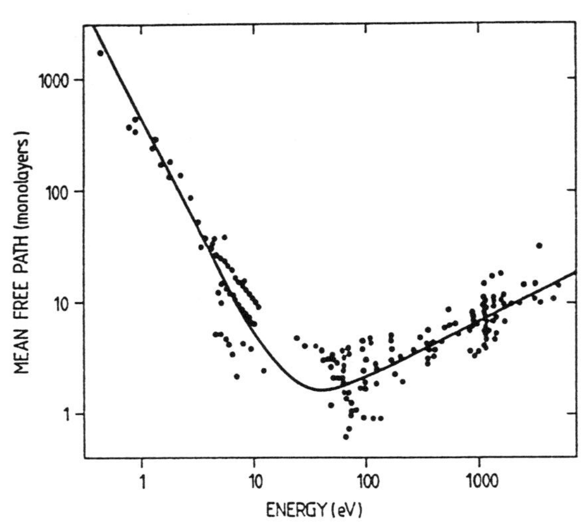

Fig. 4.1. Mean free path of electrons in solids as a function of their energy; a compilation of a variety of experimental data [4.2]. The quasi-universal dependence for a large number of different materials is due to the fact that the major interaction mechanism between the electrons and the solid is the excitation of plasmon waves whose energy is determined by the electron density in the solid

count all these effects by considering the boundary problem of matching all possible electron waves outside and inside the solid in the correct way.

Nevertheless, a simple treatment of the surface scattering process within the framework of single-scattering events, i.e. along the lines of the Born approximation, yields insights into the important features of scattering at surfaces. This approach is called the **kinematic theory**; it is applicable to all kinds of particle-surface scattering, elastic and inelastic, but cannot describe the details of the intensity distribution in low-energy electron scattering.

4.1 Kinematic Theory of Surface Scattering

We shall refer in the following to electron scattering, although other kinds of wavelike projectiles such as atoms, molecules, etc. can also be described in the same manner. The interaction of the particle with the solid – in particular with its surface – is described by a potential

$$V(\mathbf{r},t) = \sum_{\mathbf{n}} v[\mathbf{r} - \boldsymbol{\rho}_{\mathbf{n}}(t)] , \tag{4.1}$$

where $\mathbf{n}$ (= triple m,n,p) labels the atoms in a primitive lattice, and v is the interaction potential with a single near-surface atom (substrate or adsorbate) at the instantaneous position $\boldsymbol{\rho}_\mathbf{n}(t)$. $\boldsymbol{\rho}_\mathbf{n}(t)$ contains the time-independent equilibrium position $\mathbf{r}_\mathbf{n}$ and the displacement $\mathbf{s}_\mathbf{n}(t)$ from equilibrium: $\boldsymbol{\rho}_\mathbf{n}(t) = \mathbf{r}_\mathbf{n} + \mathbf{s}_\mathbf{n}(t)$. Monoenergetic particles with energy E are incident with a wave vector $\mathbf{k}$ ($E = \hbar^2 k^2/2m$) and are scattered into a state $\mathbf{k}'$. According to time-dependent quantum-mechanical perturbation theory, the scattering probability per unit time from $\mathbf{k}$ into $\mathbf{k}'$ is

$$W_{\mathbf{k}\mathbf{k}'} = \lim_{\tau \to \infty} \frac{1}{\tau} |c_{\mathbf{k}\mathbf{k}'}(\tau)|^2 \tag{4.2}$$

with the transition amplitude

$$c_{\mathbf{k}\mathbf{k}'}(\tau) = \frac{-i}{\hbar} \int d\mathbf{r} \int_0^\tau dt\, \psi_s^*(\mathbf{r},t) V(\mathbf{r}) \psi_i(\mathbf{r},t) \,. \tag{4.3}$$

For the incident and the scattered waves, ψ_i and ψ_s, we assume a plane-wave character, i.e.,

$$\psi_i(\mathbf{r},t) = V^{-1/2} e^{i(\mathbf{k}\cdot\mathbf{r} - Et/\hbar)} \,, \tag{4.4a}$$

$$\psi_s(\mathbf{r},r) = V^{-1/2} e^{i(\mathbf{k}'\cdot\mathbf{r} - E't/\hbar)} \,. \tag{4.4b}$$

The probability amplitude for scattering within a time τ is therefore

$$c_{\mathbf{k}\mathbf{k}'}(\tau) = \frac{-i}{\hbar} \sum_\mathbf{n} \int_0^\tau e^{i(E'-E)t/\hbar} \int e^{i(\mathbf{k}-\mathbf{k}')\cdot\mathbf{r}}\, v[\mathbf{r} - \boldsymbol{\rho}_\mathbf{n}(t)]\, d\mathbf{r}\, dt \,. \tag{4.5}$$

The vector $\mathbf{r} - \boldsymbol{\rho}_\mathbf{n}(t) = \boldsymbol{\xi}$ describes a position of a moving atom, whose nucleus is at the instantaneous position $\boldsymbol{\rho}_\mathbf{n}(t)$, i.e., with $d\mathbf{r} = d\boldsymbol{\xi}$,

$$c_{\mathbf{k}\mathbf{k}'}(\tau) = \frac{-i}{\hbar} \sum_\mathbf{n} \int_0^\tau dt\, e^{i(E'-E)t/\hbar} e^{i(\mathbf{k}-\mathbf{k}')\cdot\boldsymbol{\rho}_\mathbf{n}(t)} \int d\boldsymbol{\xi}\, v(\boldsymbol{\xi})\, e^{-i(\mathbf{k}'-\mathbf{k})\cdot\boldsymbol{\xi}} \,. \tag{4.6}$$

The last time-independent term is an atomic scattering factor. It describes the detailed scattering mechanism at the single atom. With the scattering vector

$$\mathbf{K} = \mathbf{k}' - \mathbf{k} \tag{4.7}$$

it becomes

$$f(\mathbf{K}) = \int d\xi\, v(\xi)\, e^{-i\mathbf{K}\cdot\xi} \; . \tag{4.8}$$

The instantaneous position of an atom near the surface can be described as

$$\boldsymbol{\rho}_{\mathbf{n}}(t) = \mathbf{r}_{\mathbf{n}} + \mathbf{s}_{\mathbf{n}}(t) = \mathbf{r}_{\mathbf{n}\parallel} + z_p \hat{\mathbf{e}}_{\perp} + \mathbf{s}_{\mathbf{n}}(t) \; . \tag{4.9a}$$

The splitting into two components parallel ($\parallel$) and perpendicular ($\perp$) to the surface $[\mathbf{n} = (\mathbf{n}\parallel, p)]$, with $\mathbf{n}_{\parallel}$ as a vector parallel to the surface, is convenient because of the translational symmetry parallel to the surface which allows a Fourier representation:

$$\boldsymbol{\rho}_{\mathbf{n}}(t) = \mathbf{r}_{\mathbf{n}\parallel} + z_p \hat{\mathbf{e}}_{\perp} + \sum_{\mathbf{q}_{\parallel}} \hat{s}(\mathbf{q}_{\parallel}, z_p) \exp\left[\pm i\mathbf{q}_{\parallel}\cdot\mathbf{r}_{\mathbf{n}\parallel} \pm i\omega(\mathbf{q}_{\parallel})t\right] . \tag{4.9b}$$

In this representation the displacement from equilibrium $\mathbf{s}_{\mathbf{n}}(t)$ is described by a harmonic wave with frequency $\omega(\mathbf{q}_{\parallel})$ and wave vector $\mathbf{q}_{\parallel}$ parallel to the surface. The most general excitation would be a superposition of modes with different $\omega(\mathbf{q}_{\parallel})$. From (4.6) it follows that

$$c_{\mathbf{k}\mathbf{k}'}(\tau) = \frac{-i}{\hbar} f(\mathbf{K}) \sum_{\mathbf{n}_{\parallel}, p} \int_0^{\tau} dt\, e^{i(E'-E)t/\hbar}$$

$$\times \exp\left\{-i\mathbf{K}\cdot\left[\mathbf{r}_{\mathbf{n}_{\parallel}} + z_p \hat{\mathbf{e}}_{\perp} + \sum_{\mathbf{q}_{\parallel}} \hat{s}(\mathbf{q}_{\parallel}, z_p) \exp\left[\pm i\mathbf{q}_{\parallel}\cdot\mathbf{r}_{\mathbf{n}\parallel} \pm i\omega(\mathbf{q}_{\parallel})t\right]\right]\right\} . \tag{4.10}$$

Because the displacements from equilibrium are small, the exponential function can be expanded:

$$\exp\left\{-i\mathbf{K}\cdot\left[\sum_{\mathbf{q}_{\|}}\hat{\mathbf{s}}(\mathbf{q}_{\|},z_p)e^{\pm i\cdots}\right]\right\}$$

$$\simeq 1 - i\mathbf{K}\cdot\left[\sum_{\mathbf{q}_{\|}}\hat{\mathbf{s}}(\mathbf{q}_{\|},z_p)\exp[\pm i(\mathbf{q}_{\|}\cdot\mathbf{r}_{\mathbf{n}_{\|}} + \omega t)]\right] + \dots \ . \tag{4.11}$$

Taking into account only the first two terms in the expansion, one arrives via (4.10) at a sum of two different (elastic and inelastic) contributions to the probability amplitude:

$$c_{\mathbf{k}\mathbf{k}'}(\tau) = \frac{-i}{\hbar}f(\mathbf{K})\sum_{\mathbf{n}_{\|},p}\int_0^{\tau} dt\exp[i(E' - E)t/\hbar]\exp[-i\mathbf{K}\cdot(\mathbf{r}_{\mathbf{n}_{\|}} + z_p\hat{\mathbf{e}}_{\perp})]$$

$$\times\left\{1 - i\sum_{\mathbf{q}_{\|}}\mathbf{K}\cdot\hat{\mathbf{s}}(\mathbf{q}_{\|},z_p)\exp[\pm i(\mathbf{q}_{\|}\cdot\mathbf{r}_{\mathbf{n}_{\|}} + \omega t)]\right\}. \tag{4.12}$$

The first "elastic" term does not contain the vibrational amplitudes $\hat{\mathbf{s}}$. In the calculation of the elastic scattering probability $W_{\mathbf{k}\mathbf{k}'}$ according to (4.2), the time limit $\tau\rightarrow\infty$ yields a delta function $\delta(E'-E)$; the summation over the two-dimensionsal (2D) set $\mathbf{n}_{\|} = (m, n)$ runs over the whole surface to infinity; it contains terms of the form

$$\sum_{m,n} e^{-i\mathbf{K}\cdot(m\mathbf{a}+n\mathbf{b})} = \sum_{m,n}(e^{-i\mathbf{K}\cdot\mathbf{a}})^m(e^{-i\mathbf{K}\cdot\mathbf{b}})^n \tag{4.13}$$

where **a** and **b** are the basis vectors of the 2D unit mesh within the surface. As in the case of a three-dimensional (3D) lattice sum, an evaluation via a geometrical series gives, in the limit $m, n\rightarrow\infty$, a nonvanishing contribution only for

$$\mathbf{K}\cdot\mathbf{a} = 2\pi h\ , \quad \mathbf{K}\cdot\mathbf{b} = 2\pi k\ ; \quad h, k \text{ integer}\ . \tag{4.14a}$$

With $\mathbf{K} = \mathbf{K}_{\|} + K_{\perp}\hat{\mathbf{e}}_{\perp}$, (4.14a) is fulfilled when

$$\mathbf{K}_{\|} = \mathbf{k}'_{\|} - \mathbf{k}_{\|} = \mathbf{G}_{\|} . \tag{4.14b}$$

The conditions (4.14) are the 2D analogs of the three Laue equations for the bulk scattering of X rays. The third Laue equation is missing, since the third summation index p in (4.12) does not run from $-\infty$ to $+\infty$ because of the boundary at the solid surface ($z = 0$).

For the *elastic scattering probability* and with N being the number of surface atoms, one finally obtains

$$W^{(el)}_{\mathbf{k}'\mathbf{k}} = \frac{2\pi}{\hbar} N \left| f(\mathbf{K}) \sum_{p} \exp(iK_{\perp} z_p) \right|^2 \delta(E'-E)\delta_{\mathbf{K}_{\|},\mathbf{G}_{\|}} , \tag{4.15}$$

where the Kronecker delta $\delta_{\mathbf{K}_{\|},\mathbf{G}_{\|}}$ expresses the condition (4.14b). With c as the periodic repeat distance perpendicular to the surface, i.e. $z_p = pc$, a summation of p up to infinity would yield the third Laue condition. The sum over p, however, extends only over those atomic layers below the surface which lie within the finite penetration depth of the incident particles. For low-energy atom and molecular scattering, p is restricted to the topmost atomic layer; for slow electrons it might extend over a few of atomic layers, whereby the exact number depends on the primary energy. Further consequences of (4.14 and 15) are considered in Chap.5.

For the second term in (4.12), the *inelastic scattering* amplitude, one obtains

$$c^{(inel)}_{\mathbf{k}\mathbf{k}'}(\tau) = \frac{-i}{\hbar} f(\mathbf{K}) \sum_{\mathbf{n}_{\|},p} \int_0^{\tau} dt \exp[i(E'-E)t/\hbar] \exp[-i\mathbf{K}\cdot(\mathbf{r}_{\mathbf{n}_{\|}} + z_p\hat{\mathbf{e}}_{\perp})]$$

$$\times(-i) \sum_{\mathbf{q}_{\|}} \mathbf{K}\cdot\hat{\mathbf{s}}(\mathbf{q}_{\|},z_p)\exp[\pm i(\mathbf{q}_{\|}\cdot\mathbf{r}_{\mathbf{n}_{\|}} + \omega t)]$$

$$= -\hbar^{-1} f(\mathbf{K}) \sum_{\mathbf{n}_{\|},p,\mathbf{q}_{\|}} \mathbf{K}\cdot\hat{\mathbf{s}}(\mathbf{q}_{\|},z_p)\exp\left[-i(\mathbf{K}_{\|} \mp \mathbf{q}_{\|})\cdot\mathbf{r}_{\mathbf{n}_{\|}}\right]$$

$$\times\exp(-iK_{\perp}z_p) \int_0^{\tau} \exp[i(E'-E\pm\hbar\omega)t/\hbar]dt . \tag{4.16}$$

Similar arguments to those in (4.13-15) yield the *inelastic scattering probability*

$$W_{\mathbf{kk}'}^{(\mathrm{inel})} = \frac{2\pi}{\hbar} N \sum_{\mathbf{q}_{\|}} \delta(E'-E \pm \hbar\omega(\mathbf{q}_{\|}))\, \delta_{\mathbf{K}_{\|} \pm \mathbf{q}_{\|}, \mathbf{G}_{\|}} \times \left| f(\mathbf{K}) \sum_{p} \mathbf{K} \cdot \hat{\mathbf{s}}(\mathbf{q}_{\|}, z_p) \exp(-iK_{\perp} z_p) \right|^2 \tag{4.17}$$

where N is the number of surface atoms. The delta function and the Kronecker symbol ensure energy conservation

$$E' = E \mp \hbar\omega(\mathbf{q}_{\|}) \, , \tag{4.18a}$$

and the conservation of the wave-vector component parallel to the surface ($\mathbf{K}_{\|} = \mathbf{k}_{\|}' - \mathbf{k}_{\|}$)

$$\mathbf{k}_{\|}' = \mathbf{k}_{\|} \pm \mathbf{q}_{\|} + \mathbf{G}_{\|} \, , \tag{4.18b}$$

i.e. in all inelastic surface scattering processes the energy lost (or gained) in scattering must be found as the quantum energy $\hbar\omega(\mathbf{q}_{\|})$ of an excited surface mode (e.g., vibrational). An extension of the present formalism can be made to include electronic excitations of the surface. Equation (4.18b) is a straightforward consequence of the 2D translational symmetry within the surface. Perpendicular to the surface the translational symmetry is broken and thus only the parallel component of the particle wave vector is conserved. The change in the wave vector $\mathbf{k}_{\|}$ upon scattering is given (to within an undetermined 2D reciprocal lattice vector $\mathbf{G}_{\|}$) by the wave vector $\mathbf{q}_{\|}$ of the excited surface mode.

The second factor in (4.17) is also of importance since it yields a selection rule for surface scattering, which helps to identify the symmetry of an excited vibrational (or electronic) surface excitation: the inelastic scattering probability vanishes if $\mathbf{K}$ is perpendicular to $\hat{\mathbf{s}}(\mathbf{q}_{\|}, z_p)$. $\hat{\mathbf{s}}$ are the Fourier components of the atomic vibration and therefore have the direction of the atomic displacement (or the electronic dipole moment). In an inelastic-scattering experiment one sees only those excitations for which the atomic displacement has a component parallel to the scattering vector $\mathbf{K} = \mathbf{k}' - \mathbf{k}$. This is illustrated in Fig.4.2 for the vibration of an adsorbed atom. For scattering in the specular direction only modes with vibrational direction normal to the surface are detectable. Vibrations parallel to the surface can only be

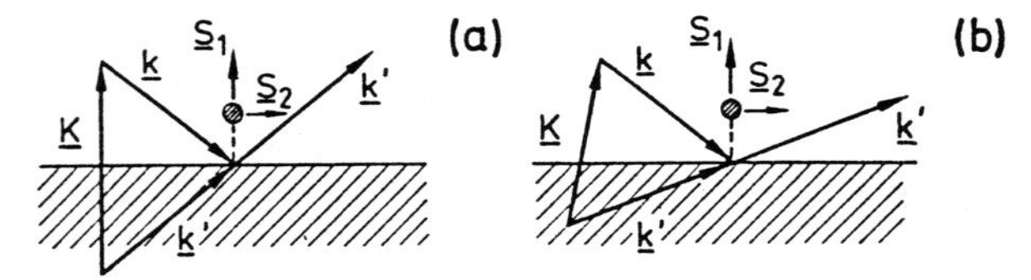

Fig.4.2. Surface scattering on an adsorbed atom (*full circle*) with detection in specular (**a**) and in off-specular geometry (**b**). The energy loss $\hbar\omega$ is assumed to be small in comparison with the primary energy E, i.e. $|\mathbf{k}'| \simeq |\mathbf{k}|$. In (a) the scattering vector **K** is perpendicular to the surface, the s_2 bending vibration of the adsorbed atom is therefore not detectable, only the stretch vibration s_1 normal to the surface. In (**b**) **K** has components parallel and normal to the surface, thus both vibrations s_1 and s_2 can be studied

studied if one observes scattering with off-specular geometry. In calculating $\mathbf{K} = \mathbf{k}'-\mathbf{k}$ one has, of course, to take into account that due to the inelastic process $|k'|$ is different from $|k|$ and that therefore, even for specular detection, there is a small component of **K** parallel to the surface. The size of this component depends on the ratio $\hbar\omega/E$.

4.2 The Kinematic Theory of Low-Energy Electron Diffraction

In almost every surface-physics laboratory Low-Energy Electron Diffraction (LEED) is used as the standard technique to check the crystallographic quality of a surface, prepared either as a clean surface, or in connection with ordered adsorbate overlayers [4.3]. In this experiment a beam of electrons with a primary energy between 50 and 300 eV is incident on the sur-

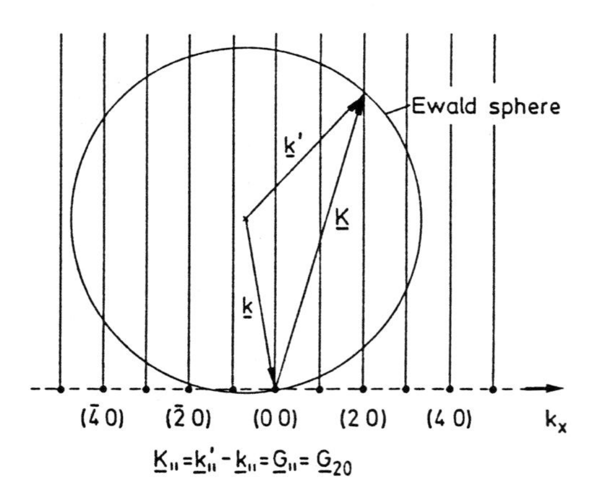

Fig.4.3. Ewald construction for elastic scattering on a 2D surface lattice. The corresponding 2D reciprocal lattice points (hk) are plotted on a cut along k_x. The scattering condition (4.14) for the plotted beams is fulfilled for the reciprocal lattice point (hk) = (20); but a number of other reflexes are also observed $(\bar{4}0)$, $(\bar{3}0)$ ...(30), $(\bar{2}\bar{2})$, ... (11) ...

face and the elastically backscattered electrons give rise to diffraction (or Bragg) spots that are imaged on a phosphorous screen (Panel VIII: Chap.4).

To understand the essential features of such an experiment, kinematic theory is sufficient. The condition for the occurrence of an "elastic" Bragg spot is given by (4.14), i.e., the scattering vector component parallel to the surface $\mathbf{K}_{\|} = \mathbf{k}'_{\|} - \mathbf{k}_{\|}$ must equal a vector of the 2D surface reciprocal lattice $\mathbf{G}_{\|}$. This condition is valid for the limiting case where only the topmost atomic layer is involved in scattering. For the component $\mathbf{K}_{\perp}$ perpendicular to the surface, no such condition applies. In order to extend the well-known Ewald construction to our 2D problem we must therefore relax the restriction of the third Laue equation (perpendicular to the surface). This is done by attributing to every 2D reciprocal lattice point (h,k) a rod normal to the surface (Fig.4.3). In the 3D problem we have discrete reciprocal lattice points in the third dimension rather than rods, and these are the source of the third Laue condition for constructing the scattered beams.

In the 2D case, the possible elastically scattered beams ($\mathbf{k}'$) can be obtained by the following construction. According to the experimental geometry (orientation of primary beam with respect to surface) the wave vector $\mathbf{k}$ of the primary beam is positioned with its end at the (0,0) reciprocal lattice point and a sphere is constructed around its starting point. As is seen from Fig.4.3, the condition $\mathbf{K}_{\|} = \mathbf{G}_{\|}$ is fulfilled for every point at which the sphere crosses a "reciprocal-lattice rod". In contrast to the 3D scattering problem in bulk solid-state physics, the occurrence of a Bragg reflection is not a singular event. No special methods like the Debye-Scherrer, or Laue techniques, etc., need to be applied to obtain a diffraction pattern. The loss of the third Laue condition in our 2D problem ensures a LEED pattern for every scattering geometry and electron energy.

These considerations are exact only in the limit of scattering from a true 2D network of atoms. In a real LEED experiment, however, the primary electrons penetrate several atomic layers into the solid. The deeper they penetrate, the more scattering events in the z-direction perpendicular to the surface contribute to the LEED pattern. In (4.15) the summation index p must run over more and more atomic layers with increasing penetration depth, and the third Laue condition becomes more and more important. This leads to a modulation of the intensities of the Bragg reflections in comparison with the case of pure 2D scattering. In the Ewald construction (Fig. 4.3) one can allow for this situation qualitatively by giving the rods periodically more or less intensity. In the extreme case of 3D scattering, where the three Laue conditions are exactly valid, the thicker regions of the rods become points of the 3D reciprocal lattice. An Ewald construction for the intermediate situation where the periodicity perpendicular to the surface enters the problem to a certain extent is shown in Fig.4.4. When the Ewald sphere crosses a "thicker" region of the rods, the corresponding Bragg spot

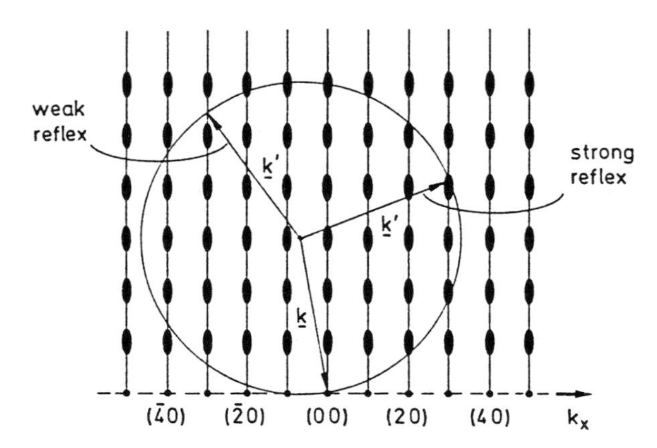

Fig.4.4. Ewald construction for elastic scattering on a quasi-2D surface lattice, as in Fig.4.3, but now not only scattering from the topmost lattice plane, but also from a few underlying planes, is taken into account. The "thicker" regions of the rods arise from the third Laue condition, which cannot be completely neglected. Correspondingly the (30) reflex has high intensity, whereas the ($\bar{3}$0) spot appears weak

has strong intensity whereas less pronounced regions of the rods give rise to weaker spots. Another important consequence is the following: if we change the primary energy of the incoming electrons the magnitude of k, i.e., the radius of the Ewald sphere, changes. As k is varied, the Ewald sphere passes successively through stronger and weaker regions of the rods, and the intensity of a particular Bragg spot varies periodically. Experimentally, this provides evidence for the limited validity of the third Laue condition (perpendicular to the surface). The effect can be checked by measuring the intensity of a particular Bragg reflex (hk) in dependence on primary energy of the incident electrons. The result of this type of measurement is known in the literature as an I-V curve (I: intensity, V: accelerating voltage of the electrons).

As is seen from Fig.4.5, there is indeed some structure found in the I-V curves measured on Ni(100) which seems to be similar to that expected from the application of the third Laue condition [4.4]. But the maxima of the observed peaks are generally shifted to lower primary energies and there are additional structures which cannot be explained in terms of the simple picture developed so far. The shift to lower energies is easily explained by the fact that inside the crystal the electrons have a wavelength different from that in the vacuum due to the mean inner potential in the crystal. The potential difference is related to the work function of the material. The shift can therefore be used to get some information about this "inner" potential. The explanation of the additional features in the I-V curves is more involved. It requires a more thorough description of the scattering process of low-energy electrons, going beyond the approximation of kine-

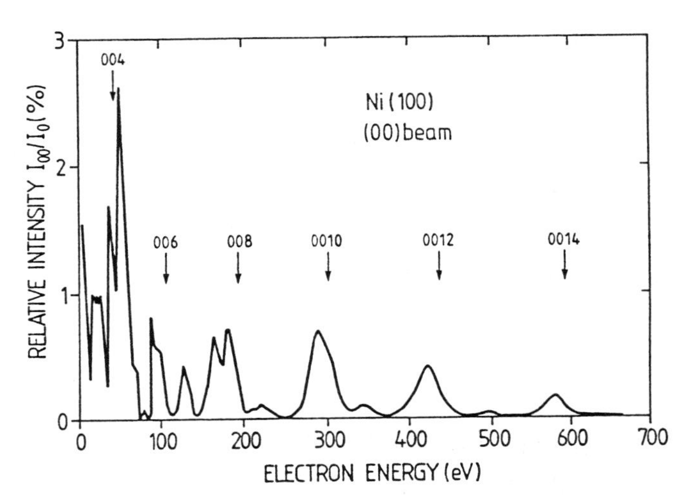

Fig.4.5. Intensity versus voltage (I-V) curve for the (00) beam from a clean Ni(100) surface. The diffracted intensity I_{00} is referred to the intensity of the primary beam I_0 [4.4]

matic theory. Due to the "strong" interaction of electrons with matter, multiple scattering processes must be taken into account. This is done in the "dynamic" theory of electron scattering to be discussed in Sect.4.4.

4.3 What Can We Learn from Inspection of a LEED Pattern?

As is clear from the discussion in the previous section, a detailed understanding of the intensity of a LEED pattern involves the complex problem of describing multiple scattering processes within the topmost atomic layers of a crystal. The evaluation of the geometric positions of surface atoms from the LEED pattern – a so-called **structure analysis** – requires detailed interpretation of the intensities (as for X-ray scattering in bulk solid-state physics). This problem is therefore beyond the applicability of kinematic theory; it will be treated in the next section. Nonetheless, simple inspection of a LEED pattern and measurement of the geometrical spot positions can yield a great deal of useful information about a surface.

At the beginning of an experiment on a crystalline surface the first step, after looking with AES for possible contamination, consists in checking the crystallographic quality of the surface by LEED. The LEED pattern must exhibit sharp spots with high contrast and low background intensity.

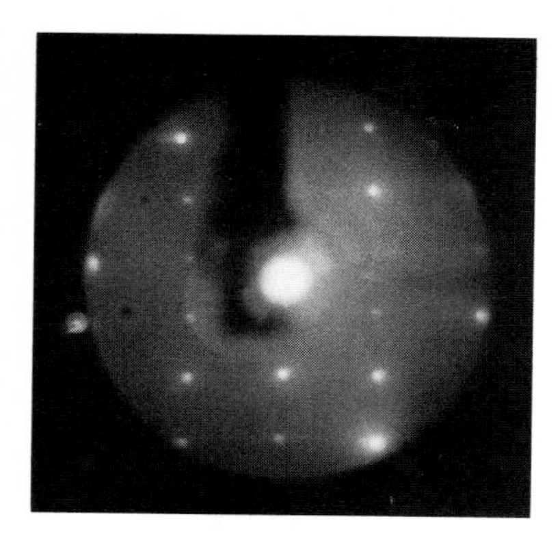

Fig. 4.6. LEED pattern of a clean cleaved nonpolar ZnO$(10\bar{1}0)$ surface. Primary voltage $U_0 = 140$ V

Random defects or crystallographic imperfections will broaden the spots and increase the background intensity due to scattering from these statistically distributed centers. One has to keep in mind, however, that the information obtained about the surface always originates from an area with a diameter smaller than the coherence width of the electron beam, i.e. for conventional LEED systems smaller than $\approx$100 Å (Panel VIII: Chap.4).

In the simplest case, the LEED pattern of the clean surface exhibits a (1×1) structure which reflects a surface 2D symmetry equal to that of the bulk. Figure 4.6 shows the (1×1) LEED pattern obtained for a clean nonpolar ZnO $(10\bar{1}0)$ surface which was prepared in UHV by cleavage along the hexagonal c-axis of the wurtzite lattice. The rectangular symmetry of this particular surface is revealed and the spot separations, on conversion from reciprocal space to real space (Sect. 3.3), give the dimensions of the 2D unit mesh. It should be emphasized that the observed (1×1) pattern does not mean that the atomic geometry is equal to that of the "truncated" bulk. Atomic positions within the 2D unit mesh might be changed and also a perpendicular relaxation towards or away from the surface is possible. Essentially the same (1×1) LEED pattern as in Fig.4.6 is also found on GaAs(110) surfaces cleaved in UHV (Sect.3.2).

More complex LEED patterns are obtained when a reconstruction with *superstructure* is found. In Fig.4.7 the well-known (2×1) reconstruction of Si(111) is displayed. This type of reconstruction is always found when a Si(111) or Ge(111) surface is prepared by cleavage in UHV at room temperature. The (111) lattice plane of these diamond-type elemental semiconductors has a sixfold symmetry which would also be found in the LEED pattern if the surface were to have the same symmetry as the truncated bulk. The LEED spots in Fig.4.7c, however, indicate that in one direction the periodicity in real space has doubled (Fig.4.7a). This causes half-order spots between the main Bragg spots thus giving a rectangular surface lattice in reciprocal space (LEED pattern) (Fig.4.7b). A further complication arises since there are three possible orientations of the (2×1) unit mesh. During cleav-

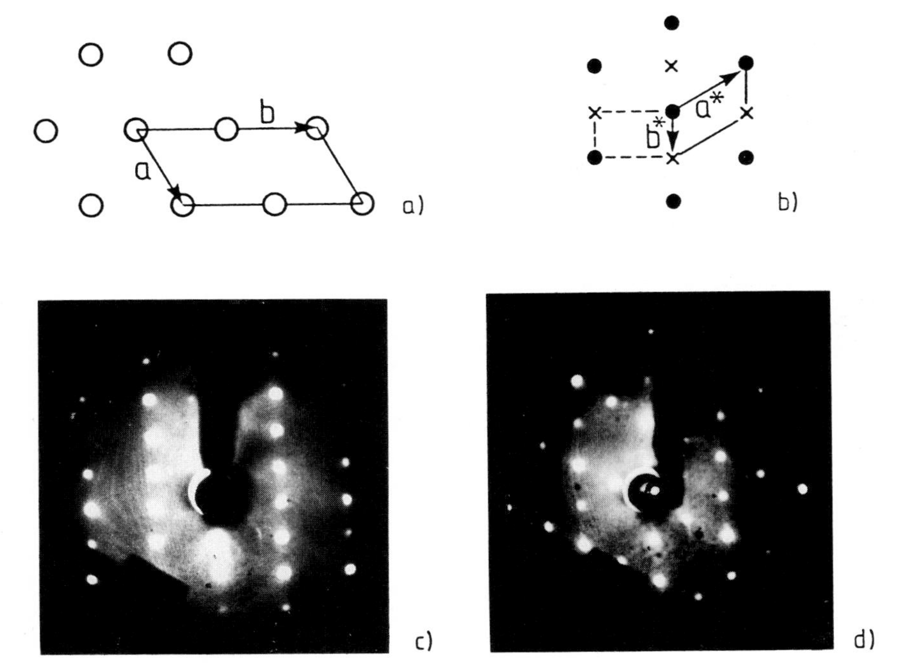

Fig.4.7a-d. (2 × 1) LEED pattern of a clean cleaved Si(111) surface. (**a**) (2 × 1) unit mesh in real space; lattice points are indicated by circles. (**b**) reciprocal lattice points of the LEED pattern; the (2 × 1) half-order spots are indicated by crosses. (**c**) LEED pattern measured with a primary energy $E_0 = 80$ eV (single domain). (**d**) LEED pattern showing the superposition of two domains rotated by 60° to one another

age the double periodicity can show up in three symmetrically equivalent directions and all three may exist in different domains. If several of these domains are hit by the primary beam, the LEED pattern will consist of a superposition of patterns rotated against each other. This situation is shown in the LEED pattern of Fig.4.7d. The nature of the (2 × 1) reconstruction, i.e. the detailed atomic positions on this surface, is discussed in Sect.3.2.

Superstructures with non-integer Bragg spots can also arise from adsorbed atoms or molecules that are positioned at certain symmetry sites of the substrate surface. In Fig.4.8 such an adsorbate superstructure is shown for the example of chemisorbed oxygen atoms on Cu(110) [4.5]. At low coverage atomic O causes half-order LEED spots in one direction of the rectangular network of (110) substrate spots. The adsorbed O atoms therefore occupy sites that are separated by twice the period of the Cu surface. Information from other experimental techniques is necessary to determine the exact adsorption sites. Care must be taken in attributing a superstructure that appears after adsorption to the adsorbate itself. An adsorbate-induced reconstruction of the substrate surface can also occur.

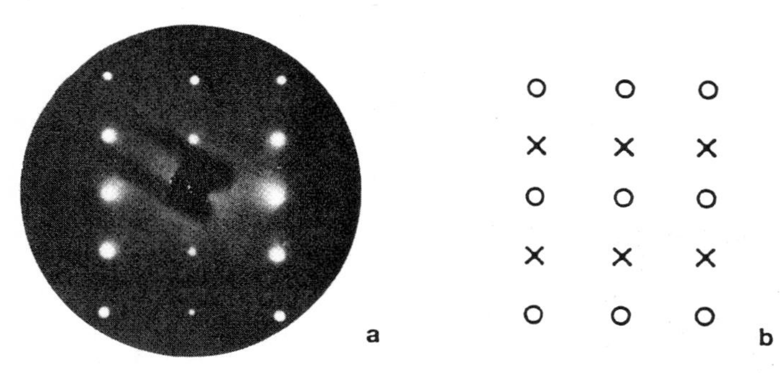

Fig.4.8a,b. LEED pattern obtained after adsorption of atomic oxygen at low coverage on Cu(110). (**a**) (2×1) superstructure as seen on the phosphorus screen. (**b**) schematic of the (2×1) pattern with substrate spots as circles and half order spots as crosses [4.5]

In some cases, the geometry of a LEED pattern can also give information about crystallographic defects on a surface. The simplest situation is a regular array of **atomic steps** on the surface. Cleaved semiconductor surfaces and poorly oriented metal surfaces often show such step arrays. Their presence can be deduced from splitting of the LEED spots. A regular step array with a definite step height d and a constant number N of atoms on the terraces between the steps means for the surface symmetry that the lattice periodicity (lattice constant a) is superimposed on a second periodicity with repeat distance Na (Fig.4.9). In the Ewald construction (Fig.4.4) a second reciprocal lattice with rod spacing $Q = 2\pi/Na$ has to be taken into account along with that of the surface (reciprocal lattice vector $\mathbf{G}_{\|}$). The two arrays of rods are inclined with respect to one another by the angle α between the stepped surface and the exposed lattice plane (Fig.4.9). According to the general principle of the Ewald construction, Bragg spots appear whenever the reciprocal lattice rods intersect the Ewald sphere. In the present case of two superimposed rod arrays this condition must be fulfilled by both rod systems simultaneously. Two situations can be distinguished. The primary wave vector $\mathbf{k}$ has a length such that the scattered wave vector $\mathbf{k}'$ ($|k'| = |k|$) reaches a point A, where the two rod systems intersect one another on the surface of the Ewald sphere (not shown in Fig.4.9). Then one single Bragg spot is observed in the LEED pattern. A second situation can occur for a larger primary wavevector $\mathbf{k}$ (this Ewald sphere is shown in Fig.4.9) where the Ewald sphere crosses a rod of the normal reciprocal lattice (point B) but the rods of the step periodicity exhibit crossing points that do not quite coincide with this. Because of the finite number of atoms on the terraces the LEED spots are not ideally sharp and this configuration leads to

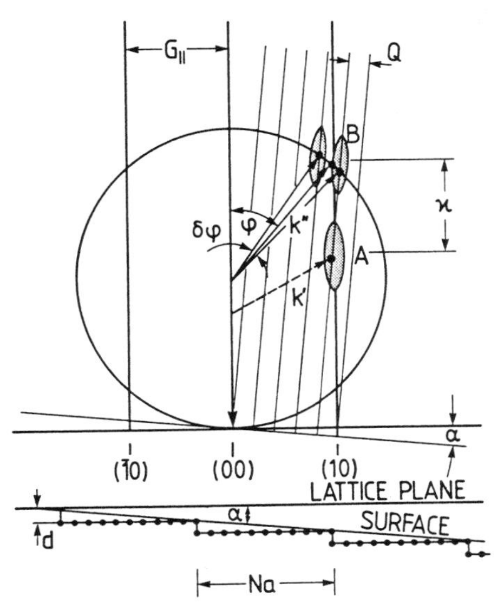

Fig.4.9. Ewald construction for a surface with a regular step array. The steps with a height d and a terrace width aN (a is the lattice constant) cause an inclination angle α between the macroscopic surface and the main lattice plane. Corresponding to the lattice distance a, the reciprocal lattice vector is $\mathbf{G}_{\|}$. The step array is described by a superimposed inclined reciprocal lattice of periodicity $Q = 2\pi/\mathrm{Na}$. The primary electrons are described by the wavevector **k** (two different primary energies with two different lengths of **k** are considered) and two different scattered beams (**k**′ and **k**") are plotted ($k = k'$ and $k = k''$)

two separate intensity maxima; the LEED spot is split into two components with an angular separation $\delta\varphi$. This angle can be directly measured on the phosphorus screen and from Fig.4.9 it follows that

$$\delta\varphi = \frac{Q}{k''\cos\varphi} = \frac{Q}{k\cos\varphi} = \frac{\lambda}{\mathrm{Na}\cos\varphi} \ . \tag{4.19}$$

The step width Na can thus be determined by finding a primary voltage (corresponding electron wavelength λ) at which a particular LEED spot at a scattering angle φ is split into two components. The angular splitting $\delta\varphi$ then determines Na according to (4.19).

According to Fig.4.9, the step height d can be determined from the two primary energies E′ and E″ at which the same LEED beam occurs as a single (point A) and as a double spot (point B), respectively. Or, equivalently, one can change the primary energy E stepwise and look for two subsequent positions at which the same spot $\mathbf{G}_{\|} = \mathbf{G}_{hk}$ occurs as a single or as a double spot. For such two subsequent energies Fig.4.9 yields

$$2\nu\kappa = \sqrt{|\mathbf{k}|^2} + k''_{\perp} \ , \quad \text{with} \quad \nu = 1, 2, 3, \ldots \ , \tag{4.20a}$$

$$2\nu\kappa = \sqrt{\frac{2mE}{\hbar^2}} + \sqrt{\frac{2mE}{\hbar^2} - |\mathbf{G}_{\|}|^2} \tag{4.20b}$$

where $k_\perp''$ is the normal component of the scattered wave vector $\mathbf{k}''$. The inclination angle α can be expressed in real space:

$$\frac{Q}{2\kappa} = \frac{2\pi}{2\kappa \mathrm{Na}} = \sin\alpha \simeq \tan\alpha = \frac{d}{\mathrm{Na}}, \tag{4.21}$$

$$2\kappa = 2\pi/d\,. \tag{4.22}$$

With (4.20b) one obtains

$$\nu\frac{2\pi}{d} = \sqrt{\frac{2mE}{\hbar^2}} + \sqrt{\frac{2mE}{\hbar^2} - |\mathbf{G}_{\|}|^2}\,, \quad \text{with} \quad \nu = 1, 2, 3, \ldots\,. \tag{4.23}$$

From two different primary-energy readings E_ν at which double or single spots are observed (4.23) allows the determination of the step height d. For the (00) LEED spot with $\mathbf{G}_{\|} = \mathbf{0}$ one has the simple relation

$$E_\nu(0,0) = \frac{\hbar^2}{2m}\left(\frac{\pi}{d}\nu\right)^2, \quad \text{with} \quad \nu = 1, 2, 3, \ldots\,. \tag{4.24}$$

Another kind of defect is also easily seen from the LEED pattern: **Facetting** is the formation of new crystal planes, which are inclined to the original surface. The effect is often observed even on clean surfaces after annealing. The cause is the tendency of a crystal to lower its surface free energy by the formation of new lower-energy planes with different crystallographic orientations. These facets give rise to a secondary LEED pattern with spot separations different from that of the normal surface. The reciprocal lattice rods of these facets are strongly inclined with respect to those of the normal surface. The (00) spot originating from the facets will therefore be found far away from the normal (00) beam. For normal incidence of the primary beam the (00) spot of the original surface is located in the center of the diffraction screen. With increasing electron energy all other spots move continuously towards the (00) spot, whose position remains fixed. By the same argument the spots originating from the facets will move towards a certain position far out from the center of the screen. A similar consideration to that of Fig.4.9 based on the Ewald construction allows the determination of the angle of inclination of the facets.

4.4 Dynamic LEED Theory, and Structure Analysis

As is seen in Fig.4.5, there are so-called **secondary Bragg peaks** in the reflected intensity versus energy (I-V) curves of a LEED spot. These structures cannot be explained by simple kinematic theory, i.e., by taking into account only single-scattering events and a limited validity of the third Laue condition. The reason is that multiple-scattering processes are taking place in the solid as a result of the large atomic scattering cross section for low-energy electrons. The multiply-scattered electrons also contribute to the LEED spots (Fig.4.10). In addition, strong inelastic-scattering processes are responsible for the fact that the electrons detected in a LEED experiment originate only from the first few atomic layers close to the surface. These complications require a more thorough theory which is usually called **dynamic theory** [4.6]. The word *dynamic* implies that the complete dynamics of the electrons, and not merely the simple lattice geometry, is included.

Two approaches may be used. The straightforward solution to the problem consists in solving the complete Schrödinger equation for a perfect semi-infinite 3D lattice, using Bloch waves which satisfy the boundary conditions at the surface. The exact solution for the diffraction by the semi-infinite solid is then obtained by matching these Bloch waves to the wave functions of the incident and the reflected electrons. In the alternative approach only the 2D periodicity of the surface net is assumed, and the field solution of the Schrödinger equation is built up from contributions of successive atomic layers.

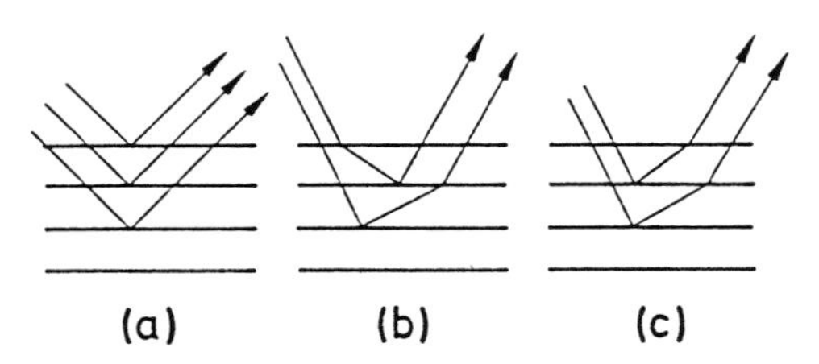

Fig.4.10a-c. Schematic representation of single and multiple scattering processes in LEED. (**a**) single-scattering events at the "lattice planes" cause a regular Bragg reflection. (**b**) Double-scattering events with forward- and subsequent back-scattering contribute to the (00) Bragg spot. (**c**) Double-scattering event with back-scattering and subsequent forward-scattering

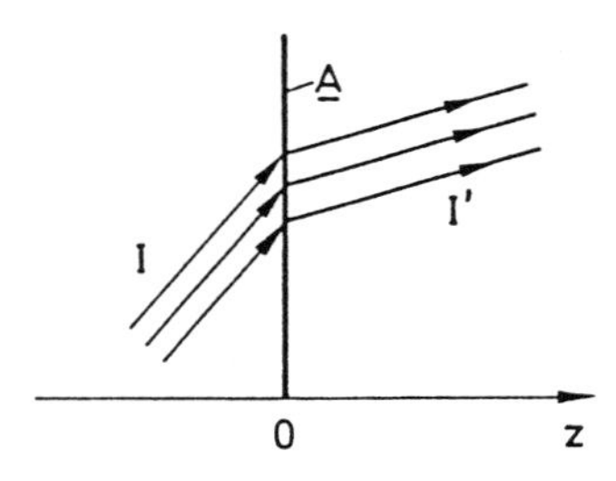

Fig.4.11. A plane-wave electron beam hits an interface at z = 0 on an area A and is refracted due to a potential step. The total current I remains constant (I = I′)

4.4.1 Matching Formalism

The *matching formulation* for the wave functions will be considered first. A particle beam is incident on a surface of a solid. For charged electrons this surface represents a potential step. Due to the potential difference refraction occurs (Fig.4.11). At the surface (z = 0) the beam diameter changes but the total current I remains unchanged, since particles do not accumulate at the interface, i.e.,

$$\mathrm{I} = \mathrm{I}' , \quad \mathbf{j}\cdot\mathbf{A} = \mathbf{j}'\cdot\mathbf{A} , \tag{4.25}$$

where **j** and **j**′ are the particle current densities on the two sides of the interface, and **A** is the area on the interface which is hit by the beam (Fig. 4.11). From the expression for the current density

$$\mathbf{j} = \frac{\hbar}{2\mathrm{im}} (\psi^* \nabla \psi - \psi \nabla \psi^*) \tag{4.26}$$

and from (4.25) we obtain the following matching conditions for the electronic wave functions ψ_o and ψ_i outside and inside the crystal

$$\psi_o \big|_{z=0} = \psi_i \big|_{z=0} , \tag{4.27a}$$

$$\left. \frac{\partial}{\partial n} \psi_o \right|_{z=0} = \left. \frac{\partial}{\partial n} \psi_i \right|_{z=0} , \tag{4.27b}$$

where $\partial/\partial n$ is the derivative in the direction normal to the surface. With $E = \hbar^2 k^2/2m$ as the incident energy, the primary electrons are described by the wave function

$$\varphi_o = \exp[\mathrm{i}\mathbf{k}_{\|}\cdot\mathbf{r}_{\|} + \mathrm{i}k_{\perp}z] , \tag{4.28}$$

with

$$\mathbf{r}_{\parallel} = (x, y) \,, \quad \mathbf{k} = (\mathbf{k}_{\parallel}, k_{\perp}) \quad \text{and} \quad E = \frac{\hbar^2}{2m}(k_{\parallel}^2 + k_{\perp}^2) \,.$$

The full wave function ψ_o outside the crystal consists of this incoming wave and the diffracted waves. The surface scattering potential has 2D periodicity. Upon scattering, therefore $k_{\parallel}$ is conserved to within a 2D reciprocal lattice vector $\mathbf{G}_{\parallel} = \mathbf{G}_{hk} = 2\pi(h, k)$ and the complete wavefunction outside becomes

$$\psi_o = \varphi_o + \sum_{hk} A_{hk} \exp[i(\mathbf{k}_{\parallel} + \mathbf{G}_{hk}) \cdot \mathbf{r}_{\parallel} - ik_{\perp,hk} z] \,. \tag{4.29}$$

A_{hk} describes the amplitudes of the scattered waves (h, k), and $k_{\perp,hk}$ are the wave-vector components normal to the surface, which are determined by energy conservation:

$$E = \frac{\hbar^2}{2m}(|\mathbf{k}_{\parallel} + \mathbf{G}_{hk}|^2 + k_{\perp,hk}^2) \,. \tag{4.30}$$

Inside the solid the wavefunctions of the electrons are Bloch waves

$$\psi_i = u_{\mathbf{k}}(\mathbf{r}) \exp(i\mathbf{k} \cdot \mathbf{r}) = \sum_{\mathbf{G}} c_{\mathbf{G}}(\mathbf{k}) \exp[i(\mathbf{k} + \mathbf{G}) \cdot \mathbf{r}] \;; \tag{4.31}$$

$u_{\mathbf{k}}(\mathbf{r})$ has the periodicity of the 3D lattice and can therefore be represented as a Fourier series in 3D reciprocal space. $\mathbf{k}$ in (4.31) can also be split up into components $\mathbf{k}_{\parallel}$ and $k_{\perp}$ parallel and normal to the surface. The coefficients $c_{\mathbf{G}}(\mathbf{k})$ and the wave vectors $k_{\perp}$ are determined by the periodic potential and the energy. They are found by substituting the above expression (4.31) into the Schrödinger equation for the periodic crystal potential. The matching conditions (4.27) imply for the wave functions (4.29 and 31) outside and inside the crystal, that the parallel wave-vector components $\mathbf{k}_{\parallel}$ and the total energy E coincide outside and inside. The matching conditions (4.27) can lead to severe restrictions on the possible diffracted beams in comparison with pure kinematic theory. For a certain energy E, which is determined by the acceleration voltage V of the primary beam, the matching conditions can only be fulfilled if there exist electronic states inside the

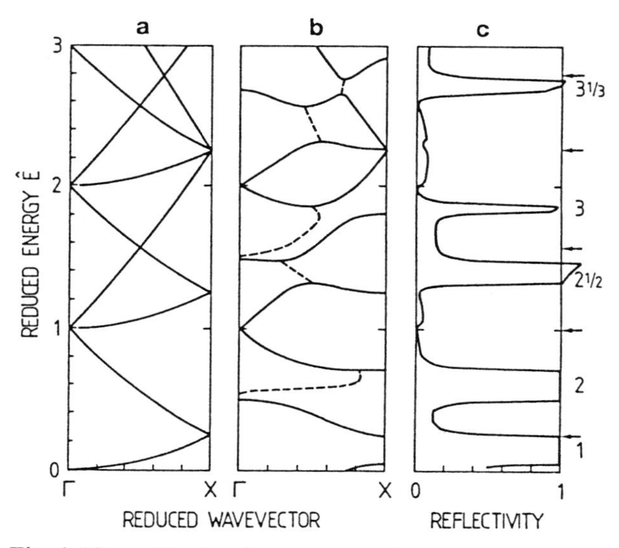

Fig.4.12a-c. The band structure and the reflectivity of a semi-infinite cubic array of s-wave scatterers with a (100) surface exposed. (**a**) Free-electron band structure along the symmetry direction Γ-X of the Brillouin zone. (**b**) Calculated band structure along Γ-X. The periodic potential causes a splitting into allowed bands (full line) and "forbidden" bands of states with complex wave vector (dotted lines). (**c**) Reflectivity of electrons at normal incidence on the (100) surface; i.e. I-V curve for the (00) spot in a LEED experiment. The arrows denote the positions of maxima from kinematic theory (3rd Laue condition). The energy scale $\hat{E}$ is in atomic units times $(2\pi/d)^2$ [4.7]

crystal at this energy. The electronic band structure with its allowed and forbidden bands is therefore of considerable importance for the intensity of a particular reflected beam (hk). If the energy E falls in a forbidden gap of the band structure, the wave functions outside can not be matched to a Bloch state inside, and a peak in the reflected LEED intensity results (Fig. 4.12). The model calculations in Fig.4.12 [4.7] have been performed on a semi-infinite cubic array (primitive lattice) of s-wave scatterers. Such a potential has essentially the same properties as an array of δ-function potentials. Normal incidence on the (100) face of this model crystal is considered, i.e. the band structure along the Γ-X symmetry line of the Brillouin zone is important. Figure 4.12 exhibits both the free-electron band structure (a) and the bands calculated under the assumption of s-wave scatterers. In Fig.4.12c the calculated reflected intensity of a normally incident electron beam is displayed. Gaps in the band structure are related to maxima in the reflectivity, since here no Bloch states are available inside the crystal for matching to the outside wave field. On the other hand, the external electronic wave functions in these regions cannot abruptly end at the surface. A detailed calculation, however, shows that, for energies within the forbidden gaps, electronic states exist at the surface. Their wave functions decay exponentially into the interior of the crystal. These states are localized at the surface (Chap.6); in contrast to the Bloch states with real wave vectors **k** in (4.31) they are characterized by a complex **k**. The imaginary part $\mathrm{Im}\{k_\perp\}$ gives the decay length of the wave functions. These decaying states are used to match the outside wave field at energies where Bloch states are forbidden. The regions of high reflectivity shown in Fig.4.12c thus correspond to the phenomenon of total reflection of electromagnetic waves on a dielectric surface. The arrows in Fig.4.12c depict the energetic positions expected for the maxima in the I-V curve on the basis of simple application of the third Laue condition (in the z-direction) in kinematic theory. When compared with an experimentally determined I-V curve (Fig.4.5) the higher-energy peaks of the calculated I-V curve are sharper and more intense. The reasons for this discrepancy might lie in the neglect of all inelastic scattering processes of electron-electron interactions.

4.4.2 Multiple-Scattering Formalism

In the second dynamic approach to understand the LEED intensity patterns, the scattered wave field is built up from scattering processes on different lattice planes below the surface. In this formalism, the first step is the calculation of the scattering amplitudes from a single atom. For different primary energies, scattering phases are calculated, usually for s, p and d scattering. Muffin-tin potentials are often used for this purpose. The next step involves the calculation of scattering processes within a single atomic layer, to give the so-called **intra-layer multiple scattering**.

The final solution is then obtained by considering scattering between different atomic planes, or the so-called **inter-layer multiple scattering**. How many layers have to be taken into account depends on the penetration depth of the primary electrons. For 1000 eV electrons, this depth is approximately 10 Å (Fig.4.1); calculations based on 8 atomic layers yield an estimated accuracy of about 1%. The wave field between the different atomic layers is composed of sets of forward and backward travelling beams, e.g. forward scattering on the first layer contributes to the amplitude backscattered at the second layer, etc. This multiple scattering approach to the LEED problem gives essentially the same results as the matching formalism. This has been shown for several model calculations. In Fig.4.13 the I-V curves of the (00) and (10) beams are plotted for the same model crystal as above (semi-infinite cubic array of s-wave scatterers). The calculations are made both on the basis of the matching method and using the multiple-scattering formalism [4.8]. Both approaches yield essentially the same features. It is worth mentioning that dynamic LEED theory has been extended

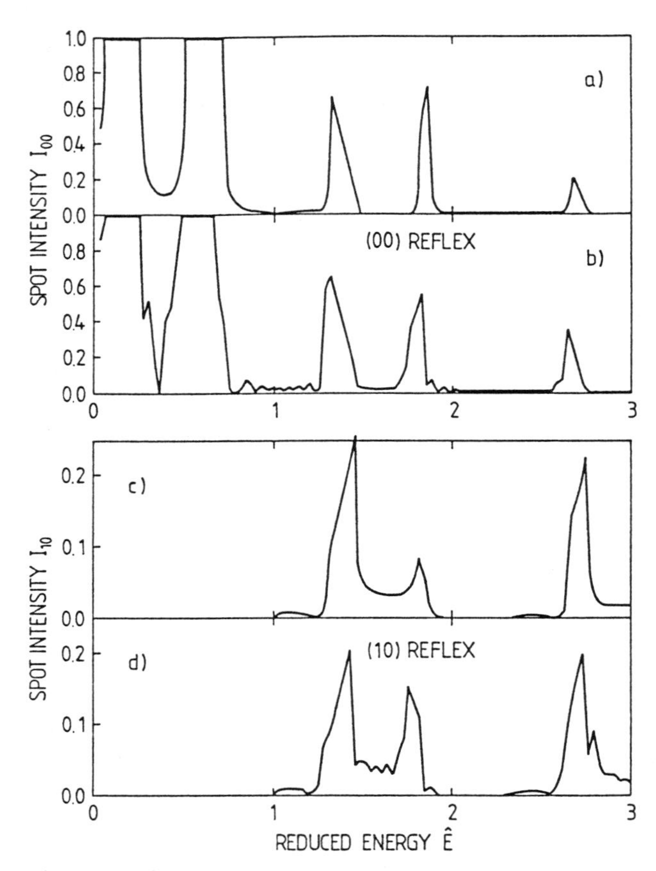

Fig.4.13a-d. Dynamic calculations of the I-V curves of the (00) and the (10) Bragg spots for a semi-infinite cubic array of s-wave scatterers. The primary electron beam is incident normal to the (100) face; (**a** and **c**) by means of the matching formalism; (**b** and **d**) by means of the multiple scattering formalism [4.7, 8]

to spin-dependent scattering [4.9]. In this case, instead of the Schrödinger equation, the relativistic Dirac equation is used to describe the dynamics of the electrons. The degree of spin polarization in the scattered electron beam is calculated as a function of primary energy and scattering angle.

4.4.3 Structure Analysis

The analysis of experimental data using the dynamic theory of elastic electron scattering is now established as one of the major tools in the determination of atomic surface structure. From simple inspection and measurement of the geometric separation of the LEED spots on the phosphorus screen, only the dimensions and the symmetry of the 2D surface lattice can be determined. In other words, the reciprocal lattice depicted in the diffraction pattern yields information only about distances and angles of the 2D unit mesh in real space (Sect.4.3). As in bulk solid-state physics, the atomic configuration, i.e. the atomic coordinates within the unit cell, can only be obtained by measuring the intensity of the Bragg spots. For X-ray scattering from bulk solids, simple kinematic theory can be used to relate structural models to the experimentally observed Bragg-spot intensities. This is not the case for a LEED experiment in surface physics; here the much more complex dynamic approach has to be applied to relate a structural model, i.e., a proposed set of atomic coordinates in the topmost atomic layers to the observed LEED spot intensities. For this purpose the intensity of a number of LEED beams is measured as a function of primary energy (I-V curves). A set of possible atomic coordinates for the atoms in the topmost layers is used as input for a dynamic calculation of these I-V curves and the results are compared with the experimental data. Depending on the quality of the fit, the structural model may be modified and a new calculation made; this "trial and error" procedure (Fig.4.14) is repeated until satisfactory agreement is obtained. One severe problem in establishing a definite model for the surface atomic structure lies in the procedure for comparing the calculated with the measured I-V curves. Simple inspection often leads to debate as to what is good and what is poor agreement. To prove a more objective assessment of the quality of a fit, so-called **reliability functions** have been introduced. For a thorough structural analysis, I-V curves for a considerable number of Bragg spots have to be measured and calculated.

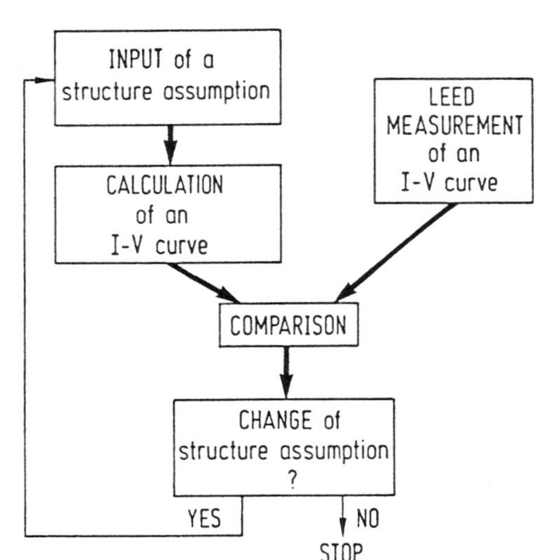

Fig.4.14. Flow diagram for a LEED structure analysis by means of dynamic theory. The calculations require big computers and the analysis must be made for a number of different Bragg spots

4.5 Kinematics of an Inelastic Surface Scattering Experiment

In almost all inelastic scattering processes on crystalline surfaces – be it with slow electrons, atoms, ions, etc. – the energy and the wave vector parallel to the surface are conserved (4.18), i.e.

$$E' - E = \hbar\omega \,, \tag{4.32a}$$

$$\mathbf{k}_{\parallel}' - \mathbf{k}_{\parallel} = \mathbf{q}_{\parallel} + \mathbf{G}_{\parallel} \,. \tag{4.32b}$$

Equation (4.32b) is a direct consequence of the 2D translational symmetry of a perfect crystalline surface. $\mathbf{G}_{\parallel}$ is an arbitrary 2D reciprocal lattice vector. If the scattering process involves irregularly distributed centers on the surface, e.g. statistically adsorbed atoms or defects, then (4.32b) breaks down and only the energy conservation (4.32a) is valid. The energy $\hbar\omega$ and wave vector $\mathbf{q}_{\parallel}$ may be transfered to collective surface excitations like phonons, plasmons, magnons, etc. to single particles like electrons in the conduction band (intra-band scattering), or to electrons excited from an occupied electronic band into an empty one (inter-band scattering). In all these cases, the characteristic excitation energy $\hbar\omega$ requires a definite $\mathbf{q}_{\parallel}$ transfer. On a periodic crystal surface collective excitations are described by a dispersion relation $\hbar\omega(\mathbf{q}_{\parallel})$ and in scattering processes within or between electronic bands the band structure $E(\mathbf{k})$ fixes the $\mathbf{q}_{\parallel}$ transfer for a particular energy transfer $\hbar\omega$:

$$E'(\mathbf{k}_{\parallel}') - E(\mathbf{k}_{\parallel}) = \hbar\omega(\mathbf{q}_{\parallel}) \,. \tag{4.33}$$

The determination of such dispersion relations or surface band structures $E(\mathbf{k}_{\parallel})$ is one of the major incentives for performing inelastic scattering on surfaces. In these experiments, therefore, the energy of the incident particles (E) and of the scattered particles (E′) has to be determined by energy analysers (Panel II: Chap. 1). The wave vectors $\mathbf{k}$ and $\mathbf{k}'$ are determined by the energy and the scattering geometry. Applying energy and wave-vector conservation (4.32) the experimental geometry uniquely relates the $\mathbf{q}_{\parallel}$ transfer to the energy transfer $\hbar\omega$. Figure 4.15 depicts the geometry of a surface scattering experiment. The incoming particles have the wave vector $\mathbf{k}$. Scattering in the specular direction with $|\mathbf{k}_s| = |\mathbf{k}|$ represents the elastic process, whereas inelastic processes generally involve scattering out of the specular direction (described by the angles ψ and φ). The scattered wave vector $\mathbf{k}'$ then differs from the specular vector $\mathbf{k}_s$ by the transfer vector $\mathbf{q}$, i.e. for the case $\mathbf{G}_{\parallel} = 0$,

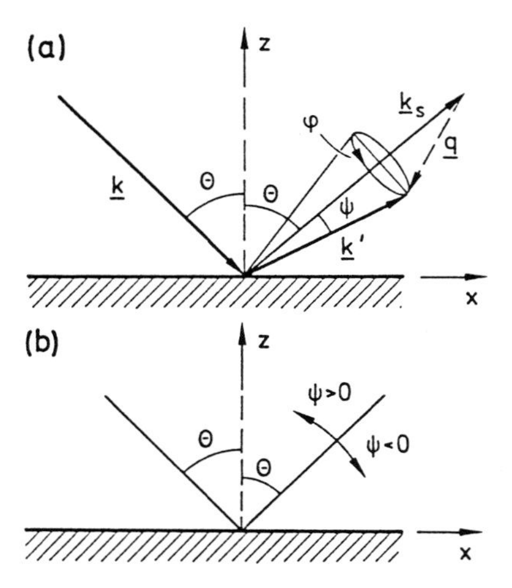

Fig.4.15. (**a**) Geometry of an inelastic scattering experiment. The plane of incidence is the xz plane of the coordinate system (x axis parallel to surface). The primary beam is described by the wave vector **k** (angle of incidence θ). Elastically scattered particles escape with a wave vector $\mathbf{k}_s$ (specular direction). Inelastically scattered particles are found with a wavevector $\mathbf{k}'$. The wave vector change is **q**. The zero of the angle φ is in the plane of incidence. (**b**) Special case of "planar scattering" where the detection direction is in the plane of incidence ($\varphi = 0$). ψ is measured from the specular direction

$$\mathbf{k}' = \mathbf{k} - \begin{pmatrix} q_x \\ q_y \\ \Delta k_z \end{pmatrix} . \tag{4.34}$$

For the transfer vector $\mathbf{q} = (q_x, q_y, \Delta k_z)$ only the components q_x and q_y parallel to the surface are fixed by a conservation law, i.e. can be found in a corresponding surface excitation. Δk_z is determined from the energy and scattering geometry only and is in no way related to any excitation. The excitations can have any wave-vector component normal to the surface. Energy conservation (4.33) is expressed as

$$\frac{\hbar^2 k'^2}{2\,m} = \frac{\hbar^2 k^2}{2\,m} - \hbar\omega , \tag{4.35a}$$

or

$$k' = k(1 - \hbar\omega/E)^{1/2} . \tag{4.35b}$$

The geometric relation between $\mathbf{k}'$ and **k** in Fig.4.15 is most readily derived in a coordinate system whose z-axis is along the specular direction, i.e., a system which is rotated by the angle θ with respect to the z-axis depicted in Fig.4.15. In this rotated system, the specular wave vector $\mathbf{k}_s$ and the inelastically scattered wave vector $\mathbf{k}'$ have the simple representations

$$\mathbf{k}_s = k \begin{pmatrix} 0 \\ 0 \\ 1 \end{pmatrix} \quad \text{and} \quad \mathbf{k}' = k' \begin{pmatrix} -\sin\psi\cos\varphi \\ \sin\psi\sin\varphi \\ \cos\psi \end{pmatrix} . \tag{4.36}$$

Their representation in the coordinate system of Fig.4.15 with the z-axis normal to the surface, is obtained by a rotation around the y-axis (parallel to surface) by means of the rotation matrix

$$\mathbf{R} = \begin{bmatrix} \cos\theta & 0 & \sin\theta \\ 0 & 1 & 0 \\ -\sin\theta & 0 & \cos\theta \end{bmatrix} . \tag{4.37}$$

We thus obtain for the specular and for the inelastically scattered beam:

$$\mathbf{k}_s = k \begin{bmatrix} \sin\theta \\ 0 \\ \cos\theta \end{bmatrix} . \tag{4.38}$$

$$\mathbf{k}' = k' \begin{bmatrix} -\sin\psi\cos\varphi\cos\theta + \cos\psi\sin\theta \\ \sin\psi\sin\varphi \\ \sin\psi\cos\varphi\sin\theta + \cos\psi\cos\theta \end{bmatrix} . \tag{4.39}$$

Combining (4.34, 38, 39) with the energy conservation (4.35b) one arrives at

$$\mathbf{k} - \mathbf{k}' = \mathbf{q} = \begin{bmatrix} q_x \\ q_y \\ \Delta k_z \end{bmatrix}$$

$$= k \left[\begin{bmatrix} \sin\theta \\ 0 \\ -\cos\theta \end{bmatrix} - \sqrt{1 - \frac{\hbar\omega}{E}} \begin{bmatrix} \cos\psi\sin\theta - \sin\psi\cos\varphi\cos\theta \\ \sin\psi\sin\varphi \\ \sin\psi\cos\varphi\sin\theta + \cos\psi\cos\theta \end{bmatrix} \right] . \tag{4.40}$$

In an inelastic scattering experiment now the orientation of the primary beam relative to the surface determines the angle of incidence θ, and the position of the entrance aperture to the energy analyser is described by the angles ψ and φ with respect to the specular direction (Fig.4.15a). An observed energy transfer $\hbar\omega$ (loss or gain) at a particular primary energy E then corresponds to a certain wave-vector transfer $\mathbf{q}$ via (4.40). The component $\mathbf{q}_{\parallel} = (q_x, q_y)$ parallel to the surface is conserved in the scattering process, i.e., it can be found as the wave vector of a particular surface excitation. The component Δk_z is determined by energy conservation, but it has no meaning for the surface excitations.

Equation (4.40) simplifies considerably for the special case of planar scattering where the scattered beam is detected in the plane of incidence

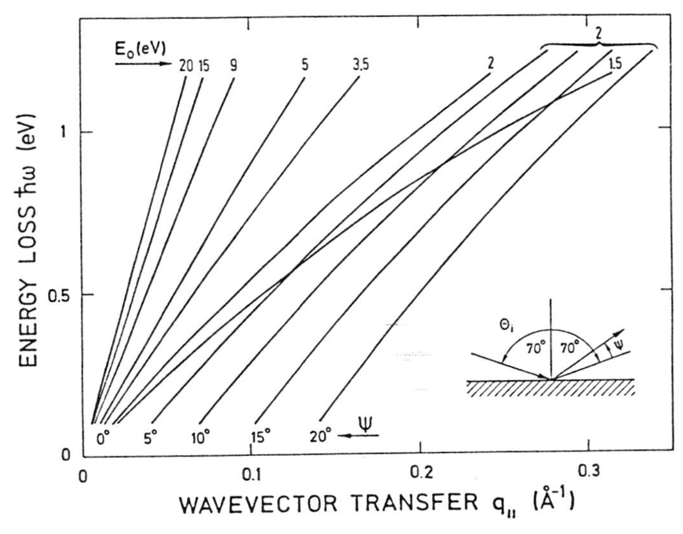

Fig.4.16. Energy transfer (loss) $\hbar\omega$ as a function of the corresponding wavevector transfer $q_{\|}$ for inelastic scattering of low energy electrons. For the specular detection ($\psi = 0$) the curves for different primary energies E = 1.5, 2, ... 20 eV all pass through the origin. The curves for off-specular detection with $\psi = 5°, 10°, 15°, 20°$ are for a fixed primary energy of E = 2 eV

(parallel to x-axis in Fig.4.15). The wave-vector transfer parallel to the surface $q_{\|} = q_x$ is then given by ($\varphi = 0, \pi$):

$$q_{\|} = k[\sin\theta + \sqrt{1 - \hbar\omega/E}\,(\sin\psi\cos\theta - \cos\psi\sin\theta)] \,. \qquad (4.41)$$

For a typical, inelastic-scattering experiment with low-energy electrons (HREELS, Panel IX: Chap.4) the dependence of the energy loss $\hbar\omega$ on the wave-vector transfer $q_{\|}$ according to (4.41) is plotted in Fig.4.16. For detection in the specular direction ($\psi = 0$) the curves originating at the zero point are calculated for different primary energies E ranging between 1.5 and 20 eV. For a fixed primary energy of 2 eV, curves are shown for detection directions deviating from the specular direction by the angles $\psi = 2°, 10°, 15°, 20°$. As is clearly seen from (4.41) and from Fig.4.16, the wave-vector transfer $q_{\|}$ in an inelastic scattering experiment can be varied by changing the ratio $\hbar\omega/E$ or by changing the detection angle ψ (off-specular detection). Figure 4.16 displays the situation for a typical HREELS experiment, but similar curves with loss energies and $q_{\|}$ values on a different scale are obtained for the scattering of atoms or ions. For He atoms scattered inelastically from a crystalline surface with loss or gain energies in the 10 meV range, $q_{\|}$ transfers in the 1 $Å^{-1}$ range can easily be obtained (Panel X: Chap.5).

A particularly simple situation occurs for specular detection ($\psi = 0$) and primary energies E that are large in comparison to the observed loss energies ($\hbar\omega/E \ll 1$). In this case (4.41) simplifies to

$$q_{\|} \simeq k \sin\theta(\hbar\omega/2E) . \qquad (4.42)$$

This case of small $\hbar\omega/2E$ and specular detection is important when low-energy electrons are scattered on a dynamic long-range potential. Such a potential, if expanded into a Fourier series, gives only components at small $q_{\|}$ values. Therefore the inelastically scattered intensity is peaked around $q_{\|} \simeq 0$, i.e. around the specularly reflected beam (Sect.4.6).

4.6 Dielectric Theory of Inelastic Electron Scattering

Among the variety of particles that can be used for surface scattering experiments, low-energy electrons play a rather important role. There are several reasons for this: slow electrons can be produced by a simple electron gun and like all charged particles can easily be dispersed by electrostatic energy analysers and detected by a channeltron or another kind of electron multiplier. On the theoretical side, a rather simple mathematical approach is possible for inelastic scattering in long-range potentials. For electrons an example of such long-range scattering potentials are oscillating dipole fields on a surface. These dipole fields may originate from collective excitations such as surface lattice vibrations (phonons) and surface plasmons (Chap.5) or from dynamic dipole moments of vibrating adsorbed molecules and atoms. Scattering cross sections can, of course, be calculated by using the dipole fields as scattering potentials within the formalism of Sect. 4.1.

It must be emphasized, however, that apart from this type of scattering by long-range dipole fields there exist other – from the theoretical standpoint more complex – scattering mechanisms for slow electrons [4.10]. Interaction with the local atomic potential of a surface atom must be described in a more localized picture since the scattering potential extends only over atomic dimensions. An expansion of such a short-range scattering potential as a Fourier series in wave vectors $\mathbf{q}_{\|}$ includes rather large $\mathbf{q}_{\|}$ values and thus leads to scattering at large angles from the specular direction (Fig.4.15a). This type of scattering is often called **impact scattering**. A possible mechanism involves the virtual excitation of an electronic state of a surface atom (in the substrate or the adsorbate); the inelastically scattered electron briefly occupies an excited state of this atom and escapes with an energy differing from the primary energy by a vibrational quantum (phonon, plasmon or adsorbate vibration).

From these qualitative arguments about the spatial range of the scattering potential and the resulting wave-vector transfers, it is evident that impact scattering and scattering in long-range dipole fields can be distinguished experimentally by the angular distribution of the inelastically scattered electrons around the specular reflection direction. Strong peaking of the scattered intensity in this direction clearly indicates scattering in long-range dipole fields. Care has to be taken in the interpretation of data when *Umklapp processes* ($\mathbf{G}_{\|} \neq 0$) are involved.

We will see in the following that these qualitative arguments follow quite naturally from the simple theoretical approach to scattering in long-range dipole fields, the so-called **dielectric theory**.

4.6.1 Bulk Scattering

The dielectric approach was first applied by *Fermi* [4.11], and *Hubbard* and *Fröhlich* [4.12] to inelastic scattering of high-energy electrons (several keV range) which penetrate thin solid films. Because of the long-range nature of the scattering potential a quasi-continuum theory is used, in which the dielectric properties of the solid are described by its complex dielectric function $\epsilon(\omega) = \epsilon_1 + i\epsilon_2$. We first consider briefly the bulk scattering process, where an electron penetrates a solid and loses part of its energy inside the bulk. This energy transfer is ascribed to the shielding of the Coulomb field of the moving electron due to the surrounding dielectric medium. The total energy transfer W is thus given by the change in the energy density of the Coulomb field inside the solid. Since in the present model calculation the bulk solid, which is penetrated by the electron, is assumed to be extended into infinity, the energy loss per unit time, i.e., the energy transfer rate is calculated as the finite quantity

$$\dot{W} = \mathrm{Re}\left\{\int d\mathbf{r}\,\mathcal{E}\cdot\dot{\mathbf{D}}\right\}. \tag{4.43}$$

If the complex representations of the $\mathcal{E}$ and $\mathbf{D}$ fields are used, only the real part of the integral describes an energy loss.

The fields are expanded in Fourier series

$$\mathcal{E}(\mathbf{r},t) = \int d\omega\, d\mathbf{q}\,\hat{\mathcal{E}}(\omega,\mathbf{q})e^{-i(\omega t + \mathbf{q}\cdot\mathbf{r})} \tag{4.44}$$

and similarly for $\mathbf{D}(\mathbf{r},t)$, whose Fourier components $\hat{\mathbf{D}}(\omega,\mathbf{q})$ are related to $\hat{\mathcal{E}}(\omega,\mathbf{q})$ via

$$\hat{\mathbf{D}}(\omega,\mathbf{q}) = \epsilon_0 \epsilon(\omega,\mathbf{q}) \hat{\mathcal{E}}(\omega,\mathbf{q}) . \tag{4.45}$$

For the calculation of (4.43) it is convenient to substitute ω by $-\omega'$ in $\hat{\mathbf{D}}(\omega,\mathbf{q})$:

$$\dot{W} = \mathrm{Re}\left\{ \int d\mathbf{r} d\omega' d\omega d\mathbf{q}' d\mathbf{q} \frac{i\omega'}{\epsilon_0 \epsilon(\omega,\mathbf{q})} \hat{\mathbf{D}}(\omega,\mathbf{q}) \cdot \hat{\mathbf{D}}(-\omega',\mathbf{q}') e^{i(\omega'-\omega)t} e^{-i(\mathbf{q}+\mathbf{q}')\cdot\mathbf{r}} \right\} . \tag{4.46}$$

With the representation for the δ-function

$$\int d\mathbf{r} e^{-i(\mathbf{q}+\mathbf{q}')\cdot\mathbf{r}} = (2\pi)^3 \delta(\mathbf{q} + \mathbf{q}') \tag{4.47}$$

one obtains

$$\dot{W} = (2\pi)^3 \mathrm{Re}\left\{ \int d\omega' d\omega d\mathbf{q} \frac{i\omega'}{\epsilon_0 \epsilon(\omega,\mathbf{q})} \hat{\mathbf{D}}(\omega,\mathbf{q}) \cdot \hat{\mathbf{D}}(-\omega',-\mathbf{q}) e^{i(\omega'-\omega)t} \right\} . \tag{4.48}$$

Since the $\mathbf{D}$ field is directly related to the moving point charge e through $\mathrm{div}\mathbf{D} = e\delta(\mathbf{r}-\mathbf{v}t)$, the Fourier components $\hat{\mathbf{D}}(\omega,\mathbf{q})$ include the time and spatial structures of the field of the moving electron. From

$$\mathbf{D} = -\frac{e}{4\pi} \nabla \frac{1}{|\mathbf{r}-\mathbf{v}t|} = \frac{e}{4\pi |\mathbf{r}-\mathbf{v}t|^3} (\mathbf{r}-\mathbf{v}t) \tag{4.49}$$

the Fourier transform is calculated by using the relations

$$\frac{e^{-\alpha r}}{r} = \int f(\mathbf{q}) e^{i\mathbf{q}\cdot\mathbf{r}} d\mathbf{q} , \tag{4.50a}$$

$$f(\mathbf{q}) = 2\pi^{-3} \int \frac{e^{-\alpha r}}{r} e^{-i\mathbf{q}\cdot\mathbf{r}} d\mathbf{r} . \tag{4.50b}$$

With $\alpha = 0$ and by use of the representation $d\mathbf{r} = d\varphi d\theta \sin\theta r^2 dr$ in spherical coordinates it follows

$$\frac{1}{r} = 2\pi^{-3} \int d\mathbf{q} \left(\frac{4\pi}{q^2}\right) e^{i\mathbf{q}\cdot\mathbf{r}} \tag{4.51}$$

and

$$\mathbf{D}(\mathbf{r},t) = e(2\pi)^{-3} \int d\mathbf{q} q^{-2} \mathbf{q} e^{-i\mathbf{q}\cdot(\mathbf{r}-\mathbf{v}t)}, \tag{4.52}$$

respectively. Using the identity

$$e^{i\mathbf{q}\cdot\mathbf{v}t} = \int d\omega e^{-i\omega t} \delta(\omega - \mathbf{q}\cdot\mathbf{v}) \tag{4.53}$$

one finally obtains

$$\hat{\mathbf{D}}(\omega,\mathbf{q}) = \frac{e}{(2\pi)^3} \mathbf{q} \frac{1}{q^2} \delta(\omega + \mathbf{q}\cdot\mathbf{v}) . \tag{4.54}$$

With $\delta(x) = \delta(-x)$ the following expression follows for the energy-loss rate (4.43)

$$\dot{W} = \frac{e^2}{\epsilon_0 (2\pi)^3} \cdot$$

$$\cdot \mathrm{Re}\left\{\int\int d\omega' d\omega \frac{i\omega'}{q^2 \epsilon(\omega,\mathbf{q})} \delta(\omega' + \mathbf{q}\cdot\mathbf{v}) \delta(\omega + \mathbf{q}\cdot\mathbf{v}) e^{i(\omega' - \omega)t} d\mathbf{q}\right\} . \tag{4.55a}$$

For a particular $\mathbf{q}$ it follows $\omega' = \omega$ and the time dependence in (4.55a) disappears:

$$\dot{W} = \frac{e^2}{\epsilon_0 (2\pi)^3} \int d\omega d\mathbf{q} \frac{\omega}{q^2} \mathrm{Im}\left\{\frac{-1}{\epsilon(\omega,\mathbf{q})}\right\} \delta(\omega + \mathbf{q}\cdot\mathbf{v}) . \tag{4.55b}$$

We decompose the wave-vector transfer $\mathbf{q}$ into the components $\mathbf{q}_{\|}$ and $\mathbf{q}_{\perp}$ parallel and perpendicular to the electron velocity (cylindrical coordinates around the electron trajectory) and find by means of

$$\mathbf{q}\cdot\mathbf{v} = q_{\|}v\,, \quad q^2 = q_{\|}^2 + q_{\perp}^2\,, \quad d\mathbf{q} = q_{\|}d\varphi dq_{\|}dq_{\perp} \tag{4.56}$$

from the δ-function in (4.55b), that maximum energy transfer takes place for

$$v = \left|\frac{\omega}{q_{\|}}\right|\,. \tag{4.57}$$

An energy loss thus results when harmonic excitations of the solid have a phase velocity $\omega/q_{\|}$ which equals the electron velocity; i.e., in order to take over energy from the moving electron, the solid excitation must propagate in phase with the electon.

Using (4.56) and the relation $\delta(\omega+q_{\|}v) = v^{-1}\delta(\omega/v+q_{\|})$ one obtains for the energy-loss rate (4.55b)

$$\dot{W} = \frac{e^2}{(2\pi)^3\epsilon_0 v}\int d\omega dq_{\perp}d\varphi\,\omega\,\frac{q_{\perp}}{(\omega/v)^2 + q_{\perp}^2}\,\mathrm{Im}\left\{\frac{-1}{\epsilon(\omega,q)}\right\}\,. \tag{4.58}$$

For a circular analyser aperture the integration over φ yields 2π and the q dependence in $\epsilon(\omega,\mathbf{q})$ can be neglected since for higher electron velocities $q_{\|}$ is negligibly small. It thus follows with $q_{\perp}dq_{\perp} = dq_{\perp}^2/2$:

$$\dot{W} = \frac{e^2}{4\pi^2\epsilon_0 v}\int \omega d\omega \mathrm{Im}\left\{\frac{-1}{\epsilon(\omega)}\right\}\int\frac{q_{\perp}dq_{\perp}}{(\omega/v)^2 + q_{\perp}^2}$$

$$= \frac{e^2}{4\pi^2\epsilon_0 v}\left\{\ell n\left[\left(\frac{\omega}{v}\right)^2 + q_{\perp}^2\right]\right\}\Bigg|_0^{q_c}\int \omega d\omega \mathrm{Im}\left\{\frac{-1}{\epsilon(\omega)}\right\}\,, \tag{4.59}$$

where the upper possible wave-vector transfer q_c can be at most the reciprocal lattice constant $1/a$ of the crystalline solid.

The total energy transfer rate (4.59) is composed of components with different angular frequencies ω. The solid can, however, offer only particular characterisic excitations with quantum energy $\hbar\omega$, to which energy can be transferred. The spectral response of the solid in an inelastic electron scattering experiment is thus described within the framework of dielectric theory by the so-called **bulk-loss function**

$$\mathrm{Im}\left\{\frac{-1}{\epsilon(\omega)}\right\} = \frac{\epsilon_2(\omega)}{\epsilon_1^2(\omega) + \epsilon_2^2(\omega)} \ . \tag{4.60}$$

Essential spectral structure in an electron energy loss spectrum therefore occurs when the nominator in (4.60), $\epsilon_2(\omega)$, i.e., essentially the optical absorption of the material exhibits peaks, e.g., due to electronic interband transitions. But the main spectral features arise from the condition $\epsilon_1(\omega) \simeq 0$ in regions, where $\epsilon_2(\omega)$ is small and monotonic. The condition $\mathrm{Re}\{\epsilon(\omega)\} \simeq 0$ determines, on the other hand, the frequencies ω of the longitudinal collective excitations of a solid as, e.g., the plasma waves of a free electron gas [4.13].

In electron energy loss experiments, where bulk scattering prevails, the essential spectral structure, therefore, is due to the excitation of the bulk plasmon, whose excitation energy is found, in accordance with bulk electron densities in the 10^{22} to 10^{23} cm^{-3} range, between 5 and 20 eV (Figs. 4.23, IX.4 and 5).

4.6.2 Surface Scattering

In surface physics the primary energy ($E < 20\,\mathrm{eV}$) used in reflection scattering experiments is so small that the electrons penetrate only a few Ångstroms into the solid (Fig.4.1). The time which they spend within the material is so short that bulk scattering according to (4.60) is a negligible process. Nevertheless, the long-range Coulomb field of the electrons penetrates into the solid as they approach the surface and on their way back after reflection (Fig.4.17). Thus, while in the vicinity of the surface, they interact with the material via their Coulomb field. The shielding of the penetrating field gives rise to surface scattering processes [4.10, 14] which can be treated mathematically in a similar way to the above bulk scattering mechanism. For this purpose the electron trajectory $\mathbf{s}(t)$ is assumed to be essentially that of an elastically scattered electron (velocity $\mathbf{v}$). The time $t = 0$ is taken to be the moment of the reflection at the solid surface ($z = 0$). The position and velocity of the electron are described according to Fig.4.17 by

$$\mathbf{s}(t) = \mathbf{v}t = \mathbf{v}_{\|}t + v_{\perp}t\hat{\mathbf{e}}_z \ , \tag{4.61a}$$

and because of the small energy losses $\hbar\omega \ll E$

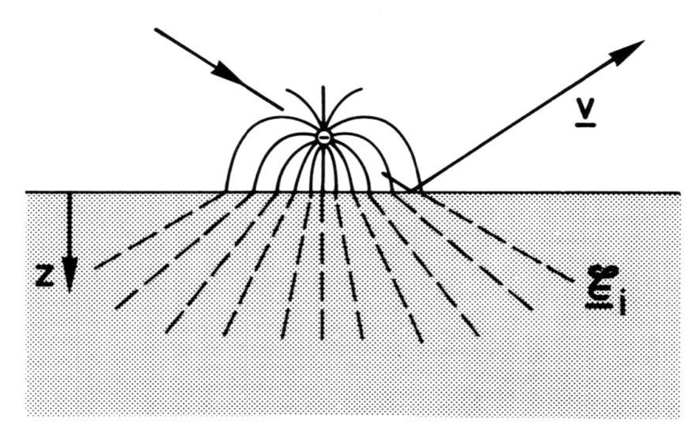

Fig.4.17. Schematic plot of the inelastic dielectric scattering process of a low-energy electron (velocity **v**) on a semi-infinite halfspace ($z > 0$). The energy transfer (energy loss) to the solid is due to shielding of the interior electric field $\mathcal{E}_i$ of the electron outside the surface. Since the energy transfer (loss) is assumed to be small in comparison with the kinetic energy of the electron, the electron trajectory is quasi-elastic

$$v_\perp(t<0) = |v_\perp| \simeq -v_\perp(t>0) , \qquad (4.61b)$$

where $\|$ and $\perp$ denote the components parallel and normal to the surface. $\hat{\mathbf{e}}_z$ is a unit vector in the z-direction. The **D** field of the moving electron far away from the surface is the same as it would be if the semi-infinite solid at $z > 0$ were not present (4.52). According to classical electrodynamics the external field of a point charge outside a semi-infinite dielectric medium can be described by an image charge inside the solid; the field $\mathcal{E}_i$ in the interior of the medium, however, has radial symmetry as if no interface were present, but it is shielded by a factor $2/(\epsilon+1)$ with respect to the field $\mathcal{E}$ of the free electron. Thus, for the Fourier components $\hat{\mathcal{E}}_i$ we have

$$\hat{\mathcal{E}}_i(\mathbf{q},\omega) = \frac{2}{\epsilon(\omega)+1}\hat{\mathcal{E}}(\mathbf{q},\omega) = \frac{\hat{\mathbf{D}}_i(\mathbf{q},\omega)}{\epsilon_0\epsilon(\omega)} . \qquad (4.62)$$

For our present problem of surface scattering, the conventional 3D Fourier series (4.44) is not a convenient representation of the electric field, since in a surface scattering experiment only the wave-vector component $\mathbf{q}_\|$ parallel to the surface is conserved rather than $\mathbf{q}$ (4.18b). The scattering cross section we are seeking should therefore be expressed as $P(\hbar\omega, \mathbf{q}_\|)$, and an expansion of the fields $\mathcal{E}$, **D** and $\mathcal{E}_i$, $\mathbf{D}_i$ in surface waves $\exp(i\mathbf{q}_\|\cdot\mathbf{r}_\| - q_\||z|)$ would be more appropriate. This expansion is indeed possible since we can integrate the 3D Fourier series (4.51) of the Coulomb potential over a coordinate normal to the surface. With

$$\mathrm{d}\mathbf{q} = \mathrm{d}\mathbf{q}_{\|}\mathrm{d}\mathbf{q}_{\perp} \quad \text{and} \quad \mathbf{q}\cdot\mathbf{r} = \mathbf{q}_{\|}\cdot\mathbf{r}_{\|} + z q_{\perp} \tag{4.63}$$

we obtain from (4.51)

$$\frac{1}{r} = \frac{1}{2\pi^2}\int \mathrm{d}\mathbf{q}_{\|}\exp(i\mathbf{q}_{\|}\cdot\mathbf{r}_{\|})\int \mathrm{d}q_{\perp}\frac{e^{iq_{\perp}z}}{q_{\|}^2 + q_{\perp}^2}$$

$$= \frac{1}{2\pi}\int \mathrm{d}\mathbf{q}_{\|}\frac{1}{q_{\|}}\exp(i\mathbf{q}_{\|}\cdot\mathbf{r}_{\|} - q_{\|}|z|)\ , \tag{4.64}$$

i.e., a representation in terms of surface waves which have wave character along the surface and decay exponentially into the semi-infinite halfspace. Since the interior fields $\mathbf{D}_i$ and $\mathcal{E}_i$ are just derivatives of the $1/r$ potential (4.52) the following representations are possible

$$\mathcal{E}_i(\mathbf{r},t) = \frac{1}{2\pi}\int \mathrm{d}\omega\mathrm{d}\mathbf{q}_{\|}\hat{\mathcal{E}}_i(\omega,\mathbf{q}_{\|})\exp(-q_{\|}|z|)\exp[i(\mathbf{q}_{\|}\cdot\mathbf{r}_{\|} - \omega t)] \tag{4.65a}$$

$$\dot{\mathbf{D}}_i(\mathbf{r},t) = \frac{1}{2\pi}\int \mathrm{d}\omega'\mathrm{d}\mathbf{q}'_{\|}(-i\omega')\hat{\mathbf{D}}_i(\omega',\mathbf{q}'_{\|})\exp(-q_{\|}|z|)\exp[i(\mathbf{q}_{\|}\cdot\mathbf{r}_{\|} - \omega t)]\ . \tag{4.65b}$$

The electric field has to be a real quantity, thus we have

$$\hat{\mathcal{E}}_i^*(\omega,\mathbf{q}_{\|}) = \hat{\mathcal{E}}_i(-\omega,-\mathbf{q}_{\|})\ , \tag{4.66}$$

and the total energy transfer to the solid can be written in analogy to (4.43) as

$$W = \mathrm{Re}\left\{\int_{-\infty}^{+\infty}\mathrm{d}t\int_{z=0}^{\infty}\mathrm{d}\mathbf{r}\,\mathcal{E}_i(\mathbf{r},t)\dot{\mathbf{D}}_i(\mathbf{r},t)\right\}\ . \tag{4.67}$$

In contrast to (4.43) the total energy transfer W rather than the rate $\dot{W}$ can be calculated since in surface scatering the scattering volume is finite and the time integral in (4.67) gives finite expressions. With the expansions (4.65) and the δ-function representation we obtain

$$W = 2\pi \mathrm{Re}\left\{ \int_0^\infty dz \int d\omega d\omega' d\mathbf{q}_{||} d\mathbf{q}'_{||} (-i\omega') \exp(-zq_{||} + zq'_{||}) \right.$$

$$\left. \times \delta(\omega + \omega') \delta(\mathbf{q}_{||} + \mathbf{q}'_{||}) \hat{\mathcal{E}}_i(\omega, \mathbf{q}_{||}) \hat{\mathbf{D}}_i(\omega', \mathbf{q}'_{||}) \right\} . \tag{4.68}$$

Relating this to the field $\hat{\mathbf{D}}$ of the free electron (without the solid) we get

$$W = 2\pi \mathrm{Re}\left\{ \int d\omega d\mathbf{q}_{||} \frac{i\omega}{2q_{||}} \frac{2}{\epsilon(\omega) + 1} \frac{2\epsilon^*(\omega)}{\epsilon^*(\omega) + 1} \frac{1}{\epsilon_0} |\hat{\mathbf{D}}(\omega, \mathbf{q}_{||})|^2 \right\} , \tag{4.69}$$

and finally

$$W = \frac{4\pi}{\epsilon_0} \int d\omega d\mathbf{q}_{||} \frac{\omega}{q_{||}} |\hat{\mathbf{D}}(\omega, \mathbf{q}_{||})|^2 \, \mathrm{Im}\left\{ \frac{-1}{\epsilon(\omega) + 1} \right\} . \tag{4.70}$$

To calculate the expansion coefficients $\hat{\mathbf{D}}(\omega, \mathbf{q}_{||})$ of the field, we use (4.52) and the expansion of the Coulomb field in surface waves (4.64)

$$\mathbf{D}(\mathbf{r}, t) = -\frac{e}{4\pi} \nabla \frac{1}{|\mathbf{r} - \mathbf{v}t|} \tag{4.71}$$

$$= \frac{e}{8\pi^2} \int d\mathbf{q}_{||} (-i\hat{\mathbf{e}}_{||}, 1) \exp(i\mathbf{q}_{||} \cdot \mathbf{r}_{||} - zq_{||}) \exp(-i\mathbf{q}_{||} \cdot \mathbf{v}_{||} t + q_{||} v_\perp t) .$$

The last exponential function of the integrand is now Fourier-transformed with respect to time:

$$\exp(-i\mathbf{q}_{||} \cdot \mathbf{v}_{||} t + q_{||} v_\perp t) = \int_{-\infty}^{+\infty} d\omega g(\omega) e^{-i\omega t} \tag{4.72}$$

to yield, with $|v_\perp| \simeq$ const and $|v_{\|}| \simeq$ const,

$$g(\omega) = \frac{1}{2\pi} \frac{2q_{\|}v_\perp}{(q_{\|}v_\perp)^2 + (\mathbf{q}_{\|}\cdot\mathbf{v}_{\|}-\omega)^2} . \tag{4.73}$$

Equation (4.73) inserted into (4.72) and finally into (4.71) gives $\hat{\mathbf{D}}(\omega,\mathbf{q}_{\|})$, and the total energy transfer (4.70) follows as

$$W = \frac{8\pi e^2}{(2\pi)^4 \epsilon_0 \hbar^2} \int d(\hbar\omega)d\mathbf{q}_{\|}\hbar\omega \frac{q_{\|}v_\perp{}^2}{[(q_{\|}v_\perp)^2 + (\mathbf{q}_{\|}\cdot\mathbf{v}_{\|}-\omega)^2]^2} \operatorname{Im}\left\{\frac{-1}{\epsilon(\omega)+1}\right\}$$

$$= \int d(\hbar\omega)d\mathbf{q}_{\|}\hbar\omega P(\hbar\omega,\mathbf{q}_{\|}) . \tag{4.74}$$

The scattering probability $P(\hbar\omega,\mathbf{q}_{\|})$ for the transfer of an energy quantum $\hbar\omega$, and a wave vector $\mathbf{q}_{\|}$ parallel to the surface is thus obtained as

$$P(\hbar\omega,\mathbf{q}_{\|}) = \frac{e^2}{2\pi^3 \epsilon_0 \hbar^2} \frac{q_{\|}v_\perp{}^2}{[(q_{\|}v_\perp)^2 + (\mathbf{q}_{\|}\cdot\mathbf{v}_{\|}-\omega)^2]^2} \operatorname{Im}\left\{\frac{-1}{\epsilon(\omega)+1}\right\} . \tag{4.75}$$

It is convenient to express the inelastic scattering cross section $P(\hbar\omega,\mathbf{q}_{\|})$ or $d^2S/d(\hbar\omega)d\mathbf{q}_{\|}$ as a differential cross section $d^2S/d(\hbar\omega)d\Omega$ for scattering into a solid angle element $d\Omega$. Measured spectra can then be calculated by integrating over the acceptance angle of the electron analyser Ω_{Apert}. For this purpose we use the relations (4.35-39) and Fig.4.15 to express the element $d\mathbf{q}_{\|} = dq_{\|x}dq_{\|y}$ in terms of a solid angle element $d\Omega$:

$$d\Omega = \sin\psi d\psi d\varphi . \tag{4.76}$$

Because of (4.32b) one has

$$d\mathbf{k}_{\|}' = d\mathbf{q}_{\|} , \tag{4.77}$$

and for small-angle scattering ($q_{\|} \ll k$) with $\psi \ll 1$ and $\varphi \simeq 0$, i.e. $\sin\varphi \simeq 0$ and $\cos\varphi \simeq 1$, one has

$$d\mathbf{q}_{\|} = dq_{\|x}dq_{\|y} = k'^2\cos\Theta\sin\psi d\psi d\varphi = k'^2\cos\Theta d\Omega . \tag{4.78}$$

For most experimental conditions the energy loss $\hbar\omega$ is small in comparison to the primary energy E, i.e.

$$\hbar\omega \ll E \,, \quad k'^2 \simeq k^2 \,, \tag{4.79}$$

and because

$$E = \frac{\hbar^2 k^2}{2m} = \frac{1}{2} m v^2 = \frac{1}{2} \frac{m v_\perp^2}{\cos^2\Theta} \tag{4.80}$$

it follows that

$$k'^2 \cos\Theta \simeq \frac{m^2 v_\perp^2}{\hbar^2 \cos\Theta} \,. \tag{4.81}$$

From (4.79) one finally obtains

$$d\mathbf{q}_{\|} = \frac{m^2 v_\perp^2}{\hbar^2 \cos\Theta} d\Omega \,, \tag{4.82}$$

and for the differential scattering cross section [from (4.75)]:

$$\frac{d^2 S}{d(\hbar\omega) d\Omega} = \frac{m^2 e^2 |R|^2}{2\pi^3 \epsilon_0 \hbar^4 \cos\Theta} \frac{v_\perp{}^4 q_{\|}}{[v_\perp{}^2 q_{\|}^2 + (\omega - \mathbf{v}_{\|} \cdot \mathbf{q}_{\|})^2]^2} \mathrm{Im}\left\{\frac{-1}{\epsilon(\omega) + 1}\right\} . \tag{4.83}$$

The reflection coefficient R allows for the fact that not every primary electron is reflected from the surface. A high percentage penetrates into the solid and can be detected as a current. Apart from a Bose occupation factor $[n(\hbar\omega) + 1]$ for the excitation $\hbar\omega$ this formula (4.83) coincides with the expression derived by *Mills* [4.10] on the basis of a quantum mechanical treatment (Sect. 4.1), where scattering is described as originating from long-range charge density fluctuations. The scattering potential is evaluated in this calculation in terms of surface waves (4.65a) and the scattering process is described in the Born approximation.

Dielectric scattering of low-energy electrons according to (4.83) is characterized by two terms, the so-called surface loss function

$$\mathrm{Im}\left\{\frac{-1}{\epsilon(\omega) + 1}\right\} = \frac{\epsilon_2(\omega)}{[\epsilon_1(\omega) + 1]^2 + \epsilon_2^2(\omega)} \,, \tag{4.84}$$

and a prefactor

$$\frac{v_\perp^4 q_\|}{[v_\perp^2 q_\|^2 + (\omega - \mathbf{v}_\| \cdot \mathbf{q}_\|)^2]^2} \tag{4.85}$$

which bears some resemblance to a resonance term.

The surface loss function (4.84) determines the essential spectral structures of a loss spectrum. Thus, as in the bulk scattering process, spectral structure is expected where $\epsilon_2(\omega) = \text{Im}\{\epsilon(\omega)\}$ exhibits strong features. Since $\epsilon_2(\omega)$ also determines the optical absorption constant of a material, optical transitions of all kinds, e.g. interband excitations, excitons, phonons etc. can be observed in inelastic low-energy electron scattering. Additionally, prominent maxima in the scattering probability occur when the condition

$$\epsilon_1(\omega) \simeq -1 \tag{4.86}$$

is fulfilled in regions of small, monotonic $\epsilon_2(\omega)$. As will be shown in the next chapter, this condition (4.86) determines the frequencies of collective excitations such as surface phonons and plasmons (polaritons) of a semi-infinite dielectric halfspace. Such excitations may therefore be conveniently studied by inelastic scattering of low-energy electrons.

The resonance-type prefactor (4.85) in the scattering probability (4.83) can most readily be discussed for grazing incidence ($v_\perp \ll v_\|$). In this situation, strong peaks in the scattering cross section occur for

$$v_\| = \frac{\omega}{q_\|}, \tag{4.87a}$$

i.e. when the electron velocity parallel to the surface $v_\|$ equals the phase velocity $\omega/q_\|$ of the surface excitation (phonon, plasmon, etc.) that is responsible for the surface scattering process. Optimal coupling of the primary electron to the excited surface mode is obtained when the electron moves as a "surf-rider" on the phase of this surface excitation. We therefore call (4.85) the "surf-rider" term, and (4.87) the "surf-rider" condition for dielectric scattering. Using $E = \hbar^2 k^2/2m = mv^2/2$ we can express (4.87a) as

$$q_\| = k\frac{\hbar\omega}{2E}. \tag{4.87b}$$

For $\Theta \simeq 90°$ (grazing incidence), (4.87b) is identical to (4.42); the condition for maximum dielectric scattering cross section restricts the $q_\|$ transfer to values that are small in comparison with the Brillouin-zone

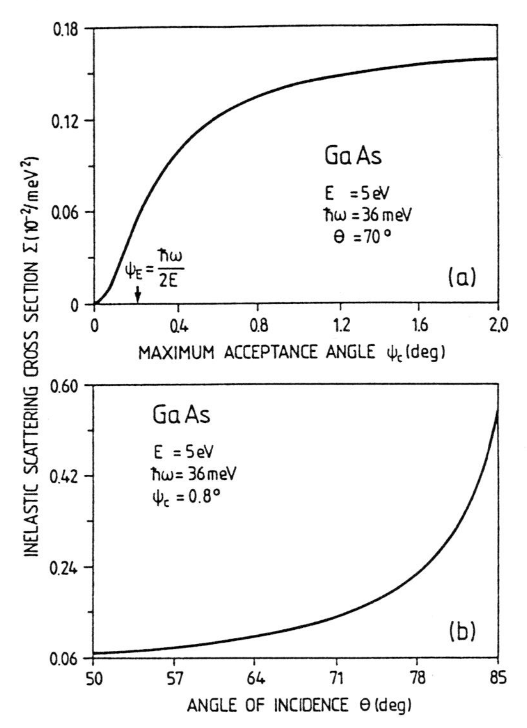

Fig. 4.18. (**a**) Total inelastic scattering cross section calculated according to dielectric theory (4.83) for scattering into an increasing angular aperture ψ_c (abscissa) around the specular direction. The parameters primary energy E, loss energy $\hbar\omega$ and angle of incidence Θ are chosen for the example of scattering from surface phonons on GaAs(110). (**b**) Total inelastic scattering cross section according to (4.83) as a function of angle of incidence. Specular detection is assumed, a primary energy of 5 eV and a loss energy of 36 meV. The integration is performed over a circular aperture (angle ψ_c = 0.8°) [4.15]

diameter. For primary energies in the 10 eV range and losses below 100 meV, $q_{\|}$ is estimated according to (4.87) to be in the range of 10^{-2} Å^{-1}. According to Fig. 4.15a such small $q_{\|}$ transfers mean scattering into small angles around the specular direction. Dielectric scattering due to long-range charge density fluctuations can therefore be distinguished experimentally from other scattering processes by measuring the angular distribution of the scattered electrons. In the case of dielectric scattering, the distribution is sharply concentrated into a lobe of angular width about 1 to 2° around the specular direction. For conventional electron spectrometers the acceptance aperture is also in the range of 1 to 2° and therefore ideally meets the requirements for studying dielectric scattering.

Figure 4.18a illustrates the calculated, total dielectric scattering cross section [angular integration over (4.83)] versus aperture angle of the detecting spectrometer ψ_c; a circular aperture is assumed. The parameters E = 5 eV (primary energy), $\hbar\omega$ = 36 meV (loss energy) and Θ = 70° are convenient for surface phonons in GaAs (Chap. 5). Figure 4.18a clearly reveals that the inelastically scattered electrons are mainly concentrated in an angle of $\psi_c \simeq 1°$. For the same scattering parameters, Fig. 4.18b displays the total scattering cross section into a circular aperture of $\psi_c = 0.8°$ versus angle of incidence Θ. The $(\cos\Theta)^{-1}$ dependence of (4.83) is clearly revealed. This

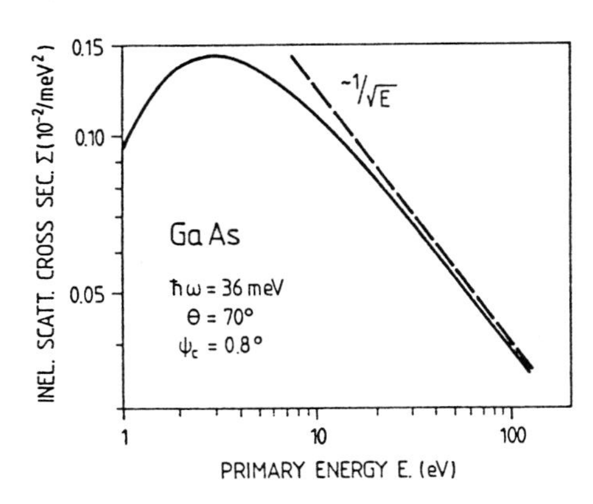

Fig.4.19. Total inelastic scattering cross section (4.83) versus primary energy. The integration of (4.83) is performed over an aperture angle of $\psi_c = 0.8°$, the angle of incidence is $\Theta = 70°$, specular detection is considered and the loss energy is assumed to be $\hbar\omega = 36$ meV [4.15]

$(\cos\Theta)^{-1}$ dependence stems from the time $\tau \sim (\cos\Theta)^{-1} E^{-1/2}$, during which the primary electron is moving in close proximity to the surface. The corresponding dependence of the scattering cross section on $E^{-1/2}$ is depicted in Fig.4.19.

4.7 Dielectric Scattering on a Thin Surface Layer

The dielectric surface scattering mechanism considered so far is restricted to the interaction of electrons with a homogeneous semi-infinite halfspace. The scattering cross section (4.75, 83) given in Sect.4.6 cannot therefore be used to describe inelastic scattering from quasi-2D excitations which are limited in space to a few Ångstroms below the surface. Physical examples of such excitations are transitions between electronic surface states (Chap. 6), excitations of a 2D electron gas in adsorbed metal layers or in tight accumulation layers of semiconductors (Chap.7), surface lattice vibrations with a finite vibrational amplitude only in the topmost atomic layers, and vibrational excitations of adsorbed molecules or atoms. In principle, a quantum mechanical approach is appropriate for the description of scattering on these systems with atomic dimensions normal to the surface. In a rough approximation, however, the elementary excitations connected with such a thin surface layer are often described in a continuum model in terms of a surface dielectric function $\epsilon_s(\omega)$. This dielectric function is used to model the dielectric response of the topmost atomic layers (thickness d). The un-

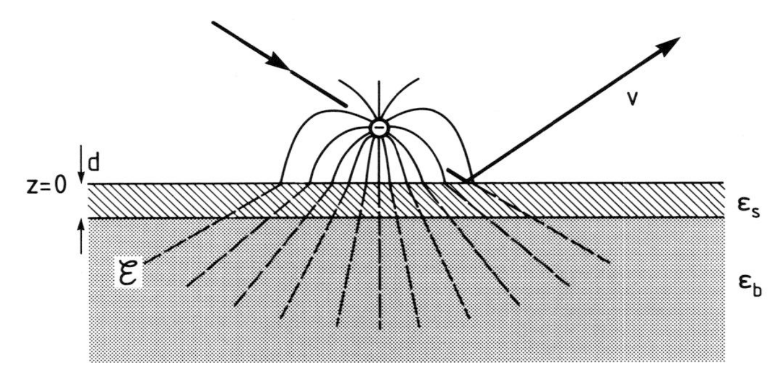

Fig.4.20. Schematic plot of the inelastic dielectric scattering process of a low-energy electron (velocity **v**) on a semi-infinite halfspace (dielectric function ϵ_b for the bulk) covered with a thin surface film (dielectric function ϵ_s). The interior electric field $\mathcal{E}_i$ of the electron outside the solid is assumed to be unchanged by the thin overlayer

derlying bulk material is described by a bulk dielectric function $\epsilon_b(\omega)$ (Fig. 4.20). As an example one might consider a thin coherent metal film with a thickness of several monolayers on a semiconductor. In this case $\epsilon_b(\omega)$ is the dielectric function of the semiconductor and $\epsilon_s(\omega)$, the dielectric function of the metal, contains as the essential part the Drude dielectric response of a free electron gas [4.16].

If we consider a spectral range of loss energies $\hbar\omega$, where substrate excitations can be neglected, the reflected electron can only transfer energy $\hbar\omega$ and wave vector $\mathbf{q}_{\|}$ to excitations within the surface layer. In the simplest approximation we assume the thickness of the surface layer to be so small ($q_{\|}d \ll 1$) that the field inside the bulk is not significantly perturbed by the surface layer. As in Fig.4.17 the field $\mathcal{E}_b$ and $\mathbf{D}_b$ inside the bulk have axial symmetry around the normal through the primary electron (Fig. 4.20). At the bulk/surface-layer interface the correct boundary conditions must be fulfilled for the components parallel and normal to the surface:

$$\mathcal{E}_b^{\|} = \mathcal{E}_s^{\|}\,, \quad \mathbf{D}_b^{\perp} = \mathbf{D}_s^{\perp}\,. \tag{4.88}$$

Since energy is only transferred to the spatial region of the overlayer, only the "surface fields" $\mathcal{E}_s(\mathbf{r},t)$ and $\mathbf{D}_s(\mathbf{r},t)$ within this layer have to be taken into account in determining the total energy transfer:

$$W = \mathrm{Re}\left\{\int_{-\infty}^{+\infty} dt \int_{z=-d}^{0} d\mathbf{r}\,\mathcal{E}_s(\mathbf{r},t)\,\dot{\mathbf{D}}_s(\mathbf{r},t)\right\}. \tag{4.89}$$

Expansion of the fields in terms of surface waves (4.65), and a calculation as in (4.68) yield

$$W = 2\pi \mathrm{Re}\left\{ \int_{-d}^{z=0} dz \int d\omega d\omega' \mathbf{dq}_{||} \mathbf{dq}'_{||} (-i\omega') \hat{\mathcal{E}}_s(\omega, \mathbf{q}_{||}) \hat{\mathbf{D}}_s(\omega', \mathbf{q}'_{||}) \right.$$

$$\left. \times \delta(\omega+\omega') \delta(\mathbf{q}_{||} + \mathbf{q}_{||}') \exp[-(q_{||} + q_{||}')z] \right\}$$

$$= 2\pi d \mathrm{Re}\left\{ \int d\omega \mathbf{dq}_{||} i\omega \hat{\mathcal{E}}_s(\omega, \mathbf{q}_{||}) \hat{\mathbf{D}}_s^*(\omega, \mathbf{q}_{||}) \right\} . \tag{4.90}$$

In order to fullfill the conditions (4.88) we decompose the fields into components normal and parallel to the surface, i.e.,

$$\hat{\mathcal{E}}_s = (\hat{\mathcal{E}}_s^{||}, \hat{\mathcal{E}}_s^{\perp}) , \quad \hat{\mathbf{D}}_s = \epsilon_s \epsilon_0 (\hat{\mathcal{E}}_s^{||}, \hat{\mathcal{E}}_s^{\perp}) . \tag{4.91}$$

Equation (4.90) is then evaluated for each field component separately:

$$W = 2\pi d \mathrm{Re}\left\{ \int d\omega \mathbf{dq}_{||} i\omega \left[\hat{\mathcal{E}}_s^{||} \hat{\mathcal{E}}_s^{||*} \epsilon_s^* \epsilon_0 + \frac{1}{\epsilon_0 \epsilon_s} \hat{\mathbf{D}}_s^{\perp} \hat{\mathbf{D}}_s^{\perp *} \right] \right\} . \tag{4.92}$$

Using the boundary conditions (4.88) one can relate this to the bulk fields

$$W = 2\pi d \mathrm{Re}\left\{ \int d\omega \mathbf{dq}_{||} \omega \left[\epsilon_0 \mathrm{Im}\{\epsilon_s |\hat{\mathcal{E}}_b^{||}|^2\} + \frac{1}{\epsilon_0} \mathrm{Im}\{-\epsilon_s^{-1} |\hat{\mathbf{D}}_b^{\perp}|^2\} \right] \right\} \tag{4.93}$$

and finally insertion of the shielding factor (4.62) for a semi-infinite half-space yields the energy transfer with respect to the "external" Coulomb field of the primary electron:

$$W = \frac{2\pi d}{\epsilon_o} \mathrm{Re}\left\{ \int d\omega d\mathbf{q}_{\|} \omega \left(\frac{4}{|\epsilon_b + 1|^2} \mathrm{Im}\{\epsilon_s |\hat{\mathbf{D}}^{\|}|^2\} \right.\right.$$

$$\left.\left. + \frac{4|\epsilon_b|^2}{|\epsilon_b + 1|^2} \mathrm{Im}\{-\epsilon_s^{-1} |\hat{\mathbf{D}}^{\perp}|^2\} \right) \right\} . \tag{4.94}$$

Using the "Fourier transforms" (4.71-73) one obtains

$$|\hat{\mathbf{D}}^{\|}|^2 = |\hat{\mathbf{D}}^{\perp}|^2 = \frac{e^2}{(2\pi)^4} \frac{q_{\|}^2 v_{\perp}^2}{[(q_{\|} v_{\perp})^2 + (\mathbf{q}_{\|} \cdot \mathbf{v}_{\|} - \omega)^2]^2} . \tag{4.95}$$

With the definition (4.74) of the probability P $(\hbar\omega, \mathbf{q}_{\|})$ for scattering from an excitation $(\hbar\omega, \mathbf{q}_{\|})$ it follows that

$$P(\hbar\omega, \mathbf{q}_{\|}) = \frac{e^2 d}{2\pi^3 \epsilon_0 \hbar^2} \frac{q_{\|}^2 v_{\perp}^2}{[(q_{\|} v_{\perp})^2 + (\mathbf{q}_{\|} \cdot \mathbf{v}_{\|} - \omega)^2]^2}$$

$$\times \left\{ \frac{1}{|\epsilon_b + 1|^2} \mathrm{Im}\{\epsilon_s\} + \frac{|\epsilon_b|^2}{|\epsilon_b + 1|^2} \mathrm{Im}\{-1/\epsilon_s\} \right\} . \tag{4.96}$$

Referring this probability to the scattering into an element of solid angle, a result similar to (4.83) is obtained. As is expected qualitatively, the inelastic scattering cross section (4.96) for low-energy electrons in a thin dielectric slab on top of the bulk substrate is proportional to the thickness d of this slab. An important result is obtained by comparing the two additive terms in (4.49 and 96), respectively: the second term related to electric field components $\hat{\mathbf{D}}^{\perp}$ normal to the surface is a factor $|\epsilon_b|^2$ larger in magnitude than the first term. For metal substrates and even on semiconductor surfaces (Si, Ge, GaAs, etc.) where $|\epsilon_b|^2$ exceeds 100, the second term originating from field components normal to the surface dominates the loss spectrum. The main structure in the loss spectrum is therefore determined by $\mathrm{Im}\{-1/\epsilon_s\}$, the bulk loss function of the thin surface layer, and because of the field direction normal to the surface $(\hat{\mathbf{D}}^{\perp})$, only those dipoles within this layer that are oriented normal to the interface contribute to the spectrum. This is the so-called **orientation selection rule**. It states that in dielectric theory the excitations (vibrations of adsorbed molecules, electronic surface state transitions, etc.) that give rise to significant loss structures are those whose dynamic dipole moment is oriented normal to the surface. A qualitative

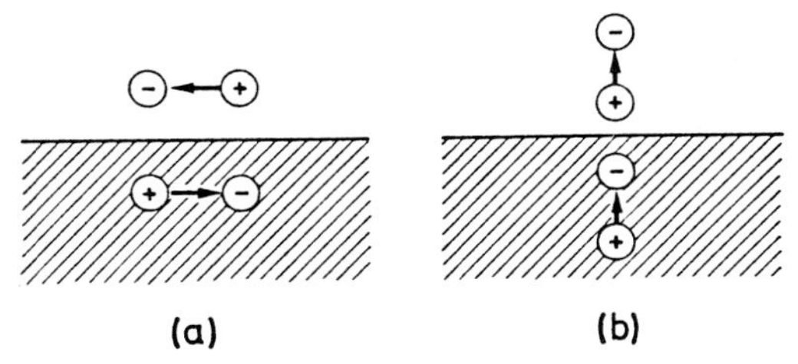

Fig.4.21a,b. Qualitative explanation of the orientation selection rule for dipole surface scattering: the image dipole within the substrate (shaded) partially compensates the effect of the adsorbed dipole for parallel orientation (**a**) but enhances the effect of dipoles with a normal orientation (**b**)

argument for this selection rule can be derived from Fig.4.21. Scattering by dynamic dipole moments on top of a metal or semiconductor substrate is considered. In addition to the dipole moments within the overlayer, image dipoles in the substrate itself contribute to the total scattering cross section. For a dipole orientation parallel to the surface (a), the image dipole has the reverse orientation and therefore partially compensates the effect of the overlayer dipole. For a dipole orientation normal to the surface the image dipole has the same direction (b) and enhances the overlayer effect. It should be emphasized that this orientation selection rule has been derived, and is only valid, within the framework of dielectric theory.

The present derivation of the scattering cross section (4.96) is an approximation valid in the limit $q_{\|}d \ll 1$. Equation (4.96) is valid for layer thicknesses d which are small in comparison to the inverse wave-vector transfer $q_{\|}$. The overlayer must not significantly perturb the Coulomb field inside the bulk substrate. A more general formula without any limitation on d has been derived by *Ibach* and *Mills* [4.10] by taking into account the correct boundary conditions both for the vacuum and for the overlayer/substrate interface. According to this derivation, the scattering cross section for surface scattering in an overlayer of thickness d is given by the formulae for the semi-infinite halfspace, (4.75, 83). But the surface loss function (4.84) has to be modified by an effective dielectric function $\tilde{\epsilon}(\omega)$, which contains both the dielectric function $\epsilon_b(\omega)$ of the bulk substrate and of the overlayer $\epsilon_s(\omega)$. The surface loss function, which has to be inserted into (4.75) or (4.83) for the continuous overlayer model, is

$$\mathrm{Im}\left\{\frac{-1}{\tilde{\epsilon}(\mathbf{q}_{\|},\omega)+1}\right\}, \tag{4.97}$$

with an effective dielectric function

$$\tilde{\epsilon}(\mathbf{q}_{\|},\omega) = \epsilon_s(\omega)\,\frac{1 + \Delta(\omega)\exp(-2q_{\|}d)}{1 - \Delta(\omega)\exp(-2q_{\|}d)}\,, \tag{4.98a}$$

where

$$\Delta(\omega) = \frac{\epsilon_b(\omega) - \epsilon_s(\omega)}{\epsilon_b(\omega) + \epsilon_s(\omega)}\,. \tag{4.98b}$$

Although both ϵ_b and ϵ_s are assumed to be independent of $q_{\|}$, the effective dielectric function $\tilde{\epsilon}(\mathbf{q}_{\|},\omega)$ becomes a function of the wave-vector transfer $q_{\|}$; this is a natural consequence of the fact that the "information depth" of the scattering experiment is dependent on $q_{\|}$ ($\propto 1/q_{\|}$) and therefore the relation between $1/q_{\|}$ and the layer thickness d determines the relative contributions of ϵ_b and ϵ_s to the total scattering cross section. For the limiting case $d \to 0$, (4.98) approaches the dielectric function ϵ_b of the semi-infinite halfspace without overlayer. For $d \to \infty$ the scattering cross section for a semi-infinite halfspace with dielectric function ϵ_s is obtained.

A further extension of this formalism to dielectric scattering from a multilayer structure on top of a semi-infinite solid has been given by *Lambin* et al. [4.17]. The model system consists of n layers (numbered by i) each described by a complex dielectric function $\epsilon_i(\omega)$ and thickness d_i. The loss function (4.97) is then calculated by using an effective dielectric function

$$\tilde{\epsilon}(\mathbf{q},\omega) = a_1(\omega) - \cfrac{b_1^2(\omega)}{a_1(\omega) + a_2(\omega) - \cfrac{b_2^2(\omega)}{a_2(\omega) + a_3(\omega) - \cfrac{b_3^2(\omega)}{a_3(\omega) + a_4(\omega) - \dots}}} \tag{4.98c}$$

instead of (4.98a).

The coefficients $a_i(\omega)$ and $b_i(\omega)$ introduce a wave-vector dependence through the relations

$$a_i(\omega) = \epsilon_i(\omega)/\tanh(q_{\|}d_i)\,, \quad b_i(\omega) = \epsilon_i(\omega)/\sinh(q_{\|}d_i)\,. \tag{4.98d}$$

For a fixed number of layers n on top of a semi-infinite substrate, the series (4.98c) is terminated by the condition $b_{n+1} = 0$. This multilayer formalism can also be used to approximate any spatially varying dielectric property $[\epsilon(\omega, z)]$ near the surface.

4.8 Some Experimental Examples of Inelastic Scattering of Low-Energy Electrons at Surfaces

In a surface scattering experiment with low-energy electrons using the reflection geometry, it is nevertheless possible to detect bulk excitations. This is because the surface loss function (4.84) for a semi-infinite halfspace contains the imaginary part $\epsilon_2(\omega)$ of the bulk dielectric function in the numerator. An experimental example is illustrated in Fig.4.22. The electron energy loss spectrum was recorded by means of a high-resolution spectrometer (Panels II, IX) with a primary energy of 21 eV on a clean InSb(110) surface, which had been prepared by cleavage in UHV. The broad loss structure with a threshold slightly below 200 meV is due to bulk interband transitions across the direct gap (E_g = 180 meV at 300 K) of InSb. The loss spectrum reflects the frequency dependence of the optical absorption constant of InSb [4.18].

Figure 4.23 displays a series of double-differentiated electron energy loss spectra, which have been measured by means of a Cylindrical Mirror Analyzer (CMA) on clean cleaved InSb(110) (*a*) after deposition of Sn on InSb(100) (*b*) and InSb(110) (*c* and *d*), and on a clean polycrystalline β-Sn foil (*e*) [4.20]. In these double-differentiated curves the negative second derivative of the loss spectrum is recorded, such that positive peak maxima correspond to peak maxima in the undifferentiated spectrum and thus indicate loss peak positions.

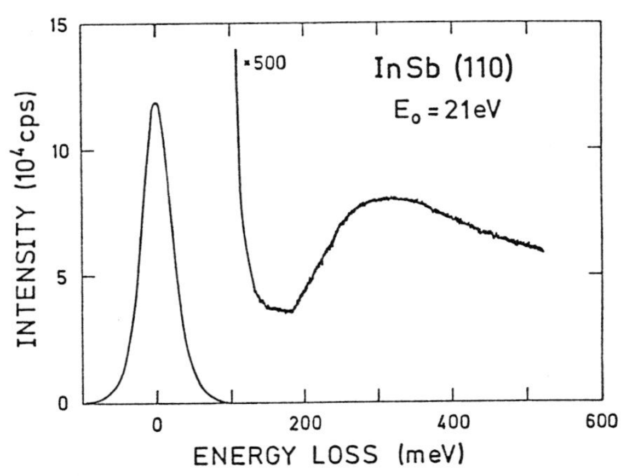

Fig.4.22. Low-energy electron loss spectrum measured for specular reflection with a primary energy E_0 of 21 eV on a clean cleaved InSb(110) surface. Because of intensity requirements, the energy resolution is not sufficient to resolve surface phonon excitations. The loss feature with an onset near 180 meV is due to electron-hole pair excitations across the forbidden gap [4.18]

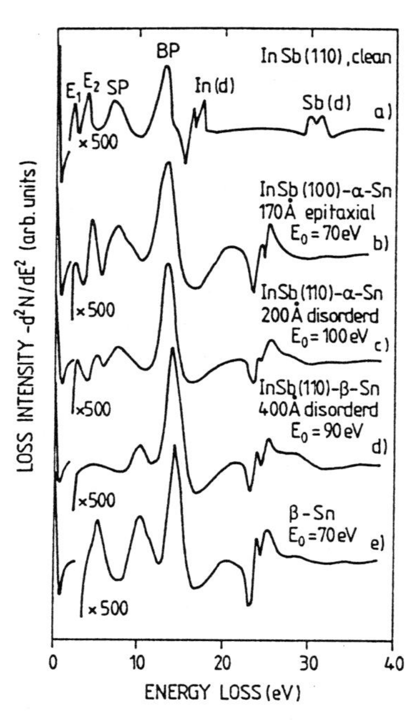

Fig. 4.23. Double differentiated electron energy loss spectra measured by means of a CMA (energy resolution ≈0.8eV, primary energy E_0) on various InSb surfaces, clean and with Sn overlayers prepared in UHV. (*a*) loss spectrum of the clean cleaved InSb(110) surface; apart from the bulk (BP) and surface plasmon (SP) losses, two losses due to bulk electronic transitions (E_1, E_2) and the d core-level transitions (spin-orbit split) of In and Sb are seen. (*b*) Loss spectrum of an InSb(100) surface cleaned by ion bombardment and annealing, after deposition of a 170 Å thick Sn overlayer; deposition at 300 K leads to epitaxial growth of α-Sn. (*c*) Loss spectrum of UHV cleaved InSb(110) surface after deposition of a 200-Å thick Sn overlayer; deposition at 300 K leads to a disordered α-Sn film. (*d*) Loss spectrum of a UHV cleaved InSb(110) surface after deposition of a 400-Å thick Sn overlayer; deposition at 300 K causes the thick Sn overlayer to be in its metallic β-modification, at least in the topmost region. (*e*) Loss spectrum of a polycrystalline metallic β-Sn sample, cleaned by ion bombardment and annealing [4.19]

The relatively high primary energies of between 70 and 100 eV give rise to both bulk and surface scattering, even though the scattering geometry is that of a reflection experiment. However, the exact relative contributions of surface and bulk scattering are not known, i.e. it is not clear to what extent the peak positions are determined by the surface (4.84) or bulk (4.60) loss function. This uncertainty is usually of minor importance since the energy resolution in these spectra is not better than 800 meV because of lack of monochromaticity of the primary beam. The prominent peaks near 14 eV in all spectra are due to the excitation of the bulk plasmon, i.e. they arise from a singularity in $\mathrm{Im}\{-1/\epsilon\}$. Because of the similarity in the atomic numbers of InSb and Sn, the bulk plasmon energies are almost identical, reflecting the very similar valence electron densities. On InSb(110) two double peaks occur near 17 and 30 eV. Comparison with core-level photoemission data allows one to interpret these structures in terms of excitations from the In and Sb d-core levels into the conduction band of the semiconductor. The doublet structure is due to spin-orbit splitting of the core levels. On the Sn-covered surfaces these structures are not observed due to shielding by the overlayer. Instead, the spin-orbit split Sn d-level excitation appears near 24 eV. The peaks near 7 eV in curves *a-c* and near 10 eV in *d* and *e* are due to the surface plasmon excitations on InSb(110), on the Sn overlayers and on polycrystalline bulk β-Sn, respectively. The two structures E_1 and E_2 below 5 eV loss energy must be interpreted in terms of bulk interband transitions, which are known from optical data for InSb and tetrahedrally bonded (semiconducting) α-Sn. The spectra in Fig.4.23 are good examples of the application of EELS (Panel IX: Chap. 4) as a fingerprinting technique for identifying the nature of a thin overlayer. Comparison of curves *d* and *e* shows that thick layers of Sn deposited on InSb(110) at 300 K are in the metallic β-modification. The presence of the two interband transitions E_1 and E_2 is indicative of a tetrahedrally bonded semiconductor. Thus in curves *b* and *c* the Sn overlayer consists essentially of tetrahedrally bonded Sn atoms. On InSb(100) (curve *b*) additional LEED investigations reveal a crystalline epitaxial Sn layer, i.e., the α-modification of Sn, usually unstable at room temperature, could be epitaxially grown on the InSb(100) substrates. On cleaved InSb(110) the Sn overlayer did not give rise to a LEED diffraction pattern and therefore the Sn must be amorphous or polycrystalline.

Figure 4.24 exhibits electron energy loss spectra that have been measured by means of a hemispherical electron energy analyzer without monochromatization of the primary electron beam on a clean Cu(110) surface and after exposure (at 90 K) to nitrous oxide N_2O [4.20]. The energy resolution is $\approx$300 meV, which is sufficient to study relatively broad electronic excitations in the energy range of several electron volts. The two loss features on the clean surface near 4 and 7 eV arise from singularities of the bulk and/or surface loss functions $\mathrm{Im}\{-1/\epsilon\}$ and $\mathrm{Im}\{-1/(\epsilon+1)\}$. They are due to bulk and surface plasmon excitations of the free electron gas coupled to d-band transitions.

After adsorption of N_2O the substrate losses appear to have changed their intensity and position, and three new loss features characteristic of the adsorbed molecules emerge near 9.6, 11 and 14.5 eV. As is seen from the bars above the spectrum, these losses correspond exactly to three electronic excitations of gaseous N_2O. The low-energy loss at 8.5 eV for the gas phase is not seen in the adsorbate spectrum. Nevertheless, the loss pattern

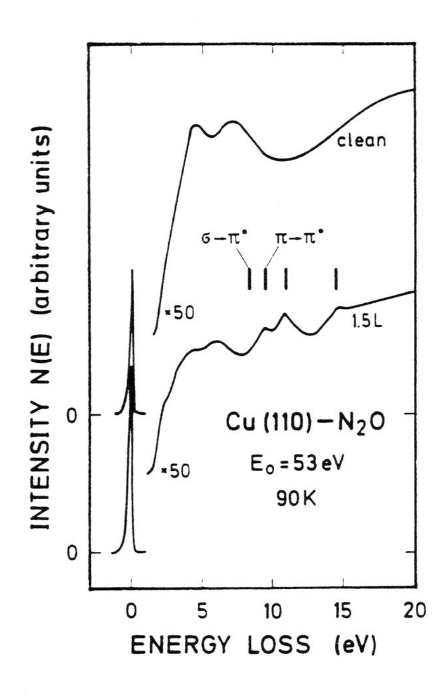

Fig.4.24. Electron energy loss spectra measured by a hemispherical energy analyser with an "unmonochromatised" primary electron beam (E_0 = 53eV, energy width $\approx$ 0.3eV) on a clean Cu(110) surface and after exposure to 1.5 L of N_2O at 90 K. The electronic transitions of gaseous N_2O are marked for comparison by bars [4.20]

of the adsorbed N_2O strongly suggests that N_2O is adsorbed on Cu(110) at 90 K as an undissociated molecule. Since the present case obviously corresponds to surface scattering on a thin adsorbate overlayer, the orientation selection rule for dipole scattering (Sect.4.7) can be applied. Accordingly, the absence of the 8.5 eV loss in the adsorbate spectrum is attributed to the particular orientation of the corresponding dynamic electronic dipole moment with respect to the surface. N_2O is a linear molecule with an atomic structure N≡N=0; the excitations at 8.5 and 9.6 eV are interpreted on the basis of molecular orbital calculations in terms of $\sigma \rightarrow \pi^*$ and $\pi \rightarrow \pi^*$ transitions. The $\sigma \rightarrow \pi^*$ transition has a dipole moment perpendicular to the molecular axis, whereas the dipole moment of the $\pi \rightarrow \pi^*$ transition is oriented parallel to the axis as is easily seen from a qualitative consideration of the dipole transition matrix element (z is the molecular axis)

$$e \int \psi^*_{\pi^*} z \psi_\pi \, d\mathbf{r} \; . \tag{4.99}$$

Since the $\pi \rightarrow \pi^*$ transition is clearly seen in the spectrum, an orientation of the adsorbed molecules with their axis parallel to the surface can be ruled out. On the other hand, the lack of the $\sigma \rightarrow \pi^*$ loss with a dipole moment normal to the molecular axis is consistent with an orientation of the N_2O molecules normal to the substrate surface.

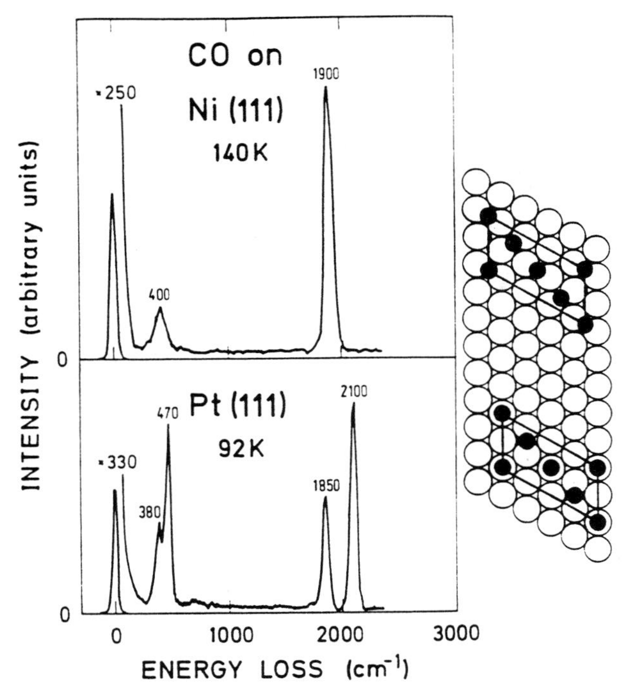

Fig.4.25. Electron energy loss spectra (HREELS) of the Ni(111) and Pt(111) surfaces, each covered with half a monolayer of CO which orders into a c(4 × 2) overlayer. On the Ni surface the vibration spectrum indicates only a single CO species in a site of high symmetry. The only possibility for positioning the two-dimensional CO lattice on the surface consistent with the single type of adsorption site is to place all CO molecules into two-fold bridges. By similar reasoning, half the CO molecules must occupy on-top sites on the Pt(111) surface. The qualitative structure analysis (depicted in the inset) is thus obtained by combining LEED and HREELS results [4.21]

This example of inelastic electron scattering from a molecular adsorbate overlayer shows how the orientation selection rule for dipoles can yield useful information about the orientation of adsorbed molecules. This selection rule is equally interesting when High-Resolution Electron Energy Loss Spectroscopy (HREELS) is used to study vibrations of adsorbed molecules [4.10]. Figure 4.25 shows the example of HREELS spectra for Ni(111) and Pt(111) surfaces [4.21], each covered with half a monolayer of CO, which orders into a c(4 × 2) overlayer (as deduced from LEED). On the Ni surface the vibrational spectrum shows only two characteristic vibrational bands. The band at a wave number of 1900 cm^{-1} is due to the C-O stretch vibration, since the corresponding gas phase excitation has a wave number of 2140 cm^{-1}. The low-energy structure at 400 cm^{-1} must be interpreted in terms of a vibration of the whole molecule against the substrate surface.

These substrate-molecule vibrations usually have energies below 1000 cm^{-1} or 100 meV. On Pt(111) the double structures with bands at 2100, 1850 cm^{-1} and 470, 380 cm^{-1} clearly indicate two types of adsorbed CO molecules, in contrast to the Ni case. Vibration frequencies due to C-O bending modes are not observed on either Ni or Pt. Applying the orientation selection rule one can infer that the CO has its molecular axis normal to the surface. In this orientation the stretching modes have a dynamic dipole moment normal to the surface, but for the bending modes the moment is parallel to the surface so that this vibration should not be observed if dipole scattering prevails, i.e. dielectric theory for a thin overlayer (Sect.4.7).

A large number of adsorption systems on clean metal surfaces have now been studied by means of HREELS. Using the orientation selection rule within the framework of dielectric theory this technique has proven to be extremely powerful in determining the chemical nature, the orientation and sometimes the adsorption site of relatively complex adsorbed organic molecules on transition metals. It should be emphasized, however, that there are also adsorption systems, e.g. atomic hydrogen on W [4.22], where impact scattering rather than dipole scattering from adsorbate vibrations has been observed. In these cases, of course, the dipole selection rule cannot be applied.

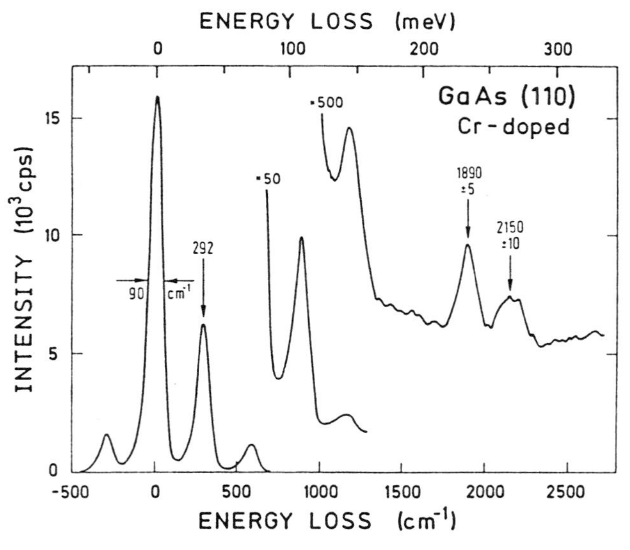

Fig.4.26. Electron energy loss spectrum (HREELS) of a cleaved GaAs(110) surface covered with a saturated atomic hydrogen adsorbate layer. Beside multiple (energetically equidistant) loss and (one) gain peaks due to Fuchs-Kliewer surface phonons (polaritons), two characteristic adsorbate losses are observed arising from Ga-H and As-H vibrations at 1890 and 2150 cm^{-1}, respectively [4.23]

HREELS on adsorbate vibrations can also give interesting information about adsorption sites on compound semiconductor surfaces. Figure 4.26 illustrates an example of a spectrum measured for a cleaved GaAs(110) surface which was exposed to atomic hydrogen (H) up to saturation [4.23]. The energetically equidistant loss (and one gain) peaks below 1500 cm^{-1} are due to multiple excitations of the Fuchs-Kliewer surface phonon (polariton) of the GaAs substrate. This kind of loss process is always observed on surfaces of InfraRed (IR) active semiconductors (Sect.5.5). The scattering process for these excitations must be described as dielectric scattering from a semi-infinite halfspace. On the other hand, the two losses at 1890 and 2150 cm^{-1} do not occur on the clean cleaved GaAs(110) surface. They are therefore due to scatttering within the thin H adsorbate overlayer. Comparison with IR data (AsH_3 and Ga-H complexes) shows that the 1890 cm^{-1} loss arises from a Ga-H, and the 2150 cm^{-1} peak from a As-H stretch vibration. Atomic H therefore adsorbs both on Ga and on As surface atoms. On IR-active materials with strong phonon losses similar to those seen in Fig.4.26, combined losses of these surface phonons and of adsorbate vibrations have also been observed [4.24]. One must therefore be careful when interpreting vibrational loss spectra of adsorbates on IR-active semiconductors.

4.9 The Classical Limit of Particle Scattering

So far, the scattering of particles, in particular of electrons, at surfaces has been described in terms of wave propagation. This picture takes into account the correct wave nature of matter, as expressed by de Broglie's relation. The underlying reason why particle scattering must be treated in this manner is that an electron of primary energy of about 150 eV has a de Broglie wavelength of 1 Å This is just the order of magnitude of interatomic distances. If the scattering potentials vary over distances comparable with the wavelength of the scattered particles, the complete wave mechanical formalism has to be applied, as was done above.

Let us now consider the extreme case of high energy atom or ion scattering which is also frequently used in the investigation of surfaces, thin overlayers and material analysis in general [4.25]. According to de Broglie's relation ($v = h/m\lambda$, $E = \frac{1}{2}mv^2$), He atoms with a kinetic energy of 2 MeV have a wavelength of 10^{-4} Å For these particles the scattering potential at a solid surface varies on a scale which is large compared to their wavelength.

Let us now compare the wave-mechanical and the classical (Newtonian) treatment of the motion of such a high-energy particle (Fig.4.27 a,b).

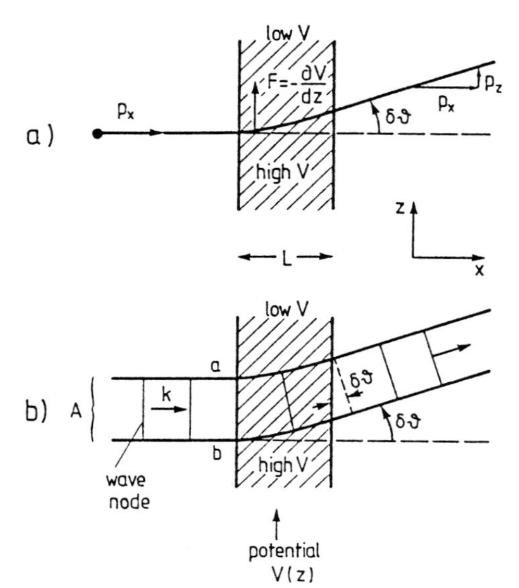

Fig.4.27. Schematic representation of (**a**) the trajectory of a classical particle (momentum p_x) traversing a region (shadowed) with changing potential V(z). The force acting on the particle is $F = -\partial V/\partial z$. (**b**) A quantum mechanical plane wave (wave vector k) traversing a region with changing potential V(z). When the potential V(z) (and also $\partial V/\partial z$) varies slowly in comparison with the wavelength $\lambda = h/p$ of the particle, both descriptions yield the same angular deflection $\delta\theta$

In wave mechanics, the wave function of a particle (mass m) moving in a potential V(z) extended over a small region L is

$$\psi \approx \exp\left[\frac{-i}{\hbar}\left(\frac{p^2}{2m} + V\right)t + \frac{i}{\hbar}\mathbf{p}\cdot\mathbf{x}\right] . \tag{4.100}$$

In the regions where V vanishes (4.100) becomes the plane wave of a freely moving particle. In the region L, where $V \neq 0$ the lines of constant phase, and thus also the node lines are given by

$$p^2/2m + V = \text{const} . \tag{4.101}$$

To find the change in angle of a wave node line after crossing the region L (Fig.4.27b) we have to consider paths (a) and (b). Corresponding to the difference $\Delta V = (\partial V/\partial z)A$ in potential there is a difference Δp in momentum according to (4.101) of

$$\Delta(p^2/2m) = p\Delta p/m = -\Delta V . \tag{4.102}$$

The wave number $p/\hbar$ is therefore different along the two paths, i.e. the phase is advancing at a different rate. The amount by which the phase φ on path (b) is "ahead" of that on path a follows as

$$\Delta\varphi = L\Delta k = \frac{L\Delta p}{\hbar} = -\frac{m}{p\hbar}L\Delta V . \tag{4.103}$$

A phase advance $\Delta\varphi$ corresponds to wave nodes advanced by a distance

$$\Delta x = \frac{\lambda}{2\pi}\Delta\varphi = \frac{\hbar}{p}\Delta\varphi \,, \tag{4.104}$$

or to an angle between incoming and outgoing wave of

$$\delta\theta = \frac{\Delta x}{A} = -\frac{m}{p^2}\frac{L}{A}\Delta V \,. \tag{4.105}$$

This derivation is valid under the condition that $V(z)$ and $\partial V/\partial z$ vary slowly in comparison with the wavelength $\lambda = h/p$ of the particle. Classically one can calculate the angular deflection $\delta\theta$ produced by the potential region L as follows (with p as initial momentum):

$$\delta\theta = \frac{p_z}{p} = \frac{FL}{pv} \tag{4.106}$$

since L/v is the time during which the force $F = -\partial V/\partial z$ acts on the particle. One thus gets

$$\delta\theta = -\frac{L}{pv}\frac{\partial V}{\partial z} \approx -\frac{m}{p^2}L\frac{\Delta V}{A} \,. \tag{4.107}$$

For the case of a potential that varies slowly in comparison to the wavelength of the particle, classical dynamics thus gives the results of wave mechanics. Interference phenomena characteristic of wave propagation can be neglected. For photons this corresponds to the limit of geometrical optics. The scattering of high-energy atoms and ions on surfaces can therefore be treated according to the kinematics of classical particle collisions. A local picture emerges for the scattering process; this is in contrast to the wave-mechanical treatment in Sect.4.1. In the kinematic theory of wave scattering a local interaction with a single surface atom is not sufficient; the whole neighbourhood, i.e. the 2D translational symmetry of the surface enters the description of the scattering process. On the other hand, in the classical limit, the interaction of the scattered particle with the surface reduces essentially to a two-body interaction of the particle with a surface target atom. This is even true for cascade processes, which may be built up from a sequence of separate two-body collisions.

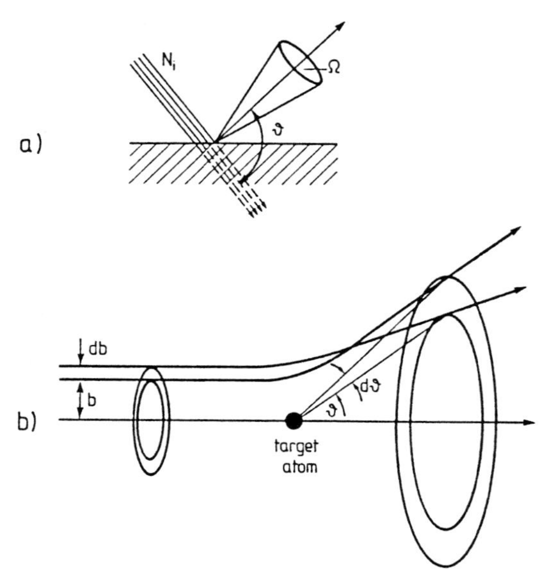

Fig. 4.28a,b. The scattering of classical particles: (**a**) Surface scattering of a primary beam of N_i particles. Some of the particles are scattered into a solid angle Ω around a direction θ counted from the direction of incidence. (**b**) Scattering of a classical particle by a heavy target particle (atom) at rest. The motion of the incoming projectile is described in terms of its impact parameter b. The scattering angle θ depends on the impact parameter b. The deflection function $\theta(b)$, which determines the scattering cross section, depends on details of the force acting between target and projectile. Projectiles arriving with impact parameters between b and b+db are scattered into directions between θ and $\theta+d\theta$

The essential measured quantity is again the differential cross section $dS/d\Omega$. For a beam of particles incident on a number N_s of surface target atoms this is defined as

$$\frac{\text{Number of particles scattered into } d\Omega}{\text{Total number of incident particles}} = \frac{dS(\theta)}{d\Omega} d\Omega\, N_s \; . \tag{4.108}$$

The average differential cross section $S(\theta)$ for scattering into a solid angle Ω in the direction θ is given by (Fig. 4.28a)

$$S(\theta) = \frac{1}{\Omega} \int_{\Omega} \frac{dS}{d\Omega} d\Omega \; . \tag{4.109}$$

For the geometry in Fig.4.28a the total number of detected particles in the solid angle Ω, i.e. the yield Y is

$$Y = S(\theta)\Omega N_i N_s \tag{4.110}$$

where N_i is the number of incident particles (time integral over incident current)

The scattering cross section can be calculated from the force that acts during the collision between the projectile and the target atom (Fig.4.28b). It is useful to introduce the so-called impact parameter b as the perpendicular distance between the incident particle trajectory and the parallel line through the target atom. Particles with impact parameters between b and b+db are scattered into angles θ and $\theta+d\theta$. For central forces there must be rotational symmetry around the beam axis. One therefore gets

$$2\pi b db = -\frac{dS}{d\Omega} 2\pi \sin\theta d\theta , \tag{4.111a}$$

or

$$\frac{dS}{d\Omega} = -\frac{b}{\sin\theta} \frac{db}{d\theta} \tag{4.111b}$$

which relate the scattering cross section to the impact parameter. The minus sign indicates that an increase in the impact parameter results in less force on the particle, i.e. in a decrease in the scattering angle. A convenient way to calculate scattering cross sections is to determine from (4.111b) the so-called **deflection function** θ(b) using the interaction force between projectile and target.

4.10 Conservation Laws for Atomic Collisions: Chemical Surface Analysis

In a collision of two classical particles, energy and momentum must be conserved. These conservation laws alone allow important conclusions concerning energy transfer from projectile to target atom without going into details of the interatomic interaction. For this purpose we assume a collision between an incident particle (mass m_1, velocity $\mathbf{v}_1$, kinetic energy E_1) and a second particle (mass m_2) initially at rest. After the collision the particles have velocities and energes v_1', v_2' and E_1', E_2', respectively (Fig.4.29a).

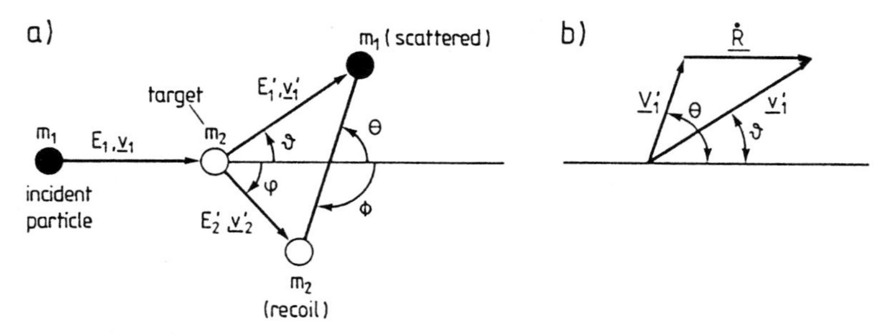

Fig.4.29a,b. Explanation of the mathematical symbols describing a classical two particle scattering event. (**a**) A particle with mass m_1 is scattered on a target with mass m_2 at rest, which receives a kinetic energy E_2' and a velocity $\mathbf{v}_2'$ after the collision. θ and φ are the scattering and recoil angles in the laboratory system respectively. Θ and Φ are scattering and recoil angles in the center of mass (CM) frame, respectively. (**b**) description of the same two particle collision in the center of mass frame: $\dot{\mathbf{R}}$ is the velocity of the center of mass (CM); $\mathbf{v}_1'$ and $\mathbf{V}_1'$ the velocities of the scattered projectile after the collision in the laboratory and in the CM frame, respectively; θ and Θ are the scattering angles in the laboratory and in the CM frame

With the scattering and recoil angles θ and φ energy and momentum conservation requires

$$\frac{1}{2}m_1 v_1^2 = \frac{1}{2}m_1 v_1'^2 + \frac{1}{2}m_2 v_2'^2 , \tag{4.112}$$

$$m_1 v_1 = m_1 v_1' \cos\theta + m_2 v_2' \cos\varphi , \tag{4.113a}$$

$$0 = m_1 v_1' \sin\theta + m_2 v_2' \sin\varphi . \tag{4.113b}$$

Equations (4.113a, b) describe momentum conservation along the directions parallel and normal to the direction of incidence ($\|\mathbf{v}_1$), respectively. After eliminating first φ and then v_2' one finds the ratio of the particle velocities of the projectile after and before collision:

$$\frac{v_1'}{v_1} = \pm \frac{(m_2^2 - m_1^2 \sin^2\theta)^{1/2} + m_1 \cos\theta}{m_1 + m_2} \tag{4.114}$$

The plus sign holds for $m_1 < m_2$. For this condition $m_1 < m_2$ the ratio of the projectile energies after and before the collision (kinematic factor) follows as

$$\frac{E_1'}{E_1} = \left[\frac{(m_2^2 - m_1^2 \sin\theta)^{1/2} + m_1 \cos\theta}{m_1 + m_2}\right]^2 . \tag{4.115}$$

The recoil energy, i.e. the energy transfer ΔE to the target (m_2) appears as an energy loss of the scattered projectile:

$$\Delta E = E_1' - E_1 = \frac{1}{2} m_2 v_2'^2 = 4 \frac{\mu}{M} E_1 \cos^2 \varphi \, , \tag{4.116}$$

where

$$M = m_1 + m_2 \, , \tag{4.117a}$$

and

$$\mu = m_1 m_2 / (m_1 + m_2) \tag{4.117b}$$

are the total and the reduced masses of the two particles, respectively. For the two simple cases of direct backscattering ($\theta = 0$) and right-angle scattering ($\theta = 90°$) (4.115) simplifies to

$$\frac{E_1'}{E_1} = \left(\frac{m_2 - m_1}{m_2 + m_1} \right)^2 \, , \tag{4.118}$$

and

$$\frac{E_1'}{E_1} = \frac{m_2 - m_1}{m_2 + m_1} \, , \tag{4.119}$$

respectively.

The dependence of the kinetic energy E_1' of the scattered particle on the mass of the target m_2 (4.115-119) clearly shows that scattering of ions from a surface can be used to identify atomic species near or at the surface. Low-energy ions ($1 \div 5$ keV) are produced by simple ion guns forming a focused beam which is directed onto the surface under study. The kinetic energy of the backscattered ions is analysed with a 127° deflector-type analyser (as used for electrons, Panel II: Chap. 1). For detection one might employ, for example, a channel electron multiplier. For high-energy scattering (50 keV $\div$ 5 MeV), accelerators are required as ion sources. Energy analysis and detection are performed with various electromagnetic energy filters and with nuclear particle detectors (e.g., solid-state Schottky-barrier-type devices), respectively.

The low-energy regime (Low-Energy Ion Scattering, LEIS) is experimentally quite accessible and offers the additional advantage of very high surface sensitivity. The ions scatter predominantly from the first atomic layer and thus a *first monolayer* analysis is possible. As an example, Fig. 4.30 shows an energy spectrum for ^{20}Ne ions scattered from the surface of

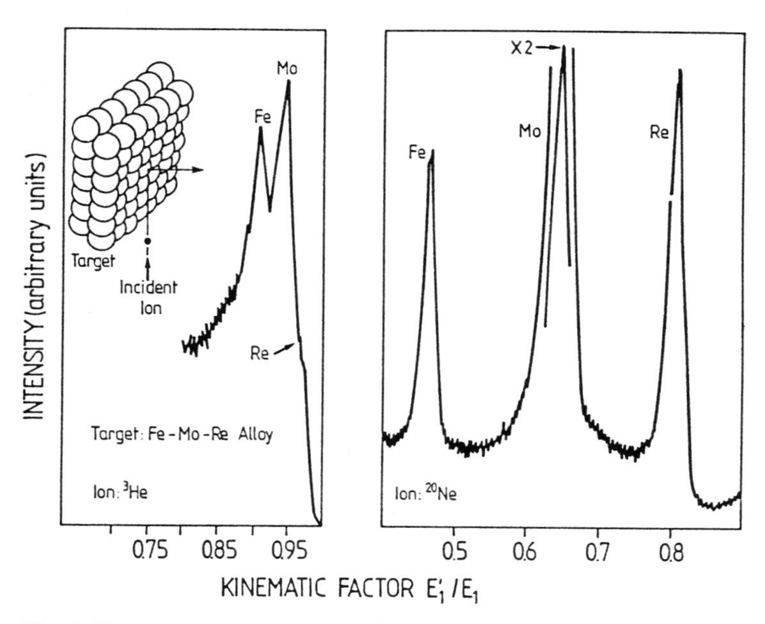

Fig.4.30. RBS energy spectra for ^{3}He scattering and ^{20}Ne scattering from the surface of an Fe-Mo-Re alloy. The energy of the incident particles was 1.5 keV. [4.26]

an Fe-Mo-Re alloy. The observed peaks at different kinematic factors E_1'/E_1 clearly reveal the presence of the different atomic species in the surface atomic layer [4.26].

A disadvantage of low-energy ion scattering is the high probability of the ions being neutralized at the surface and thus becoming invisible to any energy analyser using charge properties. At higher energies (above 50 keV), however, this neutralisation effect is less important, but ions with such high energies may penetrate hundreds and thousands of Ångstroms into the solid depending on their direction of incidence with respect to the lattice planes. Surface sensitivity can then be arranged only by utilising the concepts of *channeling*, *blocking*, and *shadowing* (Sect.4.11). These techniques rely on the fact that high-energy ions penetrating into a crystal are forced to move along *channels* between atomic rows and planes. The penetration depth and thus also the probing depth depend strongly on the geometrical factors of the single scattering event, in particular on the direction of incidence and the angular distribution of the scattering cross section $dS/d\Omega$, (4.111), for a single collision.

4.11 Rutherford BackScattering (RBS): *Channeling and Blocking*

In order to derive formulae for the scattering cross section, (4.111), for a two-body collision, details of the interatomic forces must be known. The most important case is that of Rutherford scattering, where a charged projectile (mass m_1, charge $Z_1 e$) is scattered in a Coulomb field

$$\mathcal{E} = \frac{Z_2 e}{4\pi\epsilon_0 r^2} = \frac{A}{r^2} \tag{4.120}$$

centered at a point C (Fig.4.31). The Coulomb field might be that of an ion (charge $Z_2 e$) assumed to be much heavier than the particle m_1 and thus approximately at rest. Scattering into an angle θ, i.e. a change of momentum $\Delta\mathbf{p}$ from $\mathbf{p}_1$ to $\mathbf{p}_1'$ means, according to Fig.4.31b,

$$\frac{1}{2}\frac{\Delta p}{m_1 v_1} = \sin(\theta/2)\,, \quad \Delta p = 2m_1 v_1 \sin(\theta/2)\,. \tag{4.121}$$

From Newtons law $d\mathbf{p} = \mathcal{E}dt$ we obtain the momentum change Δp in z direction (symmetry axis of scattering) according to Fig.4.30a as

$$\Delta p = \int dp_z = \int \mathcal{E}\cos\alpha dt = \int \mathcal{E}\cos\alpha \frac{dt}{d\alpha} d\alpha\,. \tag{4.122}$$

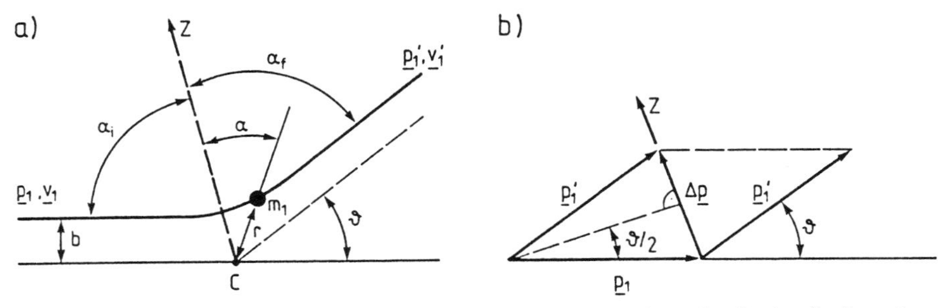

Fig.4.31. Explanation of the symbols used for the description of a Rutherford scattering event of a particle with mass m_1 from a scattering center C at rest; **(a)** trajectory of the projectile described by impact parameter b, initial momentum and velocity $\mathbf{p}_1$, $\mathbf{v}_1$ and scattering angle θ. **(b)** Relation between initial momentum $\mathbf{p}_1$, final momentum $\mathbf{p}_1'$ after collision and momentum change $\Delta\mathbf{p}$

The derivative $dt/d\alpha$ can be related to the angular momentum of the particle about the origin. Since the force is central (it acts along the line joining the particle to the origin C, Fig.4.31), there is no torque about the origin and the angular momentum is conserved. Its initial magnitude is $m_1 v_1 b$, and at a later time it can be expressed as $m_1 r^2 \, d\alpha/dt$. The conservation condition thus gives

$$m_1 r^2 \frac{d\alpha}{dt} = m_1 v_1 b \,, \quad \text{or} \quad \frac{dt}{d\alpha} = \frac{r_2}{v_1 b} \,. \tag{4.123}$$

With (4.122) we obtain the total momentum change during scattering

$$\Delta p = \frac{A}{r^2} \int \cos\alpha \frac{r^2}{v_1 b} d\alpha = \frac{A}{v_1 b} \int \cos\alpha \, d\alpha \,, \tag{4.124}$$

$$\Delta p = \frac{A}{v_1 b} (\sin\alpha_f - \sin\alpha_i) \,.$$

The initial and final angles α_i and α_f can be expressed in terms of the scattering angle θ by

$$\sin\alpha_f - \sin\alpha_i = 2\sin(90° - \theta/2) \,, \tag{4.125}$$

and the momentum change follows as

$$\Delta p = 2 m_1 v_1 \sin(\theta/2) = \frac{A}{v_1 b} 2\cos(\theta/2) \,. \tag{4.126}$$

The deflection function $\theta(b)$ is thus obtained as

$$b = \frac{A}{m_1 v_1{}^2} \mathrm{ctan}(\theta/2) = \frac{Z_1 Z_2 e^2}{8\pi\epsilon_0 E_1} \cos(\theta/2) \tag{4.127}$$

with E_1 being the kinetic energy of the incident ion. According to (4.111b) one integration yields the Rutherford scattering cross section

$$\frac{dS}{d\Omega} = - \frac{b}{\sin\theta} \frac{db}{d\theta} = \left(\frac{Z_1 Z_2 e^2}{8\pi\epsilon_0 E_1}\right)^2 \frac{1}{\sin^4(\theta/2)} \,. \tag{4.128}$$

It must be emphasized that in the above derivation the scattering center C is assumed to be at rest. Equation (4.128) can therefore be used only for collisions between a light particle m_1 and a much heavier target m_2 for which the recoil velocity v_2' can be neglected. As a numerical example we consider 180° scattering, i.e. backscattering of 2 MeV He ions ($Z_1 = 2$) from Ag atoms ($Z_2 = 47$). The distance of closest approach r_{min} of the projectile to the target may be estimated by equating the incident kinetic energy E_1 to the potential energy at r_{min}:

$$r_{min} = \frac{Z_1 Z_2 e^2}{4\pi\epsilon_0 E_1} . \tag{4.129}$$

For the present example (4.129) yields a minimum-approach distance of about $7 \cdot 10^{-4}$ Å, i.e. a distance much smaller than the Bohr radius of about 0.5 Å This result shows that, for high-energy scattering of light ions, the assumption of an unscreened Coulomb potential for the calculation of scattering cross sections is justified. The backscattering cross section follows from (4.128) as $dS(180°)/d\Omega \approx 3 \cdot 10^{-8}$ Å^2.

From Sect.4.10 it is evident that for general, two-body collisions, without the assumption of a heavy target at rest, the target will recoil from its initial position thus implying an energy loss of the projectile (4.116). An exact calculation of the Rutherford scattering cross section thus involves the treatment of a real two-body central-force problem. However, by introducing the concept of the reduced mass μ (4.117b) a reduction to a one-body problem is possible. For the two colliding particles in Fig.4.29 Newton's law is

$$m_1 \ddot{\mathbf{r}}_1 = \mathbf{F} \quad \text{and} \quad m_2 \ddot{\mathbf{r}}_2 = -\mathbf{F} . \tag{4.130}$$

Defining the Center of Mass (CM) coordinate $\mathbf{R}$ by

$$(m_1 + m_2)\mathbf{R} = m_1 \mathbf{r}_1 + m_2 \mathbf{r}_2 ,$$

$$\mathbf{R} = \frac{m_1}{M}\mathbf{r}_1 + \frac{m_2}{M}\mathbf{r}_2 \tag{4.131}$$

one obtains from (4.130) by subtraction and elimination of $\mathbf{r}_1$ and $\mathbf{r}_2$, and with the relative position vector $\mathbf{r} = \mathbf{r}_2 - \mathbf{r}_1$

$$M\ddot{\mathbf{R}} = \mathbf{0} , \quad \text{and} \quad \mu\ddot{\mathbf{r}} = \mathbf{F} . \tag{4.132}$$

The two-body problem can thus be described as a force-free motion of the CM (4.132a) and a force **F** acting between the two particles within the moving CM inertial system. Within this CM system the kinematics of the collision thus reduces to the above problem of scattering of a particle with mass μ (4.117b) on a spatially fixed center. Within the CM system the particle coordinates are

$$\mathbf{R}_1 = \mathbf{r}_1 - \mathbf{R} = \frac{m_2}{M}\mathbf{r}\,, \quad \mathbf{R}_2 = \mathbf{r}_2 - \mathbf{R} = -\frac{m_1}{M}\mathbf{r}\,. \tag{4.133}$$

In analogy with the symbols for the laboratory system ($\mathbf{v}_1$, $\mathbf{v}_1'$, $\mathbf{v}_2'$, θ, φ, see Fig.4.29a) we call the corresponding velocities and angles in the CM frame $\mathbf{V}_1$, $\mathbf{V}_1'$, $\mathbf{V}_2'$, Θ, Φ. From the vector diagrams in Fig.4.29 one thus obtains

$$\tan\theta = \frac{V_1 \sin\Theta}{(\dot{\mathbf{R}} + V_1 \cos\Theta)} = \frac{m_2 \sin\Theta}{m_1 + m_2 \cos\Theta} \tag{4.134}$$

for the transformation of the scattering angles between laboratory and CM frame. In the CM system the Rutherford scattering cross section is the same as (4.128), but only now expressed in terms of the CM scattering angle Θ, the reduced mass μ, and the solid angle element $d\tilde{\Omega}$ measured in the CM system

$$\frac{dS}{d\tilde{\Omega}} = \left(\frac{Z_1 Z_2 e^2}{4\pi\epsilon_0 \mu V_1{}^2}\right)^2 \frac{1}{\sin^4(\theta/2)}\,. \tag{4.135}$$

The transformation back into the laboratory frame by means of (4.134) yields a general expression which takes into account the recoil of the target particle: This transformation involves essentially the solid-angle element $d\tilde{\Omega} = d(\cos\Theta)d\phi'$ (ϕ' being the azimuthal angle in the CM frame). In the laboratory frame one has

$$\frac{dS}{d\Omega} = \frac{dS\,d\tilde{\Omega}}{d\tilde{\Omega}\,d\Omega} = \frac{dS\,d(\cos\Theta)}{d\tilde{\Omega}\,d(\cos\theta)} = \frac{dS\,(\sin\Theta d\Theta)}{d\tilde{\Omega}\,\sin\theta d\theta}\,, \tag{4.136}$$

i.e. the derivative $d\Theta/d\theta$ calculated from (4.134) allows direct transformation of the scattering cross sections

$$\frac{dS}{d\Omega} = \frac{dS}{d\tilde{\Omega}}\,\frac{(m_1{}^2 + m_2{}^2 + 2m_1 m_2 \cos\Theta)^{3/2}}{m_2{}^2(m_2 + m_1 \cos\Theta)}\,. \tag{4.137}$$

With (4.134, 135) the Rutherford scattering cross section in the laboratory frame follows as

$$\frac{dS}{d\Omega} = \left(\frac{Z_1 Z_2 e^2}{8\pi\epsilon_0 E_1}\right)^2 \frac{4}{\sin^4\theta} \frac{\{[1-(m_1/m_2)^2 \sin^2\theta]^{1/2} + \cos\theta\}^2}{[1-(m_1/m_2)^2 \sin^2\theta]^{1/2}} . \tag{4.138}$$

For particle masses $m_1 \ll m_2$ this expression can be expanded in a power series

$$\frac{dS}{d\Omega} = \left(\frac{Z_1 Z_2 e^2}{8\pi\epsilon_0 E_1}\right)^2 \left[\frac{1}{\sin^4(\theta/2)} - 2\left(\frac{m_1}{m_2}\right)^2 + \dots\right] . \tag{4.139}$$

The next term is of the order of $(m_1/m_2)^4$. The leading term gives exactly (4.128), i.e. the Rutherford scattering cross section for a spatially fixed scattering center. In many realistic cases the correction $2(m_1/m_2)^2$ is small, e.g. about 4% for He ions ($m_1 = 4$) incident on Si ($m_2 = 28$). Nevertheless the energy loss of the projectile (4.116) might be appreciable.

For scattering of ions with lower energy the interaction potential between the two particles must include the effect of screening. The cross section is then slightly modified in comparison with (4.139). As an example, Fig.4.32 shows a series of trajectories calculated for the scattering of He^+ ions (energy 1 keV) incident on an oxygen atom located at the origin of the diagram [4.27]. In this case an exponentially shielded Thomas-Fermi potential describes the interaction. Figure 4.32 demonstrates the existence of a **shadow cone** behind the scatterer. If another scatterer lies inside this cone, it is not "seen" by the incident ions and thus cannot contribute to the scattering. In this particular case of low-energy incident ions, a broad shadow cone of the order of half a typical interatomic distance, at about one intera-

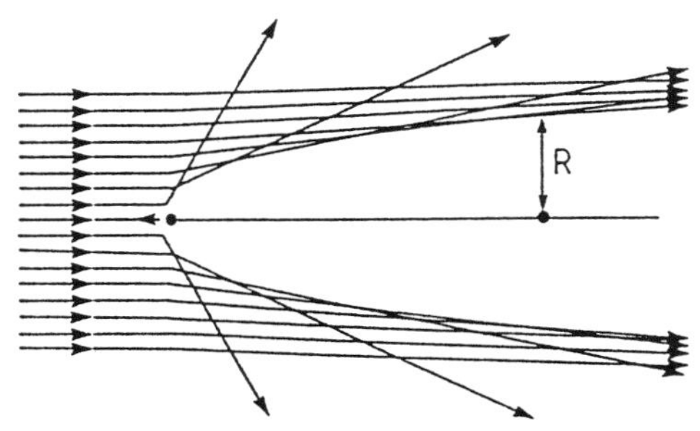

Fig.4.32. Calculated shadow cone of 1-keV He^+ ions scattered from an oxygen atom at the coordinate origin. The trajectories describe ions travelling from left to right. The calculations are based on the assumption of a Thomas-Fermi-Moliere scattering potential [4.27]

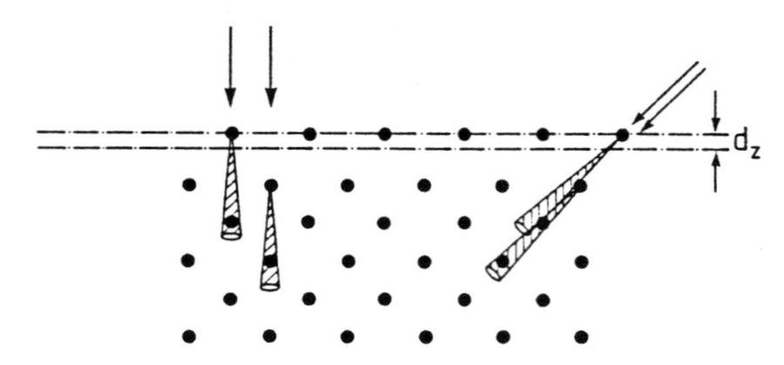

Fig.4.33. Sectional view of a surface with relaxation d_z of the topmost atomic layer showing high-energy ion scattering shadow or blocking cones for incidence along two bulk channelling directions [4.27]

tomic spacing behind the scatterer, is observed. At high energies the screening becomes less important, a simple Coulomb potential can be used, and the shadow cone becomes much narrower. This is the situation in which **channeling** and **blocking** are very useful techniques for studying surfaces and interfaces as well as thin overlayers. Figure 4.33 displays high energy ions incident on a crystalline surface at two different angles for which channeling occurs. For these incidence angles only atoms in the top two atomic layers are "visible"; all deeper lying atoms lie in the shadow cones of the topmost layers. Ions on trajectories between these atoms are "channeled" down the gaps between the atoms and penetrate relatively deep into the solid. This channeling, occurring for an ion beam that is carefully aligned with a symmetry direction of a single crystal, is shown more in detail in Fig.4.34.

Ions not hitting a surface atom are steered through the channels formed by the rows of atoms. The ions cannot get close enough to the atomic nuclei to undergo large-angle Rutherford scattering. Small-angle scattering results with oscillatory trajectories of the projectile. The interaction can easily be described in a quasi-continuous model using a continuous rate dE/dx of energy loss along the channel [4.25]. Because of the large penetra-

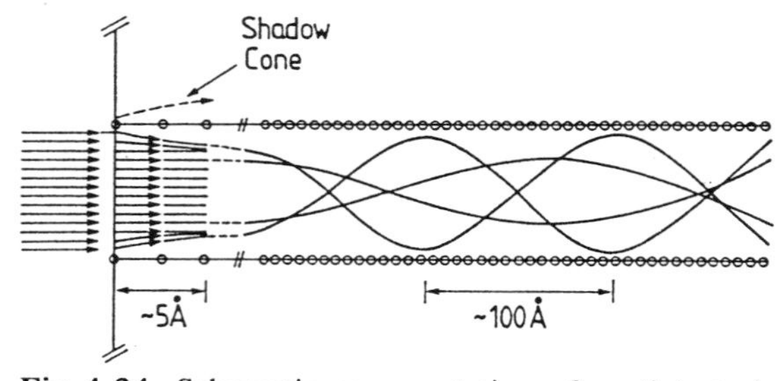

Fig.4.34. Schematic representation of particle trajectories undergoing scattering at the surface and channeling within the crystal. The depth scale is compressed relative to the width of the channel in order to display the shape of the trajectories [4.25]

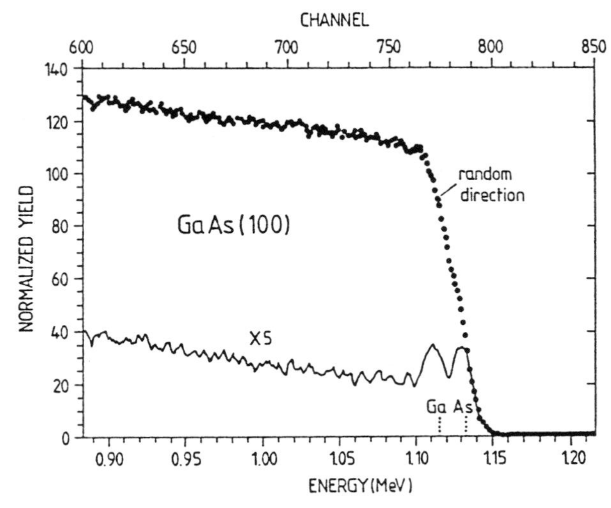

Fig.4.35. Ion backscattering yield spectra of 1.4-MeV He^+ from a GaAs(100) surface along a "random" non-channeling direction and along the surface normal ⟨100⟩, amplified by a factor of 5 [4.29]

tion depth of hundreds of Ångstroms (Fig.4.34), scattering from the substrate is drastically reduced, sometimes by a factor of 100, as compared with a non-channeling situation.

The experimental example [4.29] in Fig.4.35 of 1.4-MeV He^+ ions scattered from a GaAs{100} surface along the surface normal [100] exhibits two sharp peaks due to scattering from the Ga and As surface atomic layer. The energy losses can be calculated using (4.116). For energies lower than that of the surface peak a low-intensity plateau of backscattered ions occurs. These ions result from the few scattering events deeper in the bulk. On their channeling path the incident primary ions continuously lose small amounts of energy, mainly by electronic excitations such as the creation of plasmons, etc. The penetrating ions undergo some rare backscattering events and thus give rise to scattered energies which are reduced in proportion to their channeling depth.

The situation is different, however, when the angle of incidence does not coincide with a low index direction and channeling does not occur (Fig. 4.33). Under this *random* incidence condition many subsurface atomic layers are hit and backscattering occurs from many deeper lying atoms. The backscattering spectrum exhibits a high-intensity continuum of scattered energies up to a maximum defined by the *surface peak* due to scattering on the topmost atomic layer (Fig.4.35). Slight variations of the angle of incidence with respect to the lattice orientation can thus significantly change the backscattering yield, in particular for energies lower than the surface peak.

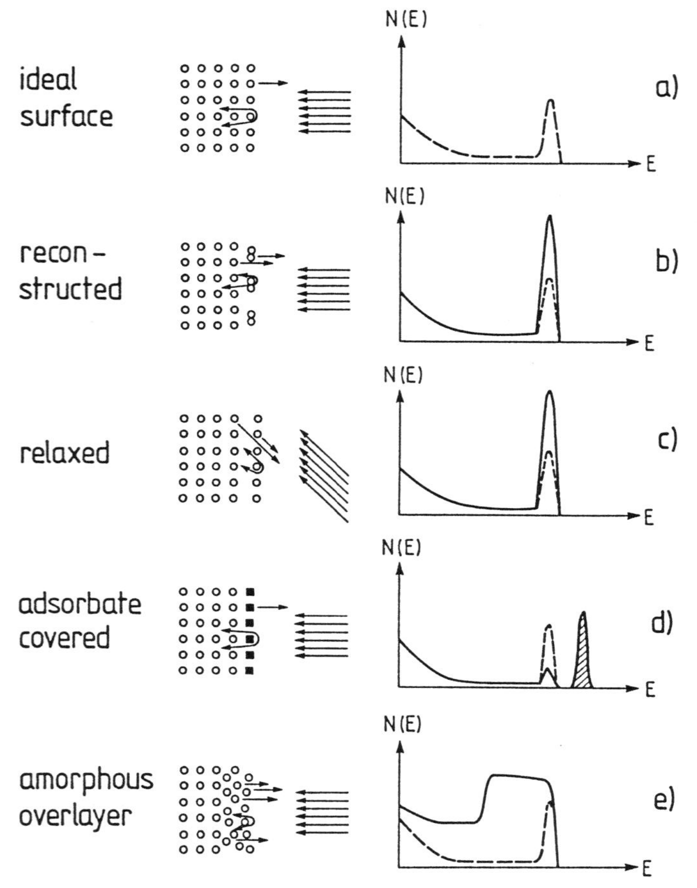

Fig.4.36a-e. Qualitative overview of some applications of ion backscattering and the corresponding spectra N(E): (**a**) Scattering from an ideal clean crystal surface under channeling conditions. The backscattering spectrum (dashed curve) consists of a *surface peak* due to scattering on the topmost atomic layer and a low intensity plateau at lower kinetic energies which arises from scattering events deep in the bulk. (**b**) Scattering on a reconstructed surface causes a considerable increase of the surface peak (up to double intensity) in comparison with the ideal surface spectrum (dashed curve). The second atomic plane contributes to the scattering events due to incomplete shadowing. (**c**) For a relaxed surface, non-normal incidence under channeling conditions causes a similar intensity increase of the surface peak due to incomplete shadowing (dashed curve: spectrum of unrelaxed ideal surface). (**d**) Scattering on an ideal surface covered with an adsorbate overlayer under channeling conditions. The adsorbate layer gives rise to a second surface peak (shadowed) due to scattering on atoms with different mass. Due to shadowing, the original surface peak (dashed curve) is reduced in intensity. (**e**) Scattering on a crystal covered with an amorphous overlayer gives rise to a broad plateau instead of a sharp surface peak. Many atoms out of registry contribute to the scattering in deeper layers. For comparison, the dashed line shows the spectrum for scattering on an ideal crystal surface. [4.25]

Because of the continuous energy loss of the primary ions on their path through the crystal there is also a simple relation between scattering depth and energy loss. These properties can be applied in a number of ways to obtain interesting information about the incorporation of impurities, relaxation effects, quality of interfaces and overlayers, etc. [4.25]. In Fig. 4.36 a qualitative overview of some examples of backscattering spectra is presented. The dashed spectra represent the scattering yield from a crystal with an ideal surface under channeling conditions. The surface peak is due essentially to scattering from only the topmost atomic layer (a). If a reconstruction is present for which the surface atoms are displaced in the plane of the surface (b), then the second atomic plane is no longer completely shadowed and therefore contributes to the backscattering yield. For a major reconstruction, the surface peak might reach an intensity of twice that of the ideal crystal. In the case of a relaxation (c), where the topmost atomic layer is displaced normal to the surface, non-normal incidence is necessary for the investigation. A geometry is used in which the shadow cones arising from the surface atoms are not aligned with the atomic rows in the bulk. In this situation both the surface and the second atomic layer contribute to the surface peak and increase its intensity. A measurement at normal incidence would yield a surface peak intensity equivalent to scattering on one atomic layer. An adsorbate layer on top of an ideal surface (d) gives rise to a new surface peak shifted in energy with respect to that of the clean surface. The sensitivity of ion scattering to atomic mass (Sect.4.10) allows discrimination between substrate and adsorbate. If the adsorbate atoms lie directly above the substrate atoms, the substrate surface peak is strongly reduced in intensity. The case of an amorphous overlayer on top of a crystalline substrate (produced, e.g., by ion bombardement or laser irradiation) is illustrated in Fig.4.36e. Since the atoms within the amorphous overlayer are out of registry, channeling is no longer possible. A broad plateau with high intensity is obtained in the backscattering yield, since numerous atoms contribute to scattering as in the case of *random* incidence on a crystalline material. The energetic width of the plateau is directly related to the thickness of the amorphous layer and the presence of the amorphous-crystalline interface is seen from the decrease of the backscattered intensity at lower energies. The intensity does not drop to the values for the completely crystalline material (a-d), since a high number of ions are deflected from their channeling direction within the amorphous overlayer. As an experimental example of such a system Fig.4.37 exhibits the channeling spectrum of an approximately 4000-Å thick amorphous Si layer on Si(100) [4.28]. 2-MeV ^{4}He ions were used in this investigation. The energy spectrum of the backscattered ions is transformed into a thickness scale (upper abscissa). For comparison, the spectrum of the clean Si(100) surface before amorphization is also shown.

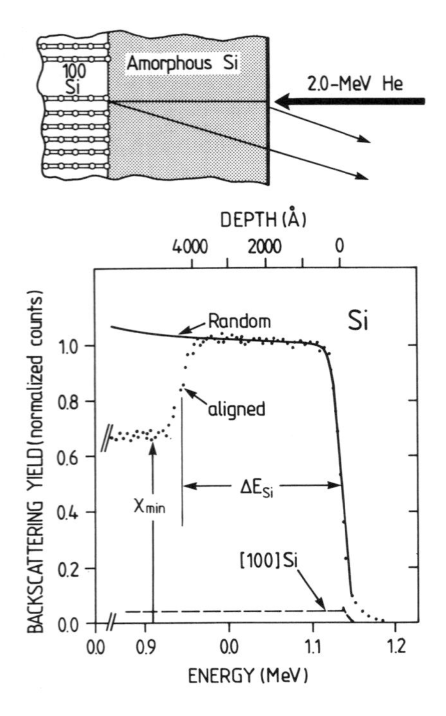

Fig. 4.37. Backscattering channeling spectrum (ion yield versus kinetic energy) of a 4000 Å thick amorphous Si-layer on top of a Si(100) surface. The spectrum was obtained under channeling conditions with 2-MeV ^{4}He ions (inset). Using a mathematical model for the Rutherford backscattering processes, the energy scale is converted into a depth scale (upper abscissa). For comparison, the spectra of the Si surface before amorphization are also given, for channeling conditions by the dashed line, and for "random" conditions by the full line. [4.28]

Further information can be obtained from RBS experiments by studying the angular dependence of the backscattering yield. In Fig. 4.38 a Si film with $1.5 \cdot 10^{21}$ cm^{-3} As atoms incorporated has been investigated [4.30]. The Si and the As backscattering yields show a very similar angular dependence. The minimum yield corresponds, of course, to the channeling direction. The similarity of the two curves clearly indicates that the pertur-

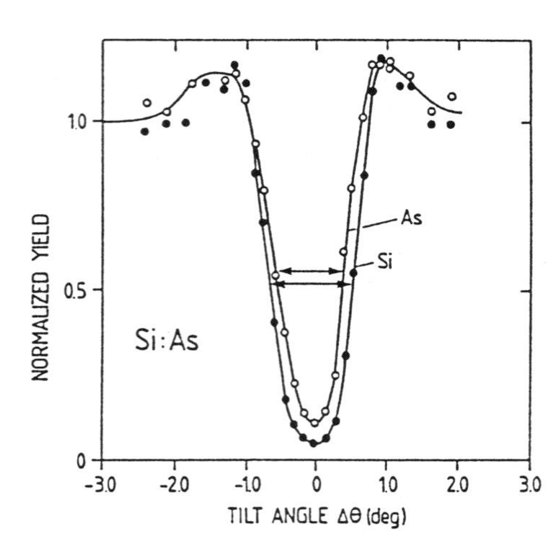

Fig. 4.38. Angular dependence of the backscattering yield around the channeling direction (tilt angle $\Delta\theta = 0$) for 1.8 MeV He ions on a Si-film, which is doped with $1.5 \cdot 10^{21}$ As atoms/cm^3. The angular dependence of the Si and of the As backscattering peak are compared. [4.30]

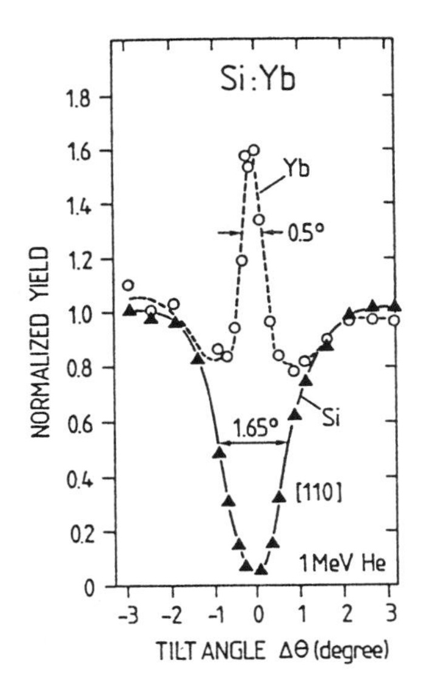

Fig. 4.39. Angular dependence of the backscattering yield (normalized) around the [110] channeling direction (tilt angle $\Delta\theta = 0$) for 1 MeV He ions on a Si-film, into which $5 \cdot 10^{14}$ Yb atoms/cm^2 are implanted (60 keV, 450 °C). The Si signal (▲) is compared with the Yb signal (○). [4.31]

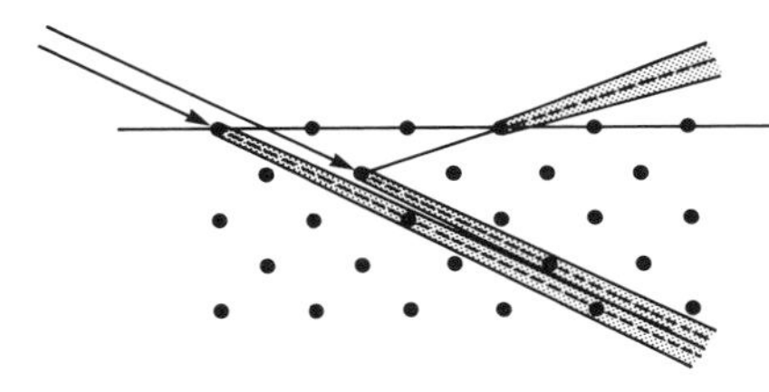

Fig. 4.40. Double-alignment scattering geometry, schematic view of a (010) scattering plane perpendicular to the (100) surface for the example of silicon

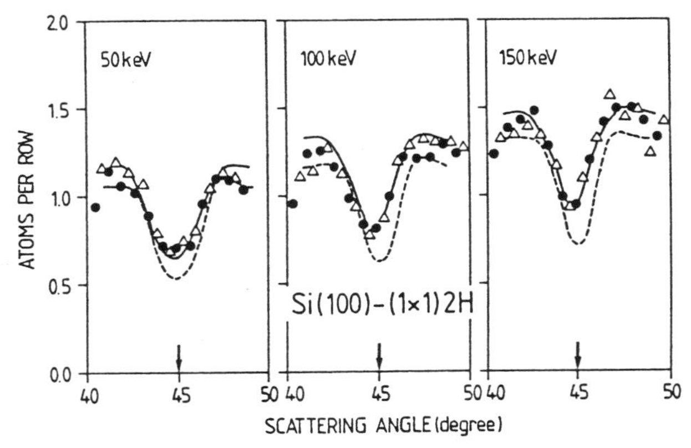

Fig. 4.41. Experimental surface blocking profiles obtained in a scattering geometry as in Fig. 4.40 for H^+ (protons) with energies of 50, 100 and 150 keV scattered on a hydrogen stabilized Si(100)-(1×1)2H surface. Circles and triangles represent results from different experimental runs. The arrows indicate the position of the blocking minimum; the dashed curves are calculated for a truncated bulk without any relaxation [4.32]

bation of the lattice due to As incorporation is negligible. As is built in as a substitutional impurity. In complete contrast are the angle-resolved spectra (Fig. 4.39) for Yb implanted into a Si film (concentration $5 \cdot 10^{14} cm^{-2}$). Exactly in the Si channeling direction (minimum of Si signal) a strong peaking of the Yb signal is observed thus demonstrating that Yb is not incorporated on Si sites but rather as an interstitial.

A further RBS technique used in surface crystallography is the so-called double alignment (Fig.4.40). The primary ion beam is aligned with a low-index crystal direction for channeling, such that only atoms in the first two atomic layers are hit. All deeper lying atoms are shadowed. An ion immediately below the surface can leave the crystal in any direction except that in which its outgoing trajectory is blocked by an atom in the topmost atomic layer. In this direction there will be a minimum in the yield of ions back-scattered from the surface, the surface blocking minimum. Deviations of the observed minimum position from that calculated for an ideal surface give information about surface reconstructions and relaxations. The Si(100) surface with adsorbed atomic hydrogen [hydrogen-stabilized Si(100)-(1×1) 2H] has been studied in this way (Fig.4.41). The observed shift of the surface blocking minimum from its calculated position (arrows and dashed curves) leads to the conclusion that the surface is relaxed inward by (-0.08 ± 0.03) Å, or $(-6 \pm 3)\%$ of the interplanar distance [4.32].

Panel VIII
Low-Energy Electron Diffraction (LEED) and Reflection High-Energy Electron Diffraction (RHEED)

Elastic scattering or diffraction of electrons is the standard technique in surface science for obtaining structural information about surfaces. The method is applied both to check the crystallographic quality of a freshly prepared surface and as a means of obtaining new information about atomic surface structure. As in all diffraction experiments, the determination of an atomic structure falls naturally into two parts: the determination of the periodicity of the system and thus the basic unit of repetition or the surface unit mesh, and the location of the atoms within this unit. The first part, the evaluation of the surface unit mesh is straightforward and involves simple measurements of symmetry and spot separation in the diffraction pattern. Since the diffraction pattern corresponds essentially to the surface reciprocal lattice (Sect.4.2), the reverse transformation yields the periodicity in real space. The second part, the determination of atomic coordinates, requires a detailed measurement of diffracted intensities. The LEED technique is frequently used for this purpose, even though the theoretical problem of deriving an atomic structure from the measured intensities is far from simple, due to the "strong" interaction of slow electrons with a solid. The analysis of the geometry and intensity of a LEED diffraction pattern is discussed in Sects.4.2-4.

The standard experimental set up for LEED consists of an electron gun to produce an electron beam with primary energies in the range of $20 \div 500$ eV and a display system for observing the Bragg diffraction spots. The energy range below 300 eV is particularly suited to surface studies since the mean-free path of these slow electrons in the solid is short enough to give good surface sensitivity (Chap.4). Furthermore, according to the de Broglie relation

$$\lambda = (150.4/E)^{1/2} \qquad \text{(VIII.1)}$$

(λ in Å and E in eV), typical LEED wavelengths are in the Ångstrom range, comparable with wavelengths used in X-ray crystallography, and of the same magnitude as the interatomic distances in a solid.

A typical three-grid LEED system is exhibited in Fig.VIII.1a. The electron gun unit consists of a directly or indirectly heated filament with a Wehnelt cylinder *W* followed by an electrostatic lens with apertures *A*, *B*,

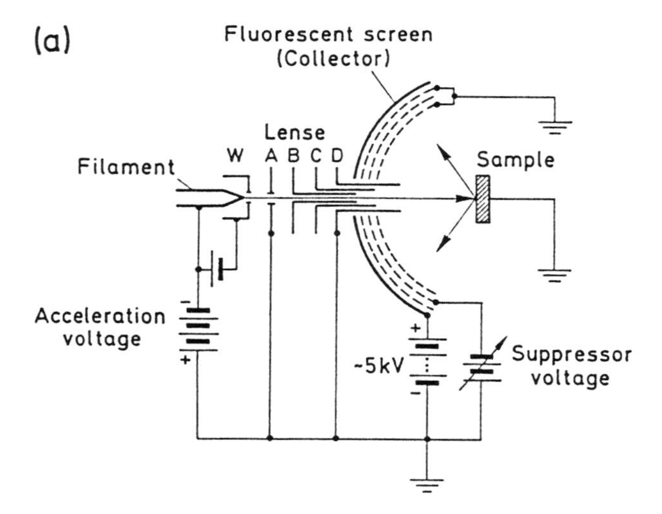

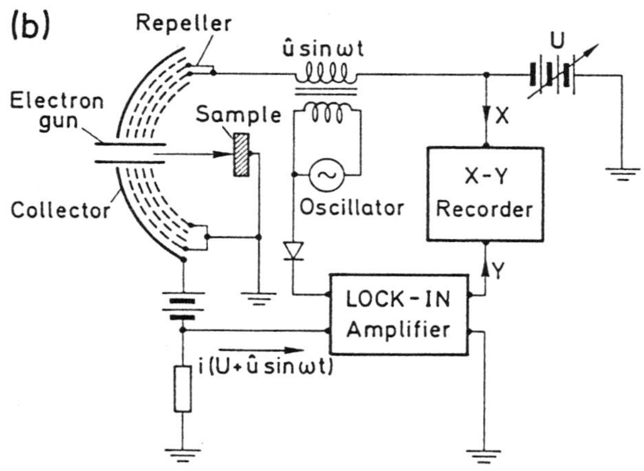

Fig. VIII. 1. (**a**) Schematic of a three-grid LEED optics for electron diffraction experiments. The integrated electron gun consists of a heated filament, a Wehnelt cylinder (W) and the electron optics containing the apertures A-D. B and C are usually held at potentials between those of A and D. (**b**) Circuit for using a four-grid LEED optics as a retarding field (RF) electron energy analyser. The retarding voltage U determines what fraction of the inelastically scattered electrons reaches the collector. Differentiation of the measured electron current i(U + usinωt) with respect to U is done by means of the superimposed AC voltage usinωt ($u \approx 1$ V) and subsequent Lock-in detection of i. For $u \ll U$ one has $i(U+u\sin\omega t) = i(U) + i'(U)u\sin\omega t + \frac{1}{2}i''(U)u^2\sin^2\omega t + \ldots$, and phase-sensitive detection at the first harmonic ω yields an output signal which is proportional to the derivative $i'(U)$

C, D. The acceleration energy ($20 \div 500\,\mathrm{eV}$) is determined by the potential between the cathode and apertures A and D. Apertures B and C have potentials intermediate between A and D and are used to focus the electron beam. Initial collimation is achieved by the Wehnelt cylinder which has a somewhat negative bias with respect to the cathode filament. The last aperture D, also called the **drift tube**, is usually at the same (earth) potential as aperture A and the sample; the same is true for the first and last grids in front of the fluorescent screen. Thus a field-free space is established between the sample and the display system through which the electrons travel to the surface and back after scattering. The fluorescent screen (collector) has to be biased positively ($\approx 5\,\mathrm{kV}$) in order to achieve a final acceleration of the slow electrons; only high-energy electrons can be made visible on the screen. Besides elastic scattering, inelastic scattering also occurs at the sample surface, thus giving rise to electrons of lower energy. These electrons are scattered through wide angles and produce a relatively homogeneous background illumination of the phosphor screen. This background illumination is suppressed by giving the middle grid a somewhat negative bias. The inelastically scattered electrons are thus prevented from reaching the collector.

The emission current of standard equipment is on the order of 1 μA; it varies with primary energy but electronic stabilization may be used to fix its value. The energy spread is about 0.5 eV, attributable mainly to the thermal energy distribution. The diameter of the primary beam is on the order of 1 mm.

As is shown in Fig. VIII.1b, standard LEED optics (four grids are preferred in this case) can also be used as an energy analyser. The electrons coming from the sample have to overcome a retarding field before they reach the collector. For a retarding voltage, U, the current reaching the collector is

$$i(U) \propto \int_{E=eU}^{\infty} N(E)\,dE \;, \qquad \text{(VIII.2)}$$

where N(E) is the energy distribution of the incoming electrons. By superposing an AC voltage $u \sin\omega t$ ($u \ll U$) and using phase-sensitive detection i(U) in (VIII.2) can be differentiated allowing N(E) to be easily obtained (Fig. VIII.1b).

The standard LEED equipment displayed in Fig. VIII.1a is used mostly for characterizing the crystalline perfection etc., of a clean surface. For investigating adsorbate layers, semiconductor surfaces and for LEED intensity studies this setup has severe disadvantages. The current density of ≈ 1 μA in the primary beam is rather high. Thus organic adsorbates and clean

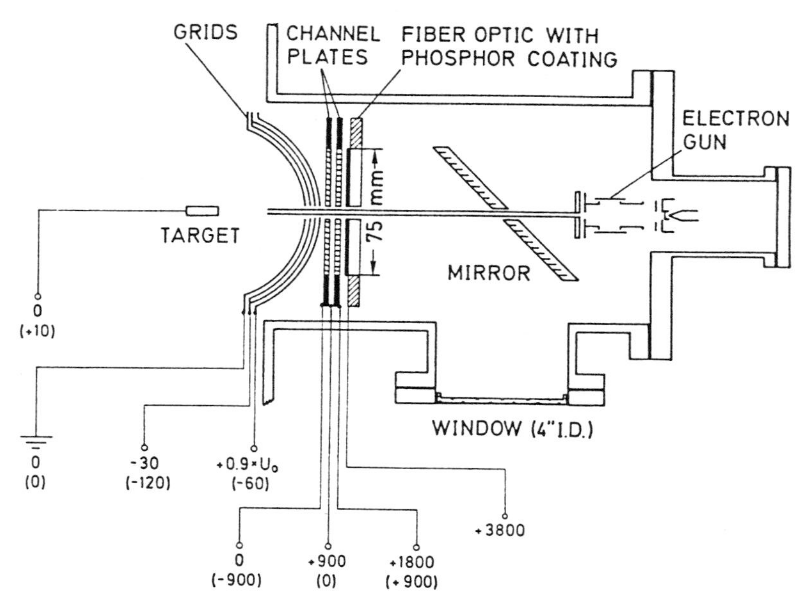

Fig. VIII.2. Schematic of a low-current optical display for LEED and ESDIAD (electron stimulated desorption ion angular distribution, Panel XIV: Chap. 9). In LEED primary beam currents in the 10^{-10} A range can be used, the channel plates enable amplification of the detected electron currents by factors in the 10^7 range. Typical bias potentials [eV] are also indicated. For ESDIAD the potentials are given in brackets; U_0 is the acceleration (primary) voltage [VIII.1]

semiconductor surfaces may suffer severe damage. Furthermore the measurement of intensity-voltage i(V) curves for a number of diffraction beams and several surface orientations requires the collection of very many data. Recent experimental developments have thus aimed to decrease the primary current and enable much faster data acquisition. Figure VIII.2 shows an advanced display system, where the primary current is decreased by four orders of magnitude to 10^{-10} A. A bright LEED pattern is then obtained by two channel plates which amplify the backscattered electron currents by a factor of about 10^7. Since inspection of the fluorescent screen (fiber optics with phosphor coating) is possible from behind via a mirror, the whole unit is extremely versatile; it can be flanged to standard ports of a UHV chamber. By changing the potentials at the grids or the channel plates, etc, (Fig. VIII.2) the same unit can be used for measurements of the angular distribution of (ionic) Electron Stimulated Desorption products (ESDIAD, Panel XIV: Chap. 9). In these experiments an electron beam is incident on an adsorbate covered surface and the angular distribution of the desorbing ions yield information about the bonding geometry of the adsorbed species (Panel XIV: Chap. 9).

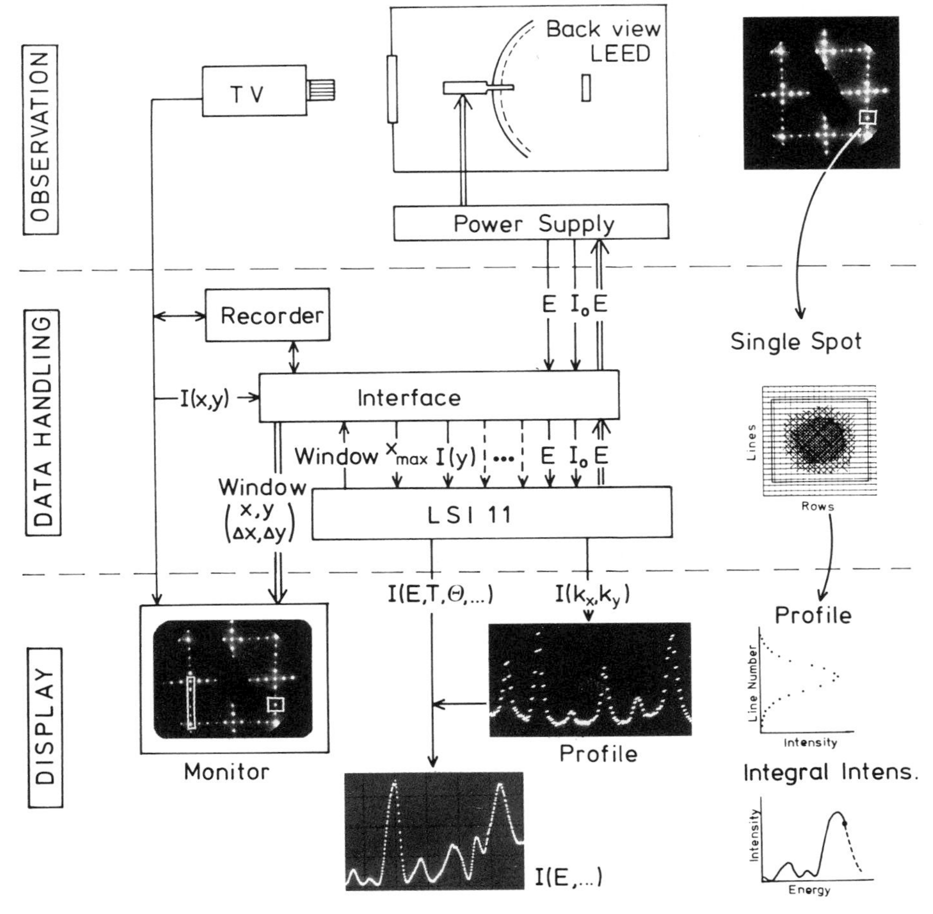

Fig. VIII.3. Information-flow diagram from observation through data handling to display of intensities by the computer controlled fast LEED system DATALEED. A video signal is generated by the TV camera viewing the LEED pattern from the back of a transparent fluorescent screen. The computer controlled system generates an electronic window of variable size and shape around a certain spot or group of spots and stores up to 10^3 pixels of digital information in a fast memory. The electronic window is made to follow the spot as it moves over the screen with varying beam energy E. The computer LSI11 calculates intensity versus energy I(E) spectra, integral intensities, FWHM of profiles, etc. Single lines carry information, double lines represent control channels [VIII.2]

Another modern improvement to the LEED equipment is to be found in so-called DATALEED (Fig. VIII.3). Here, extremely fast data acquisition is achieved by the application of an electronic video (TV camera) unit to measure the LEED spot intensities accompanied by computer-controlled

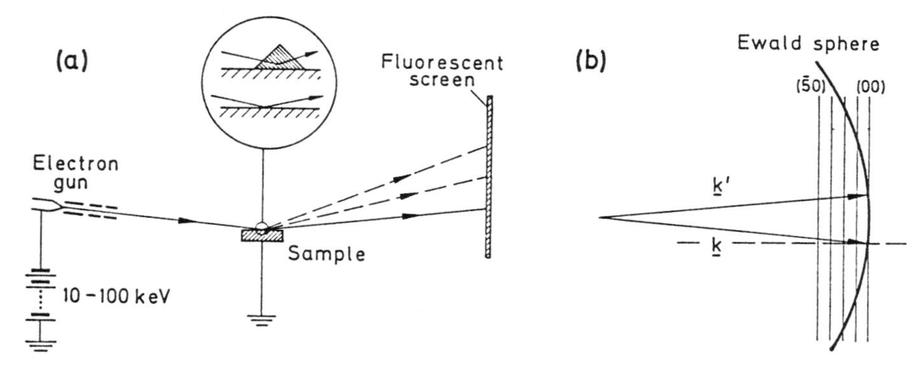

Fig. VIII.4. (**a**) Schematic of the experimental set-up for RHEED. The inset shows two different scattering situations on a highly enlarged surface area: surface scattering on a flat surface (below) and bulk scattering by a three-dimensional crystalline island on top of the surface (above). (**b**) The Ewald sphere construction for RHEED. **k** and **k**′ are primary and scattered wavevectors, respectively. The sphere radius $k = k'$ is much larger than the distance between the reciprocal lattice rods (hk). For more details, see Sect. 4.2 and Figs. 4.2,3

data handling. At present the time required to determine the integral intensity of a LEED beam at a particular energy is 20 ms including background subtraction. A whole LEED intensity spectrum is measured by this equipment in about 10 s depending on the energy steps and the energy range covered. Thus time-dependent LEED intensity analysis (Sect. 4.4), i.e. structure analysis during, e.g., surface phase transitions, becomes feasible.

Besides LEED, the second important electron diffraction technique in interface physics is RHEED (Reflection High-Energy Electron Diffraction). High energy electrons with primary energies between 10 and 100 keV are incident under grazing angles ($3° \div 5°$) onto the sample surface, and the diffracted beams are observed at similar angles on a fluorescent screen (Fig. VIII.4a). The electron guns used in RHEED are slightly more complex than those of LEED; in some cases a magnetic lens is used for focussing because of its greater efficiency at higher electron velocities. The much higher voltages used in RHEED require special power supplies and vacuum feed-throughs. The fluorescent screen needs no high-voltage source; no acceleration of the electrons is necessary since the high primary energies are sufficient to produce fluorescence. The screen is usually planar and is sometimes coated onto the inside of a window of the UHV system along with a conducting film to prevent charging. No energy filtering of inelastic and secondary electrons is necessary since the diffracted beams are much more intense than the background. Since the spatial separation of the electron gun and sample, and of sample and screen can be of the order of 50 cm the method is very flexible in relation to sample conditions. For example, one can use RHEED to study in situ surfaces at elevated temperatures during molecular beam epitaxy (Sect. 2.4).

In spite of the high primary energies, RHEED has a similar surface sensitivity to LEED: the grazing incidence and detection angles mean that a long mean-free path through the sample is associated with penetration normal to the surface of only a few atomic layers. The diffraction pattern in RHEED is rather different from that of LEED. Figure VIII.4b illustrates the Ewald construction for the conditions of a RHEED experiment. Because of the extremely high primary energies the diameter of the Ewald sphere is now much larger than a reciprocal lattice vector. The reciprocal lattice rods (Sect. 4.2) are cut at grazing angles in the region corresponding to a diffracted beam emerging close to the surface. The Ewald sphere "touches" the rods of the reciprocal lattice. Both the Ewald sphere and reciprocal lattice rods are smeared out, to some extent, due to the angular and energy spread of the primary beam, respectively, and to deviations from ideal translational symmetry in the surface (phonons, defects, etc.). Thus the diffraction pattern will usually consist not of spots, but of streaks corresponding to the sections of reciprocal lattice rod intersected. It must be emphasized, however, that on extremely flat and ideal surfaces, and with very good instrumentation, one can occasionally observe very sharp diffraction spots.

An example of RHEED patterns is shown in Fig. VIII.5. Because of the grazing-angle electron beams, very flat sample surfaces are needed in RHEED. Asperities or stronger deformations shadow part of the surface. If there are crystalline islands or *droplets* on the surface, on the other hand, bulk scattering of the grazing beam can occur and the RHEED pattern may become dominated by spots rather than streaks due to transmission electron diffraction (insert in Fig. VIII.4a). This particular property of RHEED, however, can be usefully exploited to study surface corrugation and growth modes of films during deposition and epitaxy. Stransky-Krastanov growth of islands (Chap. 3) can be readily identified through the spots in the RHEED pattern.

In both LEED and RHEED, the primary electron beam deviates from an ideal plane wave $A\exp(i\mathbf{k}\cdot\mathbf{r})$. It is actually a mixture of waves of slightly different energy and direction. These deviations from the ideal direction and energy are caused by the finite energetic width ΔE (thermal width $\approx$ 500 meV) and the angular spread 2β of the beam. The electrons impinging on the crystal surface exhibit therefore slight random variations in phase; if two spots on the surface have too large a separation, the incoming waves cannot be considered as coherent. The phases are not correlated and the outgoing waves cannot interfere to produce a diffraction pattern. There is a characteristic length called **coherence length** such that atoms in the sur-

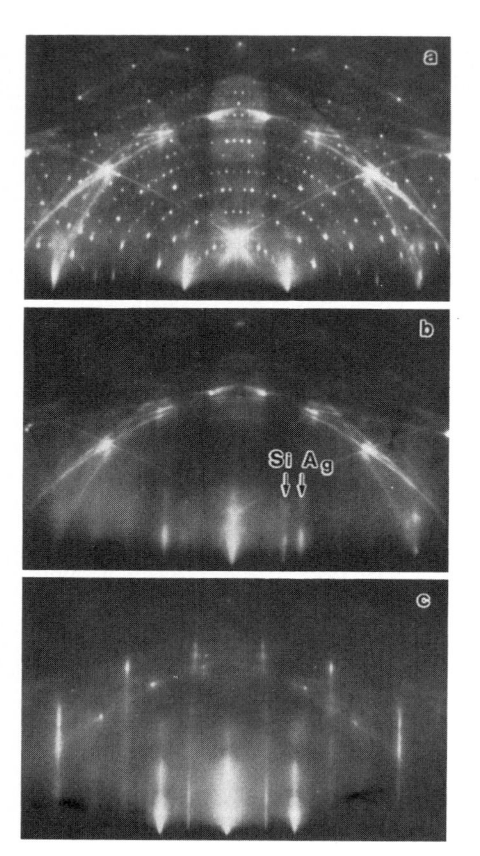

Fig. VIII.5a-c. RHEED patterns taken with a primary energy of E = 15 keV and a direction of incidence of [112] on a Si(111) surface: (**a**) Clean Si(111) fur-face with a (7× 7) superstructure. (**b**) After deposition of nominally 1.5 mono-layers (ML) of Ag streaks due to the Ag layers are seen on the blurred (7× 7) structure. (**c**) After deposition of 3ML of Ag the texture structure due to the Ag layers develops in place of the (7× 7) structure [VIII.3]

face within a coherence length (or radius) can be considered as illuminated by a simple plane wave. Waves scattered from points separated by more than a coherence radius add in intensity rather than in amplitude. Thus no surface structure on a scale larger than the coherence length forms a diffraction pattern.

The two contributions responsible for limiting the coherence are the finite energy width ΔE and the angular spread 2β, giving rise to incoherence in time and space, respectively. Since $E = \hbar^2 k^2/2m$, the energy width

$$\Delta E = \frac{\hbar^2}{m} k \Delta k \qquad \text{(VIII.3)}$$

is related to an uncertainty in wave vector (due to time incoherence)

$$\Delta k^t = k \frac{\Delta E}{2E}, \qquad \text{(VIII.4)}$$

i.e. for the component parallel to the surface (at normal incidence)

$$\Delta k_{\|}^{t} \simeq k\beta \frac{\Delta E}{E} . \tag{VIII.5}$$

The finite angular spread 2β causes an uncertainty in wave vector parallel to the surface (due to space incoherence) of

$$\Delta k_{\|}^{s} \simeq 2k\beta . \tag{VIII.6}$$

Since the two contributions are independent, the total uncertainty in $k_{\|}$ is

$$\Delta k_{\|} = \sqrt{(\Delta k_{\|}^{t})^2 + (\Delta k_{\|}^{s})^2} \tag{VIII.7}$$

and because of Heisenberg's uncertainty relation

$$\Delta r_c \Delta k \simeq 2\pi , \tag{VIII.8}$$

the coherence radius Δr_c is obtained as

$$\Delta r_c \simeq \frac{\lambda}{2\beta\sqrt{1 + (\Delta E/2E)^2}} , \tag{VIII.9}$$

where λ (VIII.1) is the wavelength of the electrons.

In a standard LEED experiment, the angular width of the primary beam is $\approx 10^{-2}$ rad with an energy spread of ≈ 0.5 eV. At a primary energy of ≈ 100 eV this leads to a coherence length of about 100 Å. In RHEED the primary energy is much higher ($\approx 5 \cdot 10^4$ eV) and the beam collimation is also better ($\beta \approx 10^{-4} - 10^{-5}$ rad), but because of the grazing incidence, $\Delta k_{\|} \simeq \Delta k$. Thus only a slightly larger coherence zone remains than in LEED. Because of the finite coherence in LEED and RHEED only limited information about the degree of long-range order on the surface can be obtained. If the long-range order is restricted to areas smaller than the coherence zone (diameter ≈ 100 Å) the diffraction pattern is weaker with a higher incoherent background. Much experimental effort has been undertaken to increase the coherence length considerably by using better electron optics. Coherence radii of several thousand Ångstroms have now been achieved enabling small deviations from long-range order to be investigated by LEED [VIII.4, 5].

References

VIII.1 C.D. Kohl, H. Jacobs: Private communication

VIII.2 K. Heinz, K. Müller: *Experimental Progress and New Possibilities of Surface Structure Determination. Springer Tracts Mod. Phys.* **91** (Springer, Berlin, Heidelberg 1982)

VIII.3 S. Hasegawa, H. Daimon, S. Ino: Surf. Sci. **186**, 138 (1987)

VIII.4 K.D. Gronwald, M. Henzler: Surf. Sci. **117**, 180 (1982)

VIII.5 M. Henzler: Appl. Phys. A **34**, 209 (1984)

Panel IX
Electron Energy Loss Spectroscopy (EELS)

Electron Energy Loss Spectroscopy (EELS) refers, in a broad sense, to every type of electron spectroscopy in which inelastic electron scattering is used to study excitations of surfaces or thin solid films [IX.1, 2]. The experiment thus involves the preparation of a more or less monoenergetic electron beam, scattering on a solid surface or within a thin solid film, and energy analysis of the electrons inelastically scattered at a certain angle by means of an electron analyser (Panel II: Chap. 1). The inelastic scattering process requires time-dependent scattering potentials (phonons, plasmons, adsorbate vibrations, electronic transitions etc.). The theory can be developed within a quite general formalism as in Sect. 4.1 or in a more restricted framework for scattering on long-range potentials (dielectric theory) as in Sect. 4.6 [IX.2, 3].

Since many different solid excitations can be studied (Fig. IX.1) over a wide energy range from some meV to more than 10^3 eV, different experimental equipment is necessary to provide the required energy resolution at very small and very high excitation energies.

Here we describe briefly the main applications of EELS and important features of the experimental set-ups but do not go into details of the scat-

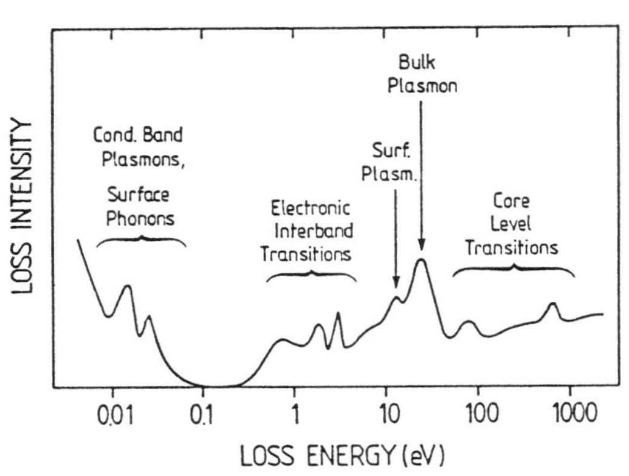

Fig. IX.1. Qualitative overview of the major excitation mechanisms which can contribute to an electron energy loss spectrum over a wide loss energy range. For investigating the different loss energy regimes, in particular below and above 1 eV, different experimental set-ups are required

tering theory. Further details of the theory are given in Chap.4. When EELS is performed with high energetic resolution at low primary energies, $E_0 < 20$ eV, it is called **High Resolution Electron Energy Loss Spectroscopy** (HREELS). The primary electron beam then has to be monochromatized, usually by means of a hemispherical or cylindrical electron analyzer (Panel II: Chap. 1). Extremely good monochromaticity is now achieved with halfwidths (FWHM) of the primary beam around 1 meV [IX.4]. At these low primary energies only reflection scattering experiments under UHV conditions are possible. The backscattered beam is analysed by means of a similar electron analyser. For dielectric scattering (Sect.4.6) with small wave-vector transfer, the scattered electrons are detected in specular direction, whereas detection at various angles is necessary when the dispersion relations of surface excitations (phonons, etc.) are measured. This techique has revealed much interesting information about surface-phonon dispersion branches on clean and adsorbate-covered metal surfaces (Chap.5).

On semiconductor surfaces typical free-carrier (electrons in the conduction or holes in the valence band) concentrations in the 10^{17} cm^{-3} range give rise to bulk and surface plasmons with energies in the $20 \div 100$ meV range. The surface plasmons can be conveniently detected by HREELS, since for primary energies below 20 eV surface scattering prevails and major loss structures occur for $\mathrm{Re}\{\epsilon(\omega)\} \simeq -1$, $\epsilon(\omega)$ being the dielectric function of the sample (Sect.5.5). One can also study the coupling of these surface plasmons to surface phonons carrying a dynamic dipole moment (**Fuchs-Kliewer surface phonons**) (Sect.5.5).

By far the widest application of HREELS is concerned with the study of vibrations of adsorbed atoms or molecules; here it is used to identify adsorbed species and to get information about adsorption sites and bonding geometry [IX.2]. The identification of an adsorbed species relies on the knowledge of its vibrational spectrum from the IR absorption or Raman measurements in the gas phase. Selection rules for IR dipole absorption and Raman scattering must be taken into account, as well as the dipole selection rule in HREELS for dipole scattering on surfaces: only dipole moments normal to the surface give rise to significant dielectric scattering. Dipole moments parallel to the surface can only be detected in off-specular geometry (Sect.4.1). The application of these selection rules allows one to draw conclusions about the adsorption geometry of a molecule: for example, chemical bonds oriented parallel to a solid surface cannot give rise to strong dipole scattering from their stretching mode in the specular direction. The adsorption site of an atom or molecule can sometimes be deduced from the occurrence of particular adsorbate-substrate atom vibrations. The occurrence of an As-H stretch vibration, for example, clearly signifies that an As surface atom on GaAs is the bonding site for a hydrogen containing adsorbate. As an example of an adsorbate vibrational spectrum Fig.IX.2 presents

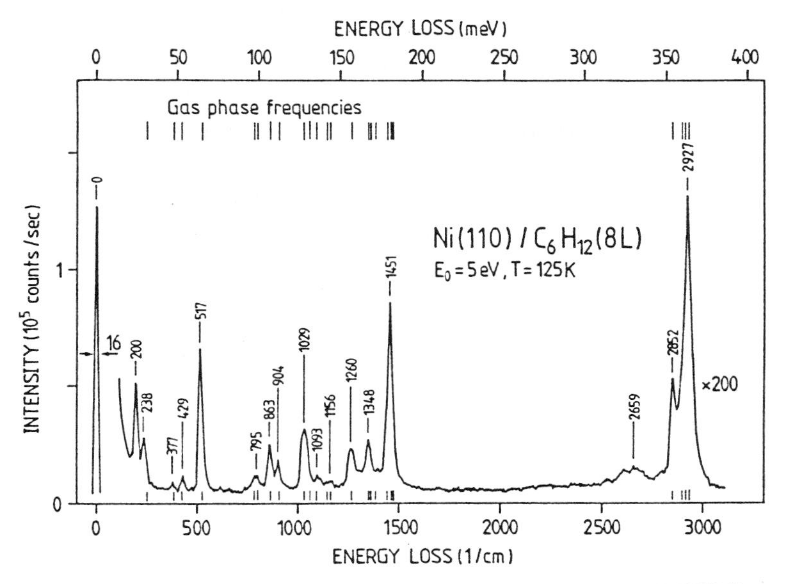

Fig. IX.2. High Resolution Electron Energy Loss Spectrum (HREELS) measured on a Ni(110) surface exposed to 8 L of C_6H_{12}. The measurement was performed with a primary energy of 5 eV under specular reflection geometry and with the sample held at a temperature of 125 K. The vibration frequencies of gas phase C_6H_{12} are given in the upper part of the figure [IX.5]

a loss spectrum measured on a Ni(110) surface exposed to 8 L of cyclohexane (C_6H_{12}) [IX, 5]. Notice the extremely good energy resolution of the electron spectrometer, yielding a FWHM of the primary peak of about 2 meV ($\approx$ 16 cm^{-1}). Nearly every observed loss peak can be explained in terms of a vibrational mode of the C_6H_{12} molecule. The frequencies of all possible modes of the gaseous C_6H_{12} species are given for comparison in the upper part of the figure. One can infer that the molecule is adsorbed non-dissociatively. There are essentially two loss features which cannot be attributed to intramolecular vibrations, the sharp peak at 200 cm^{-1} wave number and the broad loss structure near 2659 cm^{-1}. The peak at 200 cm^{-1} is due to a well-known surface phonon of the Ni(110) surface; this phonon is related to the reconstruction of the Ni surface and carries a dynamic dipole moment normal to the surface, which gives rise to strong dipole scattering [IX.5]. The broad feature near 2659 cm^{-1} is found for many hydrogen-containing molecules adsorbed on metal surfaces. It is interpreted as due to hydrogen atoms that penetrate into the surface and thus give rise to hydrogen-bonding-type vibrations with respect to the substrate atoms. From the existence of this loss feature, a planar adsorption geometry of the ring-type molecule C_6H_{12} can be concluded. The hydrogen atoms projecting away from the ring skeleton of the molecule tend to "dig" into the Ni surface. In spite of the planar adsorption geometry, all C_6H_{12} vibrations are observed with relatively high intensity since the low symmetry of the folded ring skeleton of the C_6H_{12} molecule causes dynamic dipole moments with at least one component normal to the metal surface. Nevertheless non-dipolar scattering contributions cannot be excluded in the interpretation of the spectrum.

While HREELS with primary energies typically below 50 eV involves essentially surface scattering (Sect.4.6), described in terms of the surface loss function $\mathrm{Im}\{-1/(1+\epsilon)\}$ for dielectric scattering, loss spectroscopy with higher primary energies of $100 \div 500$ eV (usually termed EELS) reveals loss structures due to both bulk and surface scattering. In dielectric theory both the bulk and the surface loss functions, $\mathrm{Im}\{-1/\epsilon\}$ and $\mathrm{Im}\{-1/(1+\epsilon)\}$, have to be considered for the interpretation of the spectra. The excitations typically observable in the energy loss range $1 \approx 100$ eV are the valence electron plasma excitations (both surface and bulk plasmons) and electronic interband transitions (Fig.IX.1). In this energy range one needs to consider bulk- and surface-state transitions of the clean surface as well as adsorbate characteristic transitions. Thus EELS with primary energies in the 100 eV range has mainly been used to study the electronic structure of clean surfaces (surface states), thin overlayers, and adsorbates. Since an energy resolution of about $0.3 \div 0.5$ eV is usually sufficient to reveal the relevant loss structures in the loss range $1 \div 50$ eV, the experiments are performed without prior monochromatization using the electron beam of an ordinary electron gun with a thermal width of about 0.3 eV. After scattering from the sample surface, energy analysis is carried out by a hemispherical analyser or a Cylindrical Mirror Analyser (CMA) as often used in AES (Panel II: Chap.1). The advantage is that standard equipment employed for AES or UPS/XPS can also be used for additional EELS studies. When hemispherical analysers with a well-defined and limited angular acceptance angle are used, EEL spectra in the direct undifferentiated mode usually display enough spectral structure to provide the necessary information. The use of CMAs involves detection of electrons within a wider acceptance angle (cone around CMA axis; Panel II: Chap.1) and thus leads to the observation of a larger variety of scattering events. The loss spectra are less well resolved, and the double differentiation technique is usually applied to discriminate the major loss structures from the background. As in AES (Panel III: Chap.2) a small AC signal with frequency ω is superimposed on the voltage between sample and analyser, and lock-in detection is performed at a frequency of 2ω. The negatively recorded $-d^2I/dE^2$ spectra show positive peaks exactly where the undifferentiated spectra would show maxima. Minima in the spectra have to be treated with care since they might originate from the double differentiation process. The energy resolution in the

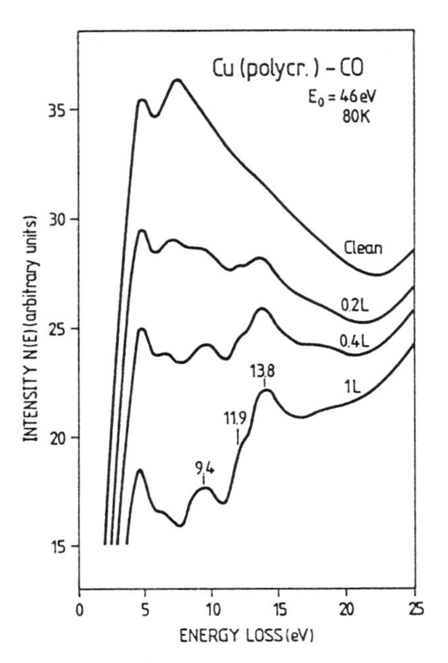

Fig. IX.3. Electron energy loss spectrum of a clean polycrystalline Cu surface and of the same surface after exposure to various amounts of CO at 80 K. The spectra were recorded at room temperature with a primary energy of 46 eV and a hemispherical analyser in the ΔE = const mode (Panel II: Chap. 1); primary peak not shown [IX.6]

double differentiation mode is determined, of course, not only by the thermal width of the primary beam, but also by the peak-to-peak value of the superimposed AC modulation (usually 1 V).

As an example of EELS measured in the undifferentiated mode by means of a hemispherical analyser (Panel II: Chap. 1) Fig. IX.3 exhibits loss spectra for a clean polycrystalline Cu surface and for the same surface after the adsorption of various amounts of CO [IX.6]. The primary peak is not shown for reasons of clarity; the loss features at 4.3 and 7.3 eV for the clean surface are most probably due to d-band transitions which are strongly coupled to surface plasmons [IX.6]. Upon adsorption of CO the spectrum gradually changes; in particular, the 7.3 eV loss is suppressed in intensity and shifts to slightly lower energies. Furthermore, new losses characteristic of the adsorbate occur at 9.4, 11.9 and 13.8 eV, which are fully developed at a CO exposure of 1 L. The losses at 11.9 and 13.8 eV are due to intramolecular Rydberg transitions of the CO molecule, already known from the literature on gaseous CO. Their occurrence thus indicates undissociated, molecular adsorption of the CO. The prominent loss at 9.4 eV could involve some contribution from intramolecular CO $5\sigma \rightarrow 2\pi^*$ transitions which are found in the free molecule at 8.5 eV; but an interpretation in terms of charge transfer transitions from occupied Cu(d) orbitals (substrate) into empty CO($2\pi^*$) adsorbate orbitals seems more likely. The latter interpretation would allow further interesting conclusions about the interaction of the adsorbed molecules with the substrate atoms, particularly since a

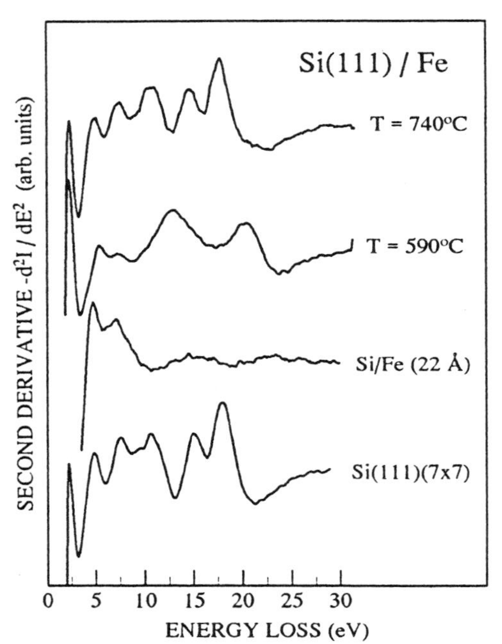

Fig. IX.4. Second-derivative electron-energy loss spectra measured by means of a CMA on a clean Si(111)-(7×7) surface, after evaporating a 22 Å thick Fe film and annealing the sample to 590 and 740 °C. The loss spectra were recorded with a primary energy of 100 eV and an AC peak-to-peak modulation voltage for differentiation of 1 eV; sample surface oriented normal to CMA axis; primary peak not shown [IX.7]

coverage dependence and surface specifity of this loss has been found [IX.6].

Figure IX.4 presents an example in which double differentiation EELS with a CMA is used to study the electronic properties of a thin overlayer system [IX.7]. When Fe is deposited onto a Si surface, annealing causes a chemical reaction involving Fe/Si interdiffusion and the formation of new compounds, the so-called Fe-silicides. Below about 500 °C a metallic silicide FeSi is formed, whereas for annealing temperatures between 550 ° and 700 °C a semiconducting overlayer of β-$FeSi_2$ results [IX.7]. The double differentiated loss spectrum of the clean Si(111)(7×7) surface (the primary peak is not shown) exhibits peaks at 2, 5, 8, 10.5, 15 and 17.8 eV. According to detailed studies of their origins [IX.8] the peak at 5 eV is due to the E_2 bulk interband transition, whereas the losses at 2, 8 and 15 eV correspond to surface state transitions of the (7×7) Si surface. The peaks near 17.8 and 10.5 eV originate from bulk and surface plasmon excitations, respectively. After deposition of 22 Å of iron the five new losses near 2.2, 5, 7, 15.5 and 23 eV are attributed to the Fe overlayer: the 23 eV feature arising from the Fe bulk plasmon and the remaining loss peaks from single-electron transitions between occupied and empty d-band states of the Fe [IX.9]. After annealing to 590 °C a 70 Å thick semiconducting β-$FeSi_2$ layer forms giving rise to characteristic losses at 2.4, 5.5, 7.5, 13.8 and 20.5 eV. The structure at 20.5 eV is most probably due to the valence-band bulk plasmon of $FeSi_2$. The 13.8 eV loss might be a superposition of surface- and interface ($FeSi_2$/Si) plasmon excitations. The remaining peaks are probably due to electronic d-band transitions within the $FeSi_2$ overlayer. Their slight shift to higher energies with respect to the corresponding peaks of the Fe overlayer is explained by the bonding shift to higher binding energies of the occupied d-orbitals in the $FeSi_2$ compound with respect to free Fe atoms. The topmost spectrum, measured after the last annealing step at 740 °C, shows a superposition of the silicide loss features with those typical for the clean Si surface. This demonstrates – in agreement with other experimental findings – that the β-$FeSi_2$ layer disintegrates thus exposing areas of free Si surface.

Panel IX

The two examples above of EELS with unmonochromatized electron beams and a hemispherical analyser or a CMA with double differentiation clearly show the usefulness of this technique for obtaining direct information about the chemical nature and the electronic structure of an adsorbate or a solid heterostructure.

An ever-growing area for the application of EELS is in connection with electron microscopy. The electron beam of a transmission or scanning electron microscope, in addition to its function in imaging, also induces all the inelastic processes given in Fig.IX.1. An electron analyser incorporated into a transmission electron microscope or into a scanning electron microprobe (Panel V: Chap.3) allows the analysis of these inelastically scattered electrons. Because of the high primary energies, 100 keV and above, used in both transmission and scanning electron microscopes, magnetic sector analysers are more effective for the energy analysis than the electrostatic analysers used in the low-energy regime (Panel II: Chap.1). The combination of EELS and electron microscopy is particularly advantageous since it combines all the spectral information from EELS with detailed spatial information. Characteristic excitation spectra can be attributed to particular spots on the sample, typically a thin film. In modern microprobes local analysis can thus be performed on spot sizes down to $10 \div 100$ Å. Characteristic core-level excitations, valence-band transitions, and plasmon losses allow the chemical identification, for example, of small embedded clusters or precipitates. Changes in the spectra indicate changes of the electronic structure within these small areas.

Figure IX.5 shows as an example a loss spectrum measured by a scanning electron microscope in transmission on a spot size with a diameter of about 50 Å on a thin Si/Fe/Si sandwich [IX.10]. The high-energy resolution, i.e., the low-energy spread of the primary peak of about 0.3 eV is obtained by the use of a field-emission cathode. A magnetic sector analyser serves as dispersive element for the energy analysis of the transmitted electron beam. Besides the Si bulk plasmon excitations (Si-PL and 2Si-PL as the double excitation), one also sees the Fe $3p_{1/2,3/2}$ transition and its combina-

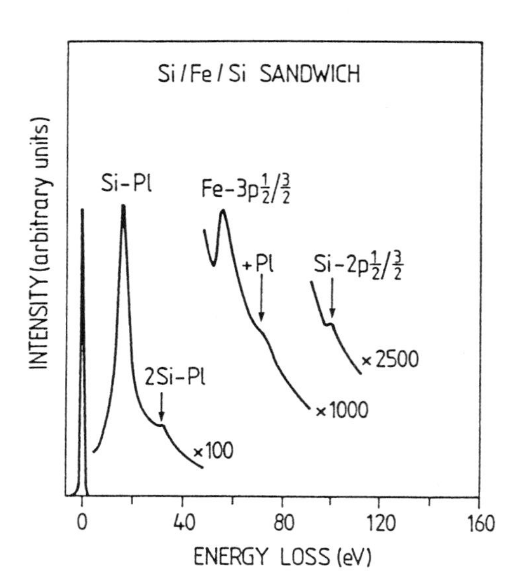

Fig. IX.5. Electron energy loss spectrum recorded by means of a scanning microprobe (VG company) in transmission on an unsupported Si/Fe/Si sandwich; primary energy 100 keV; energy width of primary peak about 0.3 eV [IX.10]

tion with a plasmon excitation (+PL). At even higher amplification the Si $2p_{1/2,3/2}$ excitation is also detected at a loss energy of about 100 eV. A clear identification of the material within spots of $10 \div 100$ Å diameter is thus possible. Scanning of the beam over different areas of the sample makes a microanalysis possible. The technique is particularly useful when applied in an UHV scanning electron microscope where surface contamination of the sample can be monitored.

References

IX.1 H. Raether: *Excitation of Plasmon and Interband Transitions by Electrons*. Springer Tracts Mod. Phys. Vol.88 (Springer, Berlin, Heidelberg 1980)

IX.2 H. Ibach, D.L. Mills: *Electron Energy Loss Spectroscopy and Surface Vibrations* (Academic, New York 1982)

IX.3 H. Lüth: Surf. Sci. **168**, 773 (1986)

IX.4 H. Ibach: *Electron Energy Loss Spectrometers*, Springer Ser. Opt. Sci. Vol.63 (Springer, Berlin, Heidelberg 1991)

IX.5 H. Ibach, M. Balden, D. Bruchmann, S. Lehwald: Surf. Sci. **269/270**, 94 (1992)

IX.6 A. Spitzer, H. Lüth: Surface Sci. **102**, 29 (1981)

IX.7 A. Rizzi, H. Moritz, H. Lüth: J. Vac. Sci. Technol. A **9**, 912 (1991)

IX.8 J.E. Rowe, H. Ibach: Phys. Rev. Lett. **31**, 102 (1973)

IX.9 E. Colavita, M. De Creszenzi, L. Papagno, R. Scarmozzino, L.S. Caputi, R. Rosei, E. Tosatti: Phys. Rev. B **25**, 2490 (1982)

IX.10 G. Crecelius (ISI, Research Center Jülich): Priv. commun. (1992)

Problems

Problem 4.1. The N_2O molecule has a linear atomic structure $N \equiv N = 0$ with σ, π and π^* molecular orbitals. Plot these orbitals qualitatively and discuss by means of symmetry considerations the orientation of the transition dipole moment

$$e \int \psi_f^* \, z \, \psi_i \, dr$$

with respect to the molecular axis z for the $\pi \to \pi^*$ transitions. ψ_i and ψ_f are the initial- and final-state wave functions, respectively.

Problem 4.2. Benzene (C_6H_6) molecules are adsorbed flat on the Pt(111) surface. Discuss, by symmetry arguments, the orientation of the electronic dipole moment for $\pi \to \pi^*$ transitions with respect to the surface normal. Are these transitions seen in an EELS measurement (reflection of a 100 eV electron beam)?

Problem 4.3. The clean, annealed Ge(111) surface exhibits a c(2×8) reconstruction. Discuss the corresponding superstructure in real and reciprocal space and plot the expected LEED pattern.

Problem 4.4. The unreconstructed (100) surface of an fcc crystal is covered by irregularly distributed adsorbate islands with an average diameter of 10 Å. The total adsorbate coverage amounts to about 50% of the surface atoms. The islands are formed by atoms, which are bonded on top of the substrate atoms and which exhibit a strong scattering probability in comparison to the substrate.

a) Discuss the LEED pattern of this surface.

b) How can one estimate the mean diameter of the islands from experimental data.

Problem 4.5. A GaAs(100) surface is covered by one monolayer of Si atoms. A Rutherford Back Scattering (RBS) experiment is made with 1.4 MeV He^+ ions along the surface normal ⟨100⟩.

Calculate the energy of the back scattered ions for the Si adsorbate peak and for the Ga and As peaks of the clean GaAs surface.

5. Surface Phonons

Classical bulk solid-state physics can, broadly speaking, be divided into two categories, one that relates mainly to the electronic properties and another in which the dynamics of the atoms as a whole or of the cores (nuclei and tightly bound core electrons) is treated. This distinction between lattice dynamics and electronic properties, which is followed by nearly every textbook on solid-state physics, is based on the vastly different masses of electrons and atomic nuclei. Displacements of atoms in a solid occur much more slowly than the movements of the electrons. When atoms are displaced from their equilibrium position, a new electron distribution with higher total energy results; but the electron system remains in its ground state, such that after the initial atomic geometry has been reestablished, the whole energy amount is transferred back to the lattice of the nuclei or cores. The electron system is not left in an excited state. The total electronic energy can therefore be considered as a potential for the movement of the nuclei. On the other hand, since the electronic movement is much faster than that of the nuclei, a first approximation for the dynamics of the electrons is based on the assumption of a static lattice with fixed nuclear positions determining the potential for the electrons. This approximation of separate, non-interacting electron dynamics and lattice (nuclear/core) dynamics is called the **adiabatic approximation**. It was introduced into solid-state and molecular physics by *Born* and *Oppenheimer* [5.1]. It is clear, however, that certain phenomena, such as the scattering of conduction electrons on lattice vibrations, are beyond this approximation.

For surface and interface physics the same arguments are valid and therefore, within the framework of the adiabatic approximation, the dynamics of surface atoms (or cores) and of surface electrons can be treated independently.

The lattice vibrations of atoms near the surface are expected to have frequencies different from those of bulk vibrations since, on the vacuum side of the surface, the restoring forces are missing. The properties of surface lattice vibrations and the conditions for their existence will be the subject of this chapter. Like the corresponding bulk excitations, surface vibrations are in principle quantized, although a classical treatment is sufficient in many cases because of the relatively high atomic masses and the small energy of the resulting quanta. The quanta of surface vibrations are called **surface phonons**.

In contrast to bulk solid-state physics, for surfaces, the distinction between surface lattice dynamics and surface electronic states is not a sufficient classification. Surface physics treats not only clean surfaces, but also surfaces with well-defined adsorbates. Surface physics therefore includes, besides the electron and lattice dynamics of the clean surface, a third important field, that of surfaces with adsorbed molecules or atoms (Chap.9). For these systems one can also apply the adiabatic approximation, i.e., the vibrations and electronic states of an adsorbed atom or molecule can be considered separately. The same is true for the interface layer between two solids, e.g., a semiconductor film epitaxially grown on a different semiconductor substrate. At the interface itself the atoms of the two "touching" materials display characteristic vibrational and electronic properties.

5.1 The Existence of "Surface" Lattice Vibrations on a Linear Chain

As in the bulk case, the essential characteristics of surface lattice dynamics can be demonstrated using the simple model of a diatomic linear chain (Fig.5.1). A model for the surface of a 3D solid is then obtained by arranging an infinite number of chains with their axes normal to the surface in a regular array, i.e. with 2D translational symmetry parallel to the surface (Fig.5.2). In the present context the chains are not extended over the whole infinite space – as in the bulk case, but they end at the surface (semi-infinite case). Nevertheless, to a rough approximation, the dynamical equations can be assumed to be unchanged with respect to those of an infinite chain:

$$M\ddot{s}_n^{(1)} = f(s_n^{(2)} - s_n^{(1)}) - f(s_n^{(1)} - s_{n-1}^{(2)}) ,$$

i.e.,

$$M\ddot{s}_n^{(1)} = -f(2s_n^{(1)} - s_n^{(2)} - s_{n-1}^{(2)}) \tag{5.1a}$$

and

$$m\ddot{s}_n^{(2)} = -f(2s_n^{(2)} - s_{n+1}^{(1)} - s_n^{(1)}) . \tag{5.1b}$$

Changes of force constants and reconstructions at the surface are not considered in this simple model.

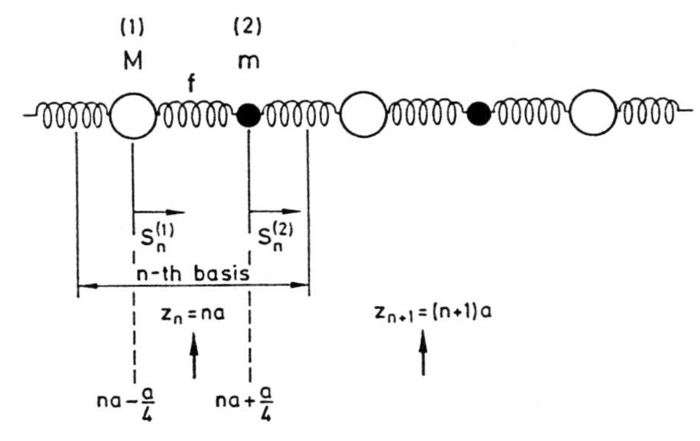

Fig.5.1. The model of a diatomic linear chain with two different atomic masses M(1) and m(2). A single restoring force f is assumed between the masses. The position of the nth unit cell is described by its geometrical centre $z_n = na$; the displacements of the two atoms in the nth unit cell from equilibrium are $s_n^{(1)}$ and $s_n^{(2)}$

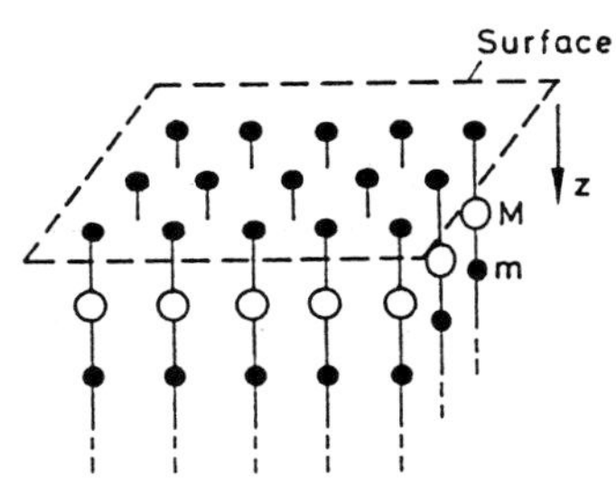

Fig.5.2. 2D arrangement of diatomic linear chains with translational symmetry in the surface. This model shows some characteristics of surface lattice dynamics

The plane-wave ansatz

$$s_n^{(1)} = M^{-1/2} c_1 \exp\{i[ka(n-\tfrac{1}{4}) - \omega t]\}, \tag{5.2a}$$

$$s_n^{(2)} = m^{-1/2} c_2 \exp\{i[ka(n+\tfrac{1}{4}) - \omega t]\}, \tag{5.2b}$$

leads to the equations

$$-\omega^2 M^{1/2} c_1 = -fc_1 M^{-1/2} + 2fc_2 \cos\frac{ka}{2} m^{-1/2}, \tag{5.3a}$$

$$-\omega^2 m^{1/2} c_2 = -fc_2 m^{-1/2} + 2fc_1 \cos\frac{ka}{2} M^{-1/2}, \tag{5.3b}$$

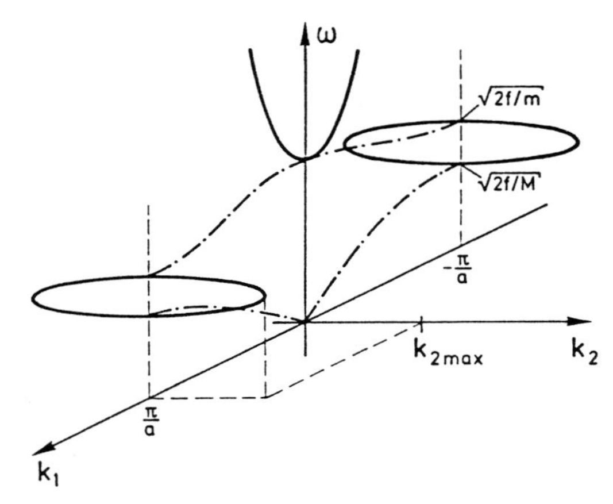

Fig.5.3. Dispersion branches for "bulk" (-··-) and "surface" (–) lattice vibrations (phonons) of a semi-infinite diatomic chain (atomic masses M,m; restoring force f). k_1 and k_2 are the real and imaginary parts of the complex wavevector, i.e., k_2 is the exponential decay constant of the "surface" phonons

which, for an infinite chain, have the solutions:

$$\omega_{\pm}^2 = \frac{f}{Mm}\left[(M+m) \pm \sqrt{(M+m)^2 - 2Mm(1-\cos ka)} \right] . \tag{5.4}$$

The frequencies $\omega_-(k)$ and $\omega_+(k)$ correspond to the well-known acoustic and optic dispersion branches of lattice waves for the infinite chain (Fig.5.3).

For surfaces, one may modify the model in the following way. The chain is terminated at one end, but extends to infinity in the other direction. Therefore, far away from the free end, approximately the same solutions exist as for the infinite chain. Furthermore, real lattice vibrations have, in any case, a finite correlation length because of anharmonic interactions. We now seek new solutions to (5.1-3) which are localized near the end of the chain, i.e., which have a negligible vibrational amplitude far away from the end of the chain in the bulk. This can be achieved by considering waves whose amplitude decays exponentially away from the end of the chain. For this purpose we make an ansatz with a complex wave vector

$$\tilde{k} = k_1 + ik_2 , \tag{5.5}$$

but we require the frequencies ω to be real. Is it possible to solve (5.4) with real $\omega_{\pm}$ but with complex $\tilde{k}$? The imaginary part k_2 would lead to exponentially decaying waves as required. Using the relations

$$\cos(iz) = \cosh(z) , \quad \sin(iz) = i\sinh(z) , \tag{5.6}$$

we can express cos(ka) in (5.4) as

$$\cos(\tilde{k}a) = \cos(k_1 a)\cosh(k_2 a) - i\sin(k_1 a)\sinh(k_2 a) . \tag{5.7}$$

Because of the reality condition on $\omega_\pm$, $\mathrm{Im}\{\cos(\tilde{k}a)\}$ in (5.7) must vanish, i.e.,

$$\mathrm{Im}\{\cos(\tilde{k}a)\} = \sin(k_1 a)\sinh(k_2 a) = 0 . \tag{5.8}$$

The solution with $k_2 = 0$ yields the bulk dispersion branches (5.4). For the surface solutions we require

$$k_2 \neq 0 \quad \text{and} \quad k_1 a = n\pi \quad \text{with} \quad n = 0, \pm 1, \pm 2, \cdots . \tag{5.9}$$

We are interested in solutions for the first bulk Brillouin zone and therefore consider the cases $n = 0, 1$, i.e.,

$$\cos(\tilde{k}a) = \cos(n\pi)\cosh(k_2 a) = (-1)^n \cosh(k_2 a) ; \quad n = 0, 1 . \tag{5.10}$$

The possible frequencies of surface solutions are therefore

$$\omega_\pm^2 = \frac{f}{Mm}\left\{(M+m) \pm \sqrt{(M+m)^2 - 2Mm[1 - (-1)^n \cosh(k_2 a)]}\right\} , \tag{5.11}$$

where the quantity under the square-root sign must be positive because we require real $\omega_\pm^2$ values. The solution with $n = 0$, i.e., $k_1 = 0$ at the Γ-point of the Brillouin zone (in k_1) is

$$\omega^2(k_1 = 0, k_2) =$$

$$= \frac{f}{Mm}\left[(M+m) + \sqrt{(M+m)^2 - 2Mm[1 - \cosh(k_2 a)]}\right] . \tag{5.12}$$

Since $[1 - \cosh(k_2 a)]$ is negative for all $k_2 a$, there is no restriction on k_2, but only the positive square root in (5.12) is a solution. The curvature of (5.12) with respect to k_2 is always positive and the value of $\omega(k_1 = 0, k_2 = 0)$ equals that of the bulk optical branch $[2f(1/M + 1/m)]^{1/2}$ at Γ. Figure 5.3 shows that at Γ ($k_1 = 0$) these surface solutions are possible with frequencies above the maximum bulk phonon frequency.

The solutions of (5.11) with $n = 1$, i.e., $k_1 = \pi/a$, are located in k-space at the Brillouin-zone boundary. The condition for a real square root now reads

$$|k_2| < \frac{1}{a} \operatorname{arccosh} \frac{M^2 + m^2}{2Mm} \equiv k_{2max} . \tag{5.13}$$

Thus there exist the solutions

$$\omega_{\pm}^2(k_1 = \pi/a, k_2) =$$

$$= \frac{f}{Mm}\left\{(M+m) \pm \sqrt{(M+m)^2 - 2Mm[1 + \cosh(k_2 a)]}\right\} \qquad (5.14)$$

only for a limited range of k_2 values (5.13). For $k_2 = 0$ the solutions are

$$\omega_+(k_2 = 0) = (2f/m)^{1/2} \quad \text{and} \quad \omega_-(k_2 = 0) = (2f/M)^{1/2} . \qquad (5.15a)$$

At the maximum value k_{2max} one obtains

$$\omega_{\pm}(k_2 = k_{2max}) = \sqrt{f(1/M + 1/m)} . \qquad (5.15b)$$

Both branches $\omega_{\pm}$ are continuous at k_{2max} and have frequencies at $k_2 = 0$ that are identical to those of the bulk acoustic and optical lattice vibrations at the zone boundary. The possible surface vibrational frequencies fill the range between the acoustic and optical branches of the bulk excitations (Fig.5.3). Boundary conditions at the surface impose further restrictions (Sect.5.2).

The possible displacements of atom i in the "surface" modes follow according to (5.2) as

$$s_n^{(i)} = C_i \exp[i(\tilde{k} z_n^i - \omega t)] , \qquad (5.16)$$

where

$$z_n^{(i)} = a(n - 1/4) \quad \text{for atom (1) = (i) ,} \quad \text{and } a(n + 1/4) \text{ for atom (2) = (i)}$$

are the corresponding atomic coordinates. From Fig.5.3 one sees that $\tilde{k}$ can have the values

$$\tilde{k} = k_1 + ik_2 = \pm\pi/a + ik_2 \qquad (5.17)$$

at the Γ- point ($k_1 = 0$) and at the boundaries of the Brillouin zone, respectively. Apart from different, constant phase factors all these solutions are vibrations of the form

$$s_n^{(i)} \propto \exp(-k_2 z_n^{(i)}) e^{-i\omega t} , \quad k_2 > 0 , \qquad (5.18)$$

whose vibrational amplitude decays exponentially away from the end of the chain, i.e. the surface at $z = 0$, into the interior of the chain.

5.2 Extension to a Three-Dimensional Solid with a Surface

Qualitatively, it is relatively easy to extend the above arguments to the case of a 3D solid with a surface. This is illustrated in Fig.5.2 where the finite solid is modelled by a regular array of parallel, semi-infinite chains. This model is only realistic in cases where the chemical bonds in directions parallel to the surface are weak, i.e., for strongly anisotropic solids. Nevertheless, we can use it to provide a qualitative idea for the features associated with general surface vibrational modes.

For every chain we have the possible vibrational modes (5.18) discussed above. However, different chains might vibrate with different phases. Due to the weak interaction between the chains the phases are correlated with each other. The phase difference can be described by a wave vector $\mathbf{k}_{\parallel}$ parallel to the surface. Since we are interested in wave propagation parallel to the surface, the wave vector of a general 3D-lattice vibration

$$\mathbf{s}_{\mathbf{k}}(\mathbf{r}) = A\hat{\mathbf{e}}_{\mathbf{k}}\, e^{i(\mathbf{k}\cdot\mathbf{r} - \omega t)} \tag{5.19}$$

can be split up into a part parallel to the surface $\mathbf{k}_{\parallel}$ describing the plane wave moving parallel to the surface, and a part $k_{\perp}$ normal to the surface. With $\mathbf{k} = \mathbf{k}_{\parallel} + \mathbf{k}_{\perp}$ and $\mathbf{r} = \mathbf{r}_{\parallel} + \hat{\mathbf{e}}_z z$, (5.19) yields

$$\mathbf{s}_{\mathbf{k}}(\mathbf{r}) = A\hat{\mathbf{e}}_{\mathbf{k}} \exp[i(\mathbf{k}_{\parallel}\cdot\mathbf{r}_{\parallel} + k_{\perp}z - \omega t) \ . \tag{5.20a}$$

Parallel to the surface plane waves with real $\mathbf{k}_{\parallel}$ are possible, but normal to the surface only solutions of the type (5.18) with imaginary $k_{\perp} = ik_2$ need be considered. The decay constant k_2 is often designated by $\kappa_{\perp}$. One obtains the following general form for a surface lattice vibration:

$$\mathbf{s}_{\mathbf{k}_{\parallel},\kappa_{\perp}} = A\hat{\mathbf{e}}_{\mathbf{k}_{\parallel},\kappa_{\perp}}\, e^{-\kappa_{\perp}z} \exp[i(\mathbf{k}_{\parallel}\cdot\mathbf{r}_{\parallel} - \omega t)] \ . \tag{5.20b}$$

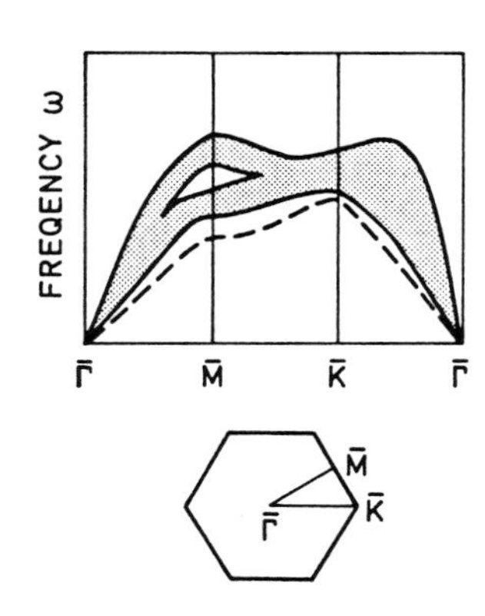

Fig.5.4. Qualitative picture of a 2D surface phonon dispersion relation (**a**) along the symmetry lines $\bar{\Gamma}\bar{M}$, $\bar{M}\bar{K}$ and $\bar{K}\bar{\Gamma}$ of the 2D surface Brillouin zone of a *hexagonal* (111) surface of a fcc lattice (**b**). The surface phonon dispersion is given by the dashed line in (**a**). The shaded area indicates the range of bulk phonon frequencies at all possible $k_{\perp}$ wave vectors for $k_{\parallel}$ values on the symmetry lines

Equation (5.20b) is only valid for primitive unit cells; if there is more than one atom per unit cell, an additional index (i) as in (5.18) describes the particular type of atom.

A surface vibrational mode is therefore characterized by its frequency ω (or quantum energy $\hbar\omega$), its wave vector $\mathbf{k}_{\parallel}$ parallel to the surface, and the decay constant $\kappa_{\perp}$, which determines the decay length of the vibrational amplitude from the surface into the interior of the crystal. These quantities are not independent of one another. They are related via the dynamical equations (as in the 3D bulk case) and via the boundary condition that no forces should act from the vacuum side on the topmost layer of surface

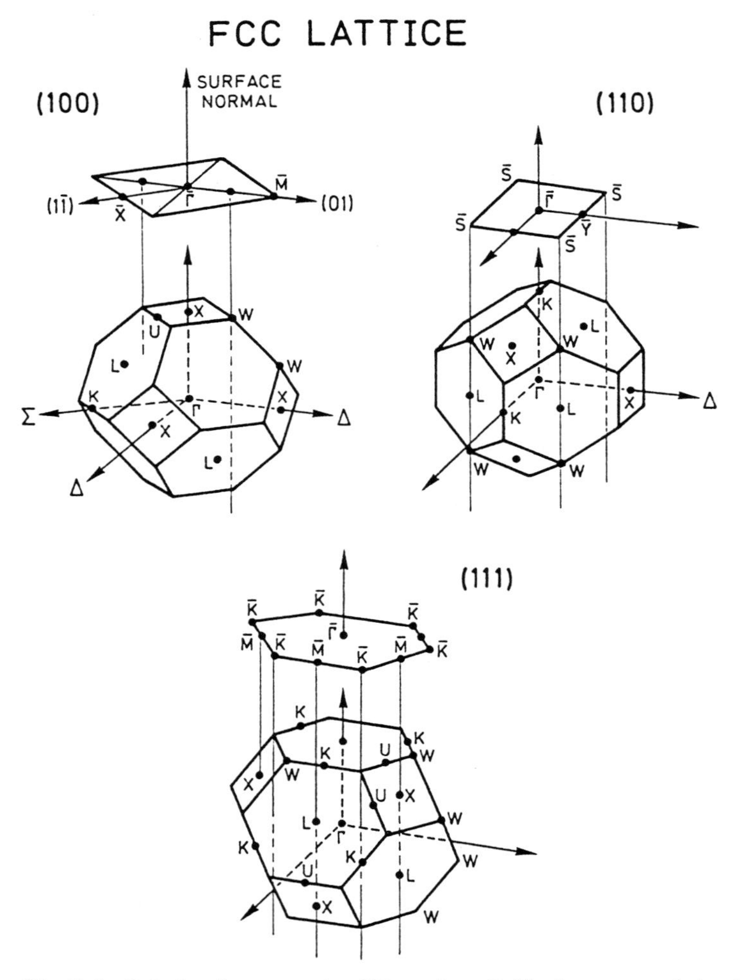

Fig.5.5. Relation between the 2D surface Brillouin zones of the (100), (111) and (110) surfaces of a fcc lattice and the bulk Brillouin zone

atoms. Thus from the "continuous" spectrum of possible surface-mode frequencies between the acoustic and optical bulk modes and above (Fig.5.3) these restrictions select (for a primitive unit cell) one particular frequency ω for each $\mathbf{k}_{\|}$ and $\kappa_{\perp}$. For a crystal with two atoms per unit cell both an acoustic and an optical surface phonon branch exist.

In analogy to the bulk case, surface phonons can therefore be described by a 2D dispersion relation $\omega(\mathbf{k}_{\|},\kappa_{\perp})$. The function $\omega(\mathbf{k}_{\|},\kappa_{\perp})$ is periodic in 2D reciprocal space. The usual way to display the dispersion relation $\omega(\mathbf{k}_{\|},\kappa_{\perp})$ is by plotting the function $\omega(\mathbf{k}_{\|})$ along certain symmetry lines of the 2D Brillouin zone (Fig.5.4). In these plots it is usual to show the bulk phonons, too, since they also contribute to the possible modes close to the surface. For a particular surface, all bulk modes with a certain $\mathbf{k}_{\|}$ have to be taken into account. The projection of the bulk modes at a fixed $\mathbf{k}_{\|}$ and

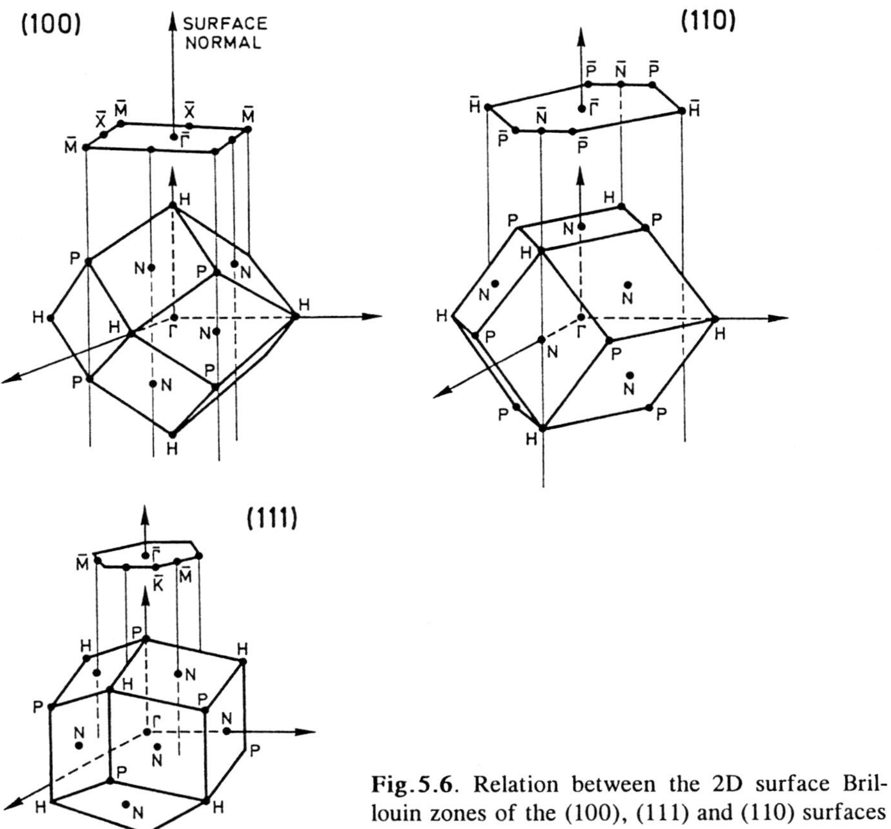

Fig.5.6. Relation between the 2D surface Brillouin zones of the (100), (111) and (110) surfaces of a bcc lattice and the bulk Brillouin zone

HCP LATTICE

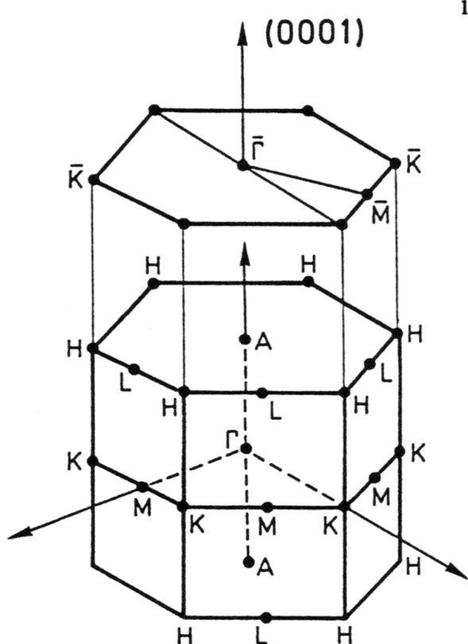

Fig.5.7. Relation between the (0001) surface Brillouin zone of a hcp lattice and the corresponding bulk Brillouin zone

for all $\mathbf{k}_\perp$ yields in the 2D plot (Fig.5.4) a continuous area of possible $\omega(\mathbf{k}_\parallel)$ values. In order to generate plots such as Fig.5.4 one has to project the 3D bulk dispersion branches onto the particular 2D surface Brillouin zone; i.e. certain bulk directions and points of high symmetry in the 3D Brillouin zone are projected onto the 2D surface zone. How this is done for some low-index faces of common 3D lattices is depicted in Figs.5.5-7.

It is worth mentioning that a more rigorous treatment of surface lattice dynamics [5.2] leads to a simple scaling rule which connects the decay length $\kappa_\perp$ of the vibrational amplitude of a surface phonon to its wave vector $k_\parallel$: in the non-dispersive regime where $d\omega/dk_\parallel$ is constant, i.e., for small wave vectors $k_\parallel$, the decay constant $\kappa_\perp$ is proportional to $k_\parallel$; the longer the wavelength of the surface vibration, the deeper its vibrational amplitude extends into the solid.

Similar considerations as applied here to the solid/vacuum interface lead, for the solid/solid interface, e.g. at an epitaxially grown semiconductor overlayer (Chap.8), to the existence of interface phonons. Their vibrational amplitude decays exponentially into each solid on both sides of the interface [5.3].

5.3 Rayleigh Waves

In the study of bulk solids the dispersionless part of the acoustic phonons was well known as sound waves long before the development of lattice dynamics [5.4]. Debye used this well-known part of the phonon spectrum to evaluate his approximation for the lattice specific heat. A similar situation holds for the dispersionless low-frequency part of the surface phonon dispersion branches. Part of these surface phonon modes were already known in 1885 as Rayleigh surface waves of an elastic continuum filling a semi-infinite halfspace [5.5, 6]. In classical continuum theory one can only describe lattice vibrations whose wavelength is long compared to the interatomic separation. A macroscopic deformation of a solid continuum can therefore be described in terms of displacements of volume elements dv whose dimensions are large in relation to interatomic distances, but small in comparison with the macroscopic body. The important variables in this sense are the displacements $\mathbf{u} = \mathbf{r}' - \mathbf{r}$ of these volume elements dv and the strain tensor

$$\epsilon_{ij} = \frac{1}{2}\left(\frac{\partial u_j}{\partial x_i} + \frac{\partial u_i}{\partial x_j}\right) . \tag{5.21}$$

In the elastic regime ϵ_{ij} is related to the stress field $\sigma_{kl} = \partial F_k / \partial f_l$ (force in k direction per area element in l direction) via the elastic compliances

$$\epsilon_{ij} = \sum_{kl} S_{ijkl} \sigma_{kl} . \tag{5.22}$$

In this continuum model the time variation and spatial structure of an elastic wave can be given in terms of the displacement field $\mathbf{u}(\mathbf{r}, t)$ which describes the microscopic movement of little volume elements containing a considerable number of elementary cells (Note that in the long-wavelength limit, neighboring elementary cells behave identically). The Rayleigh waves that are solutions of the wave equation for an elastic continuous half-space are obtained in the following way: every vector field – including the displacement field $\mathbf{u}(\mathbf{r}, t)$ – can be split up into a turbulence-free and a source-free part $\mathbf{u}'$ and $\mathbf{u}''$:

$$\mathbf{u} = \mathbf{u}' + \mathbf{u}'' \tag{5.23a}$$

with

$$\mathrm{curl}\,\mathbf{u}' = 0 \quad \text{and} \quad \mathrm{div}\,\mathbf{u}'' = 0 . \tag{5.23b}$$

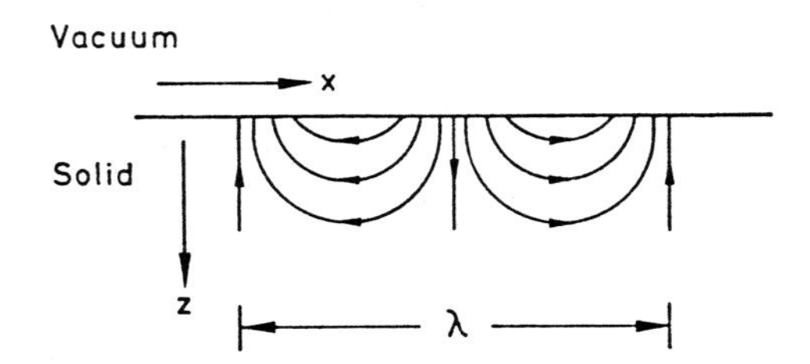

Fig.5.8. Displacement field $\mathbf{u}(\mathbf{r},t)$ (instantaneous picture) of a Rayleigh surface wave travelling in the x-direction along the boundary of a semi-infinite continuous solid halfspace

In the bulk, differential wave equations can be solved for both contributions giving longitudinal sound waves $[\mathbf{u}'(\mathbf{r},t)]$ and transverse (shear) sound waves $[\mathbf{u}''(\mathbf{r},t)]$ with the sound velocities c_l and c_t, respectively. In the present situation of a semi-infinite halfspace we assume a coordinate system as shown in Fig.5.8, and try solutions that are dependent only on x ($\|$ to the surface) and z ($\perp$ to the surface). Because of (5.23b) we can introduce two new functions ϕ and ψ which have the character of potentials

$$\mathbf{u}' = -\,\mathrm{grad}\phi \ , \tag{5.24}$$

$$u_x'' = -\frac{\partial\psi}{\partial z} \ , \quad u_z'' = \frac{\partial\psi}{\partial x} \ . \tag{5.25}$$

The definition (5.25) is possible because

$$\mathrm{div}\mathbf{u}'' = \frac{\partial u_x''}{\partial x} + \frac{\partial u_z''}{\partial z} = 0 \ . \tag{5.26}$$

Instead of treating the displacement field $\mathbf{u}(\mathbf{r},t)$ directly, we can use the functions ϕ (x, z) and ψ (x, z). The function ϕ describes longitudinal excitations, whereas ψ is related to the transverse part of the displacement field. The general equations of motion for an isotropic, 3D bulk, elastic solid can be reduced to wave equations for the generalized potentials ϕ and ψ. In analogy to the bulk problem we therefore try to solve the following wave equations for the semi-infinite halfspace:

$$c_l^{1/2}\frac{\partial^2\phi}{\partial t^2} - \Delta\phi = 0 \ , \quad c_t^{1/2}\frac{\partial^2\psi}{\partial t^2} - \Delta\psi = 0 \ . \tag{5.27}$$

In accordance with the character of ϕ and ψ Eq.(5.27) contains the longitudinal and the transverse sound velocities c_l and c_t. For the solution of (5.27) we try the ansatz of surface waves travelling parallel to the surface along x with an amplitude dependent on z ($\mathbf{u}$ must vanish for $z \to \infty$):

$$\phi(x,z) = \xi(z)e^{i(kx-\omega t)} \ , \quad \psi(x,z) = \eta(z)e^{i(kx-\omega t)} \ . \tag{5.28}$$

From (5.27) it then follows that

$$\xi'' - p^2 \xi = 0 \quad \text{with} \quad p^2 = k^2 - (\omega/c_l)^2 \ , \tag{5.29a}$$

$$\eta'' - q^2 \eta = 0 \quad \text{with} \quad q^2 = k^2 - (\omega/c_t)^2 \ . \tag{5.29b}$$

For $p^2 > 0$ and $q^2 > 0$, the amplitudes ξ and η are clearly exponential functions that decay into the bulk of the material as

$$\xi = A e^{-pz} \ , \quad \eta = B e^{-qz} \ . \tag{5.30}$$

The final solutions have the character typical of surface excitations (5.20):

$$\phi = A e^{-pz} e^{i(kx - \omega t)} \quad \text{with} \quad p = \sqrt{k^2 - (\omega/c_l)^2} \ , \tag{5.31a}$$

$$\psi = B e^{-qz} e^{i(kx - \omega t)} \quad \text{with} \quad q = \sqrt{k^2 - (\omega/c_t)^2} \ . \tag{5.31b}$$

The displacement field $(u_x, 0, u_z)$ is then derived from (5.24 and 25) by differentiation of (5.31):

$$u_x = -\frac{\partial \phi}{\partial x} - \frac{\partial \psi}{\partial z} \ , \quad u_z = -\frac{\partial \phi}{\partial z} + \frac{\partial \psi}{\partial x} \ . \tag{5.32}$$

From (5.32) we see that the displacement field of the surface excitation contains both a longitudinal and a transverse contribution; the wave is of mixed longitudinal-transverse character and its velocity must thus depend on both c_l and c_t. For the further evaluation of the Rayleigh wave phase velocity ω/k we use the boundary condition that at the very surface ($z = 0$) there is no elastic stress, i.e.,

$$\sigma_{zz}\big|_{z=0} = \sigma_{yz}\big|_{z=0} = \sigma_{xz}\big|_{z=0} = 0 \ . \tag{5.33}$$

In the subsequent somewhat tedious calculation [5.6], the elastic constants enter via (5.21 and 22). But even with the assumption of incompressibility for the semi-infinite continuum only an approximate solution is possible. One obtains the phase velocity of the Rayleigh wave as

$$c_{RW} \simeq (1 - 1/24) c_t \tag{5.34}$$

and the direct relations between its wave vector k and the parameters p and q

$$p \simeq k , \quad q \simeq k(12)^{-1/2} . \tag{5.35}$$

From (5.34) we see that the phase velocity of Rayleigh surface waves is even lower than the transverse sound velocity [5.5, 6]; this is also true for cubic crystals. Figure 5.8 illustrates qualitatively the spatial structure of the displacement field of a Rayleigh wave with wavelength $\lambda = 2\pi/k$. The mixed longitudinal-transverse character is seen from the direction of the displacements which are partially parallel and partially normal to the propagation direction x.

It should also be emphasized that the treatment in this section is based on the continuum case in which the neglect of atomic structure leads to a Rayleigh wave that shows no dispersion (like for bulk sound waves). Extending the analysis to an atomically structured medium like a crystal, the surface phonon branches will show dispersion, in particular near the Brillouin-zone boundary. In Fig.5.4 the dispersion branch indicated by the dashed line qualitatively, reflects what one can expect for Rayleigh surface phonons. Some results from experiments and more realistic calculations are presented in Sect.5.6.

5.4 The Use of Rayleigh Waves as High-Frequency Filters

Experimentally, Rayleigh waves can be excited by a variety of methods. In principle one has to induce an elastic surface strain of adequate frequency. Atomic and molecular beam scattering (Panel X: Chap.5) can be used, as can Raman scattering, in particular at low frequencies with high resolution, i.e., Brillouin scattering. For piezoelectric crystals and ceramics there is a particularly convenient way to excite Rayleigh waves. These materials are characterized by an axial crystal symmetry. Stress along such an axis produces an electric dipole moment in each unit cell of the crystal due to an unequal displacement of the different atoms in the cell. Simple examples are the wurtzite structure of ZnO which is built up along its hexagonal c-axis by double layers of Zn and O ions. Stress along the c-axis displaces the Zn and O lattice planes by different amounts and a dipole moment in the c-direction results. Other examples are the III-V semiconductors that crystallize in the zinc-blende structure with an axial symmetry along the four {111} cubic cell diagonals. For practical purposes quartz and specially designed titanate ceramics are more important. A general description of the piezoelectric effect may be given in terms of the third rank piezoelectric tensor d_{ijk} which relates a polarization P_i to a general mechanical stress ϵ_{jk}:

$$P_i = \sum_{jk} d_{ijk}\epsilon_{jk} \, , \quad \epsilon_{ij} = \sum_{k} \bar{d}_{ijk}\mathcal{E}_k \, . \tag{5.36}$$

Equation (5.36b) with the so-called inverse piezoelectric tensor $\bar{d}_{ijk}$ describes the inverse phenomenon by which an electric field $\mathcal{E}$ applied in a certain direction produces a mechanical strain ϵ_{ij} in such crystals. On surfaces of piezoelectric crystals the mechanical strain associated with Rayleigh waves can thus be induced by appropriately chosen electric fields; these are applied by evaporated metal grids (Fig.5.9). A high-frequency voltage $U_i(\omega)$ applied to the left-hand grids in Fig.5.9a gives rise via (5.36b) to a surface strain field, which varies harmonically in time with frequency ω and has a wavelength λ determined by the grid geometry. If ω and λ (i.e., ω and $k = 2\pi/\lambda$) are values which fall on the dispersion curve for Rayleigh waves of that material, such surface waves are excited. They travel along

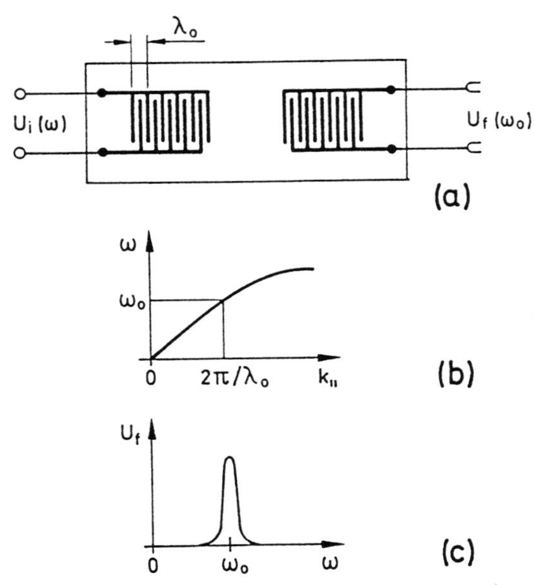

Fig.5.9. (**a**) Schematic drawing of a Rayleigh-wave high-frequency filter. Two sets of metallic grids are evaporated onto a piezoelectric plate. The rod spacing λ_0 determines the wavevector $k_0 = 2\pi/\lambda_0$ of the excited surface wave. (**b**) Through the Rayleigh-wave dispersion relation the frequency ω_0 is fixed by λ_0. (**c**) If a continuous spectrum is fed in as input voltage $U_i(\omega)$, only a sharp band $U_f(\omega_0)$ is transmitted through the device

the surface and excite a corresponding time-varying polarization which produces an electric signal in the right-hand grid structure. The grid geometry determines a fixed wavelength λ_0. Because of the single-valued dispersion relation for surface Rayleigh waves this λ_0 allows only a particular frequency ω_0 for the surface phonons. High-frequency signals $U_i(\omega_0)$ can pass the device and appear as an output signal $U_f(\omega_0)$ only if $\omega_0(2\pi/\lambda_0)$ is a particular point on the Rayleigh dispersion curve (Fig.5.9b). This dispersion curve and the geometry of the grids (equal for antenna and receiver) therefore determine the pass frequency of the filter. To give a numerical example: grids with rod distances in the 100 μm range can easily be evaporated. For a Rayleigh wave velocity of 4000 m/s we obtain a frequency of $\omega \simeq 24 \cdot 10^7 \, s^{-1}$ or $\nu \simeq 40$ MHz. Such surface-wave devices are used, e.g., in television equipment as band-pass filters for image frequencies.

5.5 Surface-Phonon (Plasmon) Polaritons

In Sect.5.3. we considered one limiting case of the surface-phonon spectrum, namely the nondispersive acoustic type of vibrations that are derived from the bulk sound waves. A similar treatment is possible for the long-wavelength optical phonons of an InfraRed (IR) active crystal. There is a certain type of optical surface phonon that is derived from the corresponding bulk TO and LO modes near $\mathbf{k} = 0$. As in the bulk, these surface modes are connected with an oscillating polarization field. Besides the dynamics of the crystal, the calculation must therefore also take into account Maxwell's equations which govern the electromagnetic field accompanying the surface vibration.

We consider a planar interface located at $z = 0$ between two non-magnetic ($\mu = 1$) isotropic media. The two media, each filling a semi-infinite halfspace, are characterized by their dielectric functions $\epsilon_1(\omega)$ for $z > 0$ and $\epsilon_2(\omega)$ for $z < 0$, respectively. The dynamics of the two media is contained in their dielectric functions; the IR activity, for example, can be expressed in terms of an oscillator type $\epsilon(\omega)$ with ω_{TO} as resonance frequency (TO denotes the transverse optical phonon at Γ). The particular case of a clean surface in vacuum is contained in our analysis for $\epsilon_2 = 1$ (or $\epsilon_1 = 1$). In general, electromagnetic waves propagating inside a non-magnetic ($\mu = 1$) medium with dielectric function $\epsilon(\omega)$ obey the dispersion law (derived from the differential wave equation):

$$k^2 c^2 = \omega^2 \epsilon(\omega) \, . \tag{5.37}$$

We look for modifications of (5.37) due to the presence of the interface. We start from the "equation of motion" for the electric field $\mathcal{E}(\mathbf{r},t)$. From Maxwell's equations we obtain for a nonmetal ($\mathbf{j} = \mathbf{0}$).

$$\mathrm{curl\,curl}\,\mathcal{E} = -\mu_0\,\mathrm{curl}\,\dot{\mathbf{H}} = -\mu_0\epsilon_0\epsilon(\omega)\ddot{\mathcal{E}}\ , \tag{5.38a}$$

i.e.,

$$-c^2\,\mathrm{curl\,curl}\,\mathcal{E} = \epsilon(\omega)\ddot{\mathcal{E}}\ , \tag{5.38b}$$

and from charge neutrality

$$\mathrm{div}[\epsilon(\omega)\mathcal{E}] = 0\ . \tag{5.39}$$

Special solutions localized at the interface should be wave-like in two dimensions (parallel to the interface) with an amplitude decaying into the two media for $z \gtrless 0$:

$$\mathcal{E}_1 = \hat{\mathcal{E}}_1 \exp[-\kappa_1 z + i(\mathbf{k}_{||}\cdot\mathbf{r}_{||}+\omega t)] \quad \text{for} \quad z>0\ , \tag{5.40a}$$

$$\mathcal{E}_2 = \hat{\mathcal{E}}_2 \exp[\kappa_2 z + i(\mathbf{k}_{||}\cdot\mathbf{r}_{||}+\omega t)] \quad \text{for} \quad z<0\ , \tag{5.40b}$$

with $\mathbf{r}_{||} = (x,y)$, $\mathbf{k}_{||} = (k_x,k_y)$ parallel to the interface and $\mathrm{Re}\{\kappa_1\}$, $\mathrm{Re}\{\kappa_2\} > 0$. From (5.39) we obtain

$$i\mathcal{E}\cdot\mathbf{k}_{||} = \mathcal{E}_z\kappa\ , \quad z\neq 0\ , \tag{5.41}$$

with $\kappa = \kappa_1$ and $\kappa = \kappa_2$ for media (1) and (2), respectively. Equation (5.41) excludes solutions with $\mathcal{E}$ normal to $\mathbf{k}_{||}$ and $\mathcal{E}_z \neq 0$, which are localized at the interface; localized waves must be sagittal with amplitudes

$$\hat{\mathcal{E}}_1 = \hat{\mathcal{E}}_1(\mathbf{k}_{||}/k_{||},\ -ik_{||}/\kappa_1) \tag{5.42a}$$

$$\hat{\mathcal{E}}_2 = \hat{\mathcal{E}}_2(\mathbf{k}_{||}/k_{||},\ +ik_{||}/\kappa_2)\ . \tag{5.42b}$$

If we insert the ansatz (5.40) together with the amplitudes (5.42) into (5.38b), we obtain dispersion laws similar to (5.37):

$$(k_{||}^2 - \kappa_1^2)c^2 = \omega^2\epsilon_1(\omega)\ , \tag{5.43a}$$

$$(k_{||}^2 - \kappa_2^2)c^2 = \omega^2\epsilon_2(\omega)\ . \tag{5.43b}$$

We now have to match the solutions $\mathcal{E}_1$ and $\mathcal{E}_2$ at the interface, i.e., we require

$$\mathcal{E}_1^{||} = \mathcal{E}_2^{||} \quad \text{and} \quad \mathbf{D}_1^{\perp} = \mathbf{D}_2^{\perp}\ . \tag{5.44}$$

This yields

$$\hat{\mathcal{E}}_1 = \hat{\mathcal{E}}_2 \quad \text{and} \quad \kappa_1/\kappa_2 = -\epsilon_1(\omega)/\epsilon_2(\omega)\,. \tag{5.45}$$

Combining (5.45) with (5.43) we get the dispersion relation for surface polaritons

$$k_{\|}^2 c^2 = \omega^2 \frac{\epsilon_1(\omega)\epsilon_2(\omega)}{\epsilon_1(\omega) + \epsilon_2(\omega)}\,. \tag{5.46}$$

Comparing this relation with the bulk polariton dispersion (5.37) one can formally define an interface dielectric function $\epsilon_s(\omega)$:

$$\frac{1}{\epsilon_s(\omega)} = \frac{1}{\epsilon_1(\omega)} + \frac{1}{\epsilon_2(\omega)}\,. \tag{5.47}$$

From the bulk dispersion relation (5.37) we obtain the frequency of the TO bulk phonon for $k \to \infty$, i.e., for k values large in comparison with those on the light curve $\omega = ck$; ω_{TO} results from the pole of $\epsilon(\omega)$. Similarly we obtain the frequency ω_s of the interface waves ($k_{\|} \to \infty$) from the pole of $\epsilon_s(\omega)$ (5.46, 47), i.e.,

$$0 = \frac{1}{\epsilon_s(\omega_s)} = \frac{\epsilon_1(\omega_s) + \epsilon_2(\omega_s)}{\epsilon_1(\omega_s)\epsilon_2(\omega_s)}\,, \tag{5.48a}$$

or

$$\epsilon_2(\omega_s) = -\epsilon_1(\omega_s)\,. \tag{5.48b}$$

If we consider the special case of a crystal in vacuum, i.e. a semi-infinite halfspace with the dielectric function $\epsilon(\omega) = \epsilon_1(\omega)$ adjoining vacuum with $\epsilon_2(\omega) = 1$, the condition determining the frequency of the surface polariton is

$$\epsilon(\omega_s) = -1\,. \tag{5.49}$$

The simplest description of an IR-active material is in terms of an undamped oscillator-type dielectric function

$$\epsilon(\omega) = 1 + \chi_{VE} + \chi_{Ph}(\omega) \tag{5.50}$$

with

$$\chi_{VE} = \epsilon(\infty) - 1$$

and

$$\chi_{Ph} = [\epsilon(0) - \epsilon(\infty)] \frac{\omega_{TO}^2}{\omega_{TO}^2 - \omega^2} ,$$

where χ_{VE} describes the valence-electron contribution in terms of the high-frequency dielectric function $\epsilon(\infty)$, $\epsilon(0)$ is the static dielectric function, and ω_{TO} the frequency of the TO bulk phonons (dispersion neglected). Inserting (5.50) into (5.46) yields the dispersion relation for surface phonon polaritons:

$$\omega^2 = \frac{1}{2}\left[\omega_{LO}^2 + \left(1 + \frac{1}{\epsilon(\infty)}\right) k_{\|}^2 c^2\right] \times \left\{1 - \sqrt{1 - 4\frac{[\omega_{LO}^2 + \epsilon(\infty)^{-1}\omega_{TO}^2] k_{\|}^2 c^2}{\{\omega_{LO}^2 + [1 + \epsilon(\infty)^{-1}] k_{\|}^2 c^2\}^2}}\right\} . \tag{5.51}$$

This dispersion is plotted in Fig.5.10 together with the dispersion branches of the bulk IR-active TO/LO polariton branches. For large $k_{\|}$ the surface-polariton branch approaches the surface phonon frequency (ω_s) which is determined by the condition (5.49). It should be emphasized that the $k_{\|}$ range shown in Fig.5.10 covers essentially the 10^{-3} part of the 2D Brillouin-zone diameter, i.e., for large $k_{\|}$ values in the remainder of the zone considerable dispersion might occur, but this is not contained in our simple approximation for small $k_{\|}$.

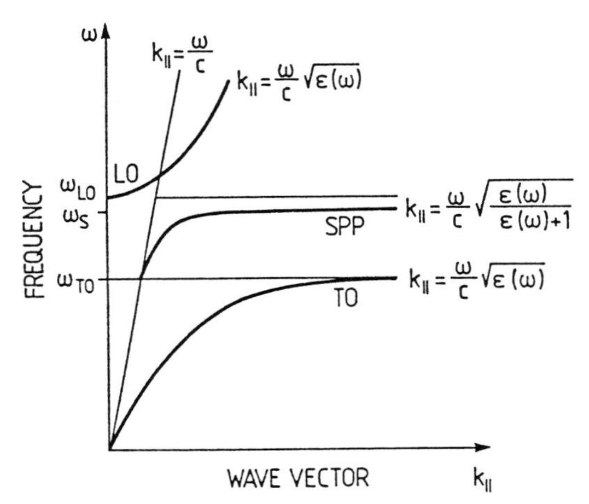

Fig.5.10. Dispersion curve of surface phonon polaritons (SPP) of an IR-active crystal together with the bulk phonon polariton curves (TO, LO) for small wave vectors parallel to the surface

It should also be noted that the analysis presented here takes into account retardation, i.e., the finite value of the light velocity c.

A much simpler derivation of the condition for the existence of optical surface phonons (5.49) and their frequency ω_s is obtained by neglecting retardation. For this purpose we ask whether there exists a wave-like solution near the interface between the IR-active crystal and the vacuum, for which both

$$\mathrm{div}\mathbf{P} = 0 \quad \text{for} \quad z \neq 0 , \tag{5.52a}$$

$$\mathrm{curl}\mathbf{P} = 0 \quad \text{for} \quad z \neq 0 , \tag{5.52b}$$

with **P** being the polarization accompanying the lattice distortion. One should remember that for the corresponding long-wavelength bulk phonons the following conditions are valid

$$\text{TO-phonon:} \quad \mathrm{curl}\mathbf{P}_{\mathrm{TO}} \neq 0 , \quad \mathrm{div}\mathbf{P}_{\mathrm{TO}} = 0 , \tag{5.53a}$$

$$\text{LO-phonon:} \quad \mathrm{curl}\mathbf{P}_{\mathrm{LO}} = 0 , \quad \mathrm{div}\mathbf{P}_{\mathrm{LO}} \neq 0 . \tag{5.53b}$$

For the surface solution we require (5.52) to hold, i.e., $\mathrm{curl}\mathcal{E} = \mathbf{0}$ and $\mathrm{div}\mathcal{E} = 0$ (for $z \neq 0$); the electric field should therefore be derived from a potential φ:

$$\mathcal{E} = - \,\mathrm{grad}\varphi \tag{5.54}$$

with

$$\nabla^2 \varphi = 0 \quad \text{for} \quad z \neq 0 . \tag{5.55}$$

For the solution of (5.55) we can make the ansatz of a surface wave

$$\varphi = \varphi_0 \mathrm{e}^{-k_x |z|} \mathrm{e}^{\mathrm{i}(k_x x - \omega t)} . \tag{5.56}$$

The coordinate system is that of Fig.5.8. The wave (5.56) already fulfills (5.55) and we simply have to demand continuity for the component $D_\perp$ at the surface, $z = 0$, i.e.,

$$D_z = - \epsilon_0 \epsilon(\omega) \left. \frac{\partial \varphi}{\partial z} \right|_{z=0-\delta} = - \epsilon_0 \left. \frac{\partial \varphi}{\partial z} \right|_{z=0+\delta} . \tag{5.57}$$

This condition (5.57) is equivalent to the condition (5.49) determining the frequency of the surface polariton. According to (5.54) the electric field $\mathcal{E} = (\mathcal{E}_x, \mathcal{E}_z)$ is derived from (5.56) by differentiation:

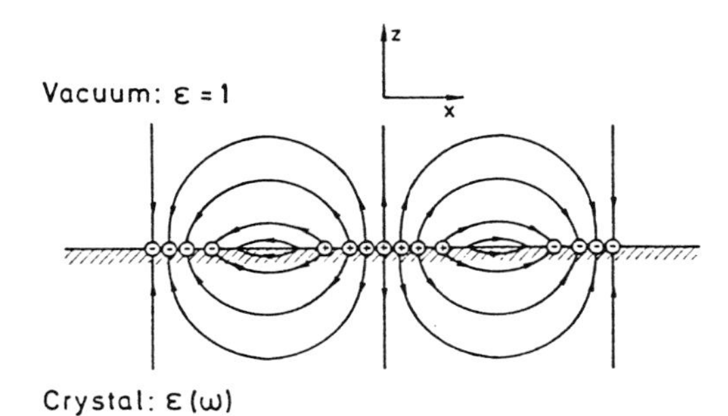

Fig.5.11. Field distribution of a (Fuchs-Kliewer) surface polariton travelling along the surface (parallel to the x-axis) of an IR-active crystal (semi-infinite halfspace z < 0) described by the dielectric function $\epsilon(\omega)$

$$\mathcal{E}_x = \hat{\mathcal{E}}_0 \sin(k_x x - \omega t)\exp(-k_x|z|) , \tag{5.58a}$$

$$\mathcal{E}_z = \pm \hat{\mathcal{E}}_0 \cos(k_x x - \omega t)\exp(-k_x|z|) . \tag{5.58b}$$

This field is depicted in Fig.5.11 with its surface polarization charges at the crystal/vacuum interface. The type of phonon polariton shown is often called **Fuchs-Kliewer phonon**.

Figure 5.12a exhibits the dielectric function $\epsilon(\omega)$ of an oscillator; this is a good approximation for long-wavelength optical phonons in an IR-active material. The frequencies ω_{TO} and ω_{LO} of the transverse and longitudinal bulk optical phonons are determined by the pole $\mathrm{Re}\{\epsilon(\omega)\}$ and the condition of $\mathrm{Re}\{\epsilon(\omega_{LO})\} \simeq 0$. According to (5.49) the frequency of the corresponding optical surface phonon ω_s is easily found as the frequency at which $\mathrm{Re}\{\epsilon(\omega)\}$ crosses the value -1 on the ordinate. If $\mathrm{Im}\{\epsilon(\omega)\}$ is not negligible in this frequency range, slight shifts in ω_s must, of course, be taken into account.

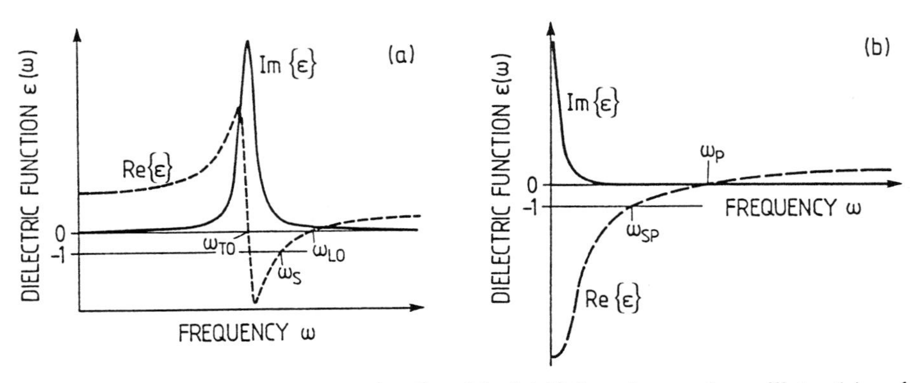

Fig.5.12. Dielectric functions $\mathrm{Re}\{\epsilon(\omega)\}$ and $\mathrm{Im}\{\epsilon(\omega)\}$ for a harmonic oscillator (**a**) and a free electron gas (**b**). (**a**) In the case of an IR-active crystal, the resonance frequency is the frequency ω_{TO} of the transverse optical (TO) bulk phonon; ω_{LO} is the frequency of the longitudinal optical (LO) bulk phonon, ω_s that of the surface phonon polariton. (**b**) ω_p is the frequency of the bulk plasmon, ω_{SP} that of the surface plasmon

A comparison of Figs.5.12a and b implies that for the free electron gas, similar arguments apply as for phonons. Indeed bulk density waves of the electron gas, i.e., **plasmon waves** are irrotational (curl$\mathbf{P}$ = 0) and their frequency ω_p follows from the condition

$$\epsilon(\omega_p) = 0 \,, \tag{5.59}$$

i.e., in the range of negligible $\mathrm{Im}\{\epsilon(\omega)\}$ for $\mathrm{Re}\{\epsilon(\omega_p)\} = 0$.

The same type of analysis as has been performed for phonons can thus be applied to a free electron gas filling a semi-infinite half space bounded by a vacuum interface. However, instead of the dielectric function of the oscillator (eigenfrequency ω_{TO}), one now has to use the dielectric function of a free electron gas (Fig.5.12b). In the simplest approximation this is a Drude dielectric function

$$\epsilon(\omega) = \epsilon(\infty) - \left(\frac{\omega_p}{\omega}\right)^2 \frac{1}{1 - 1/i\omega\tau} \,, \tag{5.60}$$

where

$$\omega_p = \sqrt{\frac{ne^2}{m^* \epsilon_0}} \tag{5.61}$$

is the **plasma frequency** (with n the carrier concentration and m^* the effective mass), and τ the relaxation time. In a better approximation one might apply a Lindhard dielectric function [5.7], or yet more sophisticated methods that take into account the special boundary conditions at a surface [5.8]. The frequency of the surface plasmon ω_{SP} is given, as in (5.49), by the condition

$$\epsilon(\omega_{SP}) = -1 \,. \tag{5.62}$$

For a Drude dielectric function, by inserting (5.60) into (5.46), one obtains the dispersion relation of **surface plasmon polaritons**

$$\omega^2 = \frac{1}{2}\left[\omega_p{}^2 + \left(1 + \frac{1}{\epsilon(\infty)}\right)k_{\parallel}c^2\right] \times \left[1 - \sqrt{1 - 4\left(\frac{\omega_p{}^2 k_{\parallel} c}{\omega_p{}^2 + [1+\epsilon(\infty)^{-1}]k_{\parallel}{}^2 c^2}\right)^2}\,\right] . \tag{5.63}$$

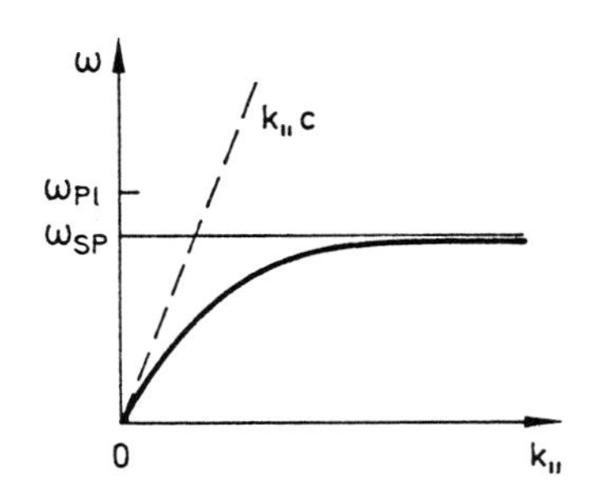

Fig.5.13. Dispersion curve $\omega(k_{\parallel})$ of surface plasmons on a semi-infinite halfspace containing a free electron gas; $k_{\parallel}$ is the wavevector parallel to surface; ω_p and ω_{SP} are the frequencies of bulk and surface plasmons for large $k_{\parallel}$ (small wavelength)

This dispersion relation is displayed in Fig.5.13. In the case of surface plasmons one has to distinguish between two different cases. In a metal the carrier concentration is on the order of 10^{22} cm^{-3}, and the corresponding plasmon energies ω_p and ω_{SP} are on the order of 10 eV. In an n-type semiconductor the plasma frequencies of the valence electrons are of the same order of magnitude ($n \approx 10^{22}$ cm^{-3}), but now we have to treat the free electrons in the conduction band separately. For a conduction electron density of typically 10^{17} cm^{-3}, the corresponding plasmon energies are in the range $10 \div 30$ meV. This is exactly the range of typical phonon energies.

An experimental example, in which one can observe both types of surface polaritons, the phonon and the plasmon, is exhibited in Fig.5.14. High-Resolution Electron Energy Loss Spectroscopy (HREELS) was used to study the clean cleaved GaAs(110) surface. In part (a) of the figure, semi-insulating GaAs, compensated by a high degree of Cr doping, was used. In this material the free-carrier concentration is negligible. Only surface phonons can be expected to occur in the low-energy range up to 200 meV loss energy. The series of energetically equidistant gain and loss peaks indicates multiple scattering on one and the same excitation. The excitation energy is derived from the spacing of the loss peaks as 36.2 ± 0.2 meV. Taking the well-known dielectric function $\epsilon(\omega)$ from IR data for GaAs [5.10] one can calculate the frequency ω_s of the surface phonon polariton by means of (5.49). This calculation yields a value of $\omega_s = 36.6$ meV in good agreement with the experimental value. A thorough quantum-mechanical theory of the scattering process (Chap.4) [5.11] predicts that the intensity of the multiple scattering events should be distributed according to a Poisson distribution, i.e.,

$$P(m) = I_m / \sum_{\nu} I_{\nu} = (m!)^{-1} Q^m e^{-Q} , \qquad (5.64)$$

where I_m is the intensity of the mth loss. Q is the one-phonon excitation probability, i.e., the squared absolute magnitude of the Fourier transform

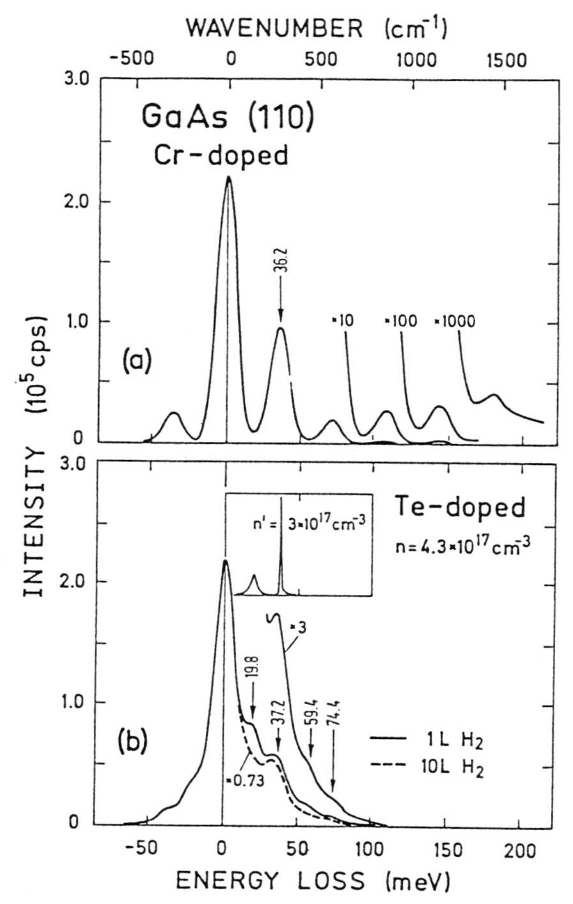

Fig.5.14. (**a**) Loss spectrum of a clean cleaved GaAs(110) surface of semi-insulating material (angle of incidence 80°). (**b**) Loss spectra measured on an n-type sample after exposure to atomic hydrogen (angle of incidence 70°; H coverage unknown) Inset: Calculated surface loss function $-\mathrm{Im}\{(1+\epsilon)^{-1}\}$ in arbitrary units; $\epsilon(\omega)$ contains contributions from the TO lattice oscillator and from the free electron gas (density $n' = 3\cdot 10^{17}\ \mathrm{cm}^{-3}$. [5.9]

of the time-dependent perturbation due to the scattered electron. This distribution law is well verified as can be seen from Fig.5.15a.

The scattered electron can not only lose energy by excitation of a surface phonon, but can also gain the same amount of energy by deexcitation of a phonon that is already thermally excited. As in Raman spectroscopy, the gain (I_{-m}) and loss (I_m) intensities (Stokes and anti-Stokes lines) are then expected to be related to each other through a Boltzman factor

$$I_{-m}/I_m = \exp(-m\hbar\omega_s/kT)\ . \tag{5.65}$$

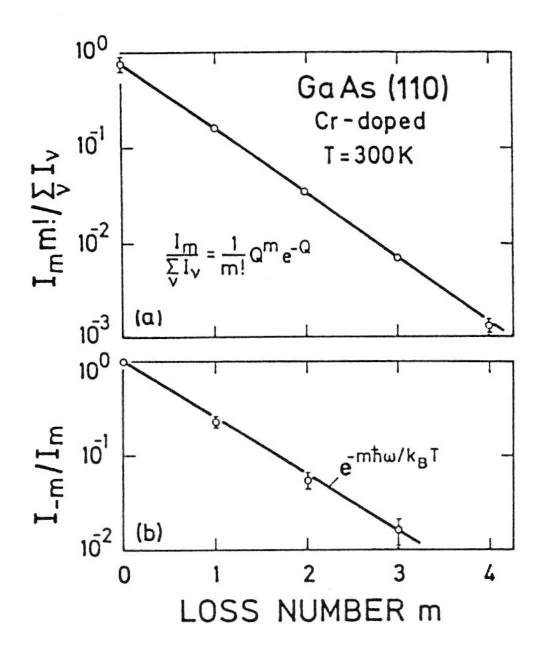

Fig.5.15. (**a**) Poisson distribution of the loss intensities I_m measured on a clean semi-insulating GaAs surface (Fig.5.14). (**b**) Intensity ratio of the mth surface phonon gain and the mth phonon loss vs. loss number m. The straight line is calculated with $\hbar\omega = 36.0$ meV [5.9]

This is also found experimentally, as is seen from Fig.5.15b. On n-doped GaAs with free electron concentrations in the conduction band of $10^{17} \div 10^{18}$ cm^{-3} loss spectra like that of Fig.5.14b are found. On clean cleaved surfaces and also after exposure to small amounts of dissociated hydrogen (or to residual gas) a series of gain and loss peaks is observed at energies $\hbar\omega_+$ (and multiples thereof) resembling those of the surface phonon ($\hbar\omega_s$). Additional gains and losses (including multiples) are also observed with a significantly smaller quantum energy $\hbar\omega_-$. The spectral position of these peaks is very sensitive to the free-carrier concentration in the bulk and to the gas treatment of the surface. An interpretation in terms of surface plasmons is therefore obvious. A quantitative description of the experimental spectra is possible by assuming a dielectric function for GaAs of the form

$$\epsilon(\omega) = \epsilon(\infty) + [\epsilon(0) - \epsilon(\infty)] \frac{\omega_{TO}^2}{\omega_{TO}^2 - \omega^2 - i\omega\gamma} - \left(\frac{\omega_p}{\omega}\right)^2 \frac{1}{1 - 1/i\omega\tau} , \quad (5.66)$$

which contains an oscillator contribution due to the TO optical phonons (ω_{TO}) and a Drude term (5.60) which takes into account the free electrons in the conduction band.

$$\omega_p{}^2 = ne^2/\epsilon_0 m_n^* \quad (5.67)$$

is the bulk plasma frequency where n is the free carrier concentration and m_n^* is their effective mass.

$$\tau = m_n^* \mu / e \tag{5.68}$$

is the Drude relaxation time with μ being the mobility. The dielectric function (5.66) is a superposition of the real and imaginary parts depicted in Figs.5.12a and b.

According to Sect.4.6 the essential structure of an electron energy loss spectrum is given within the framework of dielectric theory by the surface loss function $\mathrm{Im}\{-1/[\epsilon(\omega)+1]\}$. For monotonic and relatively small $\mathrm{Im}\{\epsilon(\omega)\}$ the maxima are found at the frequencies determined by the condition (5.49). This is also true if one inserts the more complex $\epsilon(\omega)$ of (5.66) into the surface loss function. $\mathrm{Im}\{-1/[\epsilon(\omega)+1]\}$ then exhibits two maxima at frequencies or quantum energies $\hbar\omega_-$ and $\hbar\omega_+$ that correspond to solutions of (5.49). According to (5.66-68) these two solutions $\hbar\omega_-$ and $\hbar\omega_+$ depend on the concentration n of free electrons in the conduction band. Figure 5.16 exhibits the calculated loss peak positions (full line), i.e., the energies $\hbar\omega_-$ and $\hbar\omega_+$ as functions of an effective carrier concentration n'. The lower

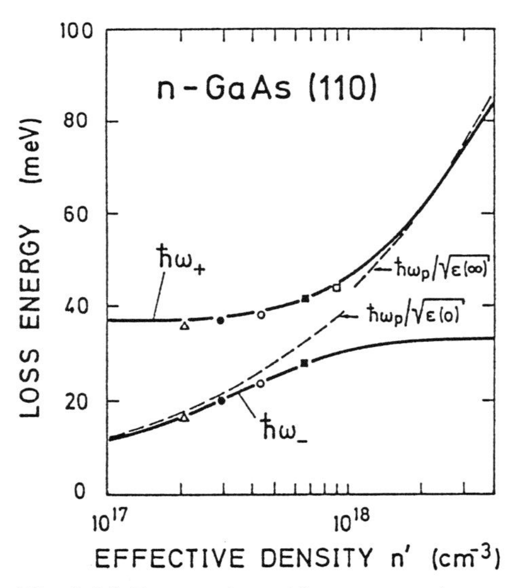

Fig.5.16. Loss peak positions $\hbar\omega_+$ and $\hbar\omega_-$ calculated from the maxima of the surface loss function with $\epsilon(\omega)$ according to (5.50) (*solid lines*). Dashed line: plasmon frequency without coupling to surface phonon. The experimental points are measured on samples with different n-type doping:
1) Te-doped (bulk density: $n = 9 \cdot 10^{17}$ cm^{-3}): (□) clean, (■) after exposure to 1L residual gas. 2) Te-doped (bulk density: $n = 4.3 \cdot 10^{17}$ cm^{-3}): (○) clean, (●) after exposure to 1L dissociated H_2. 3) Si-doped (bulk density: $n = 3 \cdot 10^{17}$ cm^{-3}): (Δ) clean [5.9]

branch $\hbar\omega_-$ has surface-plasmon-like character for small n′ whereas $\hbar\omega_+$ is surface-phonon-like. Near $n' = 10^{18}$ cm^{-3} the two branches interchange their character thus indicating a coupling between the two modes via their long-range electric fields. The two frequencies ω_- and ω_+ derive from the values ω_{SP} and ω_s in Fig.5.12 when the two dielectric functions in Figs. 5.12a and b are superimposed. The combined $\epsilon(\omega)$ exhibits two solutions of $\epsilon(\omega) = -1$ in the regime of negligible Im$\{\epsilon\}$. In Fig.5.16 some experimentally determined loss-peak positions are plotted, too. The experimentally determined $\hbar\omega_-$, $\hbar\omega_+$ values after cleavage fit the theoretical curves very well if the effective carrier concentration n′ is taken to be that of the bulk (n), as determined by Hall-effect measurements. After hydrogen treatment, however, the effective carrier concentration n′ is reduced as is seen from the positions of the loss peaks (Fig.5.16). This effect is due to the depletion of carriers in a region of some hundreds of Ångstroms below the surface due to an upwards band bending of the conduction band (space charge region; Chap.7). This so-called **depletion layer** is induced by hydrogen adsorption. It influences the loss peak position since the positions $\hbar\omega_+$ and $\hbar\omega_-$ are determined by carrier concentration within the penetration depth $1/q_{\|}$ of the electric field of the surface phonon and plasmon-like excitations. From the relation $q_{\|} \simeq \hbar\omega/2E_0$ (4.42) this penetration depth is also estimated to be about a couple of hundred Ångstroms. The measurement of surface phonon/plasmon excitations can therefore be used to investigate carrier concentrations in space charge layers at semiconductor interfaces and surfaces [5.12] (Chap.7).

5.6 Dispersion Curves from Experiment and from Realistic Calculations

When the wavelength of surface waves is comparable to the interatomic separation of the discrete crystal lattice, the continuum-type approach of the preceding sections is no longer valid. For frequencies on the order of 10^{11} s^{-1} or higher, the description of surface modes demands a lattice dynamical approach. This requires, as in bulk lattice dynamics, a detailed knowledge of the interatomic force constants. For appropriate approximations the effects of electron-lattice interactions have to be taken into account by means of shell models, in which the valence electrons are represented by a solid shell bound to the core by a spring. In even more sophisticated treatments, deformations of the electron shell itself can be taken into account by so-called ***breathing*** **shell models**. Compared to bulk lattice dynamics a fundamental new problem arises at the surface: due to reconstruction or relaxation of the topmost atomic layers, both the atomic geometry and the restoring forces may deviate near the surface from their bulk values. These changes are not generally known. They give rise to additional parameters which must be fitted to experimental data.

A variety of lattice-dynamical techniques have been applied to calculate surface phonon dispersion branches. An approach frequently used in the past is the analogy of the continuum approach (Sect.5.3): a trial solution is constructed for the semi-infinite lattice and, by means of the correct boundary conditions, dispersion curves are obtained. It is not always clear whether all possible surface modes are obtained by this calculation. Another method consists of the direct calculation of the eigenvalues and polarization vectors of a slab formed by a sufficiently large number of atomic layers. This method yields all the acoustic and optical surface modes over the entire Brillouin zone, provided that their penetration depth is less than the slab thickness. In many cases twenty layers are enough to give good results. A third method is based on the application of Green's function theory. In this approach the surface is treated as a perturbation which modifies the spectrum of bulk vibrations.

Figure 5.17 shows an example of results obtained from a slab calculation for NaCl (001) [slab orientation (001)]. Corresponding to the finite number of slabs (15) the bulk modes are obtained as a discrete set of dispersion curves. With increasing number of the slabs these bulk modes thicken to form quasi-continuous areas, i.e. bands. However, a finite number of

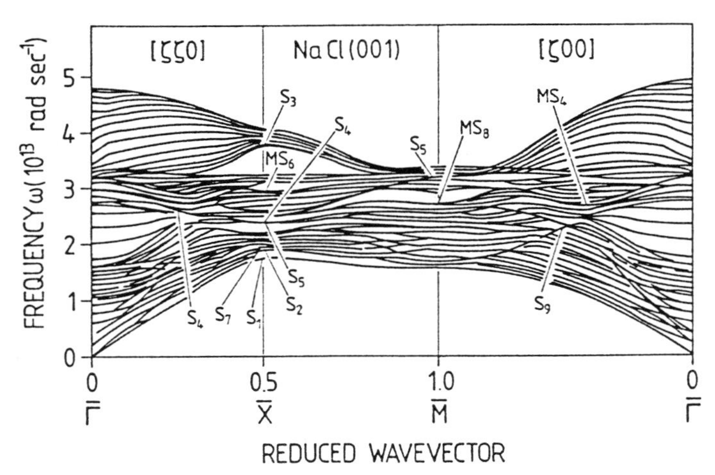

Fig.5.17. Dispersion curves of a 15 layer (001)-oriented NaCl slab obtained from a shell model calculation. The finite number of slabs gives rise to a discrete set of dispersion curves. The dispersion branches labeled by S_1, S_2... etc. belong to vibrational eigenvectors which decay into the bulk, i.e., they describe surface phonons. [5.13]

modes labeled S_1, S_2, S_3 etc., remain distinct from the bands. Their eigenvectors are found to be large near the surface and rapidly decreasing, away from the surface. These modes can obviously be identified as surface vibrations. The acoustic surface mode S_1 is localized beneath the bulk acoustic band, even in the long wavelength limit ($\mathbf{k}_{\parallel} \to \mathbf{0}$). This mode therefore represents the Rayleigh surface waves discussed in Sect.5.3. The modes S_3, S_4 and S_5 are examples of optical surface vibrations. S_4 and S_5 are the so-called **Lucas modes** with polarization normal and parallel to the surface. These modes are related to the altered force constants between the topmost atomic layers; their vibrational amplitude is therefore strongly localized near the first layer.

As an example of a calculation using the Green's function perturbation method, Fig.5.18 shows dispersion branches of surface phonons on LiF (001). The outer atomic electrons are modelled by the so-called **breathing shell model**, in which deformations of the shell are explicitly taken into account. Accordingly there is very good agreement between the calculated bulk phonon dispersion branches (shaded area) and some branches that have been determined experimentally by inelastic neutron scattering (black dots). In the plot of Fig.5.18 the bulk modes with polarization normal ($\perp$) and parallel ($\parallel$) to the (001) surface are shown separately. Some surface phonon

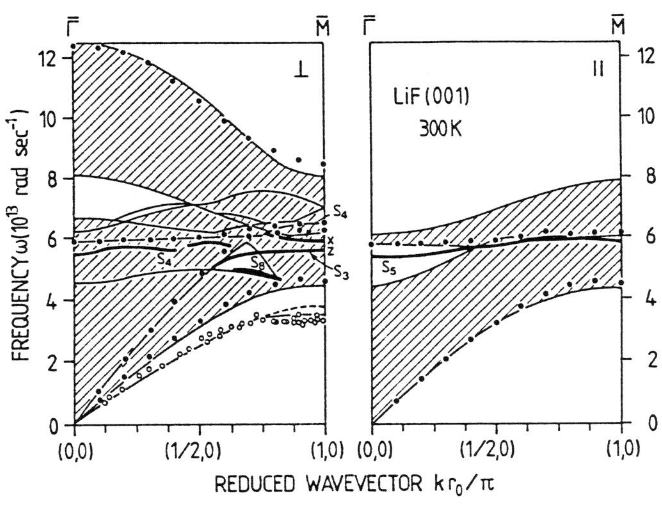

Fig.5.18. Phonon dispersion branches of LiF(001) along (100) calculated by the Green's function method [5.14]. The bulk modes (*shaded area*) with polarization normal ($\perp$) and parallel ($\parallel$) to the (001) surface are shown separately. The surface phonon modes are labeled by S_i. For comparison some experimentally determined bulk modes (from neutron scattering) are given as black dots. The open dots (near S_1) are experimental results for the Rayleigh modes, as determined by inelastic atom scattering [5.15]

bands are marked by S_3, S_4, etc. S_4 and S_5 are again the Lucas modes. They are energetically degenerate with bulk modes polarized normal to the surface and are thus called surface resonances. As is expected from continuum theory (Sect.5.3), the Rayleigh mode S_1 has frequencies (energies) below those of the bulk modes along the entire symmetry line $\overline{\Gamma M}$. This S_1 band has been calculated by both the Green's function method (full line) and by the slab method (dashed line). There is a small discrepancy near the $\overline{M}$ point which could probably be eradicated if the surface change in ionic polarizibility and/or the anharmonicity were taken into account.

The theoretical dispersion curve of the Rayleigh mode S_1 calculated using the Green's functions method is in very good agreement with the experimental results from inelastic atom scattering by *Brusdeylins* et al. [5.15] (Fig.5.18, open circles). In these experiments a supersonic nozzle beam of He atoms is inelastically scattered on the LiF(001) surface prepared in UHV, and the energy distribution of the backscattered He atoms is measured by a time-of-flight spectrometer (Panel X: Chap.5). The energy loss at the surface and the scattering angle with respect to the specular beam and the sample surface determine the phonon frequency ω and the transfer, i.e. the dispersion relation $\omega(\mathbf{q}_{\parallel})$ for the particular surface excitation mode. Rayleigh waves usually produce the strongest peaks in the time-

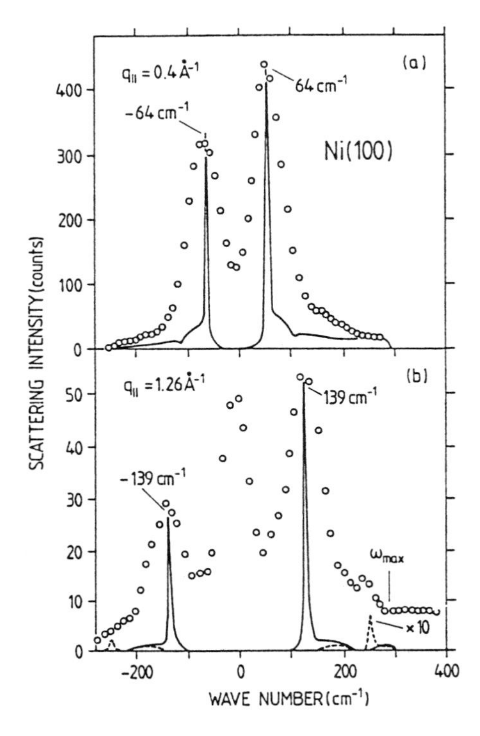

Fig.5.19a,b. Electron energy loss spectra measured by HREELS with 7 meV energy resolution in off-specular geometry between the (00) and (01) Bragg diffraction beams on the clean Ni(100) surface (geometry as in inset of Fig.5.20). (**a**) for a wavevector transfer $q_{\parallel}$ of 0.4 $Å^{-1}$; (**b**) for a wave vector transfer $q_{\parallel}$ of 1.26 $Å^{-1}$; experimental data points (*open circles*), calculated spectra (*solid lines*) [5.16]

of-flight spectra of the scattered atoms due to their large amplitude in the topmost layer.

Surface phonon dispersion branches can also be measured by the inelastic scattering of slow electrons (Panel IX: Chap.4). In order to measure $\hbar\omega(\mathbf{q}_{\|})$ throughout the whole 2D Brillouin zone, sizeable $\mathbf{q}_{\|}$ transfers have to be achieved and the measurement must thus be performed with off-specular scattering geometry. Unlike the case of optical surface phonon polaritons with $\mathbf{q}_{\|} \simeq 0$ studied in the dielectric scattering regime (Sect.5.5), the scattering is now predominantly due to short-range atomic potentials. The inelastic scattering cross section for this kind of scattering on phonons rises with increasing primary energy. The HREELS experiments on Ni(100) (Fig.5.19) were therefore performed with impact energies of 180 and 320 eV in preference to lower energies. The spectra shown in Fig.5.19 were measured with a resolution of 7 meV in off-specular scattering geometry as shown in the inset of Fig.5.20. Electrons were collected at a fixed polar angle of $\approx 72°$ along the [110] azimuth ($\overline{\Gamma X}$ direction) while the impinging beam was rotated between polar angles yielding the (01) and (00) Bragg-dif-

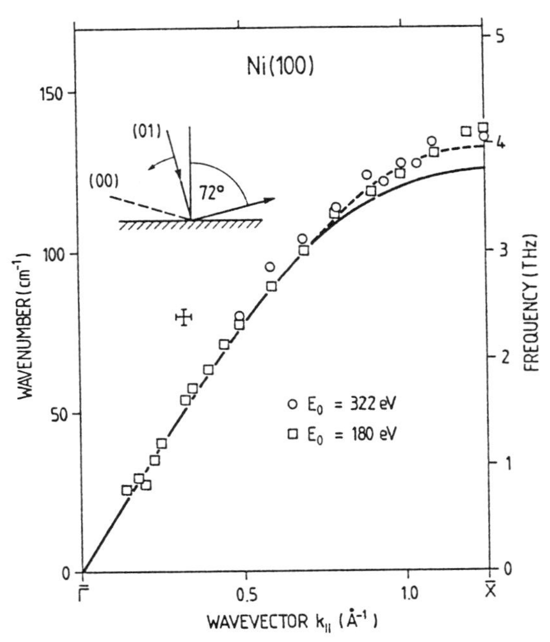

Fig.5.20. Experimentally determined (HREELS, Fig.5.19) dispersion of surface phonons on a clean Ni(100) surface (open symbols for two different primary energies E_0). The scattering geometry is shown in the inset (off-specular scattering). The full curve is the calculated dispersion according to *Allen* et al. [5.17]. The dashed line takes into account a stiffening of the force constants between the topmost and the second atomic layers [5.16]

fracted beams. The momentum resolution was $\Delta q_{\parallel} \approx 0.01$ Å^{-1}. Phonon losses were found over the entire range between the (01) and (00) positions. The particular $q_{\parallel}$ transfer of 1.26 Å^{-1} in Fig.5.19b corresponds to the $\bar{X}$ point of the 2D Brillouin zone. The experimental data (open circles) are compared with a calculation based on a nearest-neighbour central-force model in which the force constant between first and second layer is stiffened by 20%. The calculated curves show the frequency spectrum of phonons with displacements normal to the surface for the corresponding $q_{\parallel}$ wave vectors, for atoms in the outermost substrate layer. The wings on the high-frequency side of the surface phonon loss originate from the bulk phonon continuum.

The experimental peak positions as a function of $q_{\parallel}$ [calculated according to (4.41)] are plotted in Fig.5.20. The measured dispersion coincides closely with that calculated by *Allen* et al. [5.17] for a surface phonon on Ni(100) with an atomic displacement at $\bar{X}$ that is normal to the surface in the outmost layer. According to this calculation there exists a further shear-polarized surface phonon at $\bar{X}$ with displacement parallel to the surface and normal to $\mathbf{q}_{\parallel}$. Since for this phonon the displacement direction is always normal to $\mathbf{K}_{\parallel} = \mathbf{k}_{\parallel}' - \mathbf{k}_{\parallel}$ the selection rules (4.17) forbid an excitation in the present scattering geometry (inset of Fig.5.20). Indeed, this phonon is not observed in the HREELS data.

The agreement between the calculated dispersion (Fig.5.20, full line) and the data points is poorer near the $\bar{X}$ point. The agreement is improved (dotted line) if the force constant which couples atoms in the first and second layers is increased by 20%. This stiffening of the force constant mimics a modest inward relaxation of the surface atomic layer, which is indeed confirmed by other experiments.

The examples of Li(001) and Ni(100) show that measurements of surface-phonon dispersion curves and a comparison with lattice-dynamical calculations can provide interesting information about changes in force constants and atomic locations near the surface.

Panel X
Atom and Molecular Beam Scattering

Atoms and molecules such as He, Ne, H_2, D_2, impinging on a solid surface as neutral particles with a low energy (typically $<20\,eV$), cannot penetrate into the solid. Scattering experiments with neutral particle beams therefore provide a probe that yields information exclusively about the outermost atomic layer of a surface. Such experiments have now become an important source of information in surface physics. Both elastic and inelastic scattering can be studied. A schematic overview of the various scattering phenomena is given in Fig.X.1. Since He atoms, for example, with a kinetic energy of 20 meV have a de Broglie wavelength of 1 Å, scattering phenomena must be described in the wave picture (Sect.4.1). A particle approaching the surface interacts with the surface atoms through a typical interatomic or intermolecular potential $V(\mathbf{r}_{\|},z)$, $\mathbf{r}_{\|}$ being a vector parallel to the surface, and z the coordinate normal to the surface. V(z) consists of an attractive and a repulsive part (as in chemical bonding). The scattering from a two-dimensional periodic lattice of atoms (surface) is dominated by the specular quasi-elastic peak (intensity I_{00}) and elastic Bragg diffraction (intensity I_{hk}) in well-defined directions (as in LEED, Sect.4.2). This elastic scattering is adequately described in the rigid-lattice approximation with

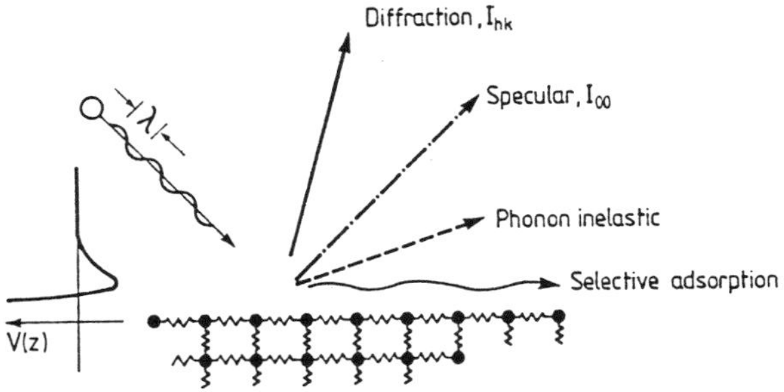

Fig.X.1. Schematic diagram showing the different collision processes that can occur in the non-reactive scattering of a light atom with the de Broglie wavelength comparable to the lattice dimensions. Since the lattice vibrational amplitudes are small, phonon inelastic scattering is expected to be weak relative to elastic diffraction (specular beam I_{00} and Bragg diffraction beams I_{hk}). Additionally, high energy losses can lead to selective adsorption of impinging atoms in the attractive part of the surface atom potential V(z) [X.1]

only an intensity correction for inelastic effects, provided by the temperature-dependent Debye-Waller factor. An incident atom or molecule can lose so much energy that it is trapped at the surface or "selectively adsorbed". This trapping of atoms in bound states on the surface can strongly modify the scattered intensities at specific angles and energies.

Inelastic scattering comes into play due to the fact that the crystal is in reality not rigid: the atoms vibrate about their average positions. The incident particle can therefore transfer part of its kinetic energy to the dynamic modes of the vibrating surface, the **surface phonons**. Similarly, it can gain energy via the annihilation of a surface phonon.

The mathematical description of the scattering is analogous to that of electron-surface scattering (Sect.4.1). The most general interaction potential $V(\mathbf{r})$ between the incident particle and the crystal surface (4.1) which enters the formula for the scattering cross section (4.17) is conveniently written as a function of $\mathbf{r}_{\|}$, a coordinate parallel to the surface, the coordinate z normal to the surface, and $\mathbf{s}_{\mathbf{n}}(t)$ the vibrational coordinate of the $\mathbf{n}$th surface atom:

$$V[\mathbf{r}_{\|}, z, \mathbf{s}_{\mathbf{n}}(t)] = V(\mathbf{r}_{\|}, z)\,|_{\mathbf{s}_{\mathbf{n}}=0} + \sum_{\mathbf{n}} (\nabla V)\cdot \mathbf{s}_{\mathbf{n}}(t) + \dots \,. \quad \text{(X.1)}$$

The first term in the potential expansion is the corrugated elastic potential, which can be determined by fitting the intensities of the elastic diffraction peaks using model potentials. Elastic scattering thus yields information about the topology of the surface and about details of the interatomic potentials. The second- and higher-order terms, which couple to the vibrations $\mathbf{s}_{\mathbf{n}}(t)$ of the surface atoms, are responsible for inelastic scattering. An understanding of these coupling terms is fundamental for an interpretation of such phenomena as sticking coefficients (Sect.9.5) and energy transfer between surface atoms and incident particles.

Before presenting some detailed examples of the application of atom scattering, the experimental set-up will be discussed briefly.

The experimental apparatus consists of a source of monoenergetic molecules or atoms which are directed as a beam towards the surface under investigation; the back-scattered distribution is recorded by a detector. Both sample and detector can be rotated around a common axis in the surface plane to allow the detection of higher diffraction orders under different angles. Since neutral particles are used, neither electric nor magnetic fields can be used as focussing or dispersive elements. A schematic diagram of a typical experimental set-up is shown in Fig.X.2. An important feature is the nozzle beam source producing the monochromatic rare-gas beam. The beam of Ne or He atoms is produced in a high-pressure expansion source. In the

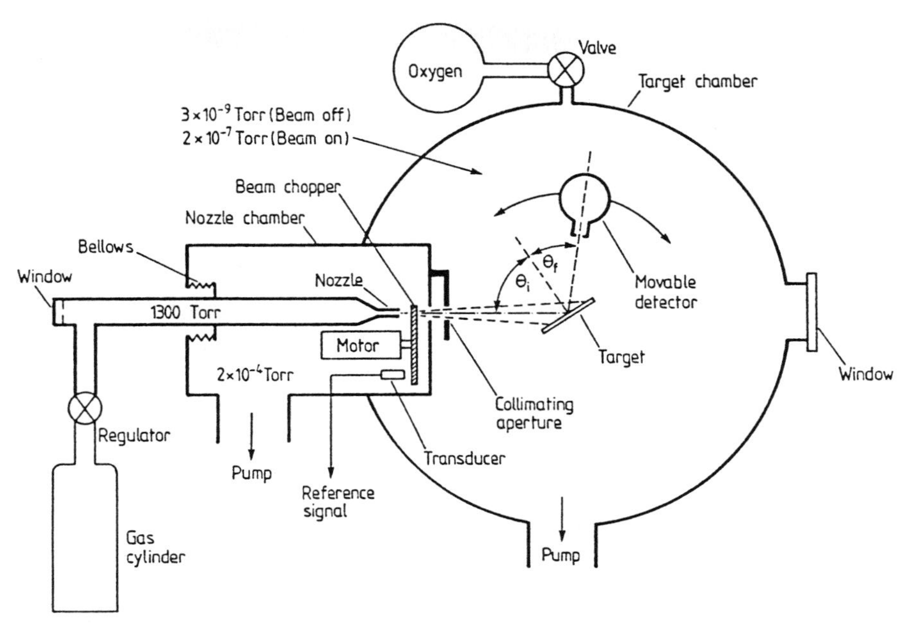

Fig. X.2. Schematic diagram of a typical low-energy molecular beam scattering apparatus [X.2]

expansion of the gas from a source pressure of about 2 atm through a thin-walled orifice (diameter $\approx 5 \cdot 10^{-2}$ mm) to a beam pressure of about 10^{-4} Torr, the random translational energy is converted into a forward mean velocity of the beam. Thus the magnitude of the random velocity component which determines the velocity spread Δv is reduced relative to the most probable velocity v. In the apparatus shown in Fig. X.2 the resultant $\Delta v/v$ is about 10%. With improved nozzle beam sources $\Delta v/v$ values on the order of 1% are achieved. *Toennies* [X.3] used a He source cooled down to 80 K. The beam is expanded from a pressure of 200 atm through a 5 μm hole into vacuum. To improve the forward velocity distribution further, the beam passes a skimmer after expansion. This funnel-shaped tube skims off atoms with insufficiently forward-directed velocity. During the expansion, chaotic thermal motion is converted into a concerted forward motion of the atoms and as a result of enthalpy conservation, the temperature in the moving gas is drastically reduced; behind a distance of about 20 mm to $\approx 10^{-2}$ K. This corresponds to a relative velocity spread of less than 1%. With modern nozzle-beam sources, primary energies from 6 meV up to 15 eV can be produced. He atoms with de Broglie wavelengths of 1 Å have an energy of about 20 meV. In Fig. X.2 the primary beam is modulated by a chopper and phase-sensitive detection is employed using a lock-in amplifier. This technique allows detection of the modulated scattered beam against a relatively high background pressure. Either standard ion gauges or more sophisticated mass spectrometers are employed as detectors.

In the following, some examples are presented of the different applications of atom and molecular scattering based on the processes of Fig.X.1. Since He atoms are essentially scattered on the almost structureless "electron sea", far from the uppermost surface lattice plane, an ideal well-ordered, close-packed metal surface gives rise to virtually no interesting scattering phenomena. But deviations from ideality, such as steps, defects or adsorbates, can affect the elastically scattered intensity in reflection direction, i.e. the specular beam intensity I_{00} quite dramatically. Figure X.3 shows the intensity variation of the specular beam of He atoms reflected from Pt(111) surfaces with differing distributions of steps and terraces. For the measurement the angle of incidence θ_i $(= \theta_r)$ is varied over a small range and the backscattered intensity I_{00} is recorded. For this purpose, of course, an experimental set-up with detection under variable observation direction is necessary. According to the differing terrace width (average values around 300 Å and above 3000 Å) an interference pattern or an essentially monotonic variation is observed in the angular region considered. The oscillations are explained in terms of constructive and destructive interferences of the He wave function reflected from (111) terraces which are separated by monatomic steps. The average terrace width, i.e., the step density, determines similarly as in an optical grid slit width and distance, the phase differences of the evading He waves under certain observation directions. The oscillation period of curve *b* allows an estimation of the step atom density of about 1%. Better preparation techniques lead to step atom densities lower than 0.1% which then give rise to a higher total reflection in-

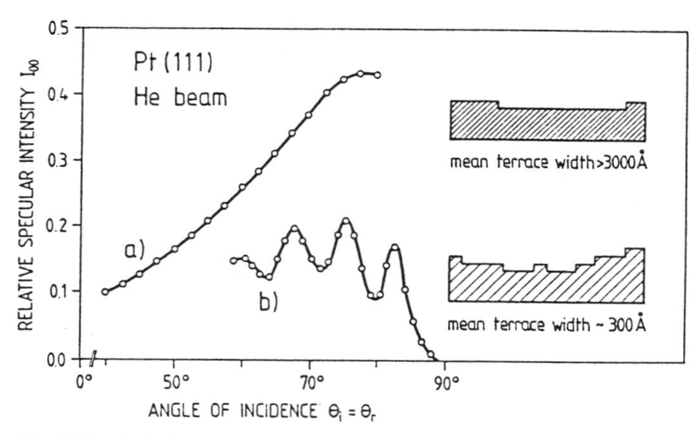

Fig.X.3. Relative specular intensity I_{00} (referred to primary beam intensity) of a low-energy He atom beam (energy E = 63meV) versus angle of incidence θ_i (= θ_r reflection angle) for two Pt(111) surfaces with differing average terrace widths [X.4]

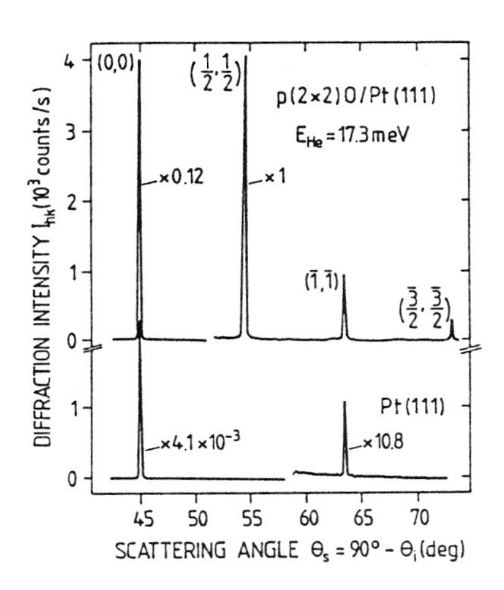

Fig.X.4. He-beam polar diffraction patterns in the [112] direction from the clean (*bottom*) and p(2×2)O/Pt(111) oxygen covered surface (*top*). The primary He energy E_{He} is 17.3 meV and the sample temperature 300 K [X.5]

tensity, and the interference oscillations are absent (curve *a*). The technique is thus useful for characterizing the degree of ideality of a clean surface after preparation.

Elastic He atom scattering, i.e. diffraction, can provide information about the structural properties of a surface. In contrast to electron scattering in LEED (Panel VIII: Chap.4), where the electrons penetrate several Ångstroms into the solid, only the outermost envelope of the electron density about the surface is probed by the He atoms. This makes the technique relatively insensitive to clean, well-ordered, densely-packed metal surfaces; but ordered adsorbate atoms or molecules whose electron density protrudes significantly from the surface, give rise to stronger scattering intensities in certain Bragg spots. This is shown for the example of a well-ordered oxygen layer with p(2×2) superstructure on Pt(111) in Fig.X.4. For the clean Pt surface the $(\bar{1},\bar{1})$ Bragg spot has ten times less intensity than on the oxygen-covered surface. The diffraction spots $(\bar{1}/2,\bar{1}/2)$ and $(\bar{3}/2,\bar{3}/2)$ due to the oxygen superlattice occur with much higher intensity. Adsorbate effects are thus clearly distinguished from substrate spots and the interpretation problems sometimes encountered for adsorbate LEED patterns (substrate vs. adsorbate superstructure) do not exist. The method of atom and molecule diffraction is therefore complementary to LEED because of its extreme sensitivity to the outermost atomic layer. In the inelastic scattering regime, atom and molecule scattering from surfaces also provides interesting advantages over other scattering techniques because of its high energy resolution. Because the possible energy and wave vector transfer are well matched throughout the whole Brillouin zone, surface-phonon dispersion branches (Chap. 5) can be measured with extremely high accuracy. Figure X.5 shows inelastic He beam spectra measured with different angles of incidence θ_i on

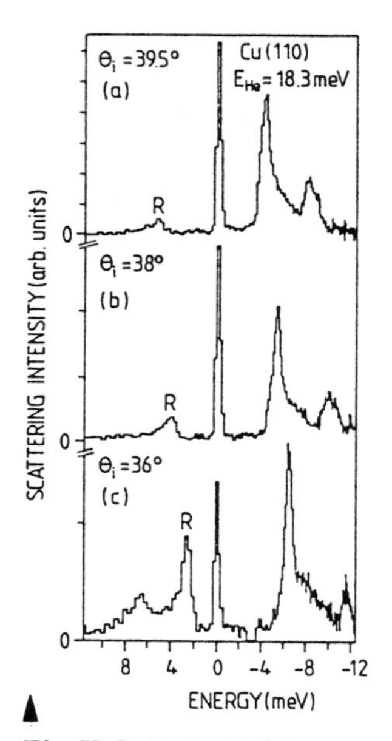

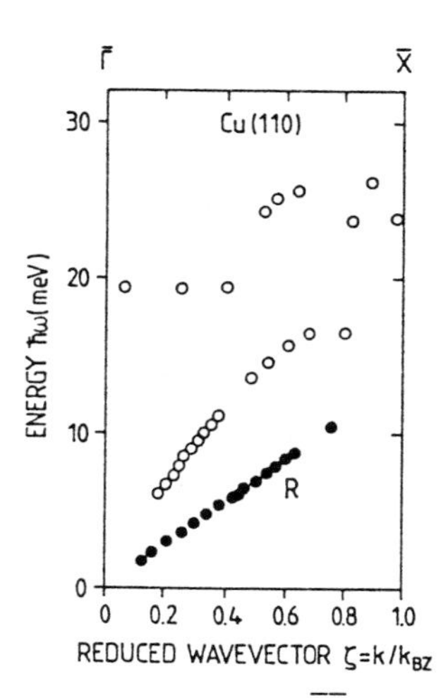

Fig.X.5. Inelastic He scattering spectra taken along the $\overline{\Gamma Y}$ direction of the surface Brillouin zone on Cu(110). The primary He beam energy is 18.3 meV [X.6]

Fig.X.6. Surface phonon dispersion curves as obtained by inelastic He scattering (primary energy E_{He} = 18.3eV) along the $\overline{\Gamma X}$ direction of the surface Brillouin zone on Cu(110). The reduced wave vector ζ is defined by $\zeta = k/k_{BZ}(\overline{X})$ with $k_{BZ}(\overline{X}) = 1.23$ $Å^{-1}$ as the Brillouin-zone dimension in the $\overline{X}$ direction [X.6]

Cu(110). The detection direction is chosen for wave-vector transfers along $\overline{\Gamma Y}$. The experimental resolution readily allows the determination of peak half-widths below 1 meV. Thus information about broadening due to phonon coupling etc. can also be derived from the experimental data. This is by no means possible from electron scattering data (HREELS, Panel IX: Chap4). where the best energy resolution is on the order of 1 meV. Surface phonon dispersion curves derived from spectra such as those of Fig.X.5 are given in Fig.X.6, but here along the $\overline{\Gamma X}$ direction of the Cu(110) surface Brillouin zone. The data denoted by R correspond to the Rayleigh surface waves (Chap.4).

References

X.1 J.P. Toennies: Phonon interactions in atom scattering from surfaces, in *Dynamics of Gas-Surface Interactions*, ed. by G. Benedek, U. Valbusa, Springer Ser. Chem. Phys. Vol.21 (Springer, Berlin, Heidelberg 1982) p.208
E. Hulpke (ed.): *Helium Atom Scattering from Surfaces*, Springer Ser. Surf. Sci., Vol.27 (Springer, Berlin, Heidelberg 1992)

X.2 S. Yamamoto, R.E. Stickney: J. Chem. Phys. **53**, 1594 (1970)

X.3 J.P. Toennies: Physica Scripta **T1**, 89 (1982)

X.4 B. Poelsema, G. Comsa: *Scattering of Thermal Energy Atoms from Disordered Surfaces*, Springer Tracts Mod. Phys., Vol.115 (Springer, Berlin, Heidelberg 1989)

X.5 K. Kern, R. David, R.L. Palmer, G. Comsa: Phys. Rev. Lett. **56**, 2064 (1986)

X.6 P. Zeppenfeld, K. Kern, R. David, K. Kuhnke, G. Comsa: Phys. Rev. B **38**, 12329 (1988)

Problems

Problem 5.1. Information can be transmitted through a solid by bulk sound waves or by Rayleigh surface waves. What phonons provide a faster transmittance velocity? Discuss the problems which arise for the signal propagation by means of short pulses when long wavelengths $\lambda \gg a$ (lattice parameter) and short wavelengths $\lambda \approx a$ are used.

Problem 5.2. The dielectric response of an infrared active, n-doped semiconductor is described in the IR spectral region by a dielectric function ϵ (5.66) which contains an oscillator contribution due to TO phonons (5.50) and a Drude-type contribution (5.60) due to free electrons in the conduction band. Calculate the surface loss function $\mathrm{Im}\{-1/[\epsilon(\omega)-1]\}$ and discuss the loss spectrum expected in an HREELS experiment as a function of carrier concentration. Flat-band situation is assumed at the surface.

Problem 5.3. Surface phonon polaritons (Fuchs-Kliewer phonons) are excited on a clean GaAs(110) surface in an HREELS experiment with a primary energy of 5 eV. Calculate from the corresponding loss peak at 36.2 meV the exponential decay length of the polarisation field of the surface phonons. Discuss the consequence for an HREELS measurement which is performed on a GaAs film which is thinner than the calculated decay length.

Problem 5.4. Calculate the frequency of a surface phonon on the (100) surface of an fcc crystal at the Brillouin-zone boundary in the [110] direction. Only central forces between next neighbour atoms are assumed. The surface phonon should have odd symmetry with respect to the mirror plane defined by the phonon wave vector $\mathbf{q}$ and the surface normal.

Why is the calculation so simple?

Does a second surface phonon exist on this surface which is localized on the first atomic monolayer.

6. Electronic Surface States

Since the surface is the termination of a bulk crystal, surface atoms have fewer neighbors than bulk atoms; part of the chemical bonds which constitute the bulk-crystal structure are broken at the surface. These bonds have to be broken to create the surface and thus the formation of a surface costs energy (**surface tension**) (Sect.3.1). In comparison with the bulk properties, therefore, the electronic structure near to the surface is markedly different. Even an ideal surface with its atoms at bulk-like positions (called **truncated bulk**) displays new electronic levels and modified many-body effects due to the change in chemical bonding. Many macroscopic effects and phenomena on surfaces are related to this change in electronic structure, for example, the surface energy (tension), the adhesion forces, and the specific chemical reactivity of particular surfaces. A central topic in modern surface physics is therefore the development of a detailed understanding of the surface electronic structure. On the theoretical side, the general approach is similar to that for the bulk crystal: In essence the one-electron approximation is used and one tries to solve the Schrödinger equation for an electron near the surface. A variety of approximation methods may then be used to take into account many-body effects.

In comparison with the bulk problem two major difficulties arise for the surface. Even in the ideal case translational symmetry only exists in directions within the plane of the surface. Perpendicular to the surface the periodicity breaks down and the mathematical formalism becomes much more complicated. Even more severe, and not generally solved up to now, is the surface-structure problem. A complete calculation of the electronic structure requires a knowledge of the atomic positions (coordinates). Because of the changed chemical bonding near the surface, however, surface relaxations and reconstructions frequently occur. This means that the atoms are displaced from the ideal positions which they would occupy if the bulk crystal were simply truncated into two parts. At present there is no general, simple and straightforward experimental technique to determine the atomic structure in the topmost atomic layers. Several relatively complex methods such as dynamic LEED analysis (Sect.4.4), SEXAFS (Panel VII: Chap.3), and atom scattering (Panel X: Chap.5) give some information, but only relatively few structures, e.g. the GaAs(110) and the Si(111) surface (Sect. 3.2) have actually been accurately established. It is interesting to mention, that in the case of the Si(111)-(7 $\times$ 7) the main contribution was obtained by means of transmission electron microscopy using simple kinematic analysis. A break through for the analysis of surface atomic structure has been achieved recently by scanning electron tunneling microscopy [6.1].

For realistic calculations of the electronic structure of surfaces one thus has to assume structural models and then to compare the calculated surface electronic band structure and other calculated physical properties, e.g. photoemission and electron energy loss spectra, with experiment.

In the following, we will first consider a simple (unrealistic) model of a surface, namely the monatomic linear chain terminated at one end (semi-infinite chain). The Schrödinger equation will be solved in the nearly-free-electron approximation and the results transferred qualitatively to the case of a real, two-dimensional (2D) surface with 2D translational symmetry. Realistic surfaces and surface states of some metals, and important semiconductors such as Si and GaAs are discussed afterwards.

6.1 Surface States for a Semi-Infinite Chain in the Nearly-Free Electron Model

If we want to make simple-model calculations of the electronic surface states on a crystalline surface, we are faced with a situation similar to that encountered in the case of surface phonons (Chap. 5). Within the surface plane we have to assume the ideal 2D periodicity. In the perpendicular direction, however, the translational symmetry is broken at the surface. Thus the most general one-electron wavefunction ϕ_{ss} for states localized near an ideal surface has plane-wave (Bloch) character for co-ordinates parallel to the surface $\mathbf{r}_{\|} = (x, y)$:

$$\phi_{ss}(\mathbf{r}_{\|}, z) = u_{\mathbf{k}_{\|}}(\mathbf{r}_{\|}, z) \exp(i\mathbf{k}_{\|} \cdot \mathbf{r}_{\|}) \; , \tag{6.1}$$

where $\mathbf{k}_{\|} = (k_x, k_y)$ is a wave vector parallel to the surface. The modulation function $u_{\mathbf{k}_{\|}}$ has the periodicity of the surface and is labelled according to the wave vector $\mathbf{k}_{\|}$. If one were to neglect the variation of the crystal potential parallel to the surface, $u_{\mathbf{k}_{\|}}(z)$ would not depend on $\mathbf{r}_{\|}$. Because of the 3D translational symmetry in the bulk one can take as the simplest model a semi-infinite chain of identical, periodically arranged atoms. The end of the chain then represents the surface. As in the phonon problem one can use this model to derive the essential properties of surface states, and can then readily generalize the results to the 2D surface of an ideal crystal. In the

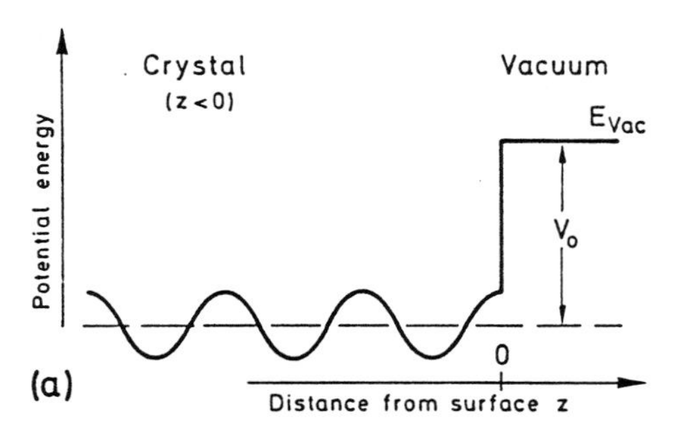

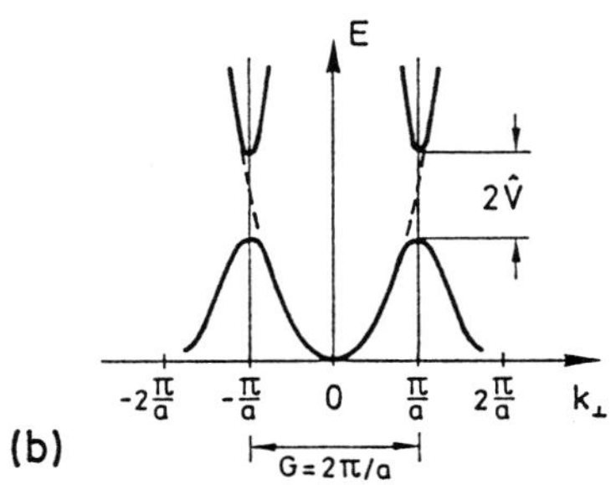

Fig.6.1a,b. Nearly-free-electron model for a cosine potential along a linear chain (z-direction). (**a**) Potential energy in the presence of a surface at z = 0. (**b**) Energy bands E ($k_\perp$) for one-electron bulk states

sense of the nearly-free electron model, we assume a cosine variation of the potential along the chain (Fig.6.1)

$$V(z) = \hat{V}\left[\exp\left(\frac{2\pi iz}{a}\right) + \exp\left(\frac{-2\pi iz}{a}\right)\right] = 2\hat{V}\cos\left(\frac{2\pi z}{a}\right), \quad \text{for } z < 0 . \qquad (6.2)$$

The surface (z = 0) is modelled by an abrupt potential step V_0 which is certainly an oversimplification for any realistic surface.

We now try to solve the Schrödinger equation

$$\left[-\frac{\hbar^2}{2m}\frac{d^2}{dz^2} + V(z)\right]\psi(z) = E\psi(z) \qquad (6.3)$$

using the potential (6.2) for V(z).

We start from regions "deep inside the crystal", i.e. at locations $z \ll 0$ far away from the surface z = 0. In this region one can consider the chain as effectively infinite and neglect surface effects. The potential V(z) can be assumed to be periodic, V(z) = V(z+na), and the well-known bulk solutions, as described in every elementary solid-state physics textbook, e.g. [6.2], are obtained. Away from the Brillouin zone boundaries $k_\perp = \pm\pi/a$ ($k_\perp$ is wave vector normal to the surface) the electronic states have plane-wave character and their energies are those of the free electron parabola

(Fig.6.1b). Near to the zone boundaries the characteristic band splitting occurs. This arises from the fact that the electronic wave function must now be taken in the lowest-order approximation, as a superposition of two plane waves. At the zone boundary an electron is scattered from a state $k_\perp = \pi/a$ into a state $k_\perp = -\pi/a$. For $k_\perp$ values near $\pi/a = G/2$ (G is reciprocal lattice vector) one therefore has in this two-wave approximation:

$$\psi(z) = Ae^{ik_\perp z} + Be^{\{i[k_\perp-(2\pi/a)]z\}} \quad . \tag{6.4}$$

Using the potential (6.2) and substituting (6.4) into the Schrödinger equation (6.3) yields the matrix equation

$$\begin{bmatrix} \frac{h^2}{2m}k_\perp^2 - E(k_\perp) & \hat{V} \\ \hat{V} & \frac{\hbar^2}{2m}\left(k_\perp - \frac{2\pi}{a}\right)^2 - E(k_\perp) \end{bmatrix} \begin{bmatrix} A \\ B \end{bmatrix} = 0 \tag{6.5}$$

which is solved by setting its determinant equal to zero. We are interested in solutions around the Brillouin-zone boundary, i.e. near $k_\perp = \pm G/2 = \pm\pi/a$. With $k_\perp = \kappa + \pi/a$, where small values of κ correspond to the interesting $k_\perp$ range, the energy eigenvalues are obtained by solving (6.5) as

$$E = \frac{\hbar^2}{2m}\left(\frac{\pi}{a}+\kappa\right)^2 \pm |\hat{V}|\left[\frac{-\hbar^2\pi\kappa}{ma|\hat{V}|} \pm \sqrt{\left(\frac{\hbar^2\pi\kappa}{ma|\hat{V}|}\right)^2 + 1}\,\right] . \tag{6.6}$$

The electronic wave functions ψ_i for spatial regions deep inside the interior crystal (subscript i) are obtained, for $z \ll 0$, by using (6.6), solving (6.5) for A and B and introducing the result into (6.4):

$$\psi_i = Ce^{i\kappa z}\left\{e^{i\pi z/a} + \frac{|\hat{V}|}{\hat{V}}\left[\frac{-\hbar^2\pi\kappa}{ma|\hat{V}|} \pm \sqrt{\left(\frac{\hbar^2\pi\kappa}{ma|\hat{V}|}\right)^2 + 1}\,\right]e^{-i\pi z/a}\right\} . \tag{6.7}$$

C is the remaining normalization constant. For regions deep inside the crystal ($z \ll 0$), the electronic energy levels form the familiar electronic bands $E(k_\perp)$ which are periodic in reciprocal $k_\perp$-space (Fig.6.1b). The "free electron" parabola splits near the zone boundaries $k_\perp = \pm\pi/a$, and allowed and forbidden energy bands arise. The $E(k_\perp = \kappa+\pi/a)$ dependence is parabolic near $\pm\pi/a$ as is seen from (6.6), and the amount of splitting, i.e. the width of the forbidden band is $2|\hat{V}|$ according to (6.6).

Our main goal in this chapter is to look for solutions of the Schrödinger equation (6.3) near a solid surface, i.e. near the end of the chain (Fig.6.1a, $z = 0$). These solutions must be composed of a part which is compatible with the constant potential $E_{vac} = V_0$ on the vacuum side ($z > 0$) and of a contribution which solves the Schrödinger equation (6.3) on the crystal side with its cosine potential (6.2). The two solutions for $z > 0$ and $z < 0$ have to be matched at the surface $z = 0$. Matching is necessary both for ψ itself and its derivation $\partial\psi/\partial z$. Any solution ψ_0 in the constant potential V_0 of the vacuum side ($z > 0$) which can be normalized must be exponentially decaying

$$\psi_0 = D\exp\left[-\sqrt{\frac{2m}{\hbar^2}(V_0 - E)}\; z\right] , \quad E < V_0 . \tag{6.8}$$

Since (6.8) contains no complex contribution $\exp(i\kappa z)$, solutions inside the crystal ψ_i of the type (6.7) can only be matched to (6.8) if a superposition of both an incoming and a reflected wave (standing wave) is taken into account. One matching condition is thus

$$\psi_0(z=0) = \alpha\psi_i(z=0,\kappa) + \beta\psi_i(z=0,-\kappa) , \tag{6.9}$$

with ψ_0 and ψ_i from (6.8) and (6.7), respectively. Equation (6.9) and the corresponding relation for the derivatives can, in fact, be fulfilled for every possible energy eigenvalue E within the allowed band. Possible surface solutions are therefore standing Bloch waves inside the crystal which are matched to exponentially decaying tails on the vacuum side (Fig.6.2a). The corresponding electronic energy levels are therefore only slightly modified from those of the infinite bulk crystal. The bulk electronic band structure therefore exists up to the very surface of a crystal with only slight alterations.

Additional surface solutions become possible if we allow complex wave vectors. Letting κ to be imaginary

$$\kappa = -\,iq \tag{6.10a}$$

and defining, for convenience,

$$\gamma = i\sin(2\delta) = -\,i\frac{\hbar^2\pi q}{ma|\hat{V}|} \tag{6.10b}$$

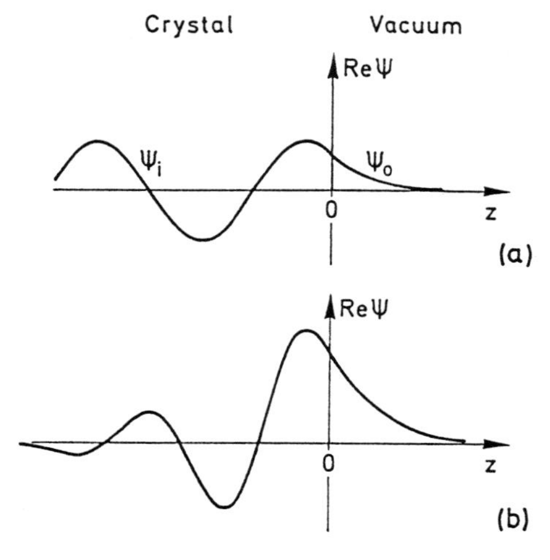

Fig.6.2. Real part of the one-electron wavefunction, Re{ψ}, for (**a**) a standing Bloch wave (ψ_i), matched to an exponentially decaying tail (ψ_0) in the vacuum; (**b**) a surface-state wave function localized at the surface (z = 0)

one can show that the energy eigenvalues (6.6) remain real for a particular range of κ, i.e. these values represent possible solutions of the Schrödinger equation. The electronic wave function for imaginary κ inside the crystal ($z<0$) results from (6.7) by using (6.10a)

$$\psi_i'(z\leq 0) = Fe^{qz}\left\{\exp\left[i\left(\frac{\pi}{a}z\pm\delta\right)\right] \mp \exp\left[-i\left(\frac{\pi}{a}z\pm\delta\right)\right]\right\}e^{\mp i\delta} \ . \tag{6.11}$$

This is essentially a standing wave with an exponentially decaying amplitude (Fig.6.2b). The energy eigenvalues are obtained from (6.6) with (6.9) as

$$E = \frac{\hbar^2}{2m}\left[\left(\frac{\pi}{a}\right)^2 - q^2\right] \pm |\hat{V}| \sqrt{1-\left(\frac{\hbar^2\pi q}{ma|\hat{V}|}\right)^2} \ . \tag{6.12}$$

The values of E remain real (as required for energies) and ψ does not diverge for large negative z if $0 < q < q_{max} = ma|\hat{V}|\hbar^2\pi$. In this q range the E versus q dependence is described by (6.12); all energies fall into the forbidden gap of the bulk electronic-band structure (Fig.6.3).

Equation (6.11) is not the complete solution for a surface electronic state, since the part of the wave function on the vacuum side, $z > 0$ is missing. To obtain the complete solution for our surface problem, we have to match the wave functions (6.11) to the exponentially decaying vacuum solu-

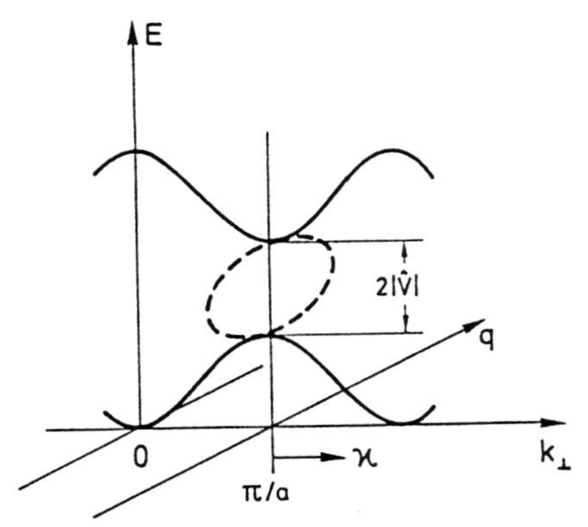

Fig.6.3. Electronic band structure (qualitative) for a semi-infinite chain of atoms. Bulk Bloch states which are little disturbed by the presence of the surface give rise to energy bands $E(k_\perp)$ that are periodic in wave vector $k_\perp$ parallel to the chain direction, i.e. normal to the surface (*solid curves*). States with a wave function amplitude exponentially decaying from the surface ($z = 0$) into the bulk ($z < 0$) are found between the bulk states for complex wave vectors $\pi/a - iq$ (*broken curve*)

tion (6.8). The matching conditions require for the wave functions and for their derivatives:

$$\psi_0(z=0) = \psi_i'(z=0) \quad \text{and} \quad \left.\frac{d\psi_0}{dz}\right|_{z=0} = \left.\frac{d\psi_i'}{dz}\right|_{z=0} . \tag{6.13}$$

For this matching procedure (two equations) there are exactly two free parameters which are thus fixed by (6.13): the energy eigenvalue E and the ratio D/F of the wave-function amplitudes in (6.8 and 11). The wave function of the resulting electronic surface state is shown qualitatively in Fig.6.2b. Its amplitude vanishes for $\pm z$ values far away from the surface. Electrons in these states are, in fact, localized within a couple of Ångstroms of the surface plane. Another important consequence of the matching conditions (6.13) is the restriction on the allowed values of E. Of the continuous range of E values within the forbidden bulk energy gap (Fig.6.3), only one single energy level E is fixed by means of the requirement (6.13). The present calculation for the semi-infinite chain therefore yields one single electronic surface state which is located somewhere in the gap of the bulk states.

6.2 Surface States of a 3D Crystal and Their Charging Character

6.2.1 Intrinsic Surface States

The generalization of the results for the one-dimensional semi-infinite chain to the 2D surface of a 3D crystal is straightforward. Because of the 2D translational symmetry parallel to the surface, the general form of a surface-state wave function is of the Bloch type (6.1) in coordinates $\mathbf{r}_{\|}$ parallel to the surface, i.e. the variation in $\mathbf{r}_{\|}$ enters through the factor $\exp(i\mathbf{k}_{\|}\cdot\mathbf{r}_{\|})$ and the energy is increased by the term $\hbar^2 k_{\|}^2/2m$. The energy eigenvalues (6.12) therefore become functions of $k_{\perp} = \pi/a - iq$ and of the wave vector $\mathbf{k}_{\|}$ parallel to the surface. The matching conditions (6.13) thus have to be fulfilled for each $\mathbf{k}_{\|}$ separately and for each $\mathbf{k}_{\|}$ a single, but in general different, energy level for the surface state is obtained. We thus arrive at a 2D band structure $E_{ss}(\mathbf{k}_{\|})$ for the energies E_{ss} of electronic surface states. The $E_{ss}(\mathbf{k}_{\|})$ bands are defined in the 2D reciprocal $\mathbf{k}_{\|}$-space of the surface. The description is analogous to that of surface-phonon dispersion branches (Chap. 5). Since bulk electronic states are also found at the surface, with only minor modifications, one has to take them into account when mapping the true surface states.

A surface state is described by its energy level E_{ss} and its wave vector $k_{\|}$ parallel to the surface. For bulk states both $k_{\|}$ and $k_{\perp}$ components are allowed. For each value of $k_{\|}$ therefore, a rod of $k_{\perp}$ values extends back into the bulk 3D Brillouin zone; bulk energy bands being cut by this rod yield a bulk state in the $E_{ss}(k_{\|})$ plane. We thus arrive at a presentation in which surface-state bands (broken lines in Fig. 6.4) are plotted together with a projection of all bulk states (hatched area in Fig. 6.4) in a particular $E(\mathbf{k}_{\|})$ plane. True surface-state bands are characterized by energy levels E_{ss} that are not degenerate with bulk bands; they lie in the gaps of the projected bulk-band structure. Surface-state bands, however, can penetrate into a part

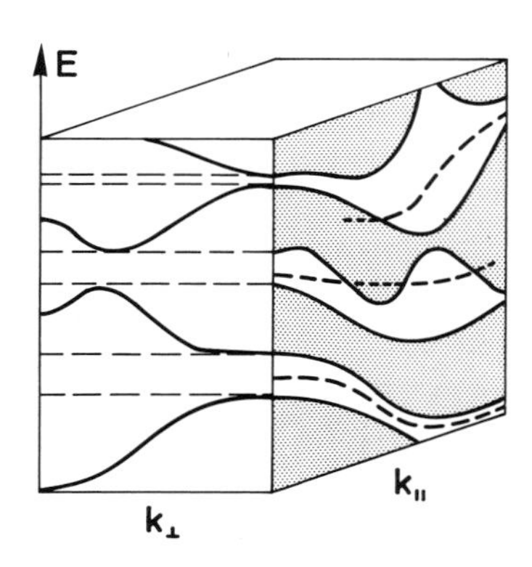

Fig. 6.4. Hypothetical electronic band structure of a crystal. The shaded areas in the $E(k_{\|})$ plane describe the projected bulk-band structure (along $k_{\perp}$). Broken lines in the $E(k_{\|})$ plane indicate surface state bands in the gaps of the projected bulk-band structure, and surface resonances where they are degenerate with bulk states (*short dotted lines*)

of the surface Brillouin zone, where propagating bulk states exist (short dotted lines in Fig.6.4). They are then degenerate with bulk states and can mix with them. Such a state will propagate deep into the bulk, similar to a Bloch state with finite $k_\perp$, but will nevertheless retain a large amplitude close to the surface. These states are known as **surface resonances**.

The discussion of realistic surface-state bands in the next chapters is based on representations such as indicated in Fig.6.4 where the energy levels E_{ss} are plotted along particular symmetry directions $\mathbf{k}_\parallel$ of the 2D surface Brillouin zone. For the most important crystal structures fcc, bcc, and hcp the relation of the low-index surface Brillouin zones to their bulk counterparts has already been described in connection with surface phonon dispersion (Figs.5.5-7). The projections of the bulk Brillouin zones on particular surfaces are very helpful for establishing whether a particular state falls into a gap of the bulk-band structure.

So far we have discussed the existence of surface states on ideal, clean surfaces in the framework of the nearly-free-electron model. For historical reasons, such surface states are often called **Schockley states** [6.3]. One can also approach the question of the existence of electronic surface states from the other limiting case of tightly bound electrons. This approximate treatment in terms of wave functions that are linear combinations of atomic eigenstates was first given by *Tamm* [6.4]. The resulting states are often called **Tamm states**, even though there is no real physical distinction between the different terms; only the mathematical approach is different. The existence of electronic surface states, whose energy is different from the bulk states, is qualitatively easy to see within the picture of tightly bound electrons (Linear Combination of Atomic Orbitals, LCAO). For the topmost surface atoms the bonding partners on one side are missing in total, which means that their wave functions have less overlap with wave functions of neighboring atoms. The splitting and shift of the atomic energy levels is thus smaller at the surface than in the bulk (Fig.6.5). Every atomic orbital involved in chemical bonding and producing one of the bulk electronic bands should also give rise to one surface-state level. The stronger the perturbation caused by the surface, the greater the deviation of the surface level from the bulk electronic bands. When a particular orbital is responsible for chemical bonding, e.g. the sp^3 hybrid in Si or Ge, it is strongly affected by the presence of the surface; bonds are broken, and the remaining lobes of the orbital stick out from the surface. They are called **dangling bonds**. The energy levels of such states are expected to be significantly shifted from the bulk values.

Beside these dangling-bond states there are other types of states, sometimes called **back bond** states, which are related to surface-induced modifications of the chemical bonds *between* the topmost layers. Thus the perturbation of the chemical bonds due to the presence of the surface is not re-

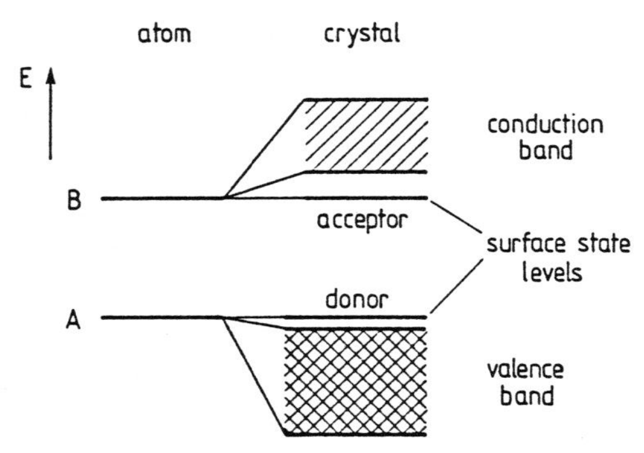

Fig.6.5. Qualitative explanation of the origin of surface states in the tight-binding picture. Two atomic levels A and B form the bulk valence and conduction bands, respectively. Surface atoms have fewer bonding partners than bulk atoms and thus give rise to electronic energy levels that are closer to those of the free atoms, i.e. surface state levels are split off from the bulk bands. Depending on their origin, these states have acceptor- or donor-like charging character

stricted to the first layer of atoms. However, back bonds are generally less disturbed than dangling bonds and the corresponding surface state levels are shifted less with respect to the bulk bands.

From Fig.6.5 it is also clear that the energetically higher-lying surface state (originating from the atomic level B) has conduction-band character, whereas the lower level which is split off from the valence band of the semiconductor is more valence-band like. The corresponding surface-state wave functions are *built up* from conduction- and valence-band wave functions which, in the absence of a surface, would have contributed to the bulk states. Therefore the charging character of the surface states also reflects that of the corresponding bulk states. A semiconductor is neutral if all conduction-band states are empty and all valence-band states are occupied by electrons. On the other hand, conduction-band states carry a negative charge if they are occupied by an electron, and valence-band states are positively charged when being unoccupied. As a consequence, surface states derived from the conduction band have the charging character

neutral (empty) $\leftrightarrow$ negative (with electron) ,

whereas surface states derived from the valence band have the character

positive (empty) $\leftrightarrow$ neutral (with electron) .

In accordance with the definition of shallow bulk impurities one calls the first type of state (conduction-band derived) an **acceptor-type** state whereas the second is a **donor-type** state.

The simple situation depicted in Fig.6.5 where one type of surface state derives solely from the conduction band and the other one from the valence band is only a limiting case. Most surface states are built up both from valence and conduction-band wave functions; their charging character is then determined by the relative valence and conduction-band contributions. Depending on their location within the gap (closer to valence or closer to conduction band) they are more donor or more acceptor like.

A good example of the simple case depicted in Fig.6.5 is that of partly ionic materials such as GaAs (III-V compounds). The bulk valence and conduction bands are derived essentially from As and Ga wave functions, respectively. Thus the As-derived surface states have more donor character, whereas the Ga-derived surface states are more acceptor like.

6.2.2 Extrinsic Surface States

The surface states discussed thus far are all related to the clean and well-ordered surface of a crystal with 2D translational symmetry. These states are called **intrinsic surface states**. They include the states arising due to relaxation and reconstruction. Because of the 2D translational symmetry of the surface these intrinsic surface states form electronic band structures in the 2D reciprocal space.

In addition to these states, there are other electronic states localized at a surface or interface which are related to imperfections. A missing surface atom causes a change in the bonding geometry for the surrounding atoms, thus giving rise to changes in the spectrum of electronic surface states. In particular, for crystals with partially ionic bonds, it is easily seen how a missing surface atom, i.e. ion, affects the electronic structure in the vicinity. If a negatively charged O^- ion in ZnO is missing in the topmost atomic layer, then a finite area of the surrounding surface contains more positive charge due to the surrounding Zn^+ ions than do other stoichiometric areas of the surface. This enhanced positive charge near the defect acts as a trap for electrons; in other words, the missing negative surface ion is related to a localized electronic defect state which can be occupied by an electron. Its charging character is acceptor-like. Similar behaviour is also known for III-V compounds. Missing As atoms in a GaAs surface are related to extrinsic electronic surface states which act as acceptors (Sect.8.4).

Similar reasoning can be applied to line defects. Atoms located on the edge of a step of a non-ideal surface are in an environment different from that of an atom on an unperturbed surface. Since usually more chemical bonds are broken at an atomic step than on the flat part of an ideal surface, atoms at the edge of the step often possess more dangling-bond orbitals. The result is a new type of surface state related to step atoms. In contrast to the intrinsic states discussed before, defect-derived states do not exhibit any 2D translational symmetry parallel to the surface. Their wave functions, of course, are localized near the defects, i.e. near the surface plane. In the particular case of a linear step, there might also be some translational symmetry along the direction of the step. Whereas intrinsic surface states are related to the existence of a perfect surface, the present **extrinsic surface** states only arise as a result of perturbations to the ideal surface.

Extrinsic surface states can also be produced by adsorbed atoms. Adsorption causes changes in the chemical bonds near the surface, thus affecting the distribution of intrinsic surface states. In addition, new electronic states are formed by the bonding and antibonding orbitals between the chemisorbed atom or molecule and the surface. Since chemisorbed atoms or molecules can form 2D lattices with translational symmetry along the surface, extrinsic electronic surface states originating from adsorbed species might form 2D band structures $E_{ss}(\mathbf{k}_{\parallel})$ as do intrinsic surface states. The special problems related to adsorption processes on surfaces are discussed more in detail in Chap. 9.

6.3 Aspects of Photoemission Theory

6.3.1 General Description

The most important and widely used experimental technique to gain information about occupied electronic surface states is photoemission spectroscopy [6.5]. The experiment is based on the photoelectric effect [6.6]. The solid surface is irradiated by mono-energetic photons and the emitted electrons are analyzed with respect to their kinetic energy. When photons in the ultraviolet spectral range are used the technique is called UPS (UV Photoemission Spectroscopy); with X-ray radiation it is called **XPS** or **ESCA** (Electron Spectroscopy for Chemical Analysis). With synchroton radiation one can cover the whole spectral range from the near-UV to the far X-ray regime.

The use of angle-integrating electron analyzers gives integrated information about large parts of the reciprocal space, i.e. essentially one obtains the densities of occupied electronic states. In order to investigate the dispersion of electronic bands $E(\mathbf{k})$ and $E(\mathbf{k}_{\parallel})$ for bulk and surface states, respectively, a determination of the electron wave vector is necessary. Besides the kinetic energy one thus also needs to know the emission direction. This can be determined by means of an electron energy analyser with small angular aperture. The method is then known as **Angle-Resolved UV Photoemission** (ARUPS). The essential geometrical parameters in an ARUPS experi-

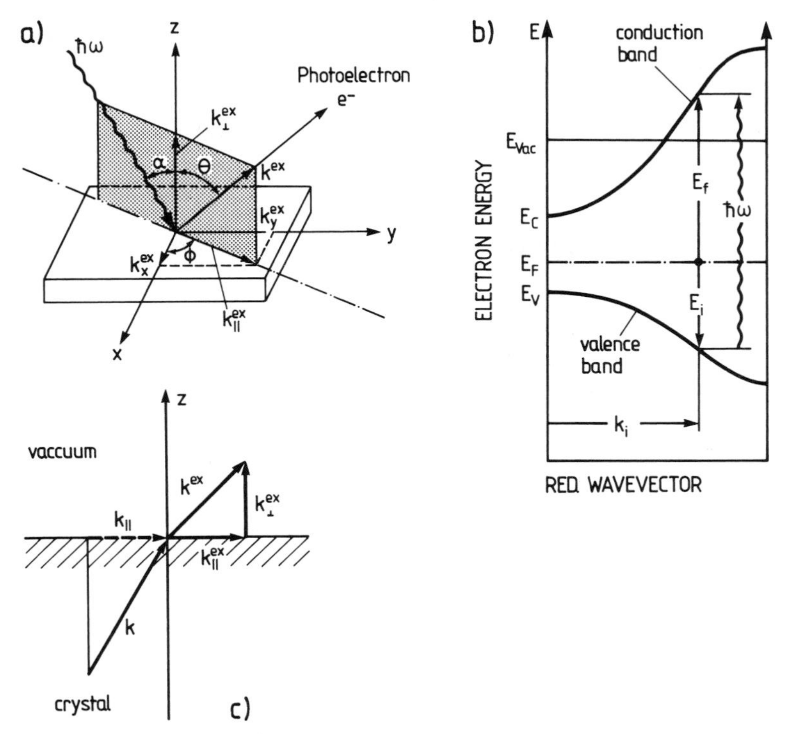

Fig. 6.6a-c. Description of a photoemission experiment. (**a**) Definition of the angles and wave vectors of the incident photon ($\hbar\omega$) and emitted electron e^-. (**b**) Representation of the photoexcitation process in the electronic band scheme $E(\mathbf{k})$ of a semiconductor. Only direct transitions with $\mathbf{k}_i \simeq \mathbf{k}_f$ are taken into account. The energies of the initial state (E_i) and final state (E_f) are referred to the Fermi level E_F. (**c**) Conservation of the wave vector component $\mathbf{k}_\parallel$ (parallel to the surface) upon transmission of the emitted electron through the surface

ment are shown in Fig. 6.6a. The angle of incidence α of the photons (energy $\hbar\omega$), their polarization and the plane of incidence, determine the electric field direction and the vector potential $\mathbf{A}$ of the photoexciting electromagnetic wave with respect to the crystal lattice. The wave vector $\mathbf{k}^{ex}$ of the emitted electrons (external of the solid) is determined by its magnitude

$$k^{ex} = (2mE_{kin}/\hbar^2)^{1/2} \tag{6.14}$$

where E_{kin} is the kinetic energy of the detected electron, and by the emission direction described by the angles ϕ and Θ.

A rigorous theoretical approach to the photoemission process requires a full quantum-mechanical treatment of the complete coherent process in which an electron is removed from an occupied state within the solid and deposited at the detector. Theoretical approaches of this kind treat the photoeffect as a **one-step process** [6.6, 7]. The less accurate but simpler and more instructive approach is the so-called **three-step model** in which the photoemission process is artificially separated into three independent parts [6.8, 9]:

(i) Optical excitation of an electron from an initial into a final electron state within the crystal (Fig. 6.6b).
(ii) Propagation of the excited electron to the surface.
(iii) Emission of the electron from the solid into the vacuum. The electron traverses the surface (Fig. 6.6c).

In principle, these three steps are not independent of each other. For example in a more rigorous treatment the final-state wave function must be considered as consisting of the excited state function together with waves scattered from neighboring atoms. Such effects are important in the method SEXAFS (Panel VII: Chap. 7). In the three-step model the independent treatment of the above three contributions leads to a simple factorization of the corresponding probabilities in the photoemission current of the emitted electrons.

The optical excitation of an electron in the first step is simply described by the golden-rule transition probability for optical excitation:

$$\begin{aligned} w_{fi} &= \frac{2\pi}{\hbar} |\langle f,\mathbf{k}| \mathscr{H} |i,\mathbf{k}\rangle|^2 \delta(E_f(\mathbf{k}) - E_i(\mathbf{k}) - \hbar\omega) \\ &= (2\pi/\hbar) m_{fi}\, \delta(E_f - E_i - \hbar\omega) \ . \end{aligned} \tag{6.15}$$

In a first approximation direct transitions with nearly unchanged $\mathbf{k}$ are taken into account between the initial and final Bloch states $|i,\mathbf{k}\rangle$ and $|f,\mathbf{k}\rangle$. The perturbation operator $\mathscr{H}$ is given by the momentum operator $\mathbf{p}$ and the vector potential $\mathbf{A}$ of the incident electromagnetic wave (dipole approximation):

$$\mathscr{H} = \frac{e}{2m}(\mathbf{A}\cdot\mathbf{p} + \mathbf{p}\cdot\mathbf{A}) \simeq \frac{e}{m}\mathbf{A}\cdot\mathbf{p} \tag{6.16}$$

with $\mathbf{B} = \mathrm{curl}\mathbf{A}$ and $\mathscr{E} = -\dot{\mathbf{A}}$. In (6.16) $\mathbf{A}$ can be assumed to commute with $\mathbf{p}$, since it is nearly constant in the long-wavelength limit (in UPS: $\lambda > 100$ Å). The δ-function in (6.15) describes energy conservation in the excitation of an electron from a state $E_i(\mathbf{k})$ into a state $E_f(\mathbf{k})$ of the electronic band structure. For the time being we consider an excitation process between two bulk bands $E_i(\mathbf{k})$ and $E_f(\mathbf{k})$; simplifications that arise for excitations between surface-state bands are explained later.

Outside the solid on the vacuum side one can only detect electrons whose energy E is above the vacuum energy E_{vac} (Fig.6.6b) and whose $\mathbf{k}$ vector in the final state is directed outwards from the surface, i.e. $\mathbf{k}_\perp > 0$.

The internal electron current density directed towards the surface with an energy E and a wave vector around $\mathbf{k}$ is therefore ($k_\perp > 0$)

$$I^{int}(E,\hbar\omega,\mathbf{k}) \propto \sum_{f,i} m_{fi}\, f(E_i)\, \delta(E_f(\mathbf{k})-E_i(\mathbf{k})-\hbar\omega)\, \delta(E-E_f(\mathbf{k})) \,. \qquad (6.17)$$

For detection at energy E (which involves adjusting the energy window of the electron analyser), the energy of the final state $E_f(\mathbf{k})$ has to equal E. The function $f(E_i)$ is the Fermi distribution function. It ensures that the initial state with E_i is occupied.

The second step in the three-step model is the propagation of the electrons described by (6.17) to the surface. A large number of electrons undergo inelastic scattering processes; they lose part of their energy E_f (or E) by electron-plasmon or electron-phonon scattering. Such electrons contribute to the continuous background in the photoemission spectrum which is called the **true secondary background**; they have lost the information about their initial electronic level E_i (Fig.6.6). The probability that an electron will reach the surface without inelastic scattering is given phenomenologically by the mean-free path λ. In general, λ depends on the energy E, the electron wave vector $\mathbf{k}$ and on the particular crystallographic direction. The propagation to the surface is thus described in a simplifying manner by the **transport probability** $D(E,\mathbf{k})$ which is proportional to the mean free path λ:

$$D(E,\mathbf{k}) \propto \lambda(E,\mathbf{k}) \,. \qquad (6.18)$$

It is this second step of propagation to the surface which makes photoemission a surface sensitive technique. The value of λ is typically between 5 and 20 Å (Fig.4.1), thus limiting the information depth to this spatial region.

The third step, transmission of the photoexcited electron through the surface can be considered as the scattering of a Bloch electron wave from the surface-atom potential with translational symmetry parallel, but not normal to the surface. One arrives at the same conclusions when the transmission through the surface is thoroughly treated by matching the internal Bloch wave functions to free-electron wave functions (LEED problem, Sect. 4.4) outside on the vacuum side. In any case, because of the 2D translational symmetry, the transmission of the electron through the surface into the vacuum requires conservation of its wave-vector component parallel to the surface (Fig. 6.6c):

$$\mathbf{k}_{\parallel}^{ex} = \mathbf{k}_{\parallel} + \mathbf{G}_{\parallel} ; \tag{6.19}$$

where $\mathbf{k}$ is the wave vector of the electron inside the crystal. Its component normal to the surface $\mathbf{k}_{\perp}$ is not conserved during transmission through the surface. For the external electron on the vacuum side, the $k_{\perp}^{ex}$ value is determined by the energy conservation requirement

$$E_{kin} = \frac{\hbar^2 k^{ex^2}}{2m} = \frac{\hbar^2}{2m}(k_{\perp}^{ex^2} + k_{\parallel}^{ex^2}) = E_f - E_{vac} . \tag{6.20}$$

With $\phi = E_{vac} - E_F$ as the work function and E_B as the (positive) binding energy referred to the Fermi level E_F (Fig. 6.6b) one also has

$$\hbar\omega = E_f - E_i = E_{kin} + \phi + E_B . \tag{6.21}$$

The wave-vector component parallel to the surface outside the crystal [from (6.20, 21)], which is determined from known experimental parameters,

$$k_{\parallel}^{ex} = \sqrt{\frac{2m}{\hbar^2}}\sqrt{\hbar\omega - E_B - \phi}\,\sin\Theta = \sqrt{\frac{2m}{\hbar^2}E_{kin}}\,\sin\Theta \tag{6.22}$$

therefore directly yields the internal wave-vector component $k_{\parallel}$ according to (6.19)

On the other hand, due to the inner microscopic surface potential V_0, the wave-vector component $k_{\perp}$ of the electron inside the crystal is changed upon transmission through the surface. The outside component is determined by energy conservation according to (6.20) as

$$k_{\perp}^{ex} = \sqrt{\frac{2m}{\hbar^2}E_{kin} - (\mathbf{k}_{\parallel} + \mathbf{G}_{\parallel})^2} = \sqrt{\frac{2m}{\hbar^2}E_{kin}}\,\cos\Theta . \tag{6.23}$$

However, without a detailed knowledge of the electronic band structure for energies above E_{vac} and of the inner microscopic potential V_0 (usually not exactly known) information about the inner wave-vector component $k_{\perp}$ cannot be obtained.

According to (6.19) the third step, transmission through the surface, can be described formally by the transmission rate

$$T(E,\mathbf{k})\delta(\mathbf{k}_{\parallel}+\mathbf{G}_{\parallel}-\mathbf{k}_{\parallel}^{ex}) . \tag{6.24}$$

In the simplest and rather naive approach one might assume that $T(E,\mathbf{k})$ is a constant $R \leq 1$ with

$$T(E,\mathbf{k}) = \begin{cases} 0 \text{ for } k_{\perp}^{ex^2} = \dfrac{2m}{\hbar^2}(E_f - E_{vac}) - (\mathbf{k}_{\parallel}+\mathbf{G}_{\parallel})^2 < 0 , \\ R \text{ for } k_{\perp}^{ex^2} = \dfrac{2m}{\hbar^2}(E_f - E_{vac}) - (\mathbf{k}_{\parallel}+\mathbf{G}_{\parallel})^2 > 0 . \end{cases} \tag{6.25}$$

This form (6.25) of $T(E,\mathbf{k})$ takes into account that only electrons with the positive wave-vector component $k_{\perp}^{ex}$ can be observed in the photoemission experiment; all others are unable to reach the vacuum side of the crystal surface and are internally reflected since their kinetic energy is not sufficient to surmount the surface barrier.

Taking together (6.15, 17, 18, 24, 25) one arrives at the following formula for the observable (external) emission current in the three-step model:

$$\begin{aligned} I^{ex}(E,\hbar\omega,\mathbf{k}_{\parallel}^{ex}) &= I^{int}(E,\hbar\omega,\mathbf{k})D(E,\mathbf{k})T(E,\mathbf{k})\delta(\mathbf{k}_{\parallel}+\mathbf{G}_{\parallel}-\mathbf{k}_{\parallel}^{ex}) \\ &\propto \sum_{f,i} m_{fi} f(E_i(\mathbf{k}))\delta\left[E_f(\mathbf{k})-E_i(\mathbf{k})-\hbar\omega)\delta(E-E_f(\mathbf{k})\right] \\ &\quad \times\delta(\mathbf{k}_{\parallel}+\mathbf{G}_{\parallel}-\mathbf{k}_{\parallel}^{ex})D(E,\mathbf{k})T(E,\mathbf{k}_{\parallel}) . \end{aligned} \tag{6.26}$$

The following discussion reveals what kind of information can be extracted from a photoemission spectrum.

6.3.2 Angle-Integrated Photoemission

If in the photoemission experiment one uses an electron energy analyzer that accepts (in the ideal case) electrons within the whole half-space above the sample surface, then the total photocurrent contains contributions with every possible $\mathbf{k}_{\parallel}$. Such an angle integrated measurement can be achieved to a good approximation by using a (LEED) retarding field analyser (Panel VIII: Chap.4). The total measured photocurrent is obtained by integrating (6.26) over $\mathbf{k}_{\parallel}^{ex}$. With the restrictions (6.22, 23), i.e. conservation of $\mathbf{k}_{\parallel}$ and determination of $k_\perp$ by energy conservation, the integration over $\mathbf{k}_\parallel^{ex}$ can be transformed into an integration over the whole $\mathbf{k}$ space, i.e.,

$$\tilde{I}^{ex}(E,\hbar\omega) \propto \int_{\text{half sphere}} I^{ex}(E,\hbar\omega,\mathbf{k}_\parallel^{ex})d\mathbf{k}^{ex} . \quad (6.27a)$$

The integration cancels the $\delta(\mathbf{k}_\parallel+\mathbf{G}_\parallel-\mathbf{k}_\parallel^{ex})$ function. Assuming furthermore that the matrix elements m_{fi} are slowly varying functions in $\mathbf{k}$ space, one arrives at an expression for the total external photoemission current:

$$\tilde{I}^{ex}(E,\hbar\omega) \propto \sum_{f,i} m_{fi} \int d\mathbf{k} f(E_i(k))\,\delta(E_f(\mathbf{k})-E_i(\mathbf{k})-\hbar\omega)\,\delta(E-E_f(\mathbf{k})) . \quad (6.27b)$$

This expression includes all possible ways in which an electron can be excited from the occupied band $E_i(\mathbf{k})$ to the band $E_f(\mathbf{k})$ with the restriction of energy and wave-vector conservations. The $\delta(E-E_f(\mathbf{k}))$ selects only those final states whose energy E_f coincides with the detection energy E. The current is therefore proportional to the joint density of states for which the final states have the energy E. The information obtained is similar to that yielded by an optical absorption experiment, but the final-state distribution enters via $\delta(E-E_f(\mathbf{k}))$.

When lower photon energies are used, i.e. UPS rather than XPS, the final states, that can be reached (in UPS), may have a considerably structured density of states giving rise to strong changes of the photoemission current $\tilde{I}^{ex}$ (6.27) with varying photon energy. In XPS, on the other hand, the photon energies are quite high and the final states are distributed quasi-continuously. The photocurrent becomes relatively insensitive against variations in photon energy. The structure obtained is then determined largely by the distribution of the initial states $E_i(\mathbf{k})$. Angle-integrated (but also non-integrated) UPS and XPS are particularly useful as a finger-printing technique to identify an adsorbed species by the characteristic emission lines from its molecular orbitals (Fig.6.7) [6.10].

6.3.3 Bulk- and Surface-State Emission

According to (6.19, 22) the externally determined wave-vector component $\mathbf{k}_\parallel^{ex}$ directly provides the internal component $\mathbf{k}_\parallel$. For 2D band structures of surface-state bands or electronic states of regularly adsorbed molecules this

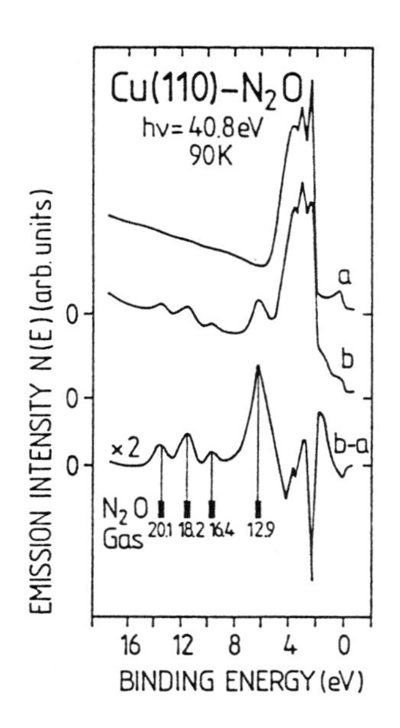

Fig. 6.7. HeII UPS spectra of a clean Cu(110) surface at 90 K (*a*), and after a 1-L exposure to N_2O (*b*). Marked under the difference curve (*b-a*), enlarged by a factor of 2, are the vertical ionization energies of gaseous N_2O (referred to vacuum level) [6.10]

is sufficient. All information about the wave vectors of the states is given. The same is true for quasi 2D crystals composed of lamellar structures (graphite, $TaSe_2$, etc.). When using ARUPS for the determination of 3D band structures of bulk electronic states one is faced with the problem of determining the $\mathbf{k}_\perp$ component inside the solid. Since $k_\perp$ is not conserved, the value measured outside $k_\perp^{ex}$ (6.23) does not yield this information. Several approaches are used to overcome this problem. One can measure the photoemission spectra for various photon energies under normal emission ($\Theta = 0$ in Fig. 6.5a) thus having vanishing wave-vector components $\mathbf{k}_\parallel$, $\mathbf{k}_\parallel^{ex}$ parallel to the surface. The wave vector components $k_\perp^{ex}$ and $k_\perp$ outside and inside the crystal are related via energy conservation and the internal potential V_0. As a simple approximation one sometimes assumes *free electron parabolae* for the final states

$$E_f(k_\perp) \simeq \hbar^2 \frac{k_\perp^2}{2m^*} . \tag{6.28}$$

The kinetic energy of the electron in the vacuum is given by

$$E_{kin} = \hbar^2 \frac{k_\perp^{ex^2}}{2m} = \frac{\hbar^2 k_\perp^2}{2m^*} + V_0 . \tag{6.29}$$

A further assumption about the inner potential, e.g. the zero of a muffin-tin potential, then allows an unequivocal determination of $E_i(k_\perp)$ from the measured outside quantities E_{kin} and $k_\perp^{ex} = k^{ex}$. If band-structure calculations are available, one can also use the theoretical final-state bands $E_f(\mathbf{k})$ instead of (6.28) to calculate the internal $k_\perp$ value from the measured value $k_\perp^{ex}$. In a trial-and-error procedure one can thus compare a progressively improved theoretical band structure, including both $E_i(\mathbf{k})$ and $E_f(\mathbf{k})$ with the measured data.

There is another direct experimental approach to the problem of $k_\perp$ determination which is shown for the example of clean Au surfaces in Fig. 6.8 [6.11]. In order to determine the electronic band structure $E(\mathbf{k})$ along a certain symmetry line, in this case along ΓL (Fig.6.8b), angle-resolved photo-

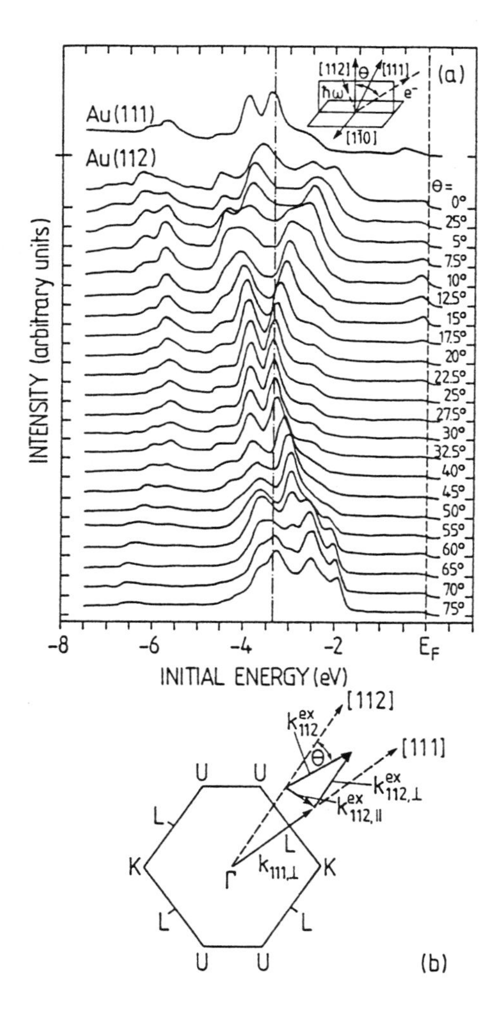

Fig.6.8. (**a**) Normal ARUPS photoemission spectra from Au(111) and a family of spectra from Au(112) obtained at different polar emission angles Θ (explanation in inset). The photon energy $\hbar\omega$ is 16.85 eV. (**b**) Cut through the Brillouin zone with the emission directions of (a) indicated. Projection of the wave vector component $\mathbf{k}_{112,\parallel}$ onto the ΓL direction yields the wave vector $\mathbf{k}_{111,\perp}$ i.e. the location of the final state in $\mathbf{k}$-space [6.11]

emission measurements normal to the surface of Au(111) are performed. Because of $\mathbf{k}_{\parallel}$ ($\simeq 0$) conservation only states described by wave vectors normal to the considered (111) surface, i.e. in the ΓL direction contribute to the normal photoemission from Au(111). The direction in **k** space is thus known for the measured spectrum on Au(111) in Fig.6.8a, but not the actual length of the internal wave vector. If ARUPS measurements from another surface of the crystal, e.g. the Au(112) surface, are now performed (Fig.6.8a) under different polar angles Θ with respect to [112], one might be able to identify the same emission bands in one of those spectra as for the normal measurement along [111]. In Fig.6.8a the spectrum with $\Theta = 25°$ best resembles the upper spectrum on Au(111). From the measured binding energies E_B, the work function ϕ on Au(112) and the angle Θ the corresponding wave vectors $\mathbf{k}_{112}^{ex}$ are calculated, and according to (6.22) also the parallel component $\mathbf{k}_{112,\parallel}^{ex}$ which is identical to the internal wave vector $\mathbf{k}_{112,\parallel}$. After projecting $\mathbf{k}_{112,\parallel}$ on the ΓL direction (Fig.6.8b) one obtains the location of the contributing final states in **k** space, i.e., the wave vector $\mathbf{k}_{111,\perp}$ (1.94 ± 0.11 Å^{-1} in Fig.6.8b).

Although the investigation of a 2D band structure is straightforward as far as the determination of $\mathbf{k}_{\parallel}$ is concerned, there is a problem in distinguishing between bulk and surface emission bands in the photoemission spectrum. Four criteria can help one to decide whether a particular band arises from surface states, i.e. is due to a 2D band structure $E_i(\mathbf{k}_{\parallel})$ localized at the very surface:

(i) Since no definite $k_{\perp}$ exists for surface-state emission, (6.22) must be fulfilled for every possible choice of photon energy, i.e. one and the same dispersion $E_i(\mathbf{k}_{\parallel})$ must be obtained by using different photon energies.

(ii) For measurements at normal emission the parallel components $k_{\parallel}^{ex}$ and $k_{\parallel}$ vanish. For emission from a surface-state band $E_{ss}(\mathbf{k}_{\parallel})$ the band structure then contributes at the Γ point ($k_{\parallel} = 0$) only, independent of the photon energy used. A surface-emission band thus occurs at the same energy in the spectrum for different photon energies. In contrast, a bulk-emission band is expected to vary in energetic position with changing photon energy.

(iii) An emission band from real surface states must fall into a bulk-band gap. Thus, if a plot of the measured $E(\mathbf{k}_{\parallel})$ dependence is not degenerate with the projected bulk-band structure (Fig.6.4), this suggests surface-state emission.

(iv) In contrast to surface states, bulk states are not affected by a surface treatment. If an emission band of the clean surface vanishes after gas adsorption, its origin is likely to be surface states. But care must be taken in applying this rule, since adsorption might also change the transmission conditions for electrons through the surface (due, e.g., to space charge layers).

6.3.4 Symmetry of Initial States and Selection Rules

According to (6.15, 16) the photocurrent in ARUPS is determined by matrix elements of the form

$$m_{fi} = \langle f,\mathbf{k}| \frac{e}{m} \mathbf{A}\cdot\mathbf{p} |i,\mathbf{k}\rangle \ , \tag{6.30}$$

where $\mathbf{A}$ is the vector potential of the incoming UV light or X rays, and $\mathbf{p}$ is the momentum operator ($\mathbf{p} = \hbar\nabla/i$). By considering special experimental geometries and the symmetry of the electronic states involved, we can derive interesting selection rules for the observability of particular initial states $|i,\mathbf{k}\rangle$. We assume that the surface has a mirror plane (Fig.6.9), and that both the direction of incidence of the exciting light and the detection direction for the emitted electrons are within that mirror plane (yz).

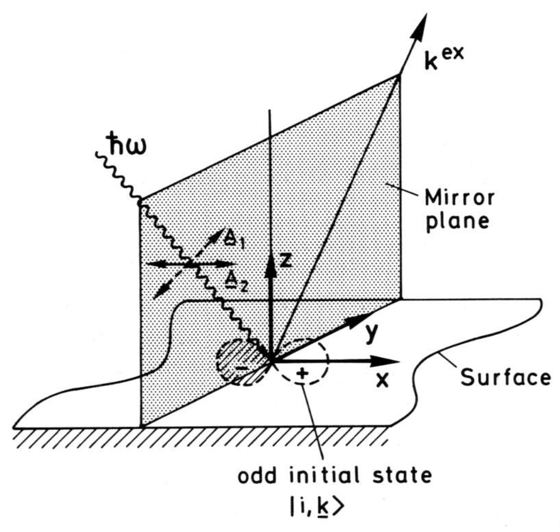

Fig.6.9. Symmetry selection rule in a photoemission experiment. The direction of the incident light ($\hbar\omega$) and the trajectory of the emitted electron (wave vector $\mathbf{k}^{ex}$) lie in a mirror plane of the crystal surface. The electron is emitted from an initial state $|i,\mathbf{k}\rangle$ that is of odd parity with respect to the mirror plane. Since the final state must be even with respect to the mirror plane, only a light polarization $\mathbf{A}_2$ (vector potential of the light wave) normal to the yz plane gives rise to a measurable emission from this initial state $|i,\mathbf{k}\rangle$

The initial electronic states $|i,\mathbf{k}\rangle$ can be classified as being even or odd with respect to reflection in the mirror plane, i.e. they retain or change their sign upon reflection. The final-state wave function must always be even, otherwise a detector located in the mirror plane would see a node of the emitted electron. We now consider two possible polarizations of the incoming exciting light. If the vector potential $\mathbf{A}_1$ is parallel to the mirror plane (yz) (Fig.6.9) the momentum operator contains only components $\partial/\partial y$ and $\partial/\partial z$ that are even with respect to reflection in the mirror plane. If $\mathbf{A}_2$ is oriented normal to the mirror plane, only $\partial/\partial x$ (odd) occurs in the perturbation operator in (6.30). In order to detect a photoemission signal for both polarizations one requires:

$$\text{for } \mathbf{A}_1 \parallel (yz): \quad \langle f,\mathbf{k}| \frac{\partial}{\partial y}|i,\mathbf{k}\rangle \neq 0 \,, \quad \langle f,\mathbf{k}| \frac{\partial}{\partial z}|i,\mathbf{k}\rangle \neq 0 \,, \tag{6.31a}$$

with $\langle f,\mathbf{k}|$, $\partial/\partial y$, $\partial/\partial z$ and $|i,\mathbf{k}\rangle$ even;

$$\text{for } \mathbf{A}_2 \perp (yz): \quad \langle f,\mathbf{k}| \frac{\partial}{\partial x}|i,\mathbf{k}\rangle \neq 0 \,. \tag{6.31b}$$

with $\langle f,\mathbf{k}|$ even and $\partial/\partial x$, $|i,\mathbf{k}\rangle$ odd.

The fact that the final states $\langle f,\mathbf{k}|$ are always even, implies, for the geometry $\mathbf{A}_1 \parallel (yz)$, that the initial state must be even with respect to reflection on the mirror plane (yz). An s-type wave function as initial state would lead to an emission signal, whereas a p-type orbital oriented along the x-axis as in Fig.6.9 would not be detected in this experimental geometry. On the other hand, an odd initial state can be observed, according to (6.31b), when the light is polarized with $\mathbf{A}_2$ normal to the mirror plane. Thus by measuring the photocurrent in a surface mirror plane for light polarized in and perpendicular to this plane, one can determine an important property of the initial state, namely its reflection symmetry. However, when spin-orbit coupling is important, odd and even states are mixed and the polarization selection rule is no longer strictly valid. An appropriate choice of the experimental geometry in ARUPS can give important information about the symmetry character (s-, p- or d-like) of electronic surface-state bands and also about molecular orbitals of adsorbates.

6.3.5 Many-Body Aspects

So far, our discussion of the photoemission experiment has been based on the one-electron states of a system of non-interacting electrons. Such a system, be it an atom, a molecule or a crystal consisting of N electrons, is described by a simple many-electron wave function, i.e.,

$$\Psi = \phi_1(\mathbf{r}_1)\phi_2(\mathbf{r}_2) \ldots\ldots \phi_N(\mathbf{r}_N) , \tag{6.32}$$

which is the product of the single one-electron functions $\phi_i(\mathbf{r}_i)$. Correspondingly the total energy E_N of such a non-interacting system is the sum of one-electron energies ϵ_i:

$$E_N = \epsilon_1 + \epsilon_2 + \epsilon_3 + \ldots\ldots + \epsilon_N . \tag{6.33}$$

In a photoemission experiment in which an electron is emitted from the energy level ϵ_ν the measured binding energy E_B in (6.21) is the difference between the initial total energy E_N of the N-electron system and E_{N-1}, that of the (N−1) electron system:

$$E_B = E_N - E_{N-1} = \epsilon_\nu . \tag{6.34}$$

This binding energy directly yields the energy of the νth electron in its one-electron state. In reality this picture is too simple since the electron-electron interaction cannot be neglected. The real many-body wave function cannot be written as a product (6.32) and the total energy E_N' of the N-electron system is not merely the sum of one-electron energies as in (6.33). In a Hartree-Fock treatment one can, of course, define single-electron energy levels ϵ_ν'. These, however, are dependent on the presence or absence of all the other N−1 electrons. The removal of an electron from such an interacting N-electron system causes the remaining (N−1) electrons to rearrange in the new potential in response to the creation of the hole. The (N−1) electron system "relaxes" into a new many-body state of minimal energy E_{N-1}'. The energy difference, termed the **relaxation energy** E_R, is passed on to the photoelectron, which then appears at higher kinetic energy, i.e. the measured binding energy E_B is not just a one-electron energy ϵ_ν' (in the Hartree-Fock sense) but includes a contribution due to this relaxation effect (or shielding of the hole):

$$E_B = \epsilon_\nu' - E_R . \tag{6.35}$$

The accuracy to which a Hartree-Fock single electron eigenvalue ϵ_ν' approximates the measured binding energy (or ionization potential) depends on how strongly the eigenvalue is influenced by the occupation with other electrons. If this effect can be neglected, the so-called **Koopmans theorem** holds, i.e. the binding energy is essentially the one-electron Hartree-Fock energy of the state.

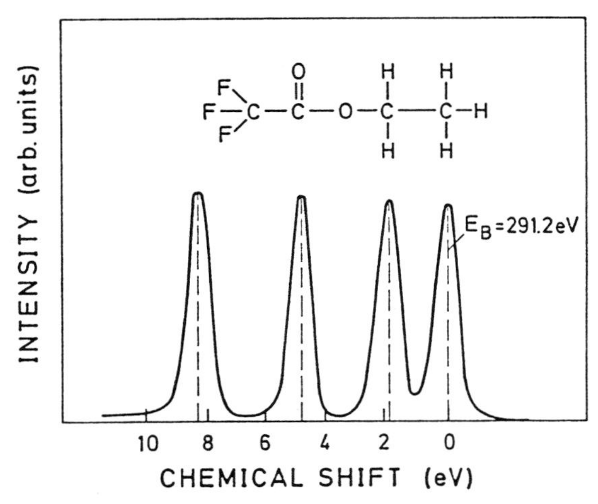

Fig.6.10. X-ray photoemission spectrum (ESCA) of the carbon core level with a binding energy of about 291 eV, obtained from the molecule $C_2H_5CO_2CF_3$. According to the different chemical surroundings of the carbon atoms in the molecule, slightly different energies are found for the core level (chemical shift). The order of the lines corresponds to the order of the C atoms in the molecule above [6.12]

The particular sensitivity of X-ray photoemission from core levels to the chemical environment of an atom (e.g., C bound to three H or to a C and two O atoms in Fig.6.10) has given XPS its alternative name ESCA (Electron Spectroscopy for Chemical Analysis). The electron configuration of the neighboring chemical bonds determines the local electrostatic potential for the core levels and thereby influences the relaxation energy E_R and the shielding of the photoexcited core hole. As seen from Fig.6.10, XPS or ESCA can be used as a fingerprinting technique to locate certain atoms in a molecule by their **core-level shifts**. The same is true for adsorbed atoms on a substrate.

So far we have assumed that the relaxation of the N-electron system leads to the ground state of the new (N−1)-electron system. The relaxation might, however, be *incomplete*, thus yielding an excited state of the (N−1)-electron system. Depending on the strength of the coupling between the surrounding electrons and the photoexcited hole (be it in the valence or core-level energy regime, UPS or XPS), collective excitations such as phonons, plasmons or interband transitions might be excited during the relaxation process. The photoemitted electron then receives only a part of the relaxation energy E_R, (6.35), the remainder having been used to excite a plasmon, interband transition, etc. The photoelectron is detected in a so-called **satellite peak** at an energy different from E_B (6.35). These satellite peaks can complicate the theoretical analysis of UPS and XPS spectra considerably if they are not identified as satellites.

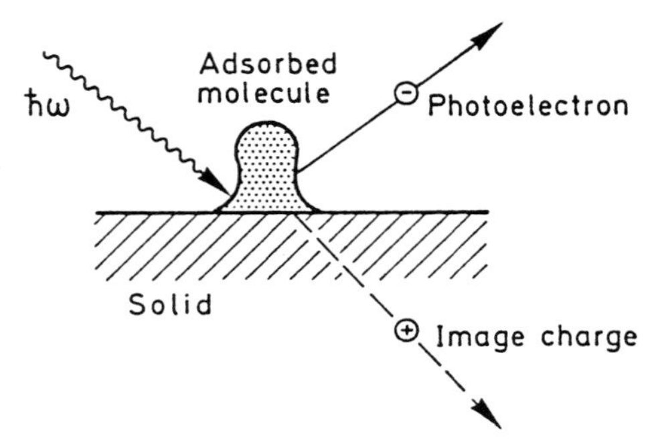

Fig.6.11. Schematic explanation of the extramolecular relaxation/polarization of a photoelectron emitted from an adsorbed molecule. The solid substrate participates in the relaxation of the many electron system after emission of an electron from the adsorbed molecule and therefore produces an additional shift of the emission line. Furthermore, the photoelectron is associated with an image charge within the substrate whose Coulomb interaction also changes its kinetic energy

The relaxation sometimes also called **relaxation/polarization effect** discussed so far concerns a single multi-electron system such as an atom, a molecule or a solid. The corresponding shifts E_R are thus more accurately termed intramolecular relaxation shifts. When we consider an atom or a molecule adsorbed on a solid surface, photo-excitation of an electron from such an adsorbate creates a valence or core hole (in UPS or XPS) in an environment different from that in the free atom or molecule. The relaxation process from the N electron ground state into the (N−1)-electron state also involves electrons in the adsorbate's chemical bond and possibly those in the substrate surface. In addition, the photoemitted electron leaving the adsorbate-surface complex is accompanied by an image charge (also a many-body shielding effect) within the substrate (Fig.6.11). The attractive interaction between the photoelectron and its image further contributes to the adsorption-induced change of E_R. This change of E_R due to adsorption is called "extramolecular" Relaxation/Polarization (R/P) shift. Experimentally this extramolecular R/P shift is evaluated by comparing photoemission spectra of an adsorbed species with those of its gas phase counterpart (Fig. 6.7). If one compares the measured binding energies referred to the same vacuum level, the gas-phase ionization potentials always exceed the binding energies of the adsorbed species by 1 to 3 eV (Fig.6.7).

As well as these so-called **final state effects** there are also "initial state" effects which might cause shifts of particular valence state emission lines with respect to their gas phase lines in UPS. These shifts arise from bonding interactions between the substrate and special adsorbate orbitals (chemical bonding shifts).

6.4 Some Surface-State Band Structures for Metals

Historically, electron surface states on metals were detected much later than on semiconductor surfaces. The reason was that ARUPS was necessary for determining the **k** vector of a particular state to see if its energy falls into a bulk-state gap. Without angular resolution a surface state can hardly be seen on a metal because of the high background of the integrated bulk-state density. In semiconductors, where an absolute forbidden gap is present, surface states located in the energy gap have a strong influence on the electronic properties of the surface (Chap. 7) and are therefore easily detected experimentally.

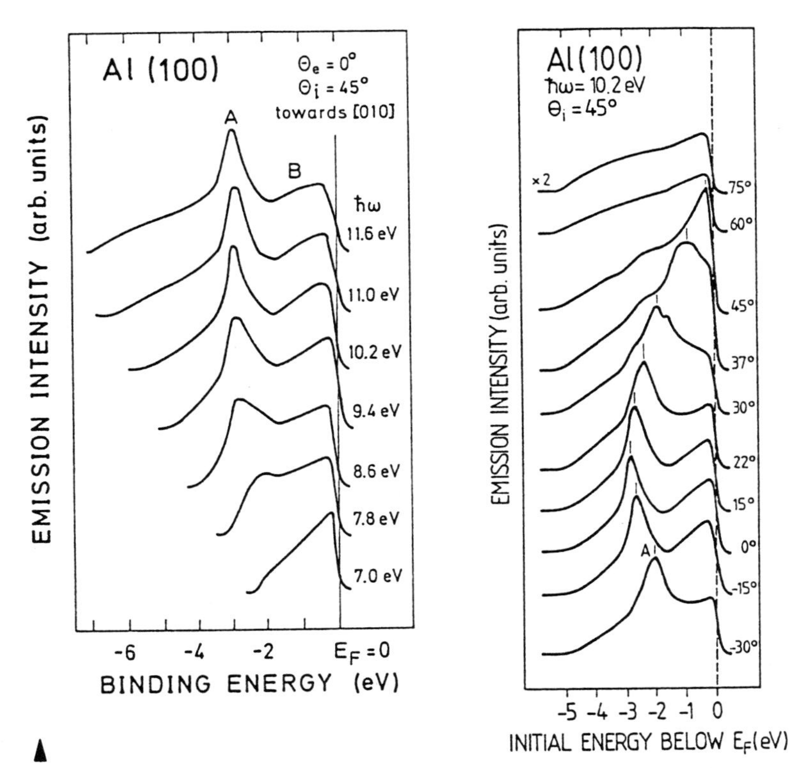

▲

Fig. 6.12. Experimental spectra of photoelectrons emitted normal to the Al(100) surface for photon energies between 7 and 11.6 eV (direction of incidence 45° to the [011] direction) [6.13]

Fig. 6.13. Photoemission spectra from the Al(100) surface with different polar angles in the $(01\bar{1})$ plane; photon energy $\hbar\omega = 10.2$ eV, direction of incidence 45° to the [011] direction [6.13]

6.4.1 s- and p-like Surface States

Simple metals such as Na, Mg and Al have an electronic band structure that strongly resembles that of the simple model of a free electron gas. The bulk bands derived from atomic s- and p-states have a nearly parabolic shape. At the Brillouin-zone boundaries and near other crossings of the "free electron" parabolae, gaps occur which also exhibit parabolic shape. Thus, in these cases one can apply surface-state theory in its simplest form based on the "nearly free electron" model (or slightly modified versions). Experimental band structures have been obtained by means of ARUPS. As an example, Fig.6.12 depicts photoemission spectra for Al(100) under normal emission ($\Theta_e = 0$) measured with various photon energies $\hbar\omega$. Figure 6.13 exhibits the results of measurements made at various angles of detection, i.e. with varying $\mathbf{k}_{\|}$ [6.13]. The sharp emission band A in Fig.6.12, which is essentially independent of photon energy but shifts with $\mathbf{k}_{\|}$ (Fig.6.13) is due to a surface-state emission. In Fig.6.14 its energetic position is plotted as a function of the $\mathbf{k}_{\|}$ vector calculated according to (6.22). The comparison with the bulk-band structure of Al along $\overline{\Gamma \mathrm{M}}$ and $\overline{\Gamma \mathrm{X}}$ reveals that the photoemission data definitely fall into a gap of the nearly free (parabolic) electron states along these symmetry lines. This is a clear indication of a surface state band. Its parabolic shape, similar to that of the bulk states, shows that it originates from free electron-like states. This band appears to be split off from the corresponding bulk band. The same bulk bands, when projected onto the (101) surface, exhibit a further parabolic gap in which another split-off surface-state band (dark data points) is found (Fig.6.15).

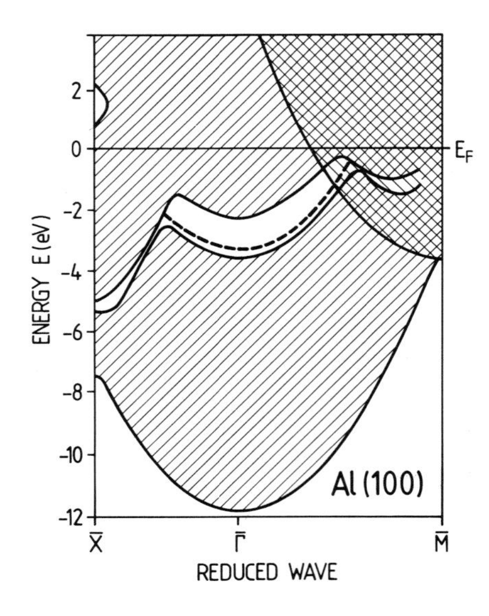

Fig.6.14. Measured surface state dispersion (*broken curve* [6.13]) and projected bulk bands for Al(100) (*shaded area* [6.15]) [6.14]

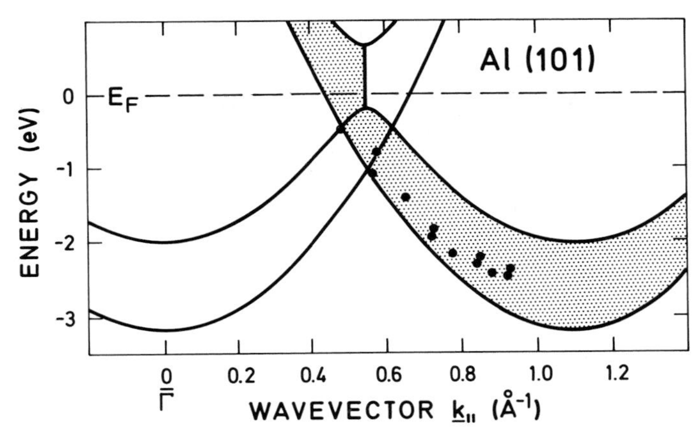

Fig. 6.15. Dispersion of sp-derived surface state band on the Al(101) surface as obtained from ARUPS. The data points fall into a gap (*shaded*) of the projected bulk-band structure [6.13]

The $\mathbf{k}_{\parallel}$ direction on the Al(101) surface is along $\overline{\Gamma X}$ of the surface Brillouin zone.

Similar sp-state-derived surface-state bands are well-known for the transition metals Au [6.16] and Cu [6.17]. In Cu the d levels are occupied and lie about 2eV below the Fermi level. The bulk states between these and the Fermi level are all sp derived. Figure 6.16 shows some photoemission

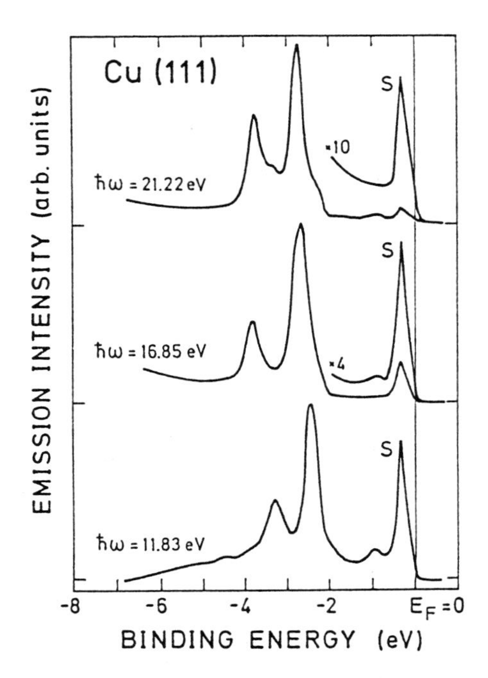

Fig. 6.16. Normal emission ARUPS data obtained from the Cu(111) surface with different photon energies. The sharp peak S is due to a surface-state band [6.17]

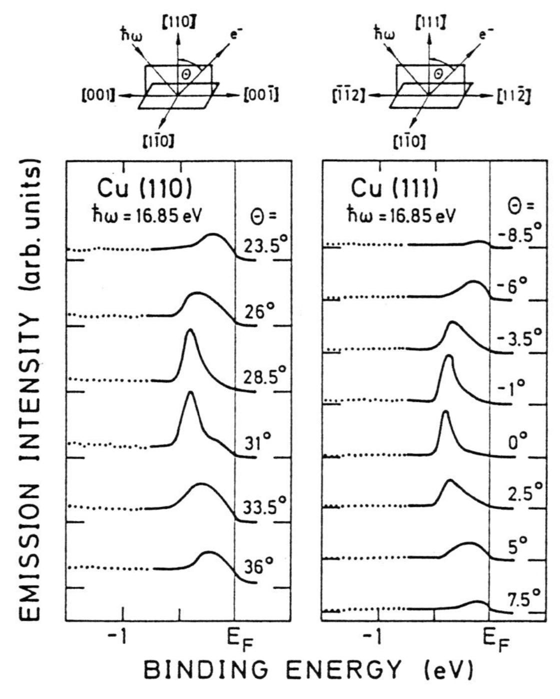

Fig.6.17. Angle-resolved UV-photoemission spectra of surface-state bands on the Cu(110) and Cu(111) surface. The peak shifts as a function of detection angle Θ indicate a significant dispersion [6.17]

spectra measured at normal emission from a Cu(111) surface with several photon energies [6.17]. The peak S visible at an amplification of 10 is definitely due to surface states. At normal emission ($\mathbf{k}_{\parallel} = \mathbf{0}$) it does not change its energetic position when the photon energy is varied. The dispersion $E(\mathbf{k}_{\parallel})$ can easily be measured by changing the detection angle (Fig.6.17). Figure 6.18 displays the dispersion of this surface-state band measured with two different photon energies (cf. the argument in favor of surface-state emission in Sect.6.3.2). The band is parabolic, as expected for sp-derived states. The points lie in a gap of the projected bulk sp states just below the Fermi level. A similar parabolic band of sp-derived surface states has also been detected on the Cu(110) surface near the $\bar{Y}$ point of the 2D Brillouin zone (Fig.6.19).

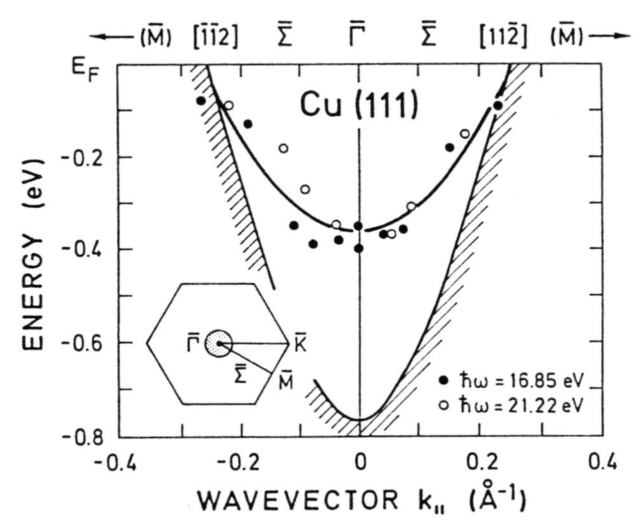

Fig. 6.18. Dispersion of the sp-derived surface state band on Cu(111) according to the ARUPS data of Fig. 6.17. Data points from measurements with two different photon energies $\hbar\omega$ are plotted in the gap of the projected bulk bands (*shaded*). The inset gives the location in reciprocal space [6.17]

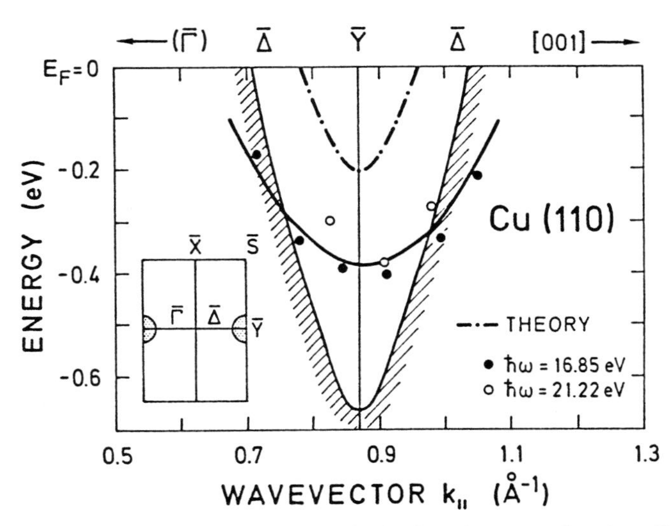

Fig. 6.19. Dispersion of the sp-derived surface state band on Cu(110) according to the ARUPS data in Fig. 6.17. Data points from measurements with two different photon energies $\hbar\omega$ are plotted in the gap of the projected bulk bands (*shaded*). The inset gives the location in reciprocal space [6.17]

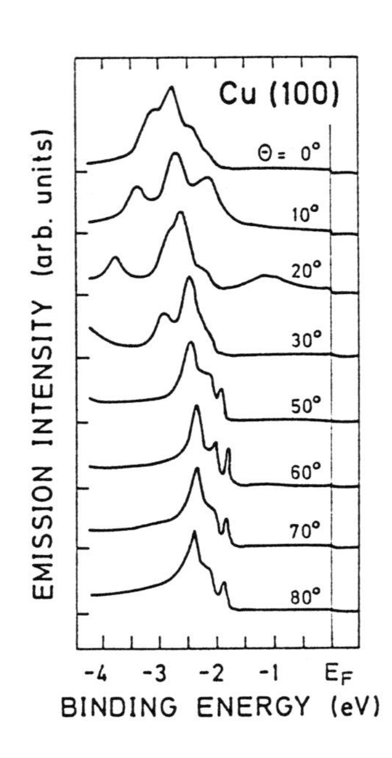

Fig.6.20. Angle-resolved UV-photoemission curves measured with a photon energy of 21.22 eV on Cu (100) at different detection angles. The emission direction is in the (001) mirror plane containing the symmetry points Γ, X, W and K of the Brillouin zone [6.18]

6.4.2 d-like Surface States

In transition metals, d orbitals of the single atoms interact in the crystal to give d bands which are typically $4 \div 10$ eV wide. Due to their localized nature the d bands show less dispersion than sp bands; they cross and hybridize with the "free electron" sp band. This hybridization introduces new gaps in the band structure where true surface states can occur.

Figure 6.20 shows ARUPS data measured on Cu(100). As a function of detection angle the d-band emission between 2 and 4eV binding energy below E_F undergoes considerable changes. In particular, a peak on top of the d-band, is recognized as being a surface-state band according to the arguments in Sect.6.32. The resulting 2D dispersion curve for this band is plotted in Fig.6.21 along the $\overline{\Gamma M}$ line of the Brillouin zone [6.18]. As expected for a surface-state band, the same dispersion is obtained with different photon energies; furthermore, the states are located in a gap of the projected bulk states. The energetic location close to the higher-lying bulk d-states clearly reveals that the band is split off from these bulk d-states.

In contrast to Cu, Ag and Au the metals Mo, W, Ni, Pt, etc. have only partially occupied d states. Their d-band is therefore cut by the Fermi energy E_F (Fig.6.22). Such relatively localized d-band states can be treated theoretically by LCAO type calculations [6.20]. In addition to yielding complete band structures mapped onto the 2D Brillouin zone, one also obtains the so-called **Local Density Of States** (LDOS). This is the total density of

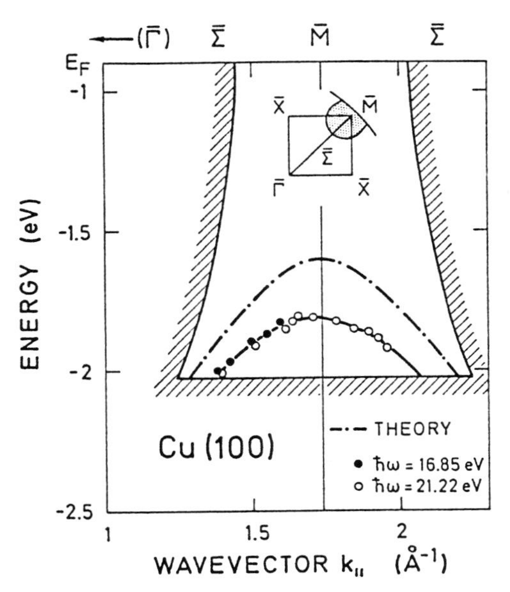

Fig. 6.21. Dispersion of the d-derived surface-state band on Cu(100) according to the ARUPS data of Fig. 6.20. Data points from measurements with two different photon energies $\hbar\omega$ are plotted in the gap of the projected bulk sd bands (bounded by the shaded region). The inset gives the location in reciprocal space. The dashed-dotted curve results from a surface-state band calculation according to [6.18, 19]

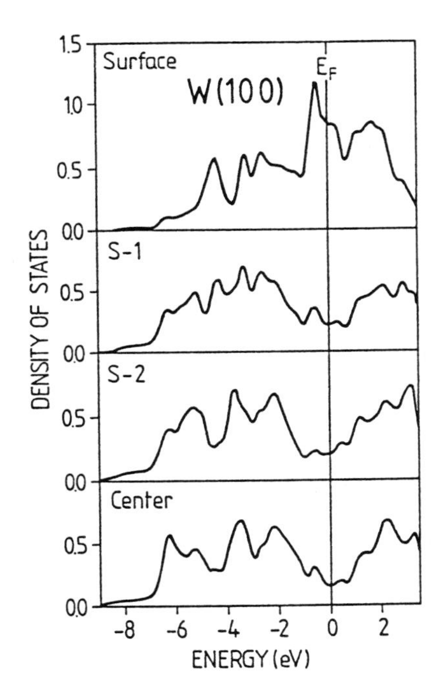

Fig. 6.22. Layer-resolved Local Density Of States (LDOS) for W(100) as obtained by a slab calculation. For the topmost surface atomic layer surface states around the Fermi level E_F produce a strong band in an energy region where, in the bulk LDOS (center layer *below*), low density is found [6.20]

states (integral over $k_{\|}$ space) within a certain layer of atoms. The LDOS of the topmost atomic layer thus reflects the density of surface states whereas the LDOS of deeper layers becomes identical to the density of bulk states. The results of such LDOS calculations are exhibited for W(100) in

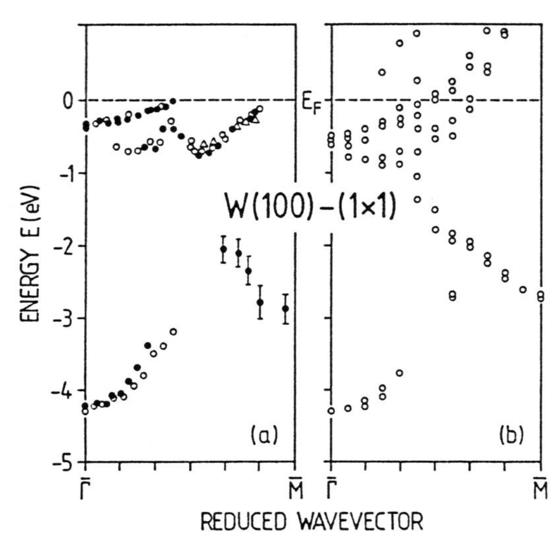

Fig.6.23a,b. Surface state bands on the W(100) surface with (1 × 1) reconstruction [6.14]. (**a**) Experimental angle-resolved UPS data [6.21,22]. (**b**) Theoretical results from a slab calculation [6.23]

Fig.6.22. Even the second- and third-layer densities very closely resemble the bulk density of states; the same valley in the state distribution can be seen around E_F. For the topmost layer, however, a relatively sharp structure, half filled, half empty is observed at E_F. The origin of this surface-state band can be described qualitatively as a splitting off of more atomic-like d-states from the bulk-state distribution. Surface atoms have fewer neighbors and their electronic structure is therefore closer to that of a free atom than for atoms deep in the bulk. Figure 6.23 displays part of the calculated surface-band structure for W(100) in comparison with experimental results from ARUPS. There are several surface-state bands of d character which contribute to the sharp structure around E_F in Fig.6.22.

The fact that surface atoms have fewer neighbors than bulk atoms leads, in a LCAO picture for the d-states, to the immediate conclusion that these states have less overlap with neighbors and that the d-LDOS for the topmost atomic layer should be sharper than for deeper layers. A calculation for the d-states on a Cu{001} slab clearly shows this effect (Fig.6.24). As for W(100) the second atomic layer already exhibits a bulk-like LDOS, whereas at the surface the d band has become considerably sharper. The effect is also clearly revealed in experimental photoemission data (Fig.6.25).

The narrowing of the d band in the LDOS of surface atoms has further interesting consequences. Because of the small screening length (high electron density) in metals, atoms at the surface tend to remain neutral. Furthermore, one and the same Fermi level must be found in the bulk and in the

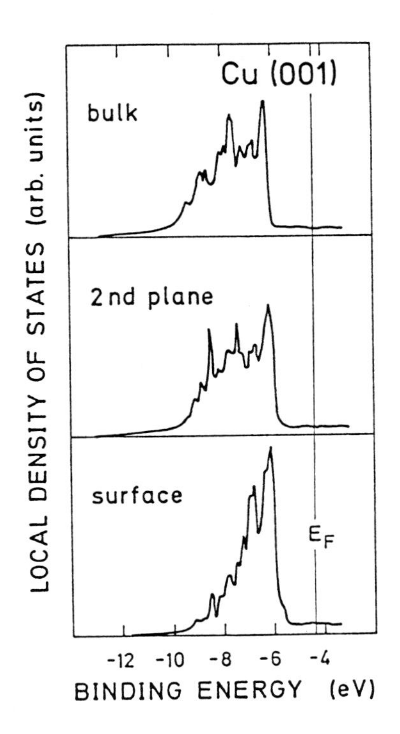

Fig. 6.24. Layer-resolved Local Density Of d-States (LDOS) for the Cu(001) surface obtained from a slab calculation [6.24]

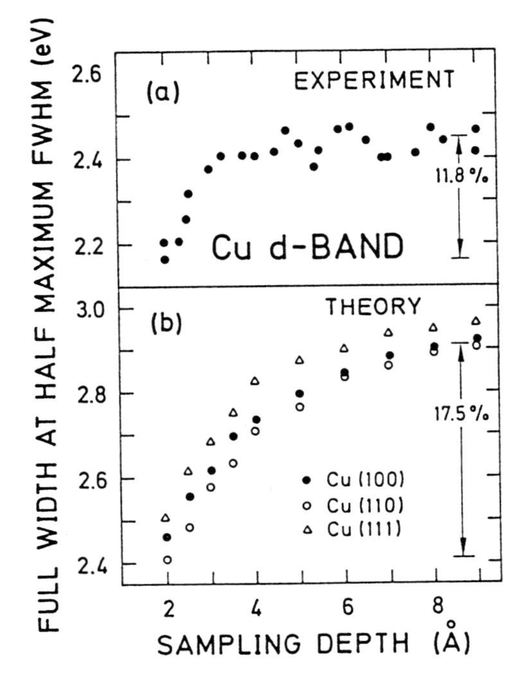

Fig. 6.25a,b. Full Width at Half Maximum (FWHM) of the Cu d-band as a function of depth into the crystal [6.25]. (**a**) Experimental data from photoemission measurements with different photon energies sampling over areas of different depth. (**b**) Results from theoretical calculations for various Cu surfaces

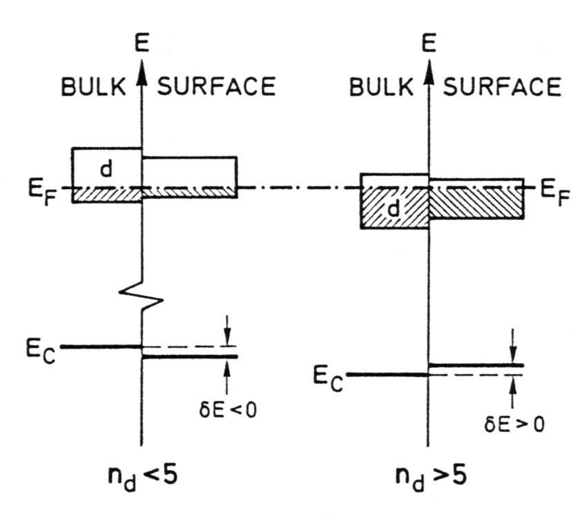

Fig.6.26. Schematic illustration of the origin of surface core-level shifts δE in d-band metals. The integer n_d denotes the number of electrons in the band. At the very surface the d-band is narrowed due to the reduced interaction with neighbors. Because of neutrality requirements this leads to shifts of the core level E_c in the topmost atomic layer [6.26]

topmost atomic layer. In order to retain local charge neutrality, shifts of the local band structure are expected and will be reflected in the surface LDOS (Fig.6.26). For metals whose d-band is less than half full, one expects a downward shift of the electronic levels at the surface. On contrast, an upward shift should occur for metals, whose d-bands are occupied by more than five electrons per atom. These characteristic shifts are also expected for the sharp core levels and should therefore be observable in XPS; but an experimental confirmation is difficult because of the many-body effects involved in interpreting the data (Sect.6.3.4).

6.4.3 Empty and Image-Potential Surface States

The importance of photoemission spectroscopy, particularly UPS, for the study of occupied surface states has its analogue in momentum resolved inverse photoemission (Panel XI: Chap.6), which is similarly useful for the investigation of empty surface-state bands located at energies above the Fermi level E_F. The experimental technique involves the capture of irradiated electrons of a certain kinetic energy in empty states with simultaneous emission of UV photons of the corresponding deexcitation energy. The energetic position of the particular empty state is obtained from the kinetic energy of the incident primary electrons and the energy of the emitted UV photon. The $\mathbf{k}_{\|}$ vector of the empty state is determined from the energy and the angle of incidence of the primary electron beam.

The underlying physical process is the inverse (time-reversed) photoeffect. The theoretical treatment is thus similar to that of UPS (Sect. 6.3). In particular, one faces the same problems as in ARUPS in distinguishing between bulk and 2D surface-state bands (Sect.6.3.2). As in ARUPS, criteria for surface-state bands are: (i) sensitivity to contamination, (ii) energetic position within a gap of the projected bulk band-structure of empty states, and (iii) dependence only on $\mathbf{k}_{\|}$ and not on $k_{\perp}$. Using these criteria inverse photoemission has meanwhile yielded a number of interesting experimental results about empty surface-state bands on solid surfaces.

Figure 6.27 shows examples of empty surface-state bands which have been measured on the (110) surfaces of Ni, Cu and Ag along the symmetry lines $\overline{\Gamma X}$ and $\overline{\Gamma Y}$ [6.27]. The energy scale is referred to the Fermi level E_F, and the vacuum energy E_{vac} is marked by an arrow. In the case of Cu(110) the occupied surface-state band around $\overline{Y}$ and slightly below E_F is the same band as appears in Fig.6.19 (Sect.6.4.1). In addition to the empty surface-state bands S_1-S_4, one also observes some empty bulk bands (B). The surface-state bands S_2-S_4 show the expected behavior, in particular they are very sensitive to contamination. They can be understood similarly to the occupied bands (Sect.6.4.1) as being split off from the empty bulk sp-bands. Correspondingly theories of the type described in Sects.6.1 and 2 are able to reproduce these surface-state bands to within an accuracy of better than 1 eV. This is not the case for the states marked S_1 in Fig.6.27. In particular, adsorption of chlorine does not lead to the disappearance of the corresponding spectral features in the inverse photoemission spectra. This is shown for a Cu(100) surface in Fig.6.28. Rather than disappearing, the spectral step near 4eV above E_F shifts by 1.1eV to even higher energies after the adsorption of Cl [c(2 × 2) superstructure in LEED]. The shift of 1.1 eV is exactly the work-function change due to the adsorbed Cl. The structure S_1, which from its $\mathbf{k}_{\|}$ dependence and its energetic location in a bulk-band gap is clearly due to surface states, thus shifts in energy as the vacuum level E_{vac} does. This is different from the behaviour of the surface states which are considered so far. All those states were "crystal derived", they are fixed (sometimes as split-off states) to the bulk-band structure rather than to the vacuum level E_{vac} which changes its position with respect to the bulk bands due to the slightest contamination.

The extraordinary behavior of the states designated S_1 in Fig.6.27 is explained in terms of so-called **image-potential states**. The physical origin of this new type of empty surface state is explained in Fig.6.29. These states are not derived in any way from bulk states or from the symmetry-breaking effect of the surface. When an electron approaches a metal surface, its charge is screened by the conduction electrons of the metal. The screening can be described in terms of a positive image charge inside the metal at the same distance from the surface as the real charge outside. This

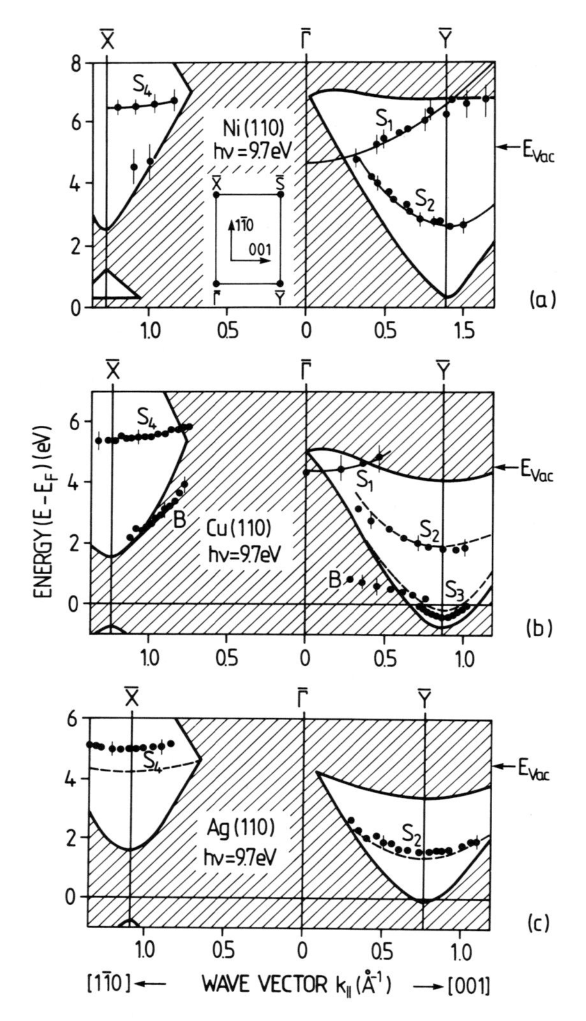

Fig. 6.27a-c. Dispersion of empty surface-state bands S_1-S_4 on the (110) surfaces of Ni, Cu and Ag along the symmetry lines $\overline{\Gamma X}$ and $\overline{\Gamma Y}$. Bulk bands are denoted by B. The energy scale is referred to the Fermi level E_F and the vacuum energy E_{vac} is marked by an arrow. The experimental data originate from inverse photoemission experiments [6.27]

leads to an attractive potential between the electron and its positive image inside the metal. In the simplest approximation this potential has the Coulomb form

$$V(z) \propto 1/z , \quad z > 0 . \tag{6.36}$$

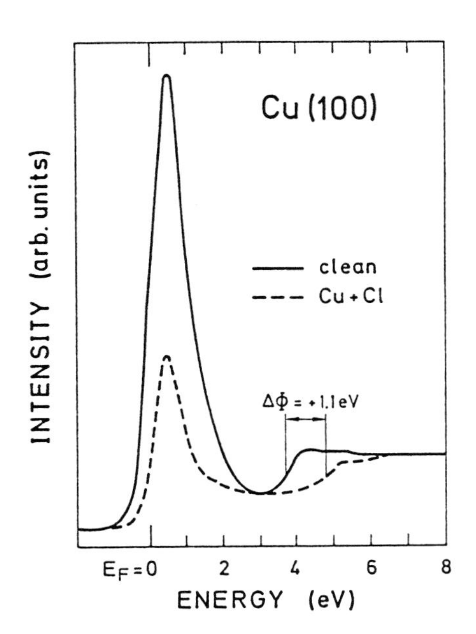

Fig. 6.28. Inverse photoemission (isochromate) spectrum on a Cu(100) surface, clean (*full line*) and after adsorption of chlorine (*broken line*). The energy scale extends from the Fermi level E_F up towards the vacuum level [6.28]

As is shown in Fig. 6.29, it is possible for bound states to exist in such a potential which can thus trap electrons in a region within a couple of Å of the surface. If the energy levels of such states (n = 1 in Fig. 6.29) fall into a gap of the bulk-band structure, the states cannot decay into bulk states and the general behavior of electrons occupying such levels is similar to that of

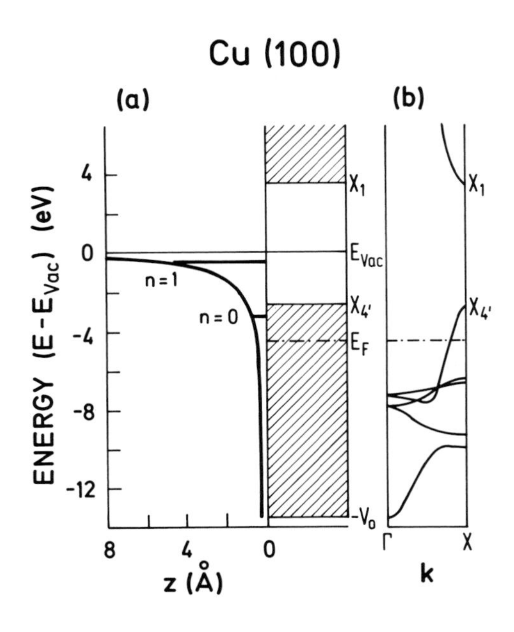

Fig. 6.29. Schematic illustration (**a**) of the origin of image potential surface states for the example of Cu(100). The bulk-band structure of Cu (**b**) gives rise to a gap between $X_{4'}$, and X_1 above the Fermi energy E_F. An electron approaching the surface can be captured by its image potential in one of the quantized states (n = 0, 1, ...). Trapping in this state occurs when, due to a gap of the empty bulk states, no decay into bulk states can occur [6.28]

ordinary crystal derived states. The energy of an electron trapped in such an image state is given by the binding energy ϵ_n of the bound state with the quantum number n. The exact value of ϵ_n is determined by the shape of the potential (6.36). Parallel to the surface the electrons are nearly free, i.e. they can move as free electrons and carry a kinetic energy $\hbar^2 k_{\parallel}^2/2m^*$, m^* being their effective mass parallel to the surface. The relevant energy reference for these states is the energy of an electron far away from the surface, i.e. the vacuum energy E_{vac}. One thus arrives at the following description for the band structure of image potential surface states:

$$E(\mathbf{k}_{\parallel}) = \frac{\hbar^2 k_{\parallel}^2}{2m^*} - \epsilon_n + e\phi \qquad (6.37)$$

where the work function term ϕ takes into account the reference energy E_{vac}. The description (6.37) is consistent with the experimental shift due to adsorption (Fig.6.28). Furthermore, the parabolic dependence on $k_{\parallel}$ (6.37) is also found in the experimental data (Fig.6.27).

It should be stressed that electrons bound in such image potential states form a quasi-2D electron gas. The very interesting properties of such 2D electron gases can be studied more easily in cases where they are formed in thermodynamic equilibrium rather than in excited states as in the present case. A further treatment of such 2D gases is given in connection with semiconductor space charge layers, where such effects are also important (Chaps. 7 and 8).

6.5 Surface States on Semiconductors

Electronic surface states or, more generally speaking, interface states were first studied on semiconductors [6.29, 30]. Their existence was derived in an indirect manner by analysing the physics underlying the rectifying action of metal-semiconductor junctions (Chap.8). An important breakthrough on the experimental front was due to optical experiments [6.30] and to the application of photoconductivity and surface photovoltage spectroscopy [6.31] (Panel XII: Chap.7). As is the case for metal surfaces, the most detailed information is obtained from UPS [6.32] and ARUPS (Chap.6.3), inverse photoemission spectroscopy (Panel XI: Chap.6) and from electron-energy loss spectroscopy (Panel IX: Chap.4). The last of these techniques can give information about excitation energies between occupied and empty surface states, as does optical absorption spectroscopy. In the following we will consider some experimental and theoretical results on low-index surfaces of elemental (Si, Ge) and III-V compound semiconductors (GaAs, InP, InSb, etc.). ZnO will also be considered briefly as an example of the II-VI compounds. The main common characteristics of these different semiconductor classes is the tetrahedral atomic bonding, i.e. the coordination of each atom by four other atoms (of the same kind for Si and Ge and of different kind in compound semiconductors). This tetrahedral bonding geometry results from the formation of the covalent sp^3 hybrids. This covalent part of the bond is the decisive factor for the crystal structure even in the presence of relatively strong ionic bonding contributions as found in II-V compounds such as ZnO. The sp^3 hybridization causes the elemental semiconductors to crystallize in the diamond structure (two fcc lattices mutually displaced by (¼, ¼ , ¼) of a unit cell) and the III-V compound materials in the zincblende structure (a diamond structure in which nearest neighbors are atoms of different kinds). Many II-VI crystals occur in the wurtzite structure, which is hexagonal but similar to a slightly deformed zincblende structure [tilted and elongated along (111)].

In Fig.6.30 the three lowest-index surfaces of the zincblende structure are displayed. The diamond lattice is obtained if all the atoms are of one species. The sp^3 hybridization leads to the formation of strongly directional bonding lobes which appear as *dangling bond* orbitals at the surface. The different dangling bond structures for the three surfaces (assumed to be non-reconstructed, truncated bulk) is shown in Fig.6.30b. The corresponding 2D Brillouin zones of the non-reconstructed surfaces are depicted in Fig.6.30c. Formation of the *truncated* (111) surface creates one half-filled dangling bond orbital per surface atom perpendicular to the surface. On the (110) surface there are two atoms in the unit mesh, each with a tilted dangling bond orbital. The unreconstructed (100) surface unit cell contains one atom with two broken bonds tilted with respect to one another. Nature, however, is not as simple as indicated in Fig.6.30: The main low-index surfaces of all important semiconductors display a variety of complicated reconstructions which are only partially understood. The few cases for which conclusive reconstruction models have been developed are discussed in the following sections.

6.5.1 Elemental Semiconductors

It is easy to see qualitatively what kind of surface-band structure would result from the local surface geometries (Fig.6.30) for the three types of unreconstructed surfaces (111), (110) and (100) of Si and Ge. Calculations for Si and Ge based on non-reconstructed surface geometries indeed show the expected behavior (for Ge see Fig.6.42). Corresponding to the one dangling

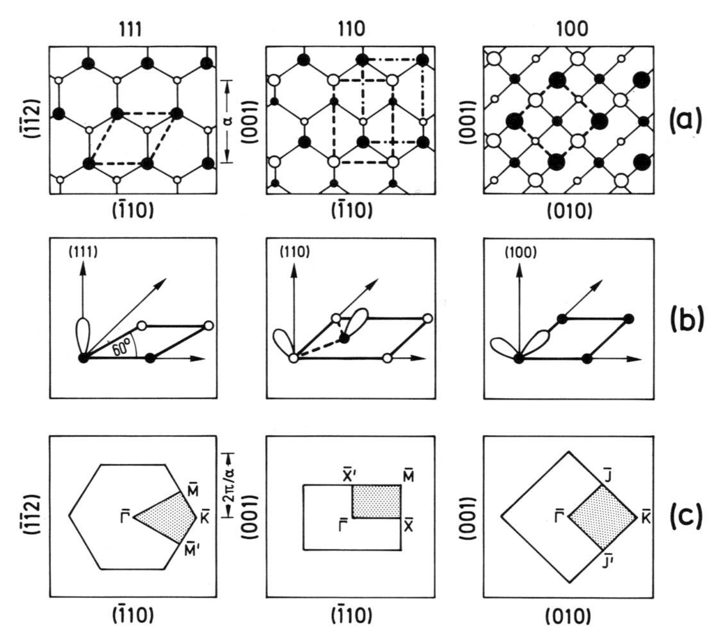

Fig.6.30a-c. Crystallography of the non-reconstructed three low-index surfaces of the zincblende lattice (sp^3-bonded). The diamond lattice of Si and Ge would be obtained if all atoms were of the same species. (**a**) Top view, smaller symbols denote deeper lying atoms. Possible unit meshs are indicated by broken lines. (**b**) Schematic plot of the dangling-bond orbitals occurring on the different surfaces. (**c**) Corresponding ideal surface Brillouin zone with conventional labeling [6.33]

bond (broken sp^3 bond) on the Ge or Si(111) surface one expects a single band of surface states within the forbidden gap. This band is split off from the bulk sp^3-states forming the valence and conduction bands. Because of the fewer neighbors at the surface, i.e. reduced orbital overlap the surface states are lowered in energy less than the bulk valence-band states and thus fall into the gap. Due to bond breaking the band is half filled since each side of a broken bond can accept one of the two electrons of the unbroken covalent bond. The creation of an ideal (110) surface leaves two dangling bonds on two different atoms in the unit cell. As a result, two dangling-bond surface-state bands are formed in the gap. Since the dangling sp^3 hybrids have only weak mutual interaction, the two bands are only slightly split and exhibit relatively little dispersion. In contrast, the two dangling-bond orbitals on one and the same atom on the (100) surface interact strongly with each other and form two gap state bands spread over a much wider energy range with higher dispersion. In addition to these gap states, all three Si surfaces give rise to so-called **back bond surface states** which lie at much higher binding energies within gaps of the projected bulk band structures. The wave functions of these states are localized between the topmost and lower lying atomic planes.

These simple conclusions concerning the band structure of the unreconstructed surfaces belie the fact that reality is much more complex. This is because of reconstructions. One of the most studied semi-conductor surfaces is the **Si(111) cleaved surface**.

If the crystal is cleaved at room temperature, a (2×1) reconstruction is found in LEED. If cleaved at very low temperature ($T < 20$K) a (1×1) LEED pattern appears. After annealing to temperatures higher than about 400°C a (7×7) superstructure occurs indicating an extremely long-range periodicity. The (7×7) structure is definitely the most stable configuration; the (1×1) and the (2×1) structures are frozen-in metastable configurations.

The Si(2×1) surface has attracted much attention in recent years. Independent of the reconstruction model considered, there is an interesting argument that the (2×1) reconstruction splits the half-filled dangling-bond surface-state band within the bulk gap into two parts: according to Fig.6.31 the 2D Brillouin zone shrinks by a factor of two in one direction for the (2×1) structure i.e., for symmetry reasons, one half of the dangling-bond band can be folded back into the new (2×1) Brillouin zone thus opening up a gap at the (2×1) zone boundary [due to the perturbation potential causing the (2×1) reconstruction]. Since the original (1×1) band was half filled, this splitting into an empty upper branch and a full lower branch leads to a total energy decrease and thus to a stabilization of the (2×1) structure. These arguments are reminiscent of a Peierls instability and no detailed assumptions about the reconstruction are necessary.

For a long time the so-called **buckling model** had been assumed to explain the double periodicity along $[\bar{2}11]$ on the cleaved Si(111)-(2×1) surface [6.34] (Figs.3.6b, 6.32). Every second row of surface Si atoms is raised with respect to the ideal lattice position and the rows of atoms in between

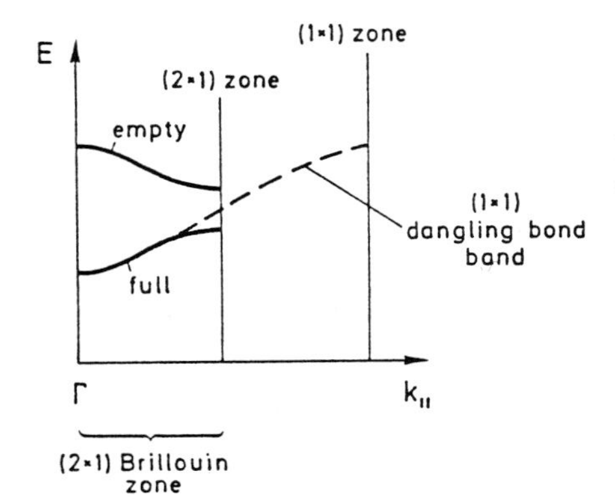

Fig.6.31. Schematic illustration of the energy gain due to the formation of a (2×1) surface reconstruction. The shrinking of the Brillouin-zone dimension by a factor of two in one direction causes the half-filled dangling bond surface state band (*broken line*) to be folded back into the new (2×1) zone and split at the zone boundary, thus leading to a reduction of the total electronic energy

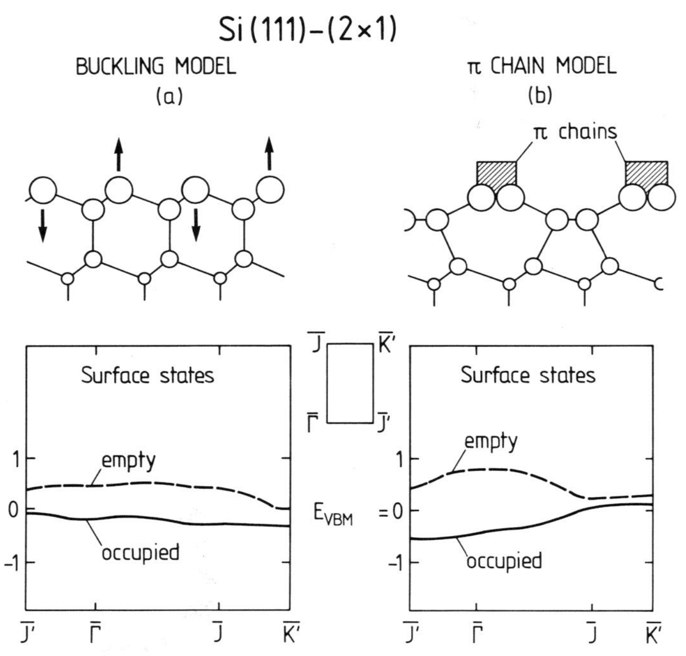

Fig.6.32. Calculated dispersion of the dangling-bond surface-state bands for Si (111)-(2 × 1) [6.36] (**a**) for the buckling model shown in the upper part; (**b**) for the π-bonded chain model shown in the upper part. See also Fig.3.6

are shifted downwards. As was shown by a number of researchers [6.35] such a reconstruction always yields an empty and an occupied surface state band (as expected, see above) with little dispersion along the $\overline{\Gamma J}$ symmetry line of the (2 × 1) surface Brillouin zone (Fig.6.32a). This is in contradiction to the experimental findings from ARUPS (Fig.6.33). A number of different groups have experimentally observed strong dispersion along $\overline{\Gamma J}$. This finding, and all other experimental data so far can be explained in terms of the so-called **π-bonded chain model** of *Pandey* [6.36] (Figs.3.6, 6.32b). Complete Si-Si bond breaking is required in the second atomic layer to induce this reconstruction. But the resulting zig-zag pattern allows the dangling p_z bonds of the topmost layer to form one-dimensional π bonds just like in a one-dimensional organic system. As has been shown theoretically, in spite of the bond breaking, the π-bonded chain model is energetically more favorable than the buckling model, due to the energy gained by the formation of the π-bonds [6.39]. The strongly dispersing, occupied and empty surface state bands are due to the bonding π-orbitals (without nodes) and the anti-bonding π^* orbitals (with nodes along the chain). If, in addition to the π-bonding, a slight buckling in the topmost atomic layers is included,

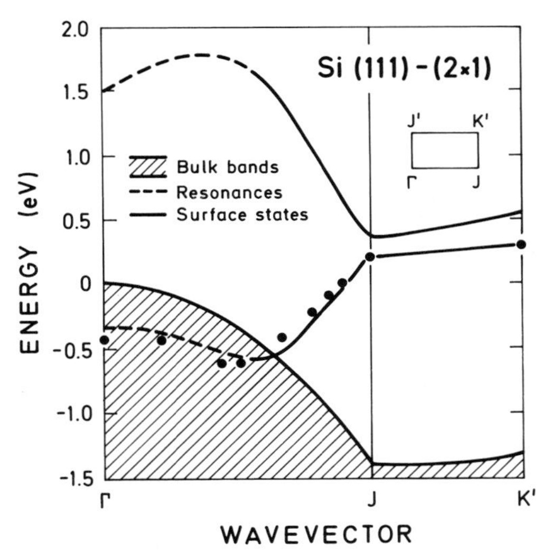

Fig.6.33. Dispersion of the dangling-bond surface-state bands for the Si(111)-(2 × 1) surface together with the projected bulk-band structure (*shaded*). The corresponding symmetry directions in k-space are explained in the inset (surface Brillouin zone). Full and broken curves are the results of theoretical calculations [6.37]. Data points were obtained by ARUPS measurements [6.38]

all available experimental data are well described. In particular, the strong dispersion measured in ARUPS is obtained (Fig.6.33). There is an absolute gap in the surface-band structure at $\bar{J}$ (Figs.6.32b, 33) with a calculated energy distance between π and π^* bands of less than 0.5eV. Indeed in High-Resolution Electron Energy Loss Spectroscopy (HREELS) (Fig.6.34) and in IR multiple internal reflection spectroscopy (Fig.6.35) the corresponding optical transitions are seen at energies of about 0.45eV. As is expected for the $\pi \rightarrow \pi^*$ optical transitions (like in an organic chain molecule) there is a strong polarization dependence of the transition matrix element. For light polarized normal to the chains the transition is not excited (Fig.6.35). The results of this reflection-absorption experiment are crucial for the π-bonded

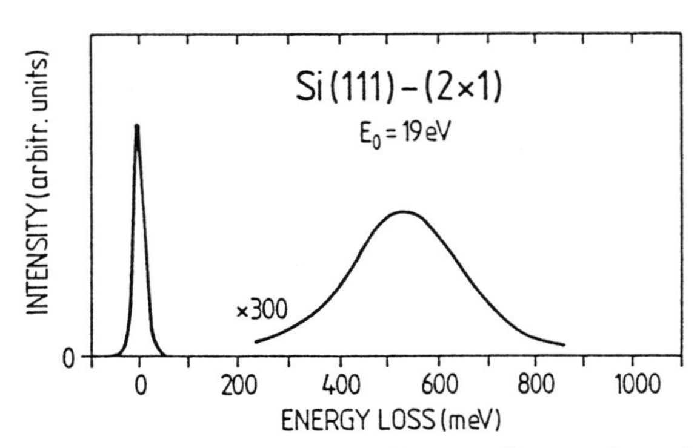

Fig.6.34. High-Resolution Electron Energy Loss Spectrum (HREELS) of a clean, cleaved Si(111)-(2 × 1) surface for specular reflection (70°) and with a primary energy E_0 of 19 eV [6.40]

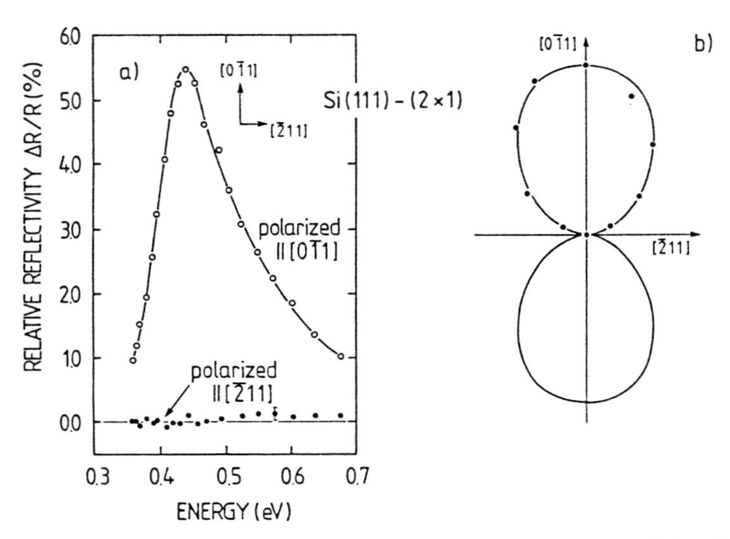

Fig. 6.35a,b. IR multiple internal reflection spectroscopy of the cleaved Si(111)-(2 × 1) surface. (**a**) Relative reflectivity change between a clean and oxygen covered surface for light polarizations parallel to [0$\bar{1}$1] (π-chain direction) and parallel to [$\bar{2}$11] (normal to π-chain direction). (**b**) Polar diagram of the relative reflectivity change as a function of light polarization direction [6.41]

chain model, since they cannot be explained by the buckling model. Moreover, the available structural data from LEED and from Rutherford backscatting are in complete agreement with a slightly buckled π-bonded chain model for the Si(111)-(2 × 1) surface. This well established model is now also used to describe the **cleaved Ge(111) surface with (2 × 1)** superstructure. Figure 6.36 shows a comparison of surface state dispersion branches along $\overline{\Gamma J}$ and $\overline{JK'}$ measured in ARUPS [6.43] with theoretical curves which have been calculated on the basis of a buckled π-bonded chain model [6.42].

The famous **Si(111)-(7 × 7) surface** with its large unit cell is the most stable surface structure on Si(111). It is prepared by annealing a clean (111) surface; the details of the annealing (and ion bombardment) depend on whether the clean surface has been obtained by cleavage (2 × 1 superstructure), sputtering or annealing. A structure model has been proposed by *Takayanagi* et al. [6.44] on the basis of transmission electron diffraction, which has found further confirmation from Scanning Tunneling Microscopy (STM) and other techniques (Fig. 6.37). This model, referred to as the DAS (Dimer-Adatom Stacking-fault) model contains 12 Si adatoms, 6 rest atoms, 9 dimers and one corner hole per surface unit cell. In one half of the unit cell there is stacking fault which explains the slight asymmetry found in STM (Fig. VI.4, Panel VI: Chap. 3).

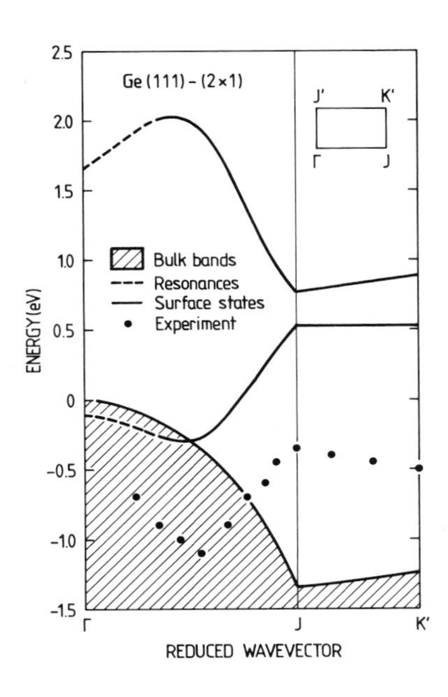

Fig. 6.36. Dispersion of the dangling-bond surface-state bands for the clean, cleaved Ge(111)-(2 × 1) surface together with projected bulk-band structure (*shaded*). The corresponding symmetry directions are explained in the inset (surface Brillouin zone). Full and broken curves are the results of theoretical calculations [6.42]. Experimental data points were obtained from ARUPS measurements [6.43]

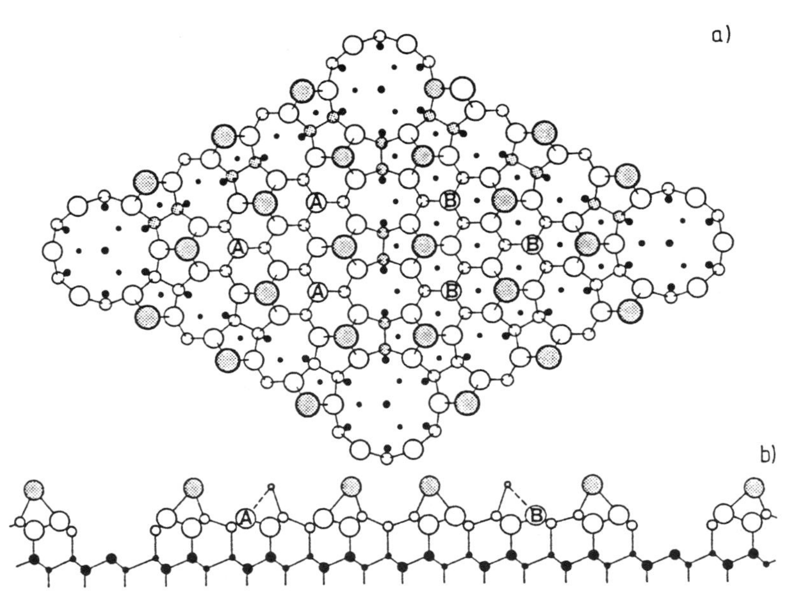

Fig. 6.37a,b. Dimer-Adatom Stacking fault (DAS) model for the Si(111)-(7 × 7) surface. (**a**) Top view: Atoms in (111) layers at increasing depth are indicated by circles of decreasing sizes. The heavy circles represent the 12 adatoms. The circles marked by A and B represent the rest atoms in the faulted and unfaulted half of the unit cell, respectively. (**b**) Side view: Atoms in the lattice plane along the long diagonal of the surface unit cell are shown with larger circles than those behind them [6.44]

To calculate the electronic structure of such a complicated reconstructed surface is an enormous task. Nevertheless some general conclusions about the surface-state density are possible. Using the arguments which were discussed in connection with the Si(111)-(2 × 1) surface (Fig.6.31) one has a 2D surface Brillouin zone for the (7 ×7) superstructure whose diameter is 1/7 of the (1 ×1) zone, i.e. because of symmetry, the one dangling-bond band of the (1 × 1) non-reconstructed surface (Fig.6.31) can be folded back seven times into the (7 ×7) Brillouin zone. This procedure is expected to yield a manifold of bands lying close to each other, i.e. a quasi-continuous distribution of states within the forbidden bulk band of Si. Thus a quasi-metallic character is expected for the Si(111)-(7 × 7) surface. A sharp Fermi edge for the surface states has indeed been found by some groups in photoemission spectroscopy [6.45,46]. In HREELS performed on very clean and well-ordered (7 × 7) surfaces, a broad and intense background is found which is attributed to a continuum of electronic transitions between continuously distributed surface states. In addition, ARUPS experiments show at least three different surface peaks above and below the upper valence-band edge (S_1, S_2, S_3 in Fig.6.38) whose dispersion plotted versus wave vector in the (1 × 1) Brillouin zone is not very significant at least in the wave-vector range considered [6.47]. The surface-state band S_1 might well show some dispersion near the zone boundary and cross the Fermi level. It thus might be responsible for the metallic character of the surface. Inverse photoemission experiments also indicate a continuous distribution of empty surface states in the upper half of the forbidden band with two maxima slightly above the conduction-band edge. Figure 6.39 shows qualitatively the density of both occupied and empty surface states, as derived from UPS and inverse photoemission measurements. The transition marked by $\hbar\omega$ (double arrow) has also been found by means of HREELS measurements [6.50].

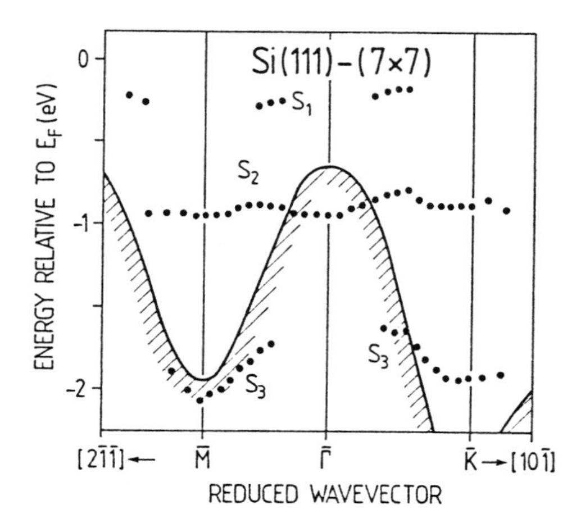

Fig.6.38. Experimentally determined dispersion of surface state bands (S_1,S_2,S_3) on the clean Si (111) -(7 ×7) surface (*points*). The projected bulk valence band is indicated by its shaded upper boundary [6.47]

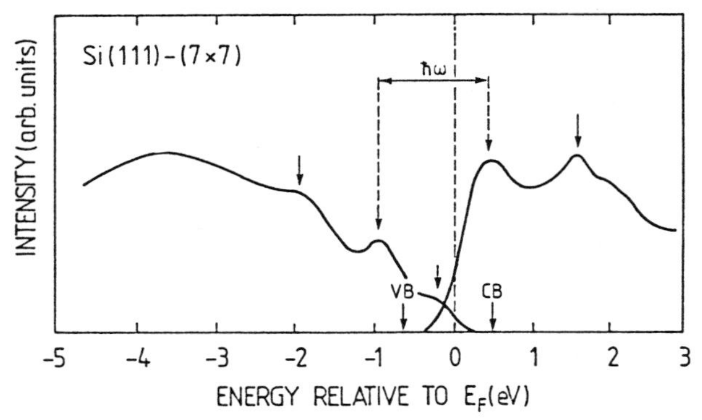

Fig.6.39. Schematic distribution of occupied and empty surface states of the Si(111)-(7×7) surface. The curves are constructed from several sets of experimental data [6.45-49]. The arrow marked $\hbar\omega$ represents an electron energy loss observed experimentally. Other tic marks show occupied and empty surface states as revealed by UPS and inverse photoemission experiments [6.50]

For semiconductor devices, the Si(100) is by far the most important surface (Sects.7.6, 7). The clean **Si(100)-(2 × 1) surface** prepared by ion bombardment and annealing has therefore also attracted much interest. The (2 × 1) superstructure could be due to missing surface atoms. But on the basis of UPS data and a calculation of the density of surface states *Appelbaum* et al. [6.51] were able to rule out such a vacancy model for the Si(100)-(2 × 1) surface (Fig.6.40a). As is expected from purely geometrical considerations, the sp^3-like dangling bonds on neighboring Si surface atoms could dehybridize into orbitals whose nature is more sp_z, p_x, p_y, and finally form Si dimers at the surface (Fig.6.40b). According to Fig.6.30 this dimerization would be in the (110) surface direction. From Fig.6.40a it is evident that such a dimer model (in contrast to Fig.6.40b a symmetric dimer) gives better agreement between measured and calculated surface state densities. More detailed investigations show that the symmetric dimer might not be the correct atomic configuration. The asymmetric dimer shown qualitatively in Fig.6.40b, which is related to a certain degree of ionicity (because of its asymmetry), leads to even lower total energy in the calculation. The dangling-bond band dispersion has been calculated both for the symmetric and the asymmetric dimers (Fig.6.41c, d). For the asymmetric dimer a total gap between the occupied and empty surface states appears, the surface is thus semiconducting – in contrast to the symmetric dimer surface. A semiconducting surface with essentially no surface-state emission near the Fermi level E_F is found in ARUPS data (Fig.6.41a). Furthermore the main surface-state band between Γ and J' agrees quite well with the calculated dispersion for the asymmetric dimer model (Fig.6.41c).

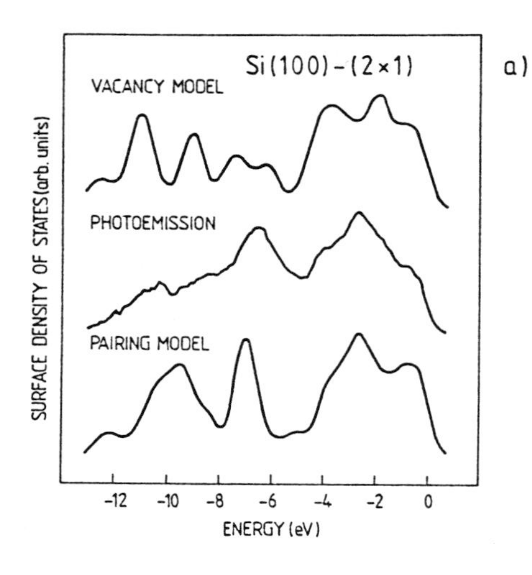

Fig.6.40a,b. Surface states on Si(100)-(2 × 1) [6.51]. (**a**) Calculated surface densities of states for the Si(100)-(2×1) surface for the vacancy and the dimer pairing models as compared to an experimental UV photoemission spectrum ($\hbar\omega$ = 21.2eV, true secondaries subtracted). (**b**) Schematic diagram of the dehybridization and dimerization leading to the non-planar dimer structure

The example of the Si(100)- (2 × 1) surface also shows how ARUPS data help to clarify structure models for semiconductor surfaces.

6.5.2 III-V Compound Semiconductors

It is instructive to compare the calculated band structures of the III-V semiconductor GaAs with its isoelectronic neighbor Ge (Fig.6.42). In the bulk-band structure the essential difference is the opening in GaAs of the so-called **ionicity gap** between −6 and −11eV. The differences in the surface-state band structure of non-reconstructed GaAs (111), (110) and (100) surfaces are easily understood in terms of the local dangling-bond geometry (Fig.6.30). Each surface state band of Ge splits into a corresponding cation- and anion-derived band in GaAs. The (110) surface is a non-polar surface with equal numbers of Ga and As atoms. On the ideal, non-reconstructed (110) surface the anion- and cation-derived surface-state bands lie in the bulk gap (Fig.6.42). According to the nature of the bulk conduction and valence-band states, the low-lying band (near E_V) is As and the high lying band (near E_C) Ga derived. The (111) and (100) surfaces are polar surfaces, which can be either Ga or As terminated. In reality, the type of termination can, to a large extent be controlled during growth by the beam flux in MBE

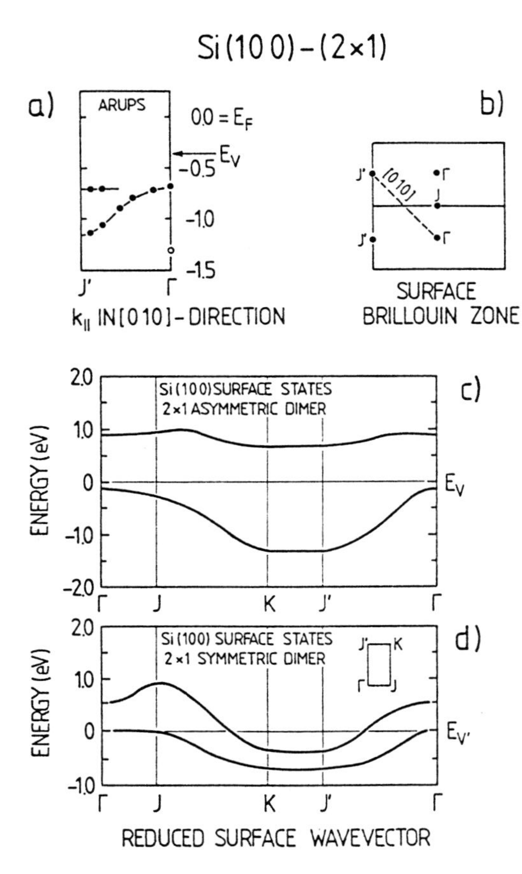

Fig.6.41a-d. Surface state dispersion on Si(100)-(2 × 1). (**a**) The experimentally determined dispersion from ARUPS [6.52]. (**b**) Surface Brillouin zone showing the orientation of the ARUPS measurement in (**a**). (**c, d**) Calculated dispersion curves for the (2 × 1) asymmetric and the (2 × 1) symmetric dimer models, respectively [6.53]

(Sect.2.4). The As-terminated surface is generally more stable, since a Ga excess easily leads to Ga aggregation and segregation. In the surface-state band schemes one observes As- or Ga-derived bands depending on the termination (Fig.6.42).

As in the case of elemental semiconductors, III-V semiconductor surfaces also reconstruct and the surface-state bands cannot be derived as easily as in Fig.6.42. In fact, the experimental investigation of surface-state band structure, e.g. by ARUPS, often helps one to construct and experimentally verify surface structure models.

Of the III-V semiconductor surfaces the **GaAs(110) surface** prepared by cleavage in UHV is certainly the most thoroughly studied. From careful measurements of the work function (Panel XV: Chap.9) for different bulk dopings [6.56] and from a number of photoemission studies, it is well known that on a perfectly cleaved surface with mirror-like finish, the Fermi level is not pinned (Sect.7.7); the bulk-band gap should essentially be free of surface states. On the other hand, according to Fig.6.42 and the calcula-

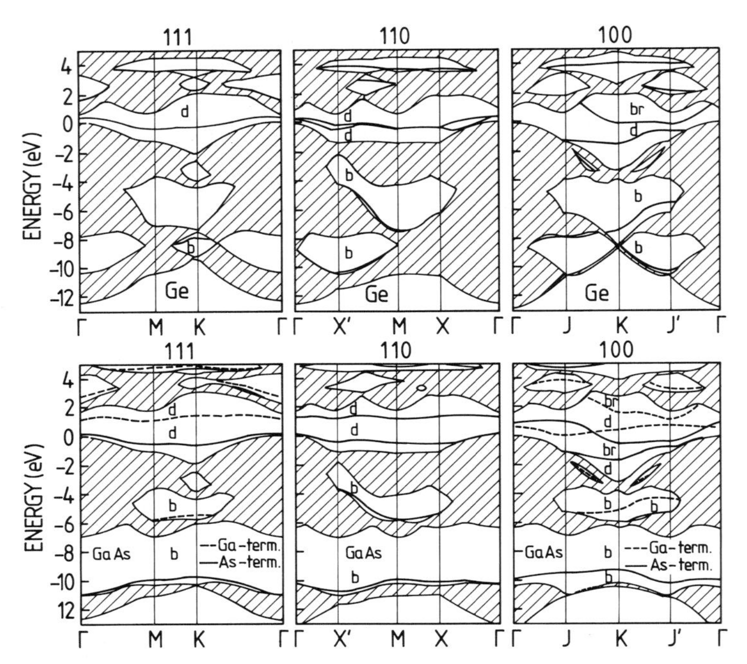

Fig. 6.42. Comparison of calculated surface state band structures of the non-reconstructed low-index surfaces of Ge and GaAs with dangling-bond surface states (d), broken-bond states (br) and back-bonding states (b). For the polar (111) and (100) GaAs surfaces Ga- and As-terminations lead to different surface state bands (*broken lines* and *solid lines*, respectively). The projected bulk band structure is shaded [6.54]

tions of a number of other groups (Fig. 6.43a) an unreconstructed (110) surface with atomic positions as in the bulk always gives Ga and As derived dangling-bond states within the gap, the Ga derived acceptor-like states near midgap and As derived donor states in the lower half of the forbidden gap (Fig. 6.43a). The latter states could not be detected in UPS. On the basis of LEED intensity analyses and Rutherford backscattering data, the reconstruction model of Fig. 6.43d was developed, in which the topmost As atoms are displaced outwards and the neighboring Ga atoms inwards with respect to their unreconstructed bulk-like position (Sect. 3.2). The best structure model achieved so far is described by a rotation of the atoms in the top layer by 27° and a contraction of the first interlayer distance by 0.05 Å. This atomic arrangement on a cleaved GaAs(110) surface indicates a dehybridization of the sp^3 tetrahedral bonds. The trivalent Ga adopts almost planar sp^2-bonds, while the As atom tends to a pyramidal $AsGa_3$ configuration with bond angles close to 90° and having more p-like character. This dehybridization is thought to be connected with a charge transfer from the Ga to the As surface atom. The Ga dangling bonds are emptied and the As

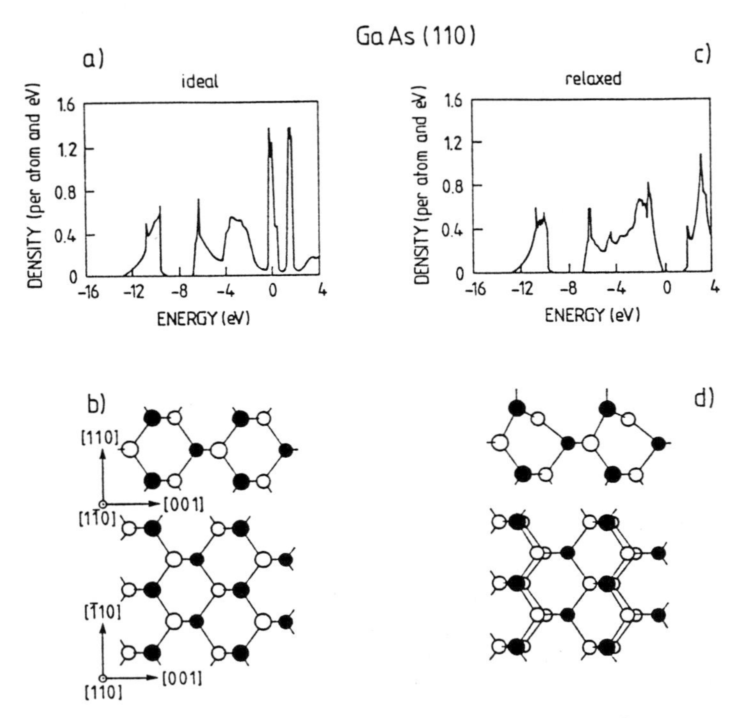

Fig.6.43a-d. Surface state densities and the corresponding structure models for the GaAs(110) surface. (**a, c**) Calculated surface state densities for the ideal, non-reconstructed (*left*) and the relaxed (*right*) surface. The zero of the energy scale is taken to be the upper valence band edge, $E_v = 0$. (**b, d**) Structure models (side and top view) for the ideal, non-reconstructed (*left*) and the relaxed (*right*) surface [6.55]

states are occupied. This picture of the charge transfer is in agreement with core level shifts observed in XPS (Fig.8.16). Note, however, that this reconstruction – a so-called **relaxed surface** with a (1×1) LEED pattern (Fig. 6.43d) – shifts the dangling bond surface states out of the bulk-band gap (Fig.6.43b).

Furthermore, there is good agreement between the calculated surface state-band structure for a relaxed surface and the dispersion of surface states measured in ARUPS (Fig.6.44). As may also be derived from Fig. 6.43a,c this agreement cannot be obtained for an unrelaxed surface. Inverse photoemission spectroscopy (isochromate mode) (Panel XI: Chap.6) was used to measure the density of empty surface states on a clean GaAs(110) surface, which was prepared by ion bombardment and annealing in UHV. The isochromate spectrum of Fig.6.45 clearly shows the empty, Ga-derived surface states to be degenerate with conduction-band states. The gap is free of states, as is required by the calculation for the relaxed surface (Fig. 6.43c).

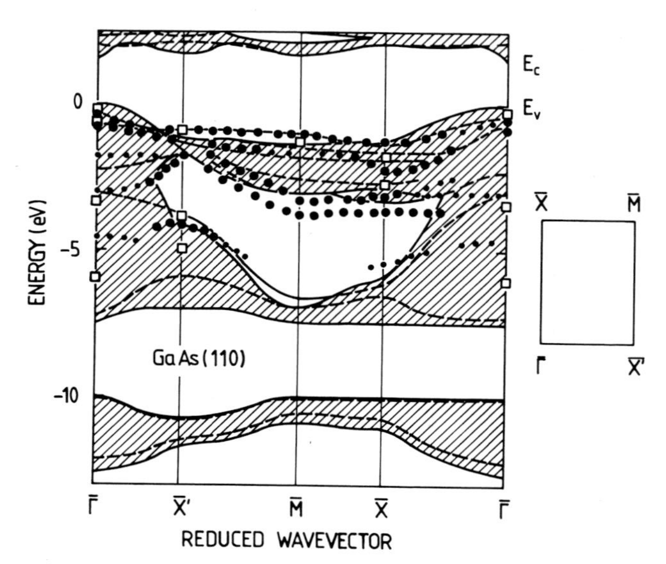

Fig. 6.44. Measured and calculated dispersion curves of surface states (*solid lines*) and surface resonances (*dashed lines*) on cleaved GaAs(110) surfaces along the symmetry lines of the surface Brillouin zone (*right-hand side*). The shaded areas represent the projected bulk band structure. (Calculation from [6.57], experimental ARUPS data from [6.58] (□) and [6.59] (●))

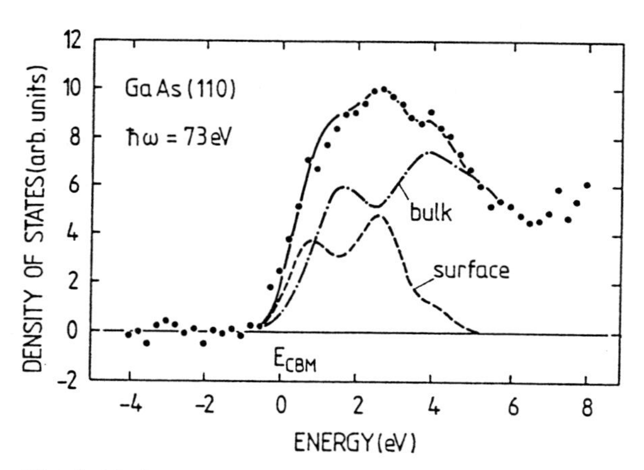

Fig. 6.45. Inverse photoemission (isochromate) spectrum of empty bulk and surface states on a GaAs(110) surface prepared by ion bombardment and annealing in UHV. The solid line is the total density of states above the conduction band minimum E_{CBM} at the surface. The surface and the bulk contributions are shown as dashed and dash-dotted lines, respectively [6.60]

In this context, it is remarkable that the same type of reconstruction (Fig. 6.43d) is obviously present on both the cleaved, as well as the ion-bombarded and annealed (110) surface. This conclusion also follows from LEED intensity analyses. Nevertheless, these differently prepared (110) surfaces differ in their electrical properties. On ion-bombarded surfaces, space charge layers are always observed due to the presence of defect-derived surface states (Sect. 7.6).

Only the (110) surface of GaAs can be prepared by cleavage. The other surfaces require ion-bombardment and annealing or MBE, the latter producing better-defined surfaces (Sects. 2.4, 5). Among the low-index surfaces **GaAs(001)** can be prepared in MBE with a number of different superstructures, depending on details of the As/Ga flux ratio, substrate temperature, etc. A particularly stable surface is the As-stabilized surface with its (2 × 4) superstructure, which is prepared at 800 K and an As_4/Ga flux ratio of about 10/1 [6.61]. The 2D surface band structure of this GaAs(001)(2 × 4) surface has been measured by ARUPS [6.61]; the bands of occupied states

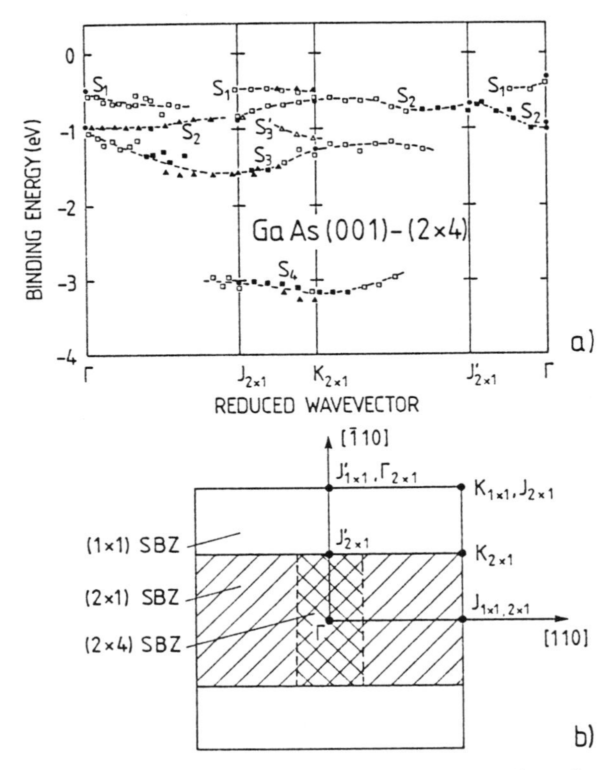

Fig. 6.46. Measured dispersion curves (**a**) of surface states on the As-stabilized GaAs(001)-(2 × 4) surface along symmetry lines of a (2 × 1) surface Brillouin zone (**b**) [6.61]

are shown in Fig.6.46a. The four-fold periodicity along the (110) direction for the (2×4) reconstruction was not observed. The experimental data are therefore plotted with respect to a (2×1) Brillouin zone in Fig.6.46. The relation of the (2×1) to the (2×4) Brillouin zone is explained in Fig.6.46b. A detailed comparison with theoretical dispersion curves having the four-fold symmetry is thus difficult. On the basis of a theoretical (2×1) reconstructed dimer model, the low lying surface state band S_4 near -3eV binding energy is attributed by the researchers to As-As dimer bonding states. From these results one can infer that As-As dimers, similar to the Si-Si dimers on Si(100)-(2×1) (Fig.6.40b), are the building blocks of the As-rich GaAs (001)-(2×4) surface.

It is assumed at present that many of the features of the GaAs(110) and (001) surfaces discussed here also apply to other III-V compound semiconductors [6.63]. A number of studies have revealed similarities between GaAs, InP and InSb. On the other hand, there are also some important differences; for example, on well-cleaved GaP(110) surfaces, the Ga derived empty surfaces states lie in the bulk gap, in contrast to those of GaAs(110).

6.5.3 II-VI Compound Semiconductors

Qualitatively, the trends in surface-state dispersion observed for III-V semiconductors are even more pronounced for the more ionic II-VI semiconductors. The best-studied example is ZnO with a direct bulk-band gap of 3.2eV at 300 K. The main low-index surfaces are the polar (0001) Zn, $(000\bar{1})$O and the non-polar hexagonal $(10\bar{1}0)$ surface (Fig.6.47). In Fig.6.47 the sp^3-like bonding character – accompanying the strong ionic contribution – is clearly revealed from the essentially tetrahedral atom surrounding. All three surfaces can be prepared by cleavage in UHV. Comparing Fig. 6.43c,d with Fig.6.47 one expects a similar reconstruction of the non-polar ZnO surface as found on GaAs(110). LEED intensity analyses indeed sug-

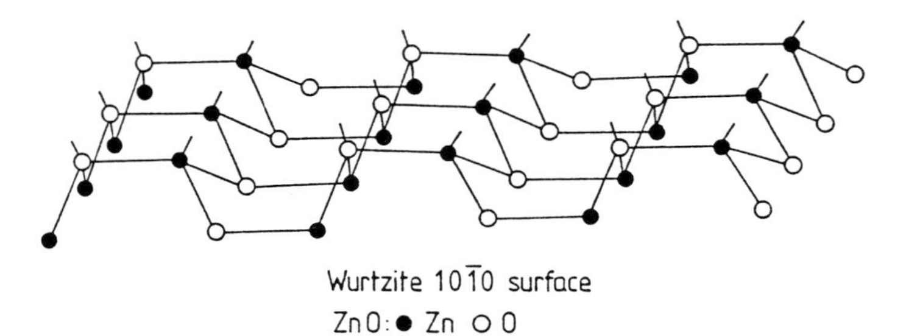

Fig.6.47. Model of the non-polar wurtzite $(10\bar{1}0)$ surface. For the example of ZnO, open circles might be attributed to oxygen and dark circles to Zn atoms

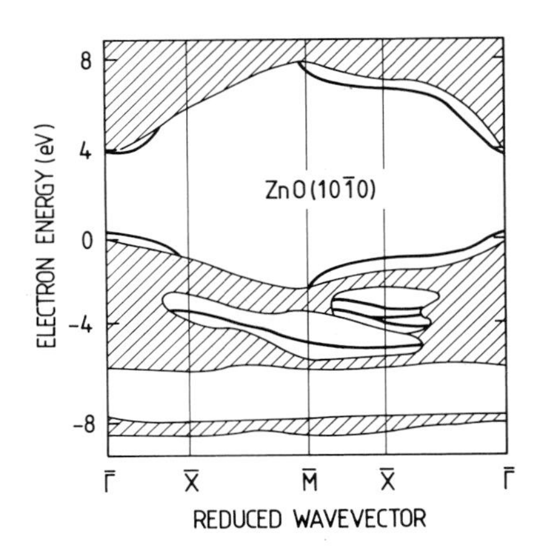

Fig.6.48. Calculated surface state dispersion (*solid lines*) within the projected bulk state gaps for the non-polar ZnO(10$\bar{1}$0) surface. The projected bulk bands are shown as shaded areas. The symmetry point $\bar{X}$ of the 2D surface Brillouin zone lies perpendicular to the c-axis, $\bar{M}$ lies on the diagonal [6.62]

gest a reconstruction in which the topmost oxygen atoms in Fig.6.47 are shifted outwards, while the neighboring Zn atoms are displaced inwards. A detailed reconstruction model has been suggested in which the total vertical shifts (with respect to bulk) are: Δz (O) = (-0.05 ± 0.1)Å and Δz (Zn) = (-0.45 ± 0.1)Å, indicating a vertical contraction of the topmost Zn-O double-layer spacing with simultaneous dehybridization (i.e., shifts out of plane) of the O and Zn atoms in the uppermost plane. In principle, such a reconstruction should have a similar effect on the Zn (empty) and O (occupied) dangling-bond orbitals as in the case of GaAs (110). For the unreconstructed surface the Zn- and O-derived dispersion branches should lie somewhere near the conduction-band E_C and valence-band E_V edge, respectively. This is indeed noted in the calculated band structure for the unreconstructed, truncated bulk-like (10$\bar{1}$0) surface (Fig.6.48). The reconstruction which is found on real cleaved surfaces is expected to shift the unoccupied Zn band to higher, and the occupied O-derived band to lower energy; i.e., if the surface-band structure of Fig.6.48 were correct in every detail for the unreconstructed surface, the reconstructed cleaved surface would have no dangling-bond surface states in the energy range of the band gap. There is indeed no indication from any experiment that the non-polar ZnO surface has intrinsic states in the forbidden band. Qualitatively, one might expect a weaker influence of the reconstruction on the surface-state band scheme than in the case of GaAs. Because of the high ionic bonding contribution in ZnO, the effect of reconstruction and even the presence of the surface (termination of the periodic potential) are relatively weak perturbations, as compared with the strong Coulomb forces. This also causes the close similarity between the bulk-band structure and surface states in Fig.6.48; the surface states follow the bulk-band structure very closely. Experiment also supports this conclusion. The double differentiated EELS data (Panel IX: Chap.4) in Fig.6.49 measured on reconstructed ZnO surfaces prepared by

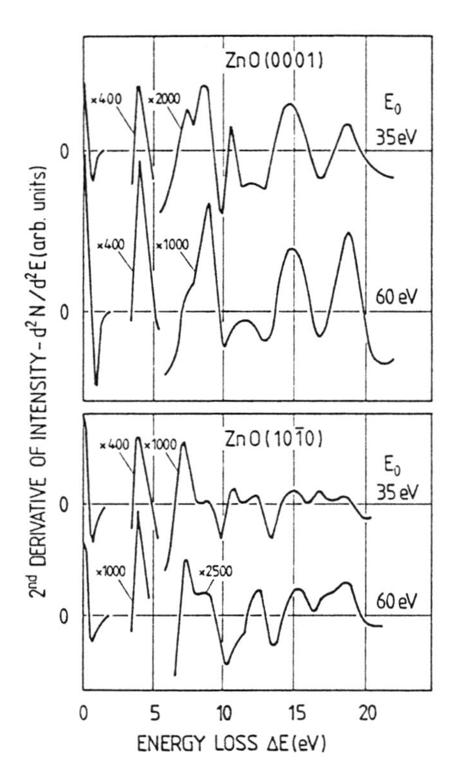

Fig.6.49. Second derivative electron energy loss spectra measured on the clean polar Zn(0001) and the non-polar hexagonal (prism) (10$\bar{1}$0) surface with two different primary energies E_0. Magnification factors are with respect to the primary peak height [6.64]

cleavage in UHV are actually consistent with the band structure (Fig.6.48) calculated for a non-reconstructed (1010) surface. Beside the bulk and surface-plasmon-like excitations near 19 and 15 eV, a number of peaks are observed due to electronic transitions at critical points of the bulk band structure. The strong transitions at 7.4 eV (on the non-polar surface) and at 11 eV (on both surfaces) are clearly due to surface states because of their characteristic dependence on primary energy E_0 (they are suppressed at higher E_0). The 7.4 eV transition obviously corresponds to excitations between occupied and empty dangling bond states from the flat $E(\mathbf{k}_{\|})$ regions around $\bar{X}$ in Fig.6.48. As is expected for a polar Zn surface, where only the Zn derived empty surface states are present, this transition is not observed in Fig.6.49a; the weak structure near 7eV is due to bulk transitions. The 11 eV transition probably originates from flat regions of the surface band structure at −4 eV near $\bar{X}$ in Fig.6.48. Surface state transitions expected from Fig.6.48 at around 4 eV from flat regions around $\bar{\Gamma}$ are likely to be contained in the strong loss feature at around 4 eV.

To conclude this chapter it should once more be emphasized that, on semiconductor surfaces in particular, the problem of atomic reconstruction is intimately connected with the dispersion of electronic surface states

[6.63, 65]. A calculation of the 2D surface band structure is impossible without the detailed knowledge of the atomic positions on the surface. Conversely, examples discussed here show how important an experimental determination of the surface band structure is for establishing a correct structure model.

Panel XI
Photoemission and Inverse Photoemission

Photoemission [XI.1] and inverse photoemission spectroscopy [XI.2] are the most important experimental techniques for studying the band structure of occupied and empty electronic states, respectively. Depending on their kinetic energy the mean-free path of electrons in a solid ranges from about 5 Å up to some hundreds of Ångstroms. These techniques are therefore suited to studying both (3D) bulk and (2D) surface band structures $E(\mathbf{k})$ and $E(\mathbf{k}_{\parallel})$, respectively. In particular, for photon energies below 100 eV the technique is highly surface sensitive and thus suitable for surface studies. *Photoemission Spectroscopy* (PS) performed with UV photons (UPS) or with X-ray photons (XPS) is based on the well-known photoelectric effect. The solid is irradiated by monochromatic photons which excite electrons from occupied states into empty states (within the solid), whence they are released into vacuum (free-electron plane-wave states) and detected by an electron-energy analyser. Thus the kinetic energy of the emitted photoelectron is determined and its wave vector $\mathbf{k}^{ex}$ outside the solid is derived from its energy and the direction of the analyser aperture with respect to the sample orientation (Sect.6.3). Since for the electron wave escaping from the crystal, the surface represents a 2D scattering potential (breakdown of translational symmetry), the wave vector $\mathbf{k}_{\perp}$ is not conserved; the internal $\mathbf{k}$ vector cannot be directly determined from the externally measured $\mathbf{k}^{ex}$ (6.19-23) in Angle-Resolved UV Photoemission Spectroscopy (ARUPS). The basic relation for the photoemission process becomes clear from Fig.XI.1, where in the upper part schematic state densities of occupied and empty states (continuum above Fermi level E_F) are plotted. Optical excitation by a fixed photon energy $\hbar\omega$ populates empty states in the crystal above the vacuum level and the corresponding energy distribution of the electrons measured outside the crystal yields a qualitative image of the distribution of occupied crystal states (valence and core level states depending on the value of $\hbar\omega$). The measured distribution of sharp peaks is superimposed on the true secondary background, which arises from electrons that have lost quasi-continuous amounts of energy due to multiple scattering in the crystal (Sect.6.3). The sharp peaks in the spectrum correspond to a kinetic energy E_{kin} of the externally detected electrons given by

$$E_{kin} = \hbar\omega - E_i - \phi \, , \tag{XI.1}$$

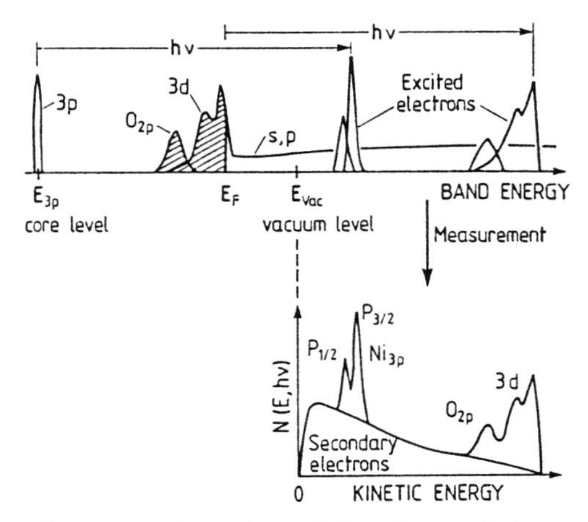

Fig. XI.1. Illustration of the photoemission process for the example of a transition metal surface (e.g., Ni) on which atomic oxygen (O 2p) is adsorbed. Shadowed areas show occupied electronic states (up to the Fermi level E_F). Photons incident with energy $h\nu$ cause electrons to be excited into unoccupied quasi-continuous electron states within the crystal. These electrons can leave the crystal and are detected in the measurement as free electrons with a kinetic energy E_{kin}. Electrons that have undergone scattering processes on their way into vacuum are detected at lower energy and form a continuous background of so-called secondary electrons

where E_i is the binding energy of the initial state (to be determined) and ϕ the work function which has to be overcome by electrons reaching vacuum states. All energies in such a photoemission experiment are conveniently referred to the Fermi level E_F of the sample, since this energy is fixed (sample at earth or other fixed potential) and can be determined from the upper emission onset in the case of a metallic sample (or metallic overlayer).

The essential parts of the experimental set-up are a monochromatic light source, the sample contained in an UHV vessel to maintain clean surface conditions, and an electron energy analyser with detector. For angle-resolved measurements hemispherical or 127° deflectors (Panel II: Chap. 1) are used since they have a limited, well-defined acceptance angle. They allow the determination of $\mathbf{k}^{ex}$. When angular resolution is not required, cylindrical mirror analysers (Panel II: Chap. 1) can be employed. For the determination of densities of occupied states (integration over $\mathbf{k}^{ex}$) retarding field analysers collecting over a large acceptance angle are convenient.

As light sources for the UV range (UPS) gas discharge lamps are used; they are flanged to the UHV chamber through a differentially pumped capillary, which supplies the UV light to the sample surface (good windows for UV spectral range are not available). The lamp and capillary are pumped at several points, such that over a capillary diameter of about 1 mm a pressure

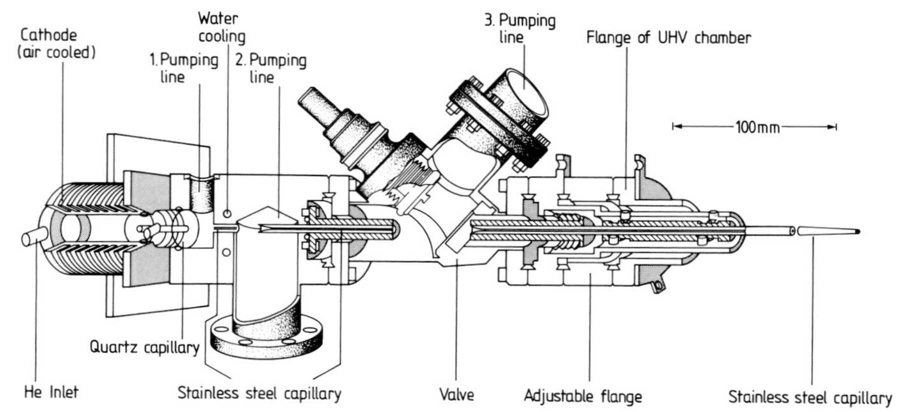

Fig. XI.2. Cross-sectional view of a UV discharge lamp for UV Photoemission Spectroscopy (UPS). The discharge quartz capillary is water cooled; three pump connections allow differential pumping; an UHV valve can interrupt the direct connection between discharge volume and UHV chamber

gradient between 1 Torr (in the lamp) and $2 \cdot 10^{-10}$ Torr (in the UHV vessel) is sustained (Fig. XI.2). The discharge burns in a water- or air-cooled compartment, which is separated from the capillary by a UHV valve, allowing closure of the lamp without breaking the vacuum in the analysis chamber. Possible filling gases together with their main spectral emission lines are listed in Table XI.1. The most important source is the He discharge, where the HeI spectral line ($\hbar\omega = 21.22$eV), originating from excitations of the neutral He atom, is extremely intense, all other spectral lines giving rise to minor background only. This line is usually employed with no UV monochromator between lamp and sample. Depending on pressure (1 Torr for HeI and 0.1 Torr for HeII) and discharge current conditions the HeII line at 40.82 eV can also be used without a monochromator. This emission originates from excited He^+ ions in the discharge.

In order to study core-level excitations, one requires higher photon energies as used in XPS (or ESCA). Conventional sources here are X-ray tubes whose characteristic emission lines (Table XI.1) are determined by the anode material (Fig. XI.3). Common anodes are composed of Mg or Al (Table XI.1). In addition, Y is an interesting anode material since it yields an emission line at 132.3 eV, just between the characteristic spectral ranges of UPS and XPS. The anodes of X-ray sources are water cooled in order to enhance the maximum emission intensity. The linewidths of the characteristic X-ray emission lines are several hundred meV (Table XI.1), such that fine-structure investigations or the analysis of chemical shifts, etc. are difficult if not impossible without the use of X-ray monochromators. Thus, for studies of core-level fine-structure, X-ray tubes are used in combination

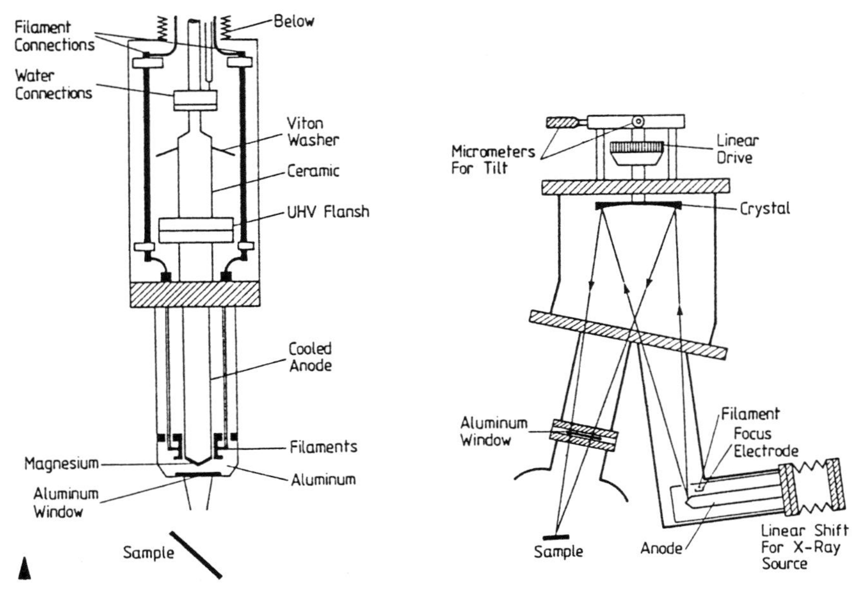

Fig. XI.3. Cross-sectional view of an X-ray source for X-ray Photoemission Spectroscopy (XPS). The anode, composed of either Mg or Al, is water cooled

Fig. XI.4. Schematic of an X-ray monochromator for high-resolution XPS. The X-ray source is flanged to an UHV chamber containing a crystal mirror which acts, by means of Bragg reflection, as a dispersive element

with an X-ray monochromator containing a crystalline mirror as a dispersive element (Fig. XI.4).

Nowadays synchrotron radiation has come to play a vital role in photoemission spectroscopy. A synchrotron yields a continuous spectrum of radiation extending from the far infrared to the hard X-ray regime. The cut-off depends on the acceleration energy (Fig. XI.5). Apart from the HeI line of a discharge lamp its spectral emission intensity exceeds considerably that of all other discharge emission lines. UV and X-ray monochromators provide an adjustable spectral resolution for the experiments. Further advantages of synchrotron radiation are its 100% polarization in the plane of the ring, its high degree of collimation (1 mrad × 1 mrad typically), its high stability and well-defined time structure (light-house effect) for time-resolved experiments. Many experimental examples involving the application of photoemission spectroscopy are given in Chap. 6.

Inverse photoemissions can be regarded as the time-reversed photoemission process [XI.5]. Electrons of well-defined energy are incident on the crystal, and are thereby injected into empty excited electronic states; from here they are deexcited into energetically lower empty states and the

Table XI.1. Commonly used line sources for photoelectron spectroscopy (UPS and XPS). In some cases (*) relative intensities of the lines depend on the conditions of the discharge. Values given are therefore only approximate. The data are compiled from several original sources [XI.1]

Source	Energy [eV]	Relative intensity	Typical intensity at the sample [photons/s]	Linewidth [meV]
He I	21.22	100	$1 \cdot 10^{12}$	3
Satellites	23.09, 23.75, 24.05	< 2 each		
He II	40.82	20*	$2 \cdot 10^{11}$	17
	48.38	2*		
Satellites	51.0, 52.32, 53.00	< 1* each		
Ne I	16.85 and 16.67	100	$8 \cdot 10^{11}$	
Ne II	26.9	20*		
	27.8	10*		
	30.5	3*		
Satellites	34.8, 37.5, 38.0	< 2 each		
Ar I	11.83	100	$6 \cdot 10^{11}$	
	11.62	80 ÷ 40*		
Ar II	13.48	16*		
	13.30	10*		
YM_t	132.3	100	$3 \cdot 10^{11}$	450
Mg $K_{\alpha 1,2}$	1253.6	100	$1 \cdot 10^{12}$	680
Satellites $K_{\alpha 3}$	1262.1	9		
$K_{\alpha 4}$	1263.7	5		
Al $K_{\alpha 1,2}$	1486.6	100	$1 \cdot 10^{12}$	830
Satellites $K_{\alpha 3}$	1496.3	7		
$K_{\alpha 4}$	1498.3	3		

corresponding deexcitation energy is released as a photon (Fig. XI.6). Thus one measures the state of an extra electron injected into the solid. The energy of the unoccupied final state is given by the energy of the incident electron eU minus the energy of the detected photon $\hbar\omega$ (both referred to the experimentally determined Fermi level E_F). The theoretical description of the processes is similar to that of the normal photoemission process (three-step approximation, Sect. 6.3). Spectroscopy of unoccupied electronic states (above E_F) can be performed by varying the primary energy of the injected electrons and detecting photons of a well-defined fixed photon energy $\hbar\omega_0$ (within a fixed spectral window). This type of inverse photoemission spec-

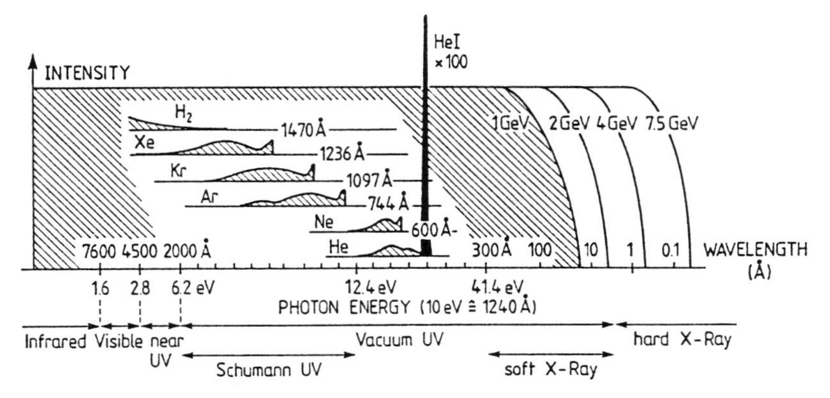

Fig. XI.5. Schematic comparison of the radiation spectrum emitted from a typical large synchrotron (e.g., DESY or BESSY) with that of classical discharge sources. Several particle energies (1GeV up to 7.5GeV) are indicated. The intensities all lie on roughly the same scale [XI.3, 4]

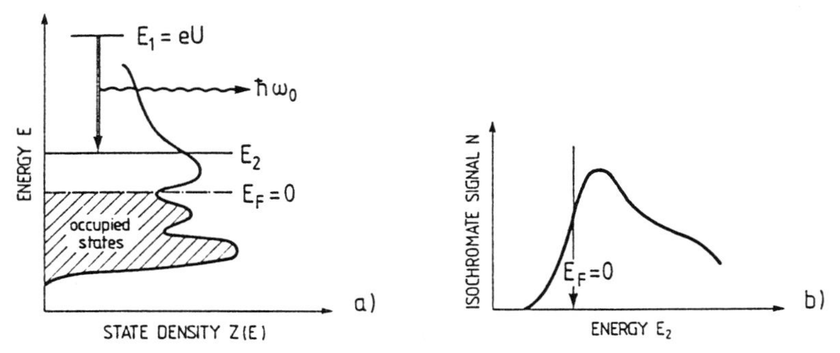

Fig. XI.6. (**a**) Schematic representation of the inverse photoemission process. An electron injected from outside the crystal enters an excited electronic state E_1 (=eU if an external voltage U accelerates the electrons onto the sample); the electron is deexcited into a state E_2 and the corresponding energy is emitted as a photon of energy $\hbar\omega_0 = E_1 - E_2$. (**b**) Schematic isochromate spectrum $N(E_2) \propto Z(E_2)$ as obtained according to (**a**)

troscopy is called **isochromate spectroscopy**. A second type of measurement uses a fixed electron energy in the primary beam and spectroscopic analysis of the emitted UV radiation by means of a UV spectrometer (**Bremsstrahlen spectroscopy**). The recorded UV spectrum (UV intensity versus photon energy) then directly yields a qualitative image of the distribution of unoccupied electronic states above E_F. Because of the involvement of electrons in this process the method is as surface sensitive as photoemission spectroscopy. The distinction between bulk and surface electronic states is performed on an experimental basis as is done in UPS or XPS

(Sect. 6.3). Angle-resolved inverse photoemission spectroscopy is possible by determining the wave vector of the injected electrons from their direction of incidence and their energy. Thus a mapping of the band structure $E(\mathbf{k})$ of unoccupied states – both 3D bulk and 2D interface states – becomes possible.

The experimental set-up in Bremsstrahlen spectroscopy consists of a high-intensity electron gun (necessary because of the low quantum yield of about 10^{-8} photons per electron) and a UV monochromator which is sometimes combined with modern multidetection units where typically 100 photon energies are recorded simultaneously with a channel plate amplifier and a position-sensitive resistive anode detector.

In isochromate spectroscopy the energy of the injected electrons is varied at the electron gun (Fig. XI.7) and the emitted UV photons are detected at a fixed energy by means of a Geiger counter equipped with a convenient band-pass filter window [XI.5, 6]. An elegant device consists of a Geiger counter filled with He (~500 mbar) and some iodine crystals; the counter is sealed with CaF_2 or SrF_2 windows (Fig. XI.7). The windows provide a high-energy cut-off for the UV radiation due to their characteristic absorption near 10.1 eV (CaF_2) or 9.7 eV (SrF_2), whereas the ionization of iodine (detection process)

$$I_2 + \hbar\omega \rightarrow I_2^+ + e^- \tag{XI.2}$$

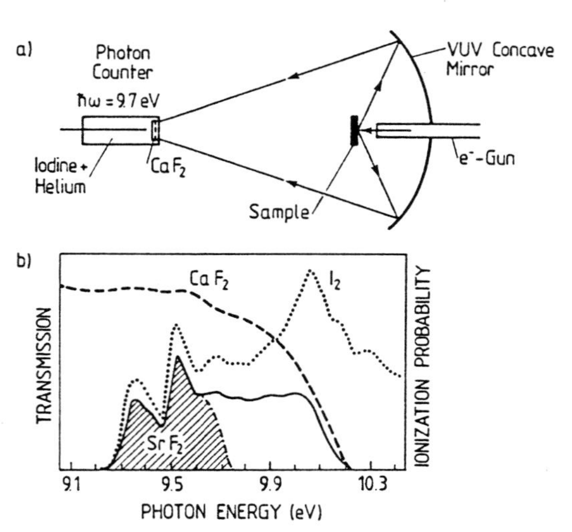

Fig. XI.7. (**a**) Inverse photoemission set-up using a Geiger counter (isochromate spectroscopy). The UV radiation emitted from the sample is focussed onto the window of a Geiger photon counter. (**b**) The spectral window of the detector is determined by the spectral transmittance of the counter window (SrF_2 or CaF_2) and by the spectral dependence of the ionization process of iodine [XI.5, 6]

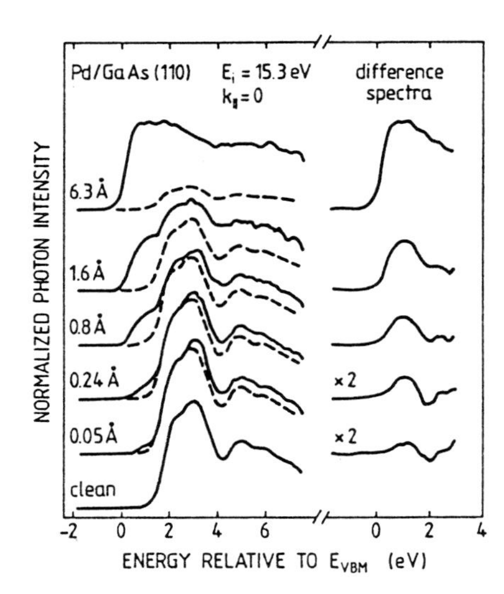

Fig. XI.8. Coverage-dependent inverse photoemission spectra (Bremsstrahlen spectra) for Pd layers of different thickness (0.05 ÷ 6.3Å) on GaAs(110). The raw data are shown together with difference curves which bring out the metal-induced interface states at about 1 eV above the valence band maximum. The clean surface spectrum is shown in each case by a dashed line [XI.7]

starts at a photon energy of 9.23 eV and thus determines the low-energy cut-off for the photon detection. One thus has a band-pass detector for UV photons around 9.5 or 9.7 eV (Fig. XI.7). This fixed photon-energy detector can be made very efficient because of a large acceptance angle, but it has limited energy resolution. On the other hand, in Bremsstrahlen spectroscopy the monochromator system achieves better energy resolution (0.3eV) and can be tuned, but it is expensive and not as efficient in the case of low signal intensities.

As an example of the application of inverse photoemission spectroscopy in the study of metal-semiconductor interfaces, Fig. XI.8 depicts Bremsstrahlen spectra measured with an electron primary energy E_i of 15.3 eV (normal injection: $k_{\|} = 0$) on cleaved GaAs(110) surfaces covered with Pd overlayers of several thicknesses [XI.7]. The energy is referred to the upper valence band edge E_{VBM} such that the onset of emission from the unoccupied states due to the conduction band on the clean surface occurs at around 1.4 eV. The forbidden band of clean GaAs(110) surfaces is free of significant densities of empty states. With increasing Pd layer thickness a remarkable structure of empty interface states occurs in the band gap. As is seen from a comparison with the spectrum of the clean surface (dashed line) and from the difference spectra, the states in the conduction band range ($E > 1.4$eV) are also modified due to the adsorbed Pd. A more detailed analysis of these results suggests that the interface states are due to d-like metal-derived levels. They are thought to be responsible for pinning the Fermi level E_F near midgap at such transition metal-semiconductor interfaces (Sect. 8.4).

References

XI.1 M. Cardona, L. Ley (eds.): *Photoemission in Solids I, II*, Topics Appl. Phys., Vols.26 and 27 (Springer, Berlin, Heidelberg 1978/79)
B. Feuerbacher, B. Fitton, R.F. Willis (eds.): *Photoemission and the Electronic Properties of Surfaces* (Wiley, New York 1978)
S. Hüfner: *Photoemission Spectroscopy, Principle and Applications*, 2nd edn., Springer Ser. Solid-State Sci., Vol.82 (Springer, Berlin, Heidelberg 1996)

XI.2 V. Dose: Momentum-resolved inverse photoemission, Surf. Sci. Rep. **5**, 337-378 (1985)

XI.3 Y. Tanaka, A.S. Jursa, F.J. Le Blank: J. Opt. Soc. Am. **48**, 304 (1958)

XI.4 E.E. Koch: In Proc. 8th All Union Conf. High Energy Particle Physics, Erevan (1975), Vol.**2**, p.502

XI.5 V. Dose: Appl. Phys. **14**, 117 (1977)

XI.6 A. Goldmann, M. Donath, W. Altmann, V. Dose: Phys. Rev. B **32**, 837 (1985)

XI.7 R. Ludeke, D. Straub, F.J. Himpsel, G. Landgren: J. Vac. Sci. Technol. A **4**, 874 (1986)

Problems

Problen 6.1. We may consider an Angle-Resolved UV Photoemission Spectroscopy (ARUPS) experiment where UV photons of energy 40.8 eV are incident on the (100) surface of a cubic transition metal with a work function of 4.5 eV. Photoemitted electrons from d states at 2.2 eV below the Fermi level are detected at an angle of 45° to the surface normal and in the [100] azimuth. Calculate the kinetic energy and the wave vector **k** of the emitted electrons. Describe the problem which arises in the derivation of the wave vector $\mathbf{k}_i$ of the initial electronic states inside the crystal.

Problem 6.2. Treatment of the surfaces of n-doped ZnO with atomic hydrogen produces accumulation layers. Discuss charging character and possible origin of the responsible surface states.

Problem 6.3. Plot a qualitative picture of the shape of the π and π^* orbitals of surface states for the π-bonded chain model of the Si(111)-2×1) surface(Sect. 6.5.1). By discussing the dipole matrix elements for the $\pi \rightarrow \pi^*$ optical transitions in a qualitative way show that optical absorption due to the transitions between the π and π^* surface-state bands can only be observed with light polarization parallel to the π-chains.

Problem 6.4. (a) For sufficiently small $\mathbf{k}_{\parallel}$ values the 2D-band of electronic surface states can be written in parabolic approximation as

$$E(\mathbf{k}_{\parallel}) = E_c + \frac{\hbar^2}{2}\left(\frac{k_x^2}{m_x} + \frac{k_y^2}{m_y}\right)$$

with m_x and m_y as positive constants. Calculate the density of states $D^{(2D)}(E)$ around the critical point E_c $(\mathbf{k}_{\parallel}=\mathbf{0})$ of the 2D band structure.

(b) Calculate the density of states in the neighborhood of a saddle point, where

$$E(\mathbf{k}_{\parallel}) = E_c + \frac{\hbar^2}{2}\left(\frac{k_x^2}{m_x} - \frac{k_y^2}{m_y}\right)$$

with positive m_x, m_y.

7. Space-Charge Layers at Semiconductor Inferfaces

If one puts a positive point charge into a locally neutral electron plasma (electrons on the background of fixed positive cores), the electrons in the neighborhood will rearrange to compensate that additional charge; they will screen it, such that far away from the charge the electric field vanishes. The higher the electron density, the shorter the range over which electrons have to rearrange in order to establish an effective shielding. In metals with free-electron concentrations of about 10^{22} cm^{-3} the screening length is short, on the order of atomic distances. On the other hand, in semiconductors the free-carrier concentrations are usually much lower, on the order of 10^{17} cm^{-3} may be, and we thus expect much larger screening lengths, of the order of hundreds of Ångstroms. These spatial regions of redistributed screening charges are called **space charge regions**.

7.1 Origin and Classification of Space-Charge Layers

A semiconductor surface which possesses electronic surface states usually represents a perturbation to the local charge balance. Depending on the type of surface states (donors or acceptors) and on the position of the Fermi level at the surface, the surface states may carry charge, which is screened by an opposite charge inside the semiconductor material.

The following table summarises the conclusions of Sect.6.2 about the charging character of occupied and unoccupied electronic surface states.

	occupied	empty
surface donors	0	+
surface acceptors	–	0 .

Depending on the position of the Fermi level at the surface, donors can carry a positive charge and acceptors a negative charge. The position of the Fermi level at the surface is determined simply by the condition of charge neutrality, namely that the charge of the surface states Q_{ss} (usually understood as a charge density per unit area) is compensated by an opposite

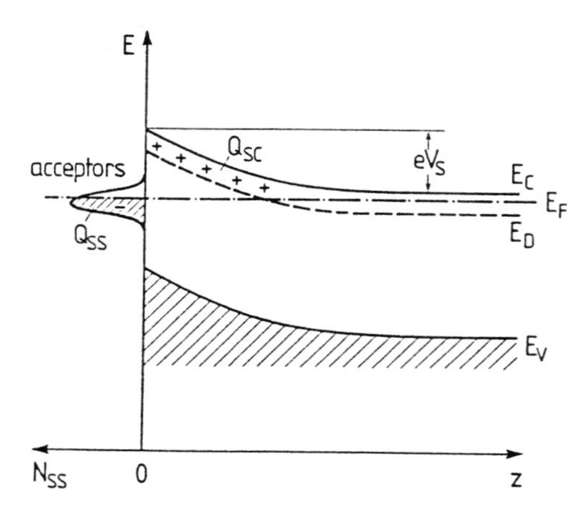

Fig. 7.1. Band scheme (band energy E versus z coordinate normal to the surface z = 0), for an n-doped semiconductor with depletion space-charge layer at low temperature (bulk donors not ionized). Partially occupied acceptor type surface states (density N_{ss}) are also indicated. Their charge Q_{ss} is compensated by the space charge Q_{sc}. E_F is Fermi energy, E_C and E_V conduction- and valence-band edges, respectively, and E_D the energy of the bulk donors. The band bending at the surface is eV_s with e being the positive elementary charge

charge inside the semiconductor. This latter charge screens the surface-state charge, and is called the **space charge** Q_{sc}. Overall neutrality requires

$$Q_{ss} = - Q_{sc} \; . \tag{7.1}$$

Figure 7.1 presents an example of a space-charge layer, in which a negative charge density Q_{ss} of occupied acceptor-type surface states near mid gap of an n-type semiconductor is compensated by ionized bulk donors (Q_{sc}). The formation of the space charge layer, i.e. the band bending (maximum value eV_s at surface) can be understood as follows: deep in the interior of the crystal, the position of the Fermi level with respect to the bulk conduction-band minimum E_C is determined by the bulk doping level. The acceptor-type surface states, however, are inherently related to the existence of the surface; their energetic position with respect to E_C is fixed and is determined by interatomic potentials (Chap. 6). Flat bands up to the very surface would therefore position the Fermi level far above the acceptor-type surface states; these states would be completely filled by electrons and a considerable uncompensated charge density would be built up. This energetically unfavorable situation cannot be stable and the result is the space charge layer depicted in Fig. 7.1. A quasi-macroscopic "deformation" of the band structure, i.e. an upwards band bending of the bands near the surface allows the surface-state band to cross the Fermi level, thereby decreasing the surface-state charge density Q_{ss}. By the same token, bulk donor states are lifted above the Fermi level and are emptied of electrons. This builds up a positive space charge Q_{sc} of immobile, ionized donor centers (E_D). The exact position of E_F at the surface within the band of surface states, and therefore also the amount of band bending, is determined by charge neutrality (7.1). The equilibrium reached in Fig.7.1 means exact compensation of the charge Q_{ss} residing in surface states by a space charge Q_{sc} within a certain depth in the semiconductor. The distribution of space charge Q_{sc} is related, via Poisson's equation, to the curvature of the electronic bands, i.e., the potential in the space-charge layer. Due to the band bending, free conduction-band electrons are pushed away from the surface, their density is lowered with respect to the bulk density n_b. This particular type of space charge layer is therefore called a **depletion layer**.

In an n-type semiconductor with bulk donors (D) (Figs.7.1,2) a depletion layer is related to a decrease in the density of free electrons (majority

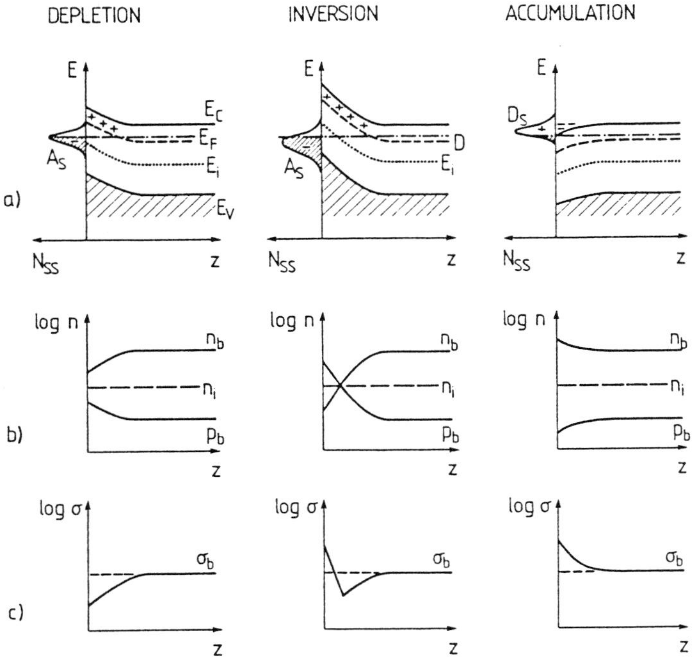

Fig.7.2a-c. Illustrating the n-type semiconductor: Schematic plots of band schemes (**a**), free carrier densities n and p on a logarithmic scale (**b**) and local conductivity σ (logarithmic scale) (**c**) for depletion, inversion and accumulation space charge layers at low temperature (bulk donors not ionized). E_C, E_V are the conduction- and valence-band edges, E_F the Fermi energy, E_i and n_i intrinsic energy and concentration, respectively. D denotes bulk donors, A_s and D_s surface acceptors and donors, respectively. The subscript b denotes bulk values

carriers) and an increase of the hole density (minority carriers) (Fig.7.2b). The local electrical conductivity $\sigma(z)$ is, of course, decreased near the surface with respect to its bulk value σ_b (Fig.7.2c).

Higher densities N_{ss} of surface acceptor states (A_s) at lower energies in the forbidden band can induce even stronger upwards band bending. The corresponding space charge layer is called an **inversion layer** for obvious reasons (Fig.7.2a). Because of the greater amount of negative surface charge Q_{ss} in the surface states A_s more bulk donors D must be ionized and the space charge layer extends deeper into the semiconductor. The band bending is so strong that the **intrinsic energy** E_i crosses the Fermi energy E_F. The intrinsic level E_i given by [7.1]

$$E_i = \frac{1}{2}(E_C + E_V) - \frac{1}{2}kT\ln(N_{eff}^c/N_{eff}^v) , \tag{7.2}$$

is a convenient quantity to describe whether a semiconductor is intrinsic, p-type or n-type. N_{eff}^c and N_{eff}^v are the effective conduction- and valence-band densities of states. If the Fermi level is identical to E_i, we have intrinsic behavior. For $E_F < E_i$ the semiconductor is p-type, whereas for $E_F > E_i$ n-type conductivity occurs with free electrons as majority carriers. According to Fig.7.2 the type of conduction changes in an inversion layer. The n-type semiconductor ($E_i < E_F$) becomes p-type near the surface ($E_i > E_F$). Correspondingly the free-electron density n decreases to below the intrinsic value n_i near the surface, whereas the hole concentration p exceeds the intrinsic value (Fig.7.2b). At the point where n and p cross on the intrinsic concentration level n_i, one finds the lowest local conductivity σ. Between this point and the very surface the conductivity σ increases again due to the presence of an enhanced density of bulk minority carriers (inversion) (Fig. 7.2c). In the charge neutrality condition (7.1) near the surface the enhanced hole concentration must also be taken into account, in addition to N_D^+, the concentration of ionized donors.

Accumulation space charge layers on n-type semiconductors, as depicted in Fig.7.2 (right column), require the presence of donor-type surface states D_s. If these states are located at high energies, they might be partially empty and thus carry a positive surface charge Q_{ss}. This charge density Q_{ss} must be compensated by an equal amount of negative space charge in the interior of the crystal. Thus free electrons accumulate in the conduction band below the surface. This electronic space charge of the so-called **accumulation layer** is related to a downward band bending. In strong accumulation layers the conduction-band minimum might even cross the Fermi level and the semiconductor becomes degenerate in the spatial region of the accumulation layer (Sect.7.5). In contrast to a depletion layer, where the positive

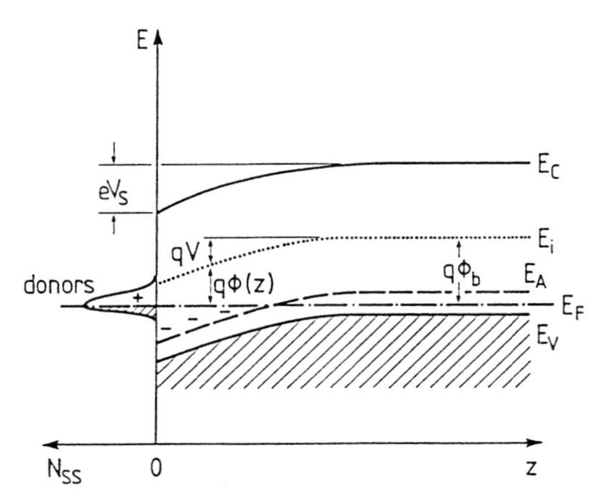

Fig.7.3. Band scheme of a hole depletion space-charge layer on a p-doped semiconductor at low temperature (bulk acceptors not ionized). eV_s is the band bending at the surface, $eV(z)$ the local band bending, $\phi(z)$ the local potential and ϕ_b the potential in the bulk. E_i is the intrinsic energy, E_F the Fermi energy, E_A the energy of bulk acceptors and E_C and E_V as in Fig.7.1

space charge originates from spatially fixed, ionized bulk donors, the accumulation layer is due to free electronic charge which is mobile. Since mobile electrons can be "squeezed", accumulation layers are, in general, much narrower than depletion layers. The qualitative dependence of the densities of majority carriers (n) and minority carriers (p) and of the local electrical conductivity σ is obvious for an n-type accumulation layer (Fig. 7.2b,c right column).

On p-type semiconductors the situations depicted in Fig.7.2 are reversed. Free holes in the valence band are the majority carriers in the bulk; an accumulation layer is therefore formed by a positively charged hole "gas". The opposite charge Q_{ss} in surface states must be negative. Thus the presence of a hole accumulation layer on a p-type semiconductor requires partially filled acceptor-type surface states and the bands are bent upwards near the surface. A depletion layer on a p-type semiconductor is shown schematically in Fig.7.3. Partially empty surface donor states D_s carry a positive charge Q_{ss}. Compensation is obtained by an equal amount of negative space charge Q_{sc} due to occupied bulk acceptor states which are "pushed" below the Fermi level E_F. This negative spatially fixed space charge is related to a downward band bending.

Figure 7.3 also explains some useful notations for describing space-charge layers. Because of the spatially varying local potential for an electron or hole, it is convenient to introduce a position-dependent potential energy that is given by the deviation of the intrinsic level E_i, (7.2), from the Fermi energy E_F:

$$e\phi(z) = E_F - E_i(z) . \tag{7.3}$$

Here, and in the following, e is the positive elementary charge. For an intrinsic semiconductor with flat bands ϕ is identically equal to zero. The values of $\phi(z)$ in the bulk and at the surface are termed ϕ_b and ϕ_s, respectively. The bulk doping determines the position of E_F with respect to the band edges and thus also the bulk potential ϕ_b. Local band deformations are described by

$$V(z) = \phi(z) - \phi_b . \tag{7.4}$$

The potential at the surface V_s (Fig.7.3) is given by

$$V_s = \phi_s - \phi_b . \tag{7.5}$$

It is convenient to define dimensionless potentials u and v via the equations

$$u = e\phi/kT , \quad v = eV/kT . \tag{7.6}$$

The values at the very surface will be denoted u_s and v_s. Using the fundamental relations for electron and hole densities in non-degenerate semiconductors [7.1]

$$n = N_{eff}^{c} \exp[-(E_C - E_F)/kT] , \tag{7.7a}$$

$$p = N_{eff}^{v} \exp[-(E_F - E_V)/kT] \tag{7.7b}$$

with N_{eff}^{c} and N_{eff}^{v} as the effective densities of states of the conduction and valence band, respectively, the following simple expressions are derived for the spatially varying carrier concentrations in a space charge layer:

$$n(z) = n_i e^{u(z)} = n_b e^{v(z)} , \tag{7.8a}$$

$$p(z) = n_i e^{-u(z)} = p_b e^{-v(z)} . \tag{7.8b}$$

Here $n_i = (np)^{1/2}$ is the intrinsic carrier concentration, and n_b and p_b are the bulk concentrations determined by the doping level. The fundamental equation which governs band bending V(z) and the form of the space-charge layers, in general, is Poisson's equation

$$\frac{d^2 V}{dz^2} = -\frac{\rho(z)}{\epsilon\epsilon_0} . \tag{7.9}$$

This equation directly relates the band curvature to the space-charge density $\rho(z)$. Usually it is sufficient to consider the dependence on the single coordinate z directed perpendicular to the surface (located at $z = 0$). The theoretical description of space-charge layers consists essentially in solving (7.9) with the appropriate boundary conditions. This is sometimes not an easy task since $\rho(z)$ itself is a function of band bending V(z). In the following we will consider some simple cases for which approximate analytic solutions of (7.9) are possible.

7.2 The Schottky Depletion Space-Charge Layer

A simple solution of the Poisson equation (7.9) is possible for strong depletion layers (Figs.7.1, 3), strong in the sense that the maximum band bending $|eV_s|$ significantly exceeds kT,

$$|eV_s| \gg kT \quad \text{or} \quad |v_s| \gg 1 . \tag{7.10}$$

Let us concentrate our attention on the depletion layer of an n-type semiconductor (Fig.7.1). The p-type case is obtained by reversing the corresponding signs for the charge. In an n-type semiconductor the positive space charge in the depletion layer is due to ionized bulk donors (density N_D, if ionized N_D^+). Because of (7.10) free electrons in the conduction band can be neglected within the space-charge layer. According to Fermi statistics, the occupation of the bulk donor levels changes from almost one to nearly zero within about 4kT. For strong band bending (7.10), the occupation change determining the sharpness of the inner boundary of the depletion layer occurs over a very short distance in comparison with the thickness d of the depletion layer (Fig.7.4). As is seen qualitatively from Fig. 7.4b, the z-dependence of the space charge density ρ can be approximated by a step function with

$$Q_{sc} = eN_D^+ d \simeq eN_D d \tag{7.11}$$

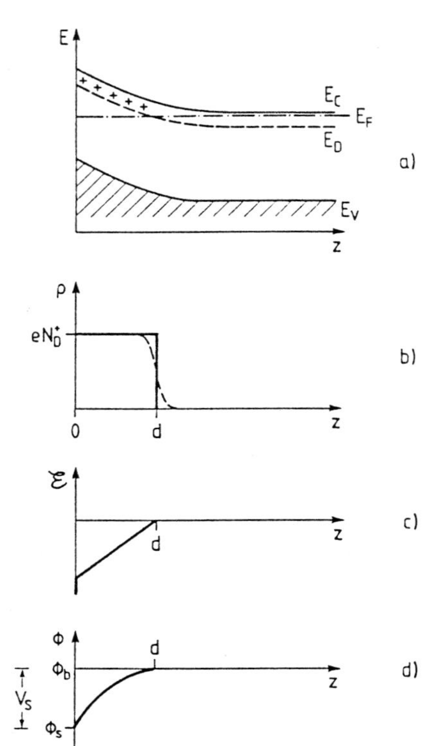

Fig. 7.4a-d. A depletion layer on an n-type semiconductor in the Schottky approximation: (**a**) Band scheme with symbols as in Figs. 7.1-3. (**b**) Volume density ρ of the space charge, realistic density (*broken line*) and Schottky approximation (*full line*). d is the thickness of the space-charge layer and N_D^+ the density of the ionized bulk donors. (**c**) Electric field $\mathcal{E}(z)$ in the space charge layer. (**d**) Electric potential $\phi(z)$ with values ϕ_b and ϕ_s in the bulk and at the surface, respectively

as the space charge (per unit area); within the space charge region the donors are assumed to be completely ionized. Poisson's equation (7.9) becomes

$$\frac{d^2\phi}{dz^2} = \frac{d^2V}{dz^2} = -\frac{d\mathcal{E}}{dz} = -\frac{\rho}{\epsilon\epsilon_0} = -\frac{eN_D}{\epsilon\epsilon_0} . \tag{7.12}$$

One integration yields the electric field $\mathcal{E}(z)$ within the space charge region (Fig. 7.4c):

$$\mathcal{E}(z) = \frac{eN_D}{\epsilon\epsilon_0}(z-d) , \quad 0 \le z \le d . \tag{7.13}$$

A further integration is necessary to obtain the potential $\phi(z)$ (Fig. 7.4d):

$$\phi(z) = \phi_b - \frac{eN_D}{2\epsilon\epsilon_0}(z-d)^2 , \quad 0 \le z \le d , \tag{7.14}$$

and the maximum potential V_s (band bending) at the surface

$$V_s = \phi_s - \phi_b = -\frac{eN_D d^2}{2\epsilon\epsilon_0} . \quad (7.15)$$

Apart from the sign, the calculation for a hole depletion layer on a p-type material (Fig.7.3) is analogous.

A simple numerical example will help to clarify the argument: on GaAs(110) surfaces, prepared by UHV cleavage one usually finds flat bands. If the cleaved surface, however, is not perfect, i.e. has a considerable number of steps and other defects, upward band bending of as much as 0.7 eV appears on n-type material with $n \simeq 10^{17}$ cm^{-3} (300K) (Sect.7.6). One evidently has acceptor-type surface states due to the defects, and these are responsible for the depletion layer on n-type material. Their density, however, is not sufficient for them to be detected by any common electron spectroscopy such as HREELS (Panel IX: Chap4) or photoemission spectroscopy (Panel XI: Chap.6). One can thus estimate that their density N_{ss} must be lower than about 10^{12} cm^{-2}. Assuming N_{ss} to be 10^{12} cm^{-2} requires a space charge density $Q_{sc}/e = N_D d$, (7.11), of equal magnitude and a thickness d of the depletion space charge layer of about 1000 Å results. By means of (7.15) the band bending at the surface $|eV_s|$ is calculated to be about 0.7 eV as is indeed found experimentally (Fig.7.19). On p-type GaAs ($p \simeq 10^{17} cm^{-3}$ at 300K) badly cleaved surfaces show p depletion layers with downward band bending of several hundred meV. A calculation similar to that for n-type material is possible.

The above numerical example shows that on common semiconductors, with dielectic constant ϵ on the order of ten, considerable band bending of about half the band-gap energy can be established by relatively low surface-state densities N_{ss} on the order of a hundredth of a monolayer. It is also concluded that typical strong depletion layers can extend up to several thousand Ångstroms into the semiconductor crystal. The surface of a semiconductor can have a long-range influence on the electronic structure. With decreasing band bending $|eV_s|$, of course, the extension d of the depletion space charge layer also decreases according to (7.15). One can also estimate by means of (7.15) that band bending on the order of 1 eV can be shielded over a length of several Ångstroms in metals with electron concentrations between 10^{22} and 10^{23} cm^{-3}. Extended space charge layers do not exist on metal surfaces. Surface charges can be screened within one or two atomic layers. On metals, therefore, interface physics is concerned only with a couple of atomic layers.

7.3 Weak Space-Charge Layers

The other limiting case of a weak space charge layer, either accumulation or depletion, also leads to a mathematically simple solution of Poisson's equation (7.9). For both weak accumulation and weak depletion where the maximum band bending $|eV_s|$ is smaller than kT ($|v_s| < 1$) the shape of the potential $\phi(z)$ is determined by mobile electrons or holes rather than by spatially fixed ionized bulk impurities. In the following we concentrate on a weak accumulation layer on an n-type semiconductor, as depicted in Fig. 7.2 (right column). The total space charge density $\rho(z)$ in Poisson's equation (7.9) is due to the free electrons n(z) in the conduction band and the density of ionized donors $N_D^+(z)$

$$\rho = -e[n(z) - N_D^+(z)] . \tag{7.16}$$

The electron concentration is given by (7.7a), whereas the density of ionized donors is determined by Fermi's occupation statistics

$$\begin{aligned} N_D^+ &= N_D - N_D \frac{1}{1 + \exp[(E_D - E_F)/kT]} \\ &\simeq N_D \exp\left(\frac{E_D - E_F}{kT}\right) \quad \text{for} \quad E_F - E_D \gg kT . \end{aligned} \tag{7.17}$$

one thus obtains the space charge density

$$\rho = -e\left[N_{eff}^c \exp\left(-\frac{E_C(z) - E_F}{kT}\right) - N_D \exp\left(\frac{E_D(z) - E_F}{kT}\right)\right] , \tag{7.18}$$

or, with E_C^b and E_D^b as bulk conduction band minimum and donor energy, respectively;

$$\rho(z) = -e\left[N_{eff}^c e^v \exp\left(-\frac{E_C^b - E_F}{kT}\right) - N_D e^{-v} \exp\left(\frac{E_D^b - E_F}{kT}\right)\right] , \tag{7.19}$$

where v(z) is the normalized band bending of (7.6b). In analogy to (7.7a) one has the general expression for the bulk electron concentration:

$$n_b = N_{eff}^c \exp[-(E_C^b - E_F)/kT] . \tag{7.20}$$

Furthermore, in the bulk, where the donors are not completely ionized in a moderately n-doped crystal, n_b is essentially given by the density of ionized donors, i.e.

$$n_b = N_D \exp[(E_D^b - E_F)/kT] . \tag{7.21}$$

One thus arrives at

$$\rho(z) = - e n_b (e^{v(z)} - e^{-v(z)}) . \tag{7.22}$$

For small band bending $|v| \ll 1$ the following approximation follows

$$\rho(z) \simeq - 2 e n_b v(z) , \tag{7.23}$$

and Poisson's equation is obtained as

$$\frac{d^2 v}{dz^2} = v/L_D{}^2 , \tag{7.24}$$

with the Debye length L_D given by

$$L_D = \sqrt{\frac{kT\epsilon\epsilon_0}{2e^2 n_b}} . \tag{7.25}$$

The solution is an exponentially decaying band bending

$$v(z) = \frac{eV(z)}{kT} = v_s e^{-z/L_D} . \tag{7.26}$$

Equations (7.23 and 26) fulfill the required conditions $v = 0$ and $\rho = 0$ for $z \to \infty$, i.e. deep in the bulk. The normalized band bending at the surface $v_s = eV_s/kT$ is determined by the condition that the total space charge

$$Q_{sc} = \int_0^\infty \rho(z)\, dz \tag{7.27}$$

is equal and opposite to the excess charge in the surface states Q_{ss}. With (7.22) one obtains

$$- Q_{ss} = Q_{sc} \approx - 2en_b v_s \int_0^\infty e^{-z/L_D} dz = - 2en_b v_s L_D \ . \tag{7.28}$$

Because of (7.1) the charge density in the surface states Q_{ss} completely determines the normalized band bending at the surface v_s when the bulk electron density, n_b, i.e. the doping level, is known. For weak band bending, i.e. weak accumulation or depletion layers, the spatial extent of the space charge is given by the Debye length L_D, i.e. essentially by the bulk carrier concentration n_b.

As a numerical example we consider n-GaAs with an electron concentration of $n_b \approx 10^{17}$ cm^{-3}. At room temperature ($kT \simeq 1/40$ eV) the Debye length (7.25) amounts to about 100 Å. Taking a band bending eV_s at 300 K of about -25 meV ($v_s \simeq -1$) as the extreme upper limit of the validity of the present simple approximation, one can estimate from (7.28) that such a band bending of $v_s \simeq -1$ requires a space charge Q_{sc} and thus also a charge density of surface states Q_{ss} of about $2 \cdot 10^{11}$ cm^{-2}. Apart from different signs the calculation for weak depletion layers is similar.

7.4 Space-Charge Layers on Highly Degenerate Semiconductors

A thorough treatment of the space-charge layer of a semiconductor that is highly degenerate in the bulk (Fig. 7.5) is complex because of the necessity to use the full Fermi distribution rather than its Boltzmann approximation. A situation such as in Fig. 7.5, where the Fermi level lies deep in the condution band, by as much as several hundred meV, occurs readily for narrow-gap semiconductors such as in InSb (gap $\simeq$ 180 meV) with a very low density of states in the conduction band. In this case of strong degeneracy with $|eV_s| \gg kT$ and $|e\psi_F| \gg kT$ one can assume, as for a free electron gas in a square-well potential [7.1]

$$n_b = N_D \propto \psi_F^{3/2} \tag{7.29}$$

for the bulk free electron concentration n_b and the donor concentration N_D.

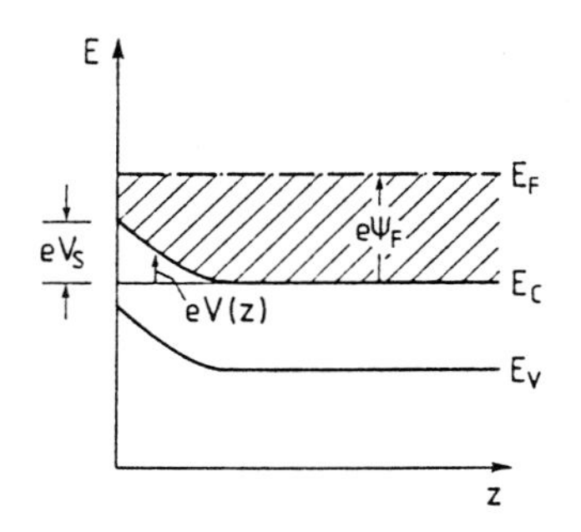

Fig. 7.5. Band scheme of a depletion space-charge layer in a highly degenerate n-type semiconductor. $e\psi_F$ describes the position of the Fermi energy E_F with respect to the lower conduction-band edge E_C in the bulk

The actual concentration n(z) at a distance z below the surface can then be written as

$$n(z) \propto [\psi_F - V(z)]^{3/2} , \tag{7.30}$$

from which the space charge $\rho(z)$ follows as

$$\rho(z) = - eN_D [1 - (1-V/\psi_F)^{3/2}] . \tag{7.31}$$

Poisson's equation (7.9) is obtained in this approximation as

$$\frac{d^2V}{dz^2} = \frac{1}{2}\frac{d}{dV}\left(\frac{dV}{dz}\right)^2 = \frac{-\rho(z)}{\epsilon\epsilon_0} = \frac{eN_D}{\epsilon\epsilon_0}\left[1 - (1-V/\psi_F)^{3/2}\right] . \tag{7.32}$$

Since $\mathcal{E}(z) = -dV/dz$ one integration of (7.32) yields the electric field $\mathcal{E}$ in the space charge layer:

$$\mathcal{E}^2 = \frac{2eN_D}{\epsilon\epsilon_0}\int [1 - (1-V/\psi_F)^{3/2}]dV . \tag{7.33}$$

With the boundary condition

$$\mathcal{E}^2(V = 0) = 0$$

one obtains for the space-charge field:

$$\mathcal{E}(z) = \sqrt{\frac{2eN_D}{\epsilon\epsilon_0}}\left\{V(z) - \frac{2}{5}\psi_F + \frac{2}{5}\psi_F[1-V(z)/\psi_F]^{5/2}\right\}^{1/2} . \tag{7.34}$$

With $\mathcal{E}_s = \mathcal{E}(z=0)$ as the electric field at the very surface the space-charge density follows as

$$Q_{sc} = \epsilon\epsilon_0 \mathcal{E}_s$$

$$= \sqrt{2eN_D \epsilon\epsilon_0}\left[V_s - \frac{2}{5}\psi_F + \frac{2}{5}\psi_F(1-V_s/\psi_F)^{5/2}\right]^{1/2} . \tag{7.35}$$

In order to determine the spatial dependence of the potential V(z) a second integration of (7.34) is necessary. It is easy, however, to calculate the space charge capacitance (per unit area) C_{sc} by differentiating (7.35):

$$C_{sc} = \frac{dQ_{sc}}{dV_s} = \frac{(2eN_D\epsilon\epsilon_0)^{1/2}}{2}\frac{1-(1-V_s/\psi_F)^{3/2}}{\left[V_s - \frac{2}{5}\psi_F + \frac{2}{5}\psi_F(1-V_s/\psi_F)^{5/2}\right]^{1/2}} . \tag{7.36}$$

This capacitance C_{sc} describes the change in space charge Q_{sc} due to a band bending change. In a metal-semiconductor junction (Schottky barrier, Sect. 8.6) V_s might be controlled by an external bias V, and $C_{sc}(V)$ has to be taken into account for the dynamic behavior of such a junction in an electric circuit.

7.5 The General Case of a Space-Charge Layer

The general case of a space-charge layer cannot be treated in a closed mathematical form. According to *Many* et al. [7.2] the following formalism is convenient for non-degenerate semiconductors.

In terms of the reduced potential v(z), (7.6-8), Poisson's equation can be written as

$$\frac{d^2 v}{dz^2} = -\frac{e^2}{kT\epsilon\epsilon_0}(n_b - p_b + p_b e^{-v} - n_b e^{v}) . \tag{7.37}$$

With the shorthand

$$L = \sqrt{\frac{\epsilon\epsilon_0 kT}{e^2(n_b + p_b)}} \tag{7.38}$$

as an effective Debye length [different from L_D of (7.25)] and using (7.8) Poisson's equation (7.37) can be expressed as

$$\frac{d^2 v}{dz^2} = \frac{1}{L^2}\left[\frac{\sinh(u_b + v)}{\cosh(u_b)} - \tanh(u_b)\right], \tag{7.39}$$

where u, v, u_b, etc. are defined in (7.3-6). Multiplication of both sides by 2(dv/dz) as in (7.32) and one integration with the condition dv/dz = 0 at v = 0 leads to

$$\frac{dv}{dz} = \mp \frac{F(u_b, v)}{L}, \tag{7.40}$$

(minus sign for v > 0, plus sign for v < 0), where

$$F(u_b, v) = \sqrt{2}\left[\frac{\cosh(u_b + v)}{\cosh(u_b)} - v\tanh(u_b) - 1\right]^{1/2}. \tag{7.41}$$

In order to calculate the band bending v(z) as a function of the z coordinate, (7.40) has to be integrated once more, i.e.,

$$\frac{z}{L} = \int_{v_s}^{v} \frac{dv}{\mp F(u_b, v)}. \tag{7.42}$$

This integration generally has to be performed numerically. Further approximate solutions can be found in [7.2].

Figure 7.6 shows the results of a numerical integration of (7.42). The normalized potential barrier $|v|$ is plotted as a function of the distance from the surface z divided by the effective Debye length L (7.38). The bulk potential $u_b = (E_F - E_i^b)/kT$ (7.6) is used as a parameter. For accumulation layers the curves with $|u_b| \lesssim 2$ resemble that of an intrinsic semiconductor with $u_b = 0$. Inversion and depletion layers with higher $|u_b|$ values extend much deeper into the bulk. The qualitative reason for this difference is the fact that, in accumulation layers, mobile free carriers, electrons or holes, are responsible for the shape of v(z). These free carriers can be "squeezed" and therefore form a narrower space-charge layer than the spatially fixed ionized impurities in a depletion layer. In Fig. 7.6 the band bending at the surface $|v_s|$ has been assumed to be 20, but in fact every barrier having a

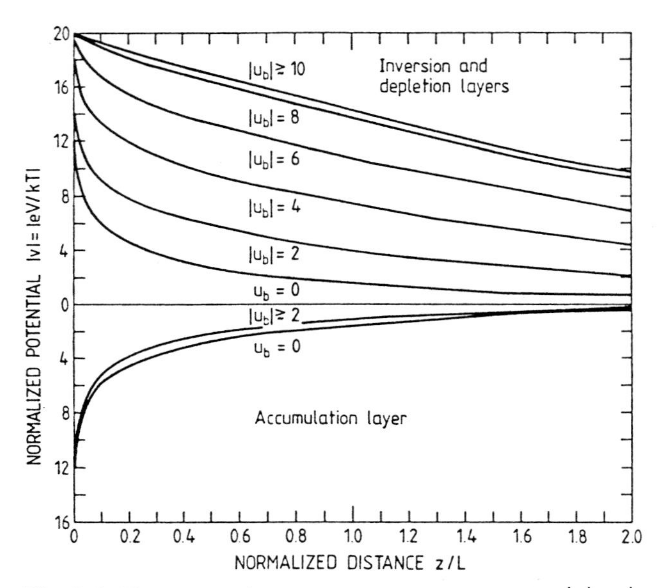

Fig. 7.6. The shape of the normalized band bending $|v| = |eV(z)/kT|$ as a function of the normalized distance from the surface z/L for various values of the bulk potential $|u_b|$. L is the effective Debye length (7.38). The potential at the surface $|v_s|$ has been taken equal to 20 [7.2]

value $|v_s|$ smaller than 20 can be evaluated simply by a translation along the z/L axis. This follows from rewriting (7.42) as

$$\frac{z'}{L} + \frac{z - z'}{L} = \int_{v_s}^{v_s'} \frac{dv}{\mp F} + \int_{v_s'}^{v} \frac{dv}{\mp F} . \tag{7.43}$$

For the new barrier height v_s' one has to measure z/L from the point where the curve for the particular $|u_b|$ intersects the horizontal line $|v| = |v_s'|$.

Another numerical calculation plotted in Fig. 7.7 shows the absolute band bending $|eV_s|$ at the surface of a semiconductor whose bulk characteristics are those of GaAs (e.g., $E_g = 1.4$ eV, etc.). Acceptor-type (A_s) and donor-type (D_s) surface states are responsible for the formation of depletion layers [7.3]. The states are described by single levels at separations E_{ss} from the conduction-band minimum and valence band maximum, respectively. The band bending at the surface eV_s is calculated as a function of the density of these surface state levels. An interesting result can be seen: below a surface-state density of $5 \cdot 10^{11}$ cm^{-2} band bending is quite small, but then, at a density of about 10^{12} cm^{-2} or a thousandth of the surface-atom density, the band bending begins to change very rapidly with increasing surface-state density. The final saturation-band bending, which is

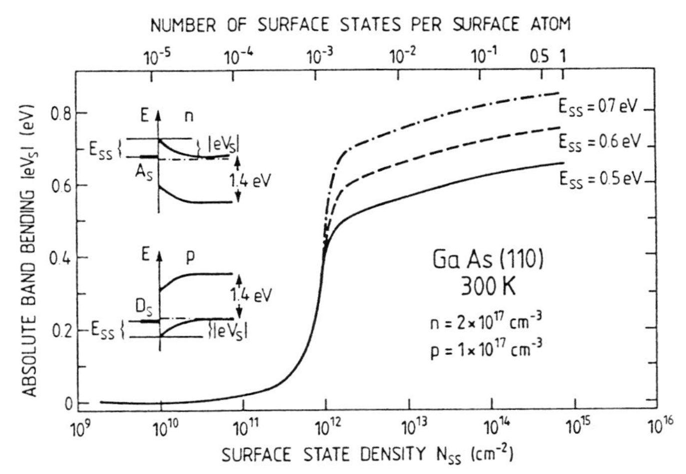

Fig.7.7. Calculated absolute band bending $|V_s|$ due to an acceptor surface-state level A_s and a donor level D_s for n- and p-type GaAs. $|V_s|$ is plotted versus the surface state density N_{ss} (*lower scale*) and related to the number of surface states per surface atom (*upper scale*). With the different definition of the energetic position E_{ss} for n- and p-type crystals (insets) the calculated curves for n- and p-type material are not distinguishable on the scale used [7.3]

directly related to the position E_{ss} of the surface-state level, is approached at an occupation of $5 \cdot 10^{12}$ cm^{-2} or a couple of hundredths of the surface-atom density. Even if the density of states increases by a further one or two orders of magnitude, the band bending does not increase any more. The reason is that the surface states are now located energetically close to the Fermi level E_F. Each increase of their density N_{ss} causes an infinitesimal increase in band bending $|eV_s|$ and simultaneous decharging of states, thus leading to a stabilization of the E_F. This effect is sometimes called **pinning of the Fermi level**. A sharp surface-state band with a density of at least 10^{12} cm^{-2} is able to establish a band bending that is similar to the maximum achievable by extremely high densities of states ($N_{ss} \approx 10^{15}$ cm^{-2}). Roughly speaking, the band bending saturates when the Fermi level crosses the surface state band; at sufficiently high density the Fermi level becomes locked or *pinned* near the surface states. Further band bending or movement of E_F relative to the band edges occurs only very slowly with increasing N_{ss}.

This effect is also evident from the following simple consideration: For a semiconductor, even with a high bulk doping level of 10^{18} cm^{-3}, the corresponding carrier density at the surface amounts to about 10^{10} cm^{-2}. For a total surface-state density as low as 10^{12} cm^{-2}, typically spread over an energy range of 100 meV, the surface-state density per energy reaches values around 10^{13} (eV)$^{-1}$cm^{-2}. The charge in these surface states being necessary to compensate the dopant surface charge thus causes a shift of

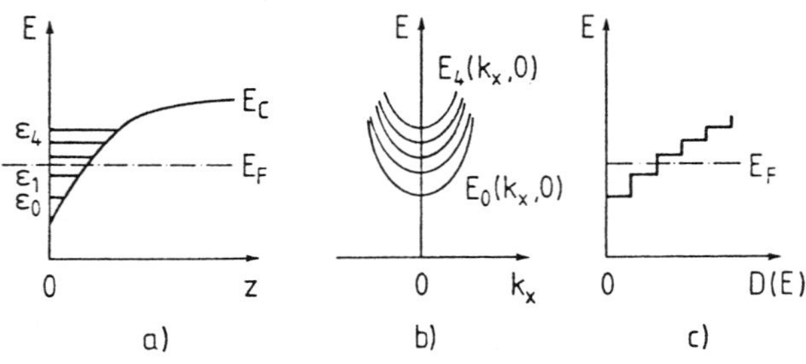

Fig.7.8a-c. Scheme of a quantized electron accumulation layer; E_C conduction-band edge, E_F Fermi level: (**a**) Conduction band edge E_C versus z coordinate normal to surface $z = 0$; $\epsilon_0, \epsilon_1, \dots, \epsilon_i \dots$ are the minima of the subbands resulting from z-quantization. (**b**) Subband parabolas $E_i(k_x, 0)$ versus wave vector k_x parallel to the surface. The minimum of E_i is ϵ_i in a). (**c**) Density of states D(E) resulting from the subband structure in (**a**) and (**b**)

the Fermi level at the surface of about $10^{10}\,cm^{-2}$ / $10^{13}\,(eV)^{-1}cm^{-2}$, i.e., 10^{-3} eV. This negligible shift means essentially a pinning of E_F.

7.6 Quantized Accumulation and Inversion Layers

According to Fig.7.6 space-charge layers, in particular those corresponding to the steepest band slope, can be very thin ($\approx$10 Å). Such narrow accumulation and inversion layers are related to strong band bending. In the case of an accumulation layer on an n-type semiconductor, whose Fermi level E_F is close to the lower conduction band edge, the conduction-band minimum might cross E_F (Fig.7.8a) such that the accumulation layer contains a degenerate free-electron gas. This electron gas will exhibit metallic behavior. The excess electron concentration at the surface (per unit area) is

$$\Delta N = \int_0^\infty [n(z) - n_b]\,dz \,. \tag{7.44}$$

This can be measured by means of the Hall effect and is essentially independent of temperature, just like the electron density in a metal. As an experimental example, Fig.7.9 shows surface electron densities ΔN measured by the Hall effect in strong accumulation layers. The layers were prepared on UHV-cleaved hexagonal $(10\bar{1}0)$ surfaces of n-type ZnO by treatment with large doses of atomic hydrogen. For the highest hydrogen dosages the downwards band bending can reach values of up to 1eV (ZnO band gap: 3.2eV) and the electron gas becomes degenerate. This degenerate electron

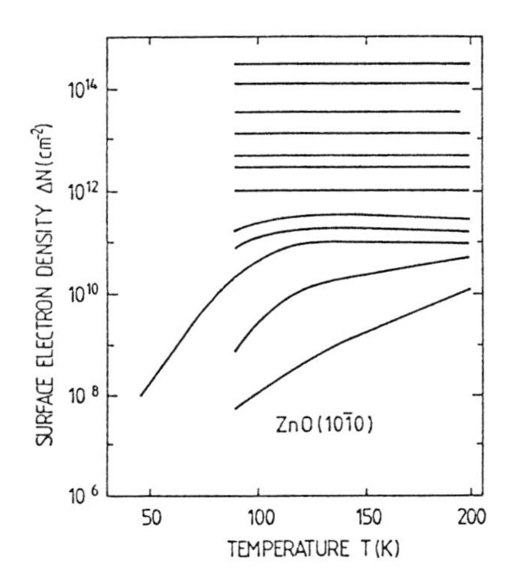

Fig.7.9. Surface density ΔN of free electrons as a function of temperature for a nonpolar ZnO (10$\bar{1}$0) surface in UHV. The accumulation layers were generated by adsorption of atomic hydrogen. ΔN was measured using the Hall effect; the different curves correspond to increasing total carrier density, i.e. to increasing band bending [7.4]

gas in a narrow accumulation layer can extend perpendicular to the surface no further than about 10 to 30 Å, i.e., in the z-direction the electron wave function cannot be the normal Bloch state found in a 3D periodic crystal. A similar situation was mentioned previously for electrons trapped in image-potential states (Sect.6.4.3). 2D periodicity is retained parallel to the surface, but along the z-direction this periodicity breaks down. The formal quantum mechanical description in the effective mass (m^*) approximation, starts from a Schrödinger equation for an electron in the accumulation layer of the type

$$\left[-\frac{\hbar^2}{2}\left(\frac{1}{m_x^*}\frac{\partial^2}{\partial x^2}+\frac{1}{m_y^*}\frac{\partial^2}{\partial y^2}+\frac{1}{m_z^*}\frac{\partial^2}{\partial z^2}\right)-eV(z)\right]\psi(\mathbf{r})=E\psi(\mathbf{r}) \qquad (7.45)$$

where the potential V(z) is only a function of z (e.g., linear in z for a "triangular" potential well, Fig.7.8a). The solutions of (7.45) are thus free electron waves for x,y parallel to the surface, and the wave function can be expressed by the ansatz

$$\psi(\mathbf{r})=\phi_i(z)\,e^{ik_x x+ik_y y}=\phi_i(z)\,e^{i\mathbf{k}_{\|}\cdot\mathbf{r}}\;. \qquad (7.46)$$

The Schrödinger equation (7.45) then separates into two equations:

$$\left[-\frac{\hbar^2}{2m_z^*}\frac{\partial^2}{\partial z^2}-eV(z)\right]\phi_i(z)=\epsilon_i\phi_i(z) \qquad (7.47a)$$

and a second equation which simply describes 2D free motion parallel to the surface

$$\left[\frac{-\hbar^2}{2m_x^*}\frac{\partial^2}{\partial x^2}-\frac{\hbar^2}{2m_y^*}\frac{\partial^2}{\partial y^2}\right]e^{ik_x x+ik_y y}=E_{xy}\,e^{ik_x x+ik_y y}\,. \tag{7.47b}$$

In (7.47a) the potential V(z) is the electrostatic potential of an electron (7.4) with the boundary conditions $V(z=0)=V_s$ and $V(z\to\infty)=0$ (7.5). In principle, V(z) has to be determined from Poisson's equation (7.9) in which the space charge $\rho(z)$ is calculated self-consistently in terms of wave functions $\phi_i(z)$ that are solutions of (7.47a) itself

$$\rho(z)=e\left[-\sum_i N_i|\phi_i(z)|^2+N_D^+-N_A^-\right]. \tag{7.48}$$

N_i is the concentration of electrons possessing the i-th energy eigenvalue ϵ_i in (7.47a). The appropriate approach is therefore a self-consistent solution of (7.47a) together with (7.9), where these equations are related to one another through (7.48). Rather than applying this complex procedure, one often approximates the space charge potential near the "bottom" of the conduction band by a linear function of z (Fig. 7.8a)

$$V(z)=-\mathcal{E}_{sc}z\,,\quad \mathcal{E}_{sc}<0\,, \tag{7.49}$$

where $\mathcal{E}_{sc}$ is the space charge field, assumed to be independent of z. This triangular potential well contains bound electronic states whose wavelength is determined by the width of the well (Fig. 7.10). The eigenvalues ϵ_i of (7.47a) are thus discrete energies, whereas E_{xy}, the solutions of (7.47b), are the energies of free electron waves propagating in the x,y-plane:

$$E_{xy}=\frac{\hbar^2}{2m_x^*}k_x^2+\frac{\hbar^2}{2m_y^*}k_y^2\,. \tag{7.50}$$

The total energy of an electron in the accumulation or inversion layer can thus be written, with $\mathbf{k}_{\|}=(k_x,k_y)$ as

$$E_i(\mathbf{k}_{\|})=\frac{\hbar^2}{2m_x^*}k_x^2+\frac{\hbar^2}{2m_y^*}k_y^2+\epsilon_i\,,\quad i=0,1,2,\dots\,. \tag{7.51}$$

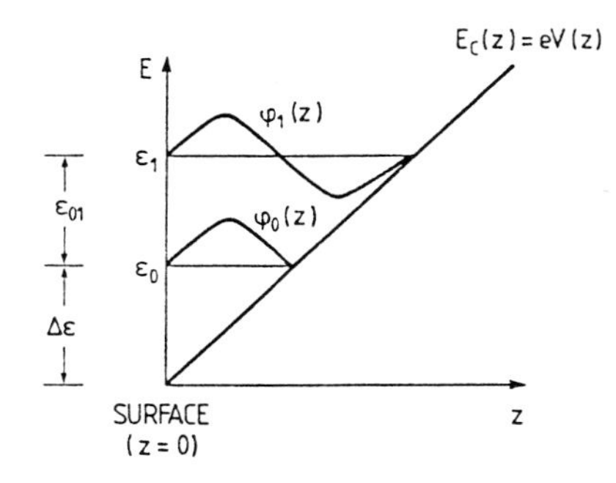

Fig.7.10. A triangular potential well used to approximate narrow electron accumulation or inversion space-charge layers. $E_C(z)$ is the conduction-band minimum as a function of z (coordinate normal to the surface). The lowest quantized states ϵ_0 and ϵ_1 are indicated with the corresponding wave functions ϕ_0 and ϕ_1 (schematic)

To each ϵ_i ($i = 0, 1, 2 \ldots$) there belongs a 2D parabolic band, such that along k_x a series of parabolas is obtained (Fig.7.8b). These different 2D bands are called **subbands**. A 2D parabolic band has a constant density of states $D(E) = dZ/dE$ as is easily shown: the number of states dZ per unit area within a ring of thickness dk and radius k in (k_x, k_y) space is

$$dZ = \frac{2\pi k dk}{(2\pi)^2} . \tag{7.52}$$

Since $dE = \hbar^2 k dk/m^*$ one obtains with spin degeneracy (a factor 2):

$$D = dZ/dE = m^* / \pi\hbar^2 = \text{const} . \tag{7.53}$$

The total density of states of the subband series in Fig.7.8b thus consists of a superposition of constant contributions, each belonging to one subband $E_i(\mathbf{k}_\parallel)$ (Fig.7.8c).

The position of ϵ_0 with respect to the bottom of the conduction band ($\Delta\epsilon$ in Fig.7.10) is easily estimated by means of the uncertainty principle

$$d_0 p_0 \simeq \hbar , \tag{7.54}$$

where d_0 is the width of the potential well for the particular wave function whose momentum is p_0. With the space charge field $\mathcal{E}_{sc}$ one has, according to Fig.7.10,

$$\epsilon_0 = \frac{{p_0}^2}{2m_\perp^*} \simeq ed_0 |\mathcal{E}_{sc}| . \tag{7.55}$$

With $p_0 = (2m_\perp^* \epsilon_0)^{1/2}$ and using (7.54) the lowest subband energy, referred to the conduction-band minimum, follows as

$$\epsilon_0 \simeq \frac{(\hbar e)^{2/3}}{(2m_\perp^*)^{1/3}} |\mathcal{E}_{sc}|^{2/3} , \qquad (7.56)$$

where $m_\perp^*$ is the principal effective mass normal to the surface.

With typical values for $|\mathcal{E}_{sc}|$ and $m_\perp^*$ of 10^5 V/cm and $0.1 m_0$, respectively, one estimates ϵ_0 to be about 30 meV and the "size" d_0 of the corresponding state to be approximately 30 Å. The exact solution of the Schrödinger equation (7.47a) with a triangular potential well (7.49) (Fig.7.10) yields the following spectrum of energy eigenvalues

$$\epsilon_i = \left[\frac{3}{2}\pi\hbar e\right]^{2/3} \frac{|\mathcal{E}_{sc}|^{2/3}}{(2m_\perp^*)^{1/3}} \left[i + \frac{3}{4}\right]^{2/3} , \quad i = 0,1,2,3,\dots . \qquad (7.57)$$

Comparing this with (7.56) we see that our estimate for $\epsilon_0 (i = 0)$ was actually rather good.

It should be emphasized that our simple considerations up to this point are most applicable to inversion layers, where the carriers in the quasi-2D gas are separated from the free carriers in the bulk by a sharp zone of depletion (Fig.7.2b, second column). In a strong accumulation layer there is no such barrier containing the bulk carriers. Both the electrons bound to the surface in the subbands and those whose motion normal to the surface is not quantized, contribute to the self-consistent potential and must be considered in (7.48).

Another complication arises in indirect-gap semiconductors such as Si, Ge and GaP, for which the conduction band minima are not in the center of the Brillouin zone. For Si, for example, the surfaces of minimal constant energy in **k**-space are ellipsoids around the k_x, k_y, k_z axes, thus yielding two different effective electron masses m_ℓ (longitudinal) and m_t (transversal). Different lattice planes cut the energy ellipsoids differently. On a (100) surface the projections of two ellipsoids are circles whereas the other four are ellipses. According to (7.56, 57) the effective mass component normal to the surface enters into the subband energies. On the Si(100) surface one thus has two subband series according to (7.57) which differ in their effective masses normal to (100). The same is true for the Si(110) surface. A (111) surface, however, cuts all six ellipsoids in the same way. Thus only one mass component normal to (111) has to be regarded, and a single series of subbands exists.

For both accumulation and inversion layers, there is a remarkable difference between the free electron density n(z) calculated in a completely

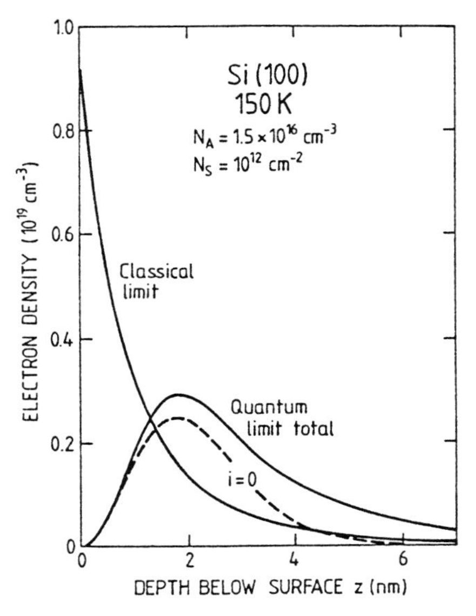

Fig.7.11. Classical and quantum-mechanical charge density for a Si(100) inversion layer at 150 K with 10^{12} electrons per cm^2 and a bulk acceptor doping of $1.5 \cdot 10^{16}$ cm^{-3}. The dashed curve shows the contribution of the lowest subband to the quantum-mechanical charge density [7.5]

classical way (Sects.7.2-4) and that, for which z-quantization has been taken into account. In the classical description, the charge density depends only on the local separation of the band edge $E_C(z)$ from the Fermi level E_F; it must thus have a maximum at the very surface, where this separation is smallest and largest, respectively, for accumulation and inversion layers (Fig.7.11). On the other hand, in the quantum-mechanical description, the free electron density n(z) is given by $\Sigma_i N_i |\phi_i(z)|^2$ (7.48), and the wave functions $\phi_i(z)$ for each subband i must have a node at the surface (Fig. 7.10); i.e. n(z) must vanish at the surface. The calculated electronic charge density n(z) for Si(100) in Fig.7.11 clearly shows this behavior [7.5]. For this particular example, most of the total charge is contained in the lowest subband ($i = 0$). The average distance of the charge from the surface is greater when calculated quantum mechanically than when calculated classically.

7.7 Some Particular Interfaces and Their Surface Potentials

As was pointed out in Sect.7.5, surface or interface densities of states in the forbidden bulk band exceeding about 10^{12} cm^{-2} are usually high enough to pin the Fermi level E_F at the surface or interface, i.e. the position of E_F becomes locked with respect to the bulk band edges E_C and E_V. For a particular interface with high density of states in the gap, there is a fixed sur-

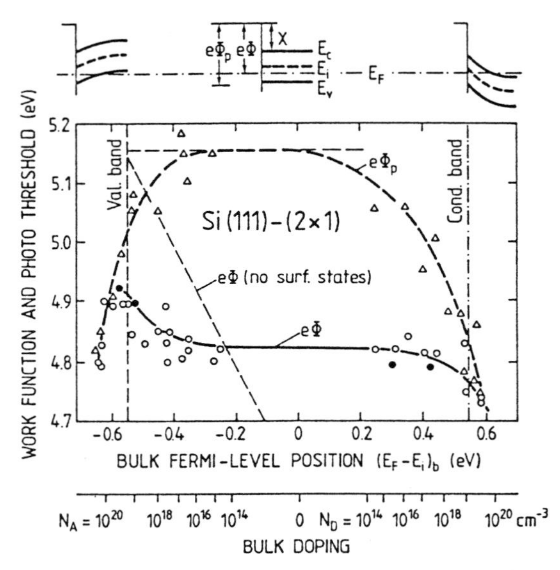

Fig.7.12. Work function $e\phi$ and photo-threshold $e\phi_p$ (explanation in the band scheme *above*) of the cleaved Si(111)-(2 × 1) surface as a function of bulk doping (p- and n-type) and bulk Fermi level position $(E_F - E_i)_b$ [7.6]

face potential $\phi_s = e^{-1}[E_F - E_i(0)]_{surf}$ that is characteristic for the particular surface. The band bending $eV_s = e(\phi_s - \phi_b)$, (7.5), is then determined by ϕ_b, i.e. by the bulk doping. This is clearly the case for a clean **Si(111) surface with the (2 × 1) superstructure**, obtained by cleavage in UHV. Figure 7.12 shows the classic results of *Allen* and *Gobeli* [7.6] who measured the work function $e\phi = E_{vac} - E_F$ and the photothreshold $e\phi_P = E_{vac} - E_V$ on Si(111)- (2 × 1) after cleavage in UHV. As is expected for Fermi level pinning both ϕ and ϕ_p remain nearly constant over a wide range of bulk doping levels. Consistent with this, the surface potential ϕ_s is also independent of temperature between 130 and 350 K. The situation for Si(111)-(2 × 1) is more clearly explained in Fig.7.13. The bulk doping level determines the energetic distance between E_F and E_C deep in the bulk ($z \gg 0$). According to Sect.6.5.1, there is a significant density of dangling-bond surface states reaching into the forbidden band. Two bands, one of occupied π states (donors), and the other of empty π^* states (acceptors) give rise to densities of states of the form shown qualitatively in Fig.7.13; they are separated by an absolute gap of about 0.4 eV width (see also Fig.6.35). The surface density of dangling-bond orbitals corresponds to a monolayer and is thus higher than 10^{14} cm^{-2}. This is large enough to pin the surface Fermi level E_F at an energy between the two dangling bond bands. From experiment, the pinning position $(E_F - E_V)_{surf}$ is found to be about 0.35 eV [7.6, 8] (Fig.7.13). The band bending must adjust so that on

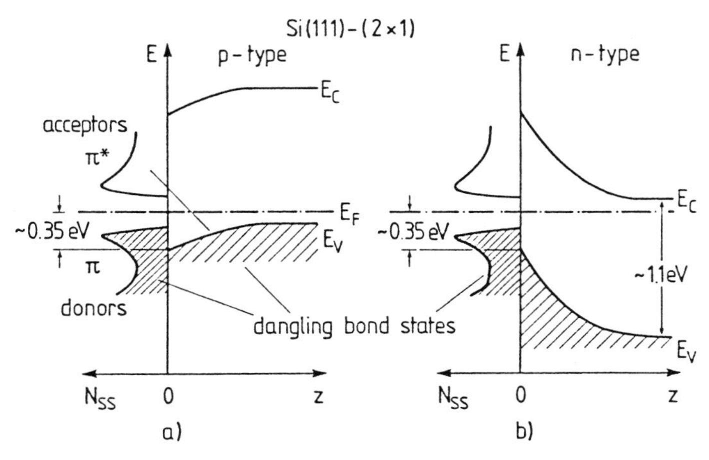

Fig. 7.13. Qualitative surface state density N_{ss} and band scheme of the cleaved Si (111)-(2×1) surface for p-doped (**a**) and n-doped (**b**) material. π and π^* dangling-bond states are derived from the π-bonded chain model [7.7]

p-type Si a hole depletion is formed (Fig. 7.13a) whereas on highly n-doped material a p-inversion layer $(E_i - E_F)_{surf} > 0$, appears (Fig. 7.13b).

The **Ge(111)-(2×1)** surface prepared by cleavage in UHV exhibits properties similar to the Si(111)-(2×1) surface with respect to its surface states (Sect. 6.5.1). Two bands of occupied and empty π and π^* state are responsible for pinning E_F at about 40 meV above the valence band edge [7.9]. For a comparison with Si one has to take into account the smaller band gap for Ge of about 0.7 eV at 300 K.

Si(100)-(2×1) surfaces, for which the reconstructions are described by dimer models (Sect. 6.5.1), are prepared by argon ion bombardment and annealing in UHV, or by only annealing. The distribution of surface states on this surface is not as well established as for the Si(111)-(2×1) surface, but according to Sect. 6.5.1 (Fig. 6.41a,c) the surface is semiconducting similar to Si(111)-(2×1). Accordingly, the pinning position of the Fermi level at the surface is not very different. $(E_F - E_V)_{surf}$ is found to be about 0.4 eV, i.e. comparable to that on Si(111)-(2×1) [7.10, 11]. The properties of space charge layers are thus similar to those described in Fig. 7.13.

The **Si(111)-(7×7)** surface is the most stable surface. It can be produced by annealing a cleaved (2×1) reconstructed surface to above 350°C or by argon ion bombardment and annealing. Details of the geometrical and electronic structure are known (Sect. 6.5.1). Photoemission (Fig. 6.38) and inverse photoemission studies [7.12] indicate that the distribution of electronic surface states roughly resembles that in Fig. 7.14. The surface probably possesses a nearly continuous distribution of surface states (metal-like surface) with a peak at about 0.7 eV above the valence-band edge E_V

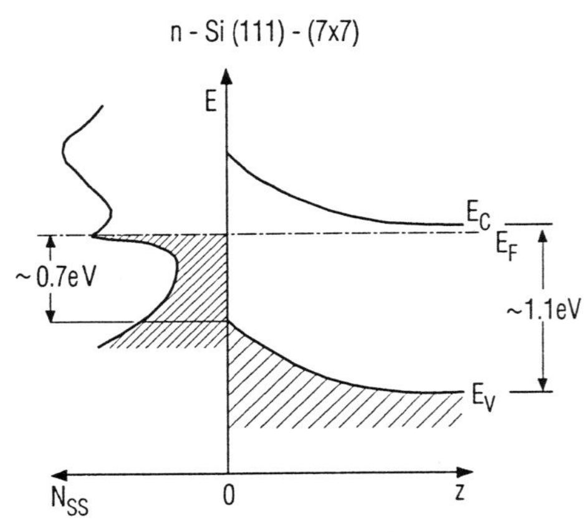

Fig.7.14. Qualitative surface state density N_{ss} and band scheme of the Si (111)-(7 × 7) surface (n-doped material) [7.12]

(surface-state band S_1 in Fig.6.38). The density of states is high enough to effectively pin the Fermi level at about 0.70 eV above the valence band edge, i.e. near midgap [7.12]. One therefore expects electron depletion layers on n-type Si and hole depletion on p-type material.

With respect to semiconductor device technology the most important Si interface is the **Si/SiO$_2$ interface** [7.13]. Extensive studies have therefore been dedicated to achieving a deeper understanding of the oxidation process of Si surfaces. On the basis of such studies, preferably performed on surfaces prepared in UHV, detailed models of the oxidation process have been developed. High-Resolution Electron Energy Loss Spectroscopy (HREELS) [7.14] clearly shows the different stages of oxidation of a Si(111)-(7 × 7) surface. The data suggest the following model: Exposure to oxygen at 700 K first leads to monolayer adsorption of atomic oxygen (Fig.7.15a) with the oxygen atoms bonded at bridge sites to two surface Si atoms. Higher exposures in the 10^3 L range cause a penetration of oxygen into the second atomic layer. The HREELS spectra can be rationalized by assuming oxygen at bridge sites with a bonding configuration as in Fig.7.15b [7.14]. For even higher exposures, the loss spectrum develops into that of Fig.7.15c, which, by comparison with optical data, clearly indicates the presence of a "thick" oxide layer of SiO_2. This native oxide layer on Si is amorphous, no long-range order is found in any scattering technique. But the layer is of very high quality, both in the sense of being homogeneous and closed and of having an extremely low defect level density at the Si/SiO_2 interface. In high-resolution electron micrographs, this interface appears sharp to within one or two atomic layers. The perfection of the interface is especially important for its electrical properties. Oxidation leads to the saturation of dangling bond orbitals, the surface states of Figs.7.13,14 disappear from the forbidden band and new chemisorption bonding and antibonding states are formed at energies far below E_V and above E_C, respectively. These states

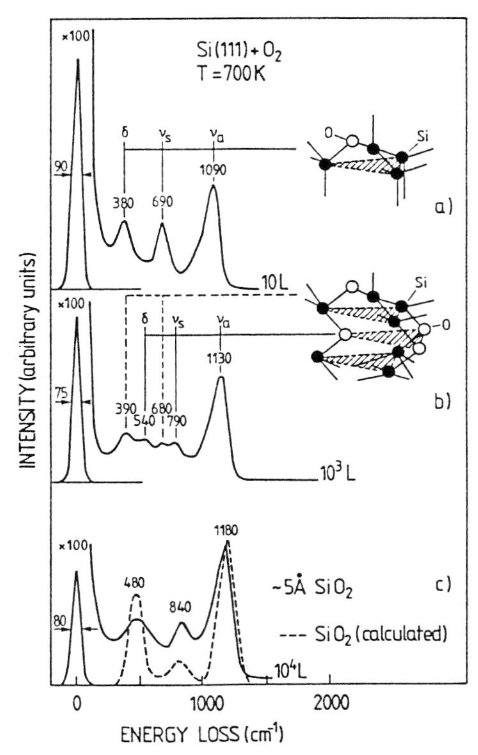

Fig. 7.15a-c. Energy loss spectra (HREELS) of the Si(111)-(7×7) surface after oxygen exposure at 700 K (primary energy E_0 = 15eV, specular reflection geometry). The interpretation of the observed vibrational losses is given on the right-hand side in terms of structure models. For clarity, Si atoms have been assumed to lie at bulk-like positions. (**a**) In the first fast adsorption stage (10 L) atomic oxygen bonds between Si atoms within the topmost Si layer. (**b**) In the subsequent slow chemisorption stage (10^3 L) oxygen penetrates into the Si lattice and is additionally bonded on bridge sites between the first and second Si(111) double layers. (**c**) Spectrum for even higher exposures (>10^4 L oxygen). A comparison with a spectrum calculated from optical data of vitreous SiO_2 (broken line) indicates the presence of a SiO_2 overlayer. The thickness (≈5 Å) was measured by AES [7.14]

are always occupied and empty, respectively; they are no longer of interest for the electrical properties.

As expected from this simple picture, no significant density of interface states is found in the forbidden band. Recent preparation and measurement techniques show density values N_{IS} as low as 10^8 cm^{-2}eV^{-1} near midgap. Over the whole forbidden band one estimates a total areal density of less than 10^{10} cm^{-2}. This is such a low value that it does not affect the position of the Fermi level at the interface. Not enough charge is trapped in these interface states to cause any pinning of E_F. The Fermi level moves

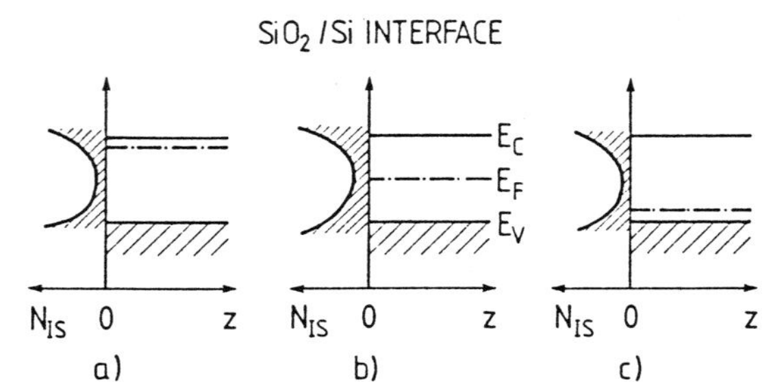

Fig. 7.16a-c. Qualitative plot of the interface state density N_{IS} and the band scheme near the SiO_2/Si interface for different bulk dopings, i.e. different Fermi level positions: (**a**) n-type Si; (**b**) nearly intrinsic Si; (**c**) p-type Si

freely with changes of doping and temperature over the entire forbidden gap (Fig. 7.16). Nevertheless, a major effort is being made to understand the nature of the remaining electronic interface states at this nearly perfect SiO_2/Si interface and to further decrease their density. As will be shown in the next chapter, the performance of important Si devices can be improved by lowering the density of these interface states. As is qualitatively plotted in Fig. 7.16, the distribution of these interface states is U-shaped; it has increasing tails towards the band edges E_V and E_C, and a minimum near midgap. Figure 7.17 depicts a characteristic example of interface-state densities measured some years ago [7.15]. Compared with recent results, a relatively high density of states is found. Today it is possible to prepare inter-

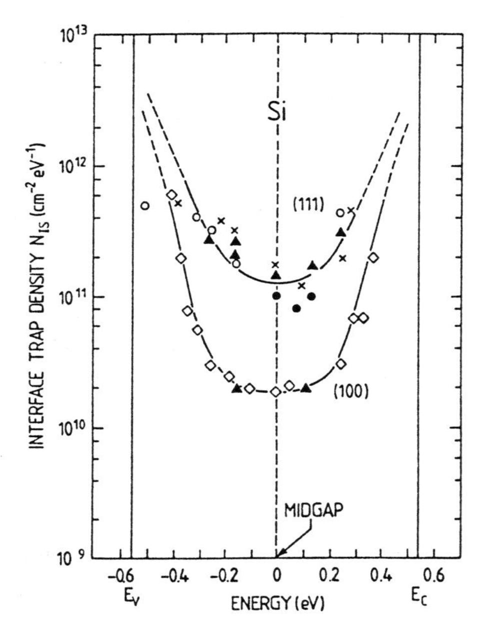

Fig. 7.17. Interface state density in thermally oxidized Si for two different surface orientations. E_V and E_C are the valence- and conduction-band edges [7.15]

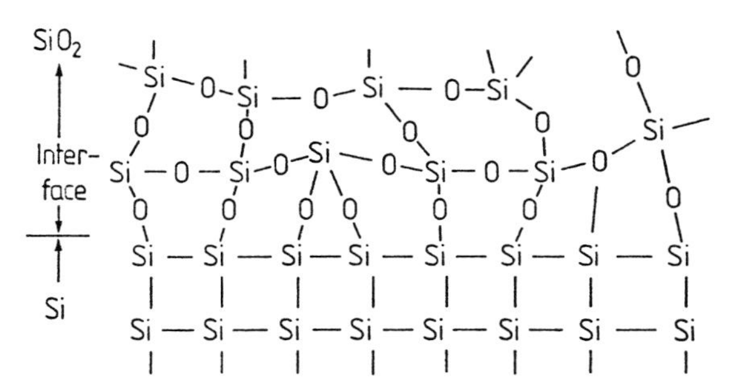

Fig. 7.18. Schematic model of the microscopic Si/SiO_2 interface in which bonds of silicon and oxygen are distorted to match the silicon lattice. An unsaturated Si dangling bond is also shown

faces whose trap densities are about two orders of magnitude lower. But the U-shape of the distribution is clearly resolved. There is a clear orientation dependence; on (111) surfaces the trap density is about an order of magnitude higher than on the (100) surface. This result can be correlated with the number of available bonds per unit area on the Si surface. The orientation dependence is the reason why all modern Si field effect transistors (Sect. 7.8) are fabricated on (100)-oriented substrates.

The models presented thus far to describe the origin of interface states, have been based on the assumption of unsaturated *dangling* bonds at the interface. Figure 7.18 depicts a possible model for such interface defects. In addition to the free Si dangling bonds, silicon and oxygen bonds in the oxide are distorted at the interface to match the silicon lattice. In the chemist's language: near the interface, the oxidation is incomplete. Instead of having only Si^{+4} as in SiO_2, at the interface Si is also found in its Si^{+1}, Si^{+2} and Si^{+3} states. This can be concluded from XPS studies [7.16, 17].

Surfaces of *III-V compound semiconductors* display completely different behavior to the Si surfaces just discussed. Let us consider, as an example the much studied compound GaAs.

The **GaAs(110) surface** can be prepared by cleavage in UHV. As has been shown by numerous UPS and work-function measurements, no pinning of the Fermi level occurs on good mirror-like cleaves of either n- or p-type GaAs. The variation of the Fermi level at the surface as a function of bulk doping is similar to that shown in Fig. 7.16 for the Si/SiO_2 interface. To within the experimental accuracy flat bands are found, independent of the type of bulk doping; the Fermi level at the surface can move over the entire forbidden band as the bulk doping is varied from high n-type (n^+) to high p-type (p^+). This means that the density of surface states in the gap must be extremely low on well-cleaved GaAs(110) surfaces. As was previously discussed in Sect. 6.5.2, this surface displays a kind of buckling (1×1)

reconstruction with the As surface atoms shifted outwards and the Ga atoms inwards in respect to their bulk position (Figs.6.43). This reconstruction is connected with a surface-state distribution, in which the As and Ga dangling-bond surface states are essentially degenerate with the bulk-valence and conduction-band states, respectively (Fig.6.44). The gap is free of intrinsic states and merely contains a few defect-induced extrinsic surface states. In this sense the good, clean, UHV-cleaved GaAs(110) surface more closely resembles the Si/SiO_2 interface, than the good, cleaved Si(111) surface.

It has been shown theoretically by a number of different approaches (Sect.6.5) that an "ideal" GaAs(110) surface, whose atomic configuration is that of the truncated bulk, is related to a different distribution of surface states (Fig.6.43). For this surface the As- and Ga-related dangling-bond states shift into the gap, the Ga-derived empty states probably down to a position near midgap, whereas the occupied As states are probably located in the lower half of the forbidden band. This configuration leads to pinning of the Fermi level. The acceptor-type Ga states (near midgap) cause depletion layers on n-doped GaAs, whereas the donor-type As states are responsible for hole depletion on p-type material.

A qualitatively similar situation is also obtained when a perfectly cleaved GaAs(110) surface with initially flat bands is ion bombarded, etched or exposed to one of a number of gases (oxygen, atomic hydrogen etc.). For all these treatments the Fermi level is pinned somewhere between 0.65 and 0.8 eV for n-type GaAs, and between 0.45 and 0.55 eV for p-type material, above the valence band maximum E_V (Fig.7.19). Similar pinning positions are also found after deposition of a variety of metals on the GaAs (110) surface (Chap.8). The fact that the GaAs(110) surface responds similarly to the different surface treatments suggests that in each case defect surface states (acceptors on n-type and donors on p-type) are responsible for the formation of the depletion layers. As and Ga vacancies and also antisite defects As_{Ga} and Ga_{As}, have been proposed as likely agents (Sect.8.4).

The idea that a certain degree of nonstoichiometry is responsible for the surface states on GaAs(110) is also supported by the observation of space-charge layers and Fermi-level pinning on **GaAs(001) surfaces** grown by Molecular Beam Epitaxy (MBE) [7.18]. In MBE (Sect.2.4) the stoichiometry of a growing surface can be varied within certain limits by controlling the As/Ga supply ratio. Surfaces with different stoichiometry are usually characterized by different LEED or RHEED patterns, i.e. superstructures. For GaAs(001) the observed (4×6), $c(2 \times 8)$, $c(4 \times 4)$ patterns correspond to increasing surface As concentrations, as revealed by AES or XPS. From Fig.7.19 it is evident that in all cases of MBE-grown GaAs(001) surfaces, similar Fermi-level pinning is obtained as for GaAs(110) surfaces, although a slight shift of the pinning energies towards the conduction-band

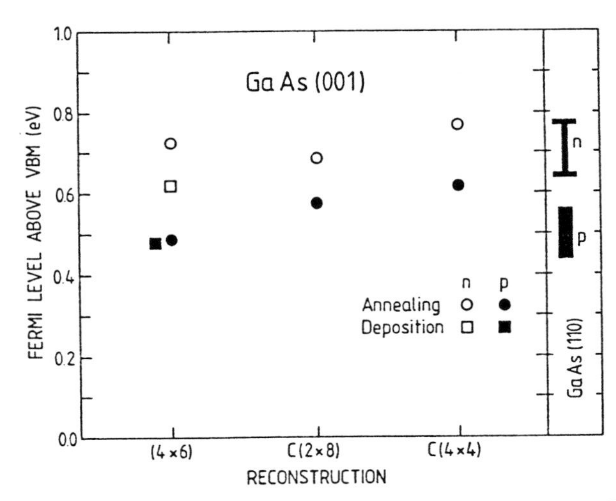

Fig. 7.19. Pinning of the Fermi level above the Valence Band Maximum (VBM) for n- and p-type GaAs (001) with three different surface reconstructions produced in MBE [7.18]. For comparison the Fermi level-pinning positions on ion bombarded, etched, and O- and H-covered GaAs(110) surfaces are indicated by dark (p-type) and open (n-type) bars

minimum E_C occurs with increasing As content. The same type of surface states as on GaAs(110), probably defect states, are likely to be responsible for this effect.

A somewhat different situation has been found for the cleaved (110) surfaces of the **narrow gap III-V semiconductors** InAs (gap $\approx$ 340 meV) and InSb (gap $\approx$ 180 meV). In both cases the clean surfaces appear to have flat bands, independent of the level and type of doping, i.e. a negligible density of surface states exists within the forbidden band. However, adsorption of gases, ion bombardment, or deposition of metals pins the Fermi level at energies within the bulk conduction-band range, i.e. above the lower conduction-band edge E_C [7.19]. For InSb(110) the pinning position at the surface is $(E_F - E_C)_{surf} \approx$ 100 meV, and for InAs(110) it amounts to about 150 meV [7.19, 20].

ZnO surfaces are typical both of many metal oxides and of II-VI compound semiconductors. ZnO is always found as an n-type semiconductor with a band-gap energy of $\approx$3.2 eV (300 K). Three types of surfaces can be prepared by cleavage in UHV: the non-polar $(10\bar{1}0)$ prism face as well as the polar (0001)Zn and $(000\bar{1})$O faces. After cleavage in UHV all three surfaces show flat bands or weak electron depletion layers. For the polar O face the situation is depicted in Fig. 7.20 [7.21]. The band bending seen here is established by a low density of surface states at 0.38 ÷ 0.45 eV below the conduction-band edge. Treatment with atomic hydrogen produces

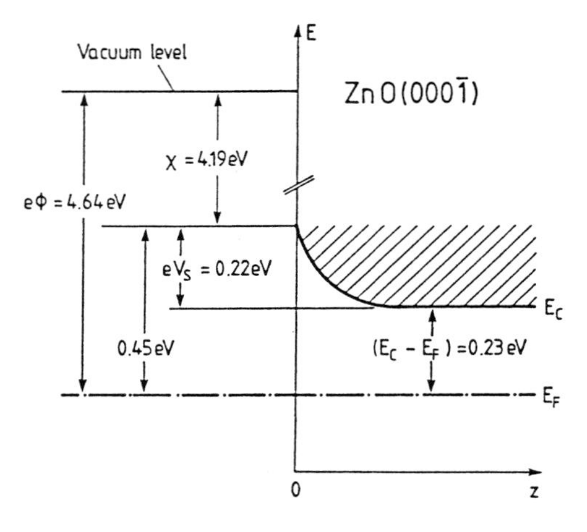

Fig. 7.20. Energy-band scheme near the conduction band edge for the polar (000$\bar{1}$)O face of ZnO after cleavage in UHV. After subsequent annealing in UHV at temperatures up to 800 K the same scheme remains valid [7.21]

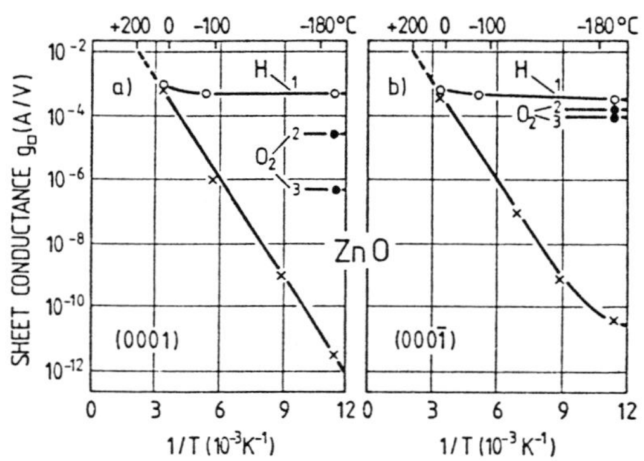

Fig. 7.21a,b. Sheet conductance $g_{\text{sheet}} = \int_0^d dz\mu n(z)$ versus temperature measured after cleavage in UHV for polar surfaces of ZnO: (**a**) (0001) Zn surface; (**b**) (000$\bar{1}$)O surface. After cleavage, the clean surface (x) was exposed to atomic H (curve *1*), then O_2 was admitted at $3\cdot10^{-3}$ Torr for 10 min (curve *2*) and subsequently at 10^{-2} Torr for 10 min (curve *3*) [7.23]

strong accumulation layers on all three types of surface. Downward band bending of as much as 1 eV has been observed. In contrast to the usual exponential dependence of the bulk conductivity on 1/T (Fig. 7.21) the sheet conductance containing essentially the effect of the degenerate electron gas in the accumulation layer is nearly independent of temperature T. Hydrogen acts to reduce the surface and probably produces a Zn excess, which is related to the formation of donor-like defect surface states [7.22]. From the observed band bending one can deduce that these states lie at least 0.4 eV above the conduction-band edge, i.e. they are degenerate with bulk conduction-band states.

Oxygen has the opposite effect on ZnO surfaces (Fig. 7.21). For both polar surfaces oxygen can at least partially remove the effect of hydrogen adsorption. The removal of the accumulation layer proceeds faster on the Zn surface, but the saturation values are the same for the two polar faces. On all three types of clean ZnO surfaces chemisorbed oxygen forms electron depletion layers with upward band bending. The oxygen adsorbs as a negatively charged species which binds electrons and therefore acts as a surface acceptor.

To conclude this section, let it be once more emphasized that there is an important and probably general difference between the clean surfaces of most elemental semiconductors and those of compound semiconductors. For the latter class of materials, defect surface states appear to play the major role in determining the Fermi-level position at the surface. On the other hand, for Si and Ge, it is intrinsic dangling bond states that are important.

7.8 The Silicon MOS Field-Effect Transistor

The most important direct application of space-charge layers is in field-effect devices. **Field-effect transistors** are basically voltage- or electric-field-controlled resistors. Since their conduction process involves essentially one kind of carrier in a space-charge layer, field-effect transistors are called **unipolar**, in contrast to p-n junction derived devices, such as the npn transistor. There, the current is carried by both electrons and holes, and consequently the device is known as a **bipolar transistor**. There is a whole family of Field Effect Transistors (FETs) of which two major classes are the Metal-Semiconductor Field-Effect Transistors (MESETs) and the Metal-Oxide Semiconductor Field-Effect Transistors (MOSFETs). The function of a MESFET is based on the external control of a Schottky depletion space-charge layer beneath a metal semiconductor contact (Sect.8.6). In contrast, in the MOSFET the space-charge layer beneath an insulating surface oxide is controlled by an external voltage. By far the most advanced MOSFET, also with respect to its possible degree of circuit integration, is the Si MOSFET. Its basic structure is illustrated in Fig.7.22. For this so-called n-channel device the substrate is p-type Si, into which two n^+ - doped (n^+ means high level of n-type doping) regions, the source and drain, are formed (e.g., by ion implantation). Metal films on top serve as contacts. These two n^+ regions are separated from each other by the so-called channel (length L). In operation this channel carries the current between source and drain. It is a space-charge layer which is separated from the third metal contact, the gate, by a SiO_2 insulating film. This film consists of native

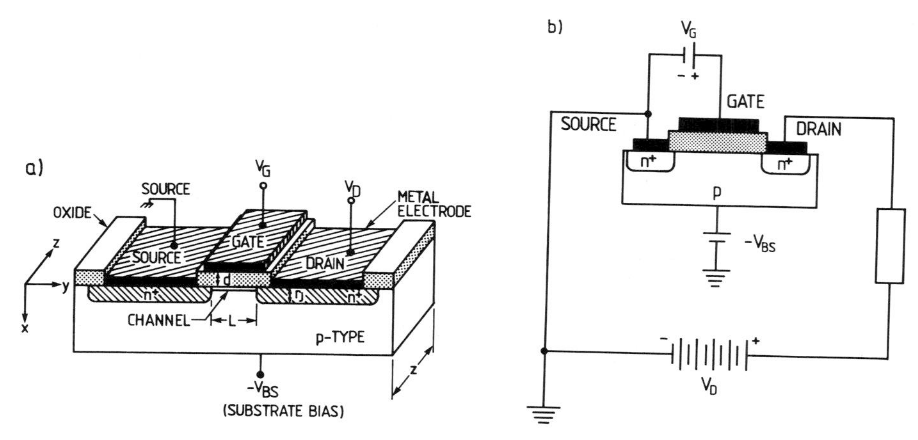

Fig. 7.22a,b. Si metal-oxide-semiconductor field effect transistor (MOSFET). (**a**) Schematic diagram of the device [7.24]. (**b**) Schematic circuit showing the MOSFET and the voltages relevant to its performance

oxide which is formed by heat treatment during the fabrication process. In particular the Si/SiO_2 interface which was discussed briefly in Sect. 7.7 has an important influence on the performance of the MOSFET. In contrast to the studies discussed in Sect. 7.7, however, the oxide barrier in MOS devices is not usually prepared in UHV systems but, for practical reasons, in stream reactors. On a MOSFET there is a fourth contact, the substrate contact, on the back surface of the Si substrate. It is usually used as a reference. Figure 7.22b shows schematically the simplest circuit for operating a MOSFET. When no voltage is applied to the gate ($V_G = 0$), the Si substrate is p-type up to the Si/SiO_2 interface below the gate (Fig. 7.23). The source-to-drain electrodes correspond to two p-n junctions connected back-to-back. The channel has an extremely high resistance. When a sufficiently high positive bias is applied to the gate, negative charge is induced across the oxide, electrons build up in the channel to the level where inversion occurs (Fig. 7.23b). An electron inversion layer (or channel) has been formed between the two n^+ regions. Source and drain are then connected by a conducting n-type channel and a current, controlled by V_G, can flow between source and drain. With source and drain at the same potential, i.e. $V_D = 0$ in Fig. 7.22b, the band scheme of Fig. 7.23b would be valid over the entire gate length. In operation, however, a controllable current flows between source and drain, and the drain contact is thus biased with respect to the source. A complete description of this nonequilibrium condition requires a two-dimensional band scheme $E(x, y)$ with x being the coordinate normal to the Si/SiO_2 interface and y parallel to the channel (Fig. 7.24). The band scheme depicted in Fig. 7.23b thus corresponds to Fig. 7.24c and qualitatively also to spatial regions in the channel close to the source contact in

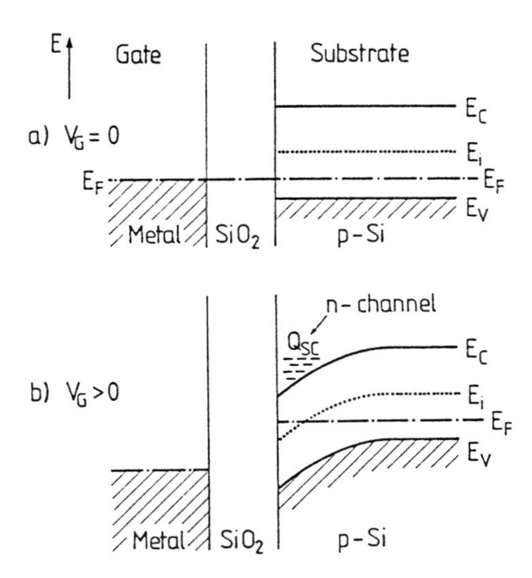

Fig. 7.23a,b. Band diagram of a MOSFET along a line normal to the gate electrode (x direction in Fig. 7.22a). Interface states between Si and SiO_2 and oxide traps within the SiO_2 layer are neglected. (**a**) At zero gate voltage ($V_G = 0$) the bands are essentially flat. (**b**) For sufficiently high positive gate voltage ($V_G > 0$) an electron inversion layer is established and the channel is opened

Fig. 7.24d. Note that in the non-equilibrium case (Fig. 7.24d) the Fermi level splits into quasi-Fermi levels E_{Fp} and E_{Fn} along the channel. The voltage required for inversion at the drain is larger than in equilibrium, since the applied drain bias lowers the electronic quasi Fermi level E_{Fn}; an inversion

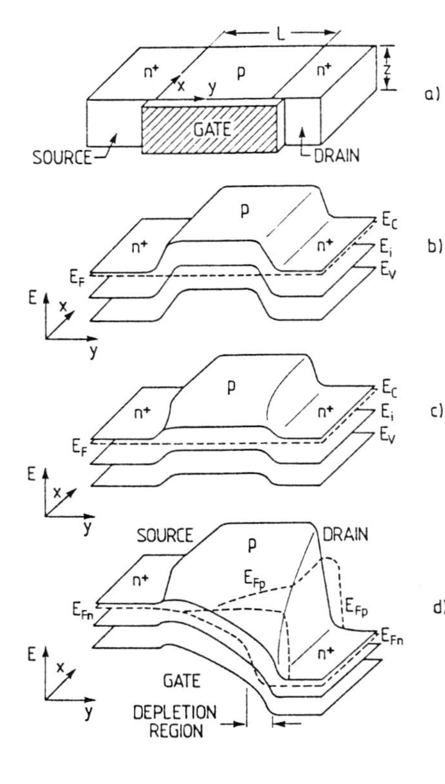

Fig. 7.24a-d. Two-dimensional band diagram of an n-channel MOSFET as in Fig. 7.22. The x,y plane of (E, x, y) coordinate system is the same as in Fig. 7.22. The intrinsic energy E_i, Fermi energy E_F and conduction- and valence-band edges E_C and E_V are plotted versus coordinates x and y normal and parallel to the gate electrode. (**a**) Device configuration. (**b**) Flat-band zero-bias equilibrium condition as in Fig. 7.23a. (**c**) Equilibrium condition under a gate bias ($V_G \neq 0$) as in Fig. 7.23b. (**d**) Non-equilibrium condition under both gate and drain biases. A current flows through the channel and in the region of current flow the Fermi-level splits into quasi-Fermi levels for electrons E_{Fn} and holes E_{Fp} [7.25]

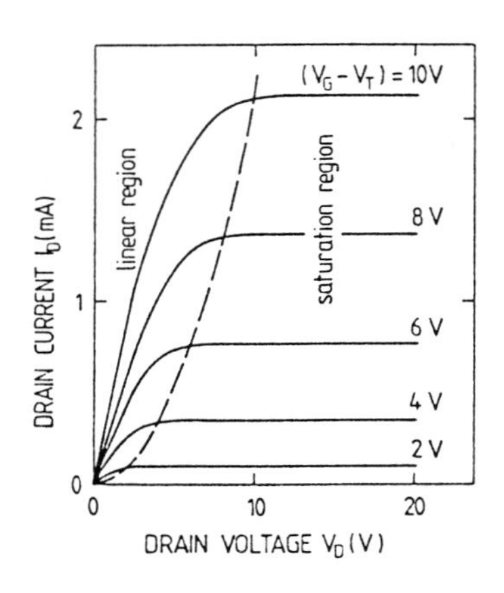

Fig.7.25. Idealized drain current characteristics (I_D versus V_D) of a MOSFET. V_T is the threshold voltage at which channel conductance appears. The dashed line describes the locus of the saturation drain voltage [7.26]

layer can be formed only when the potential at the surface crosses E_{Fn}. A detailed mathematical treatment of the MOSFET has been given by *Sze* [7.26]. Based on a simple description of the space-charge layer, as in Sects. 7.2-4, one can derive the drain current I_D as a function of drain voltage V_D (Fig.7.25). A certain threshold voltage V_T is necessary to open the n-channel; for increasing gate voltages above the threshold voltage ($V_G - V_T > 0$) increasing drain currents are obtained because the carrier concentration in the inversion layer increases. For a given V_G the drain current first increases linearly with V_D (linear region), then gradually levels off and approaches a saturation value (saturation region) which is determined by the maximum available carrier density in the channel at the particular gate voltage V_G. The dashed line indicates the locus of the drain voltage at which saturation is reached.

So far, our treatment has completely neglected the existence of interface states at the Si/SiO_2 interface, and of bulk trap states within the SiO_2 layer itself. Such interface states, if their density is too high, cause the functioning of a MOSFET to deteriorate considerably. If in Fig.7.23 a continuous distribution of states with high density in the Si gap has to be taken into account, these states have to cross the Fermi level during the formation of the band bending in Fig.7.23b (formation of inversion layer). A large number of interface states must be charged because they have shifted below the Fermi level. The charge required to fill these states is missing in the inversion layer, and is not effective for increasing the channel current. The charge Q_G supplied to the gate-metal contact by the action of an applied gate voltage V_G has to compensate both the charge in interface states Q_{IS} and that in the space charge region Q_{SC} (channel):

$$Q_G = Q_{IS} + Q_{SC} \quad . \tag{7.58}$$

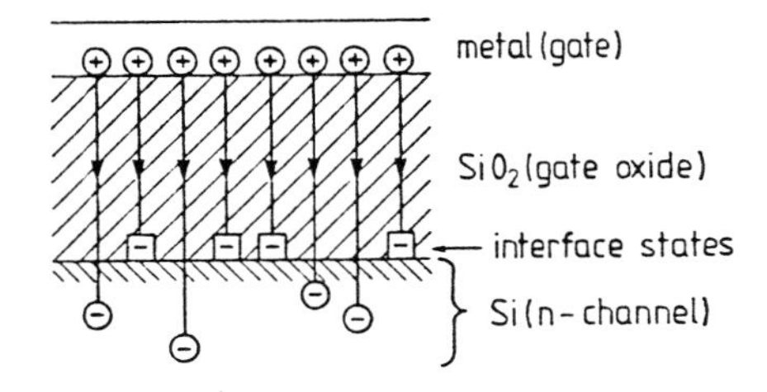

Fig.7.26. Schematic plot of electric field across the gate oxide of a MOSFET. Field lines connect the charge on the metal gate electrode with free carriers in the channel and with charge trapped in interface states

Electric field lines connect charges on the gate electrode with electrons in the channel and in interface traps (Fig.7.26). For a given gate voltage V_G the input capacitance C_i determines the gate charge $Q_G = C_i V_G$; but if the interface state density is high enough, Q_{IS} completely compensates Q_G and Q_{SC} must be negligibly small. Consequently the applied gate bias cannot induce enough charge in the channel and the bands are not bent (Fig.7.23b). To achieve a satisfactory MOSFET performance one thus requires a semiconductor/oxide interface that has a very low density of interface states. According to Sect.7.6 this is precisely the case for the Si/SiO_2 interface. But in contrast, the GaAs-oxygen interface is characterized by a high density of interface states (probably defect states, Sect.7.6); MOSFET structures based on GaAs can therefore not be fabricated in a simple manner as with Si. A possible solution is related to Schottky barriers at metal-semiconductor junctions; this will be discussed in more detail in Chap.8.

7.9 Some Experimental Results on Narrow Inversion and Accumulation Layers

Together with III-V heterostructures (Chap.8) Si MOSFET's have often been used in recent years to study the unusual quantum effects that arise in narrow inversion and accumulation layers. Since the energetic seperation ϵ_i between subbands (Sect.7.5) is typically on the order of 10 meV, optical excitations between occupied (below E_F) and empty subbands should be observable in IR spectroscopy. Such experiments have been performed on Si MOS arrangements and the subband structure has been verified. In Fig.7.27 the principle of the experimental setup is shown [7.27]. Monochromatic IR radiation with a photon energy in the 10 meV range is supplied from an IR laser to a Si MOS structure which can be biased by a variable external gate voltage V_G. The optical absorption of the Si substrate is measured by an IR bolometer. At low bias, the inversion layer is less squeezed and the subband separations are small (Fig.7.27b); the transition energy $\epsilon_0 \rightarrow \epsilon_1$ does not

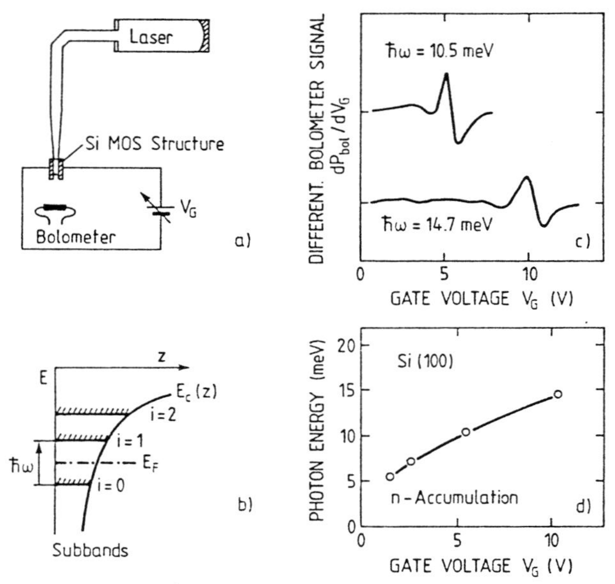

Fig.7.27a-d. Determination of subbands in the space charge layer of a Si(100) MOS structure by means of infrared (IR) absorption. (**a**) Experimental setup. (**b**) Subband minima plotted together with the potential well formed by the bent conduction-band edge $E_V(z)$. (**c**) Differentiated IR absorption signal versus gate voltage V_G. (**d**) Energetic position of resonance in the differentiated absorption (**e**) as a function of gate voltage [7.27]

match the photon energy of the laser. With increasing gate voltage V_G the channel becomes narrower, the energy splitting $E_{01} = \epsilon_1 - \epsilon_0$ increases and suddenly equals the incident photon energy; optical excitation, i.e. absorption, occurs, showing up as a characteristic absorption band in the differentiated bolometer signal (Fig.7.27c). For higher gate voltages the energy spacing E_{01} no longer matches the photon energy; because of the step-like subband density of states (Fig.7.8c) the differentiated absorption decreases again. For a higher photon energy (14.7meV in Fig.7.27c) the absorption structure occurs at a higher gate voltage. The dependence of the subband separation E_{01} on gate voltage is measured by means of different laser photon energies in Fig.7.27d. The experiment clearly reveals the quantized nature of the electronic states in the Si MOS channel.

Further clear evidence for the existence of subbands comes from tunneling experiments on MOS structures. The principle of such a measurement is illustrated for a PbTe-oxide-Pb structure in Fig.7.28a [7.28]. If the Pb metal electrode is positively biased with respect to the semiconductor, unoccupied metal states occur at energies $E > E_F$ opposite the occupied conduction-band states. Thus electrons from these states can elastically tunnel into the metal states. The tunnel current increases, first continuously

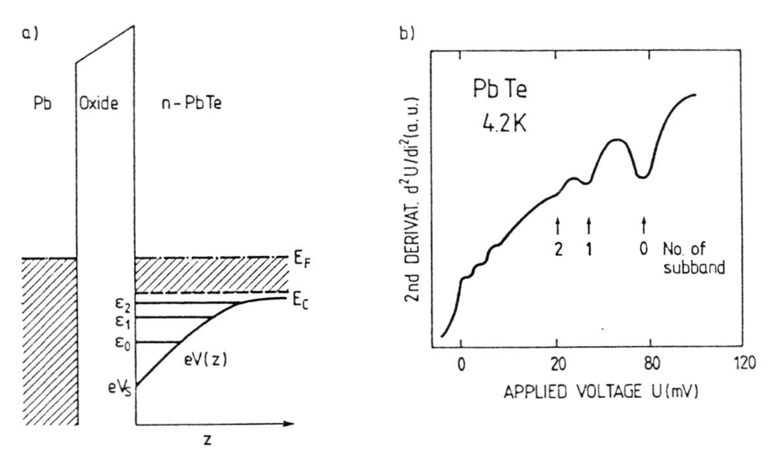

Fig.7.28a,b. Determination of subbands in an electron accumulation layer at the PbTe oxide interface by means of elastic tunneling spectroscopy. (**a**) Band diagram of a n-PbTe/oxide/Pb tunnel sandwich. The semiconductor is degenerate such that an electron continuum exists at energies higher than the subband minima $\epsilon_0, \epsilon_1, \ldots$. (**b**) Doubly differentiated tunnel characteristics d^2U/di^2 where U is the voltage and i the current between the Pb counterelectrode and the semiconductor substrate. The structures marked by arrows originate from the three lowest subbands 0,1,2 [7.28]

due to the degenerate electron gas in the conduction band, but beyond a certain point a further increase of bias produces a step in the tunnel current since the subband ϵ_2 reaches the Fermi energy of the metal. Subsequent steps occur when ϵ_1 and ϵ_0 cross the Fermi level. Since the tiny step-like increases in the tunnel characteristics are not well resolved, second derivatives of the tunnel characteristics d^2U/di^2 or d^2i/dU^2 are usually recorded. These curves clearly show the presence of the discrete subbands (Fig. 7.28b).

A whole class of experiments on MOS structures is concerned with the interesting transport properties of quasi-2D electron gases. A typical set of mobility data for a Si(100) MOS structure is shown in Fig.7.29 [7.29]. The measurements were made at 4.2 K using the Hall effect. As expected, higher trap densities in the oxide (oxide charge density given at each curve in units of $10^{11}\,cm^{-2}$) cause lower mobilities due to scattering on charged impurities. One should note here that bulk defects (donors or acceptors) play a negligible role since the thin space-charge layer ($d \simeq 10^{-7}$ cm) contains only 10^{10} cm^{-2} traps, even at high doping concentrations of 10^{17} cm^{-3}. This is comparable with the areal densities of charged states at the Si/SiO_2 interface. But acceptors such as boron induce interface charges and states by migration into the oxide. The charge on such impurities in the oxide or at the interface is screened by the mobile electrons. When more electrons are available the screening is more effective and the mobility

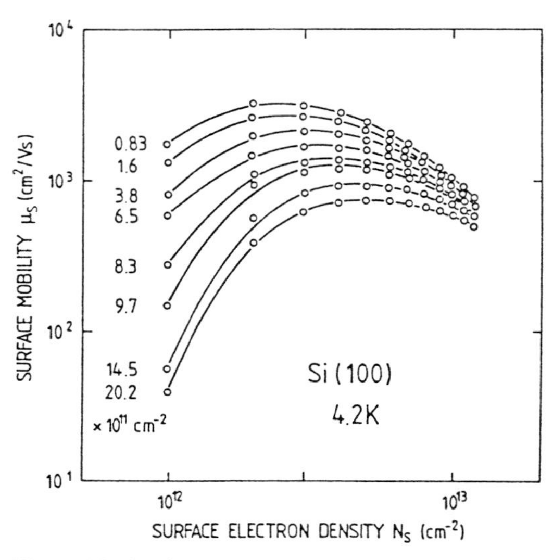

Fig. 7.29. Surface mobilities measured at 4.2 K on a Si(100) MOS sandwich in which the oxide charge (given on each curve in units of $10^{11}\,cm^{-2}$) was varied by causing Na^+ ions to drift through the SiO_2 [7.29]

should increase. This is indeed observed at the left flanks of the curves in Fig. 7.29.

Even more complex dependencies on surface electron density, namely *oscillatory* mobilities, are observed for the Hall mobility of electrons in accumulation layers on the polar O(000$\bar{1}$) face of ZnO [7.30]. The surface conductivity of the O face prepared by cleavage in UHV is raised in these experiments by controlled photolysis. By means of defect formation this procedure can establish strong accumulation layers on ZnO; the effect is similar to that of hydrogen adsorption (Sect. 7.7). Measurements of surface conductivity and Hall effect yield surface mobilities and surface electron concentrations separately. The oscillatory behavior of the surface mobility as a function of electron concentration (Fig. 7.30) is observed only on the polar ZnO surfaces, not on the hexagonal (10$\bar{1}$0) faces [7.30]. At electron concentrations greater than 10^{11} cm^{-2} and at temperatures higher than 125 K, the oscillations disappear even at the polar faces. They are suppressed at high electron concentrations by the onset of degeneracy and the occupation of higher subbands. Annealing studies reveal that the location of the maxima in Fig. 7.30 depends on the donor density at the surface. As yet, there is no detailed theoretical explanation for this interesting phenomenon.

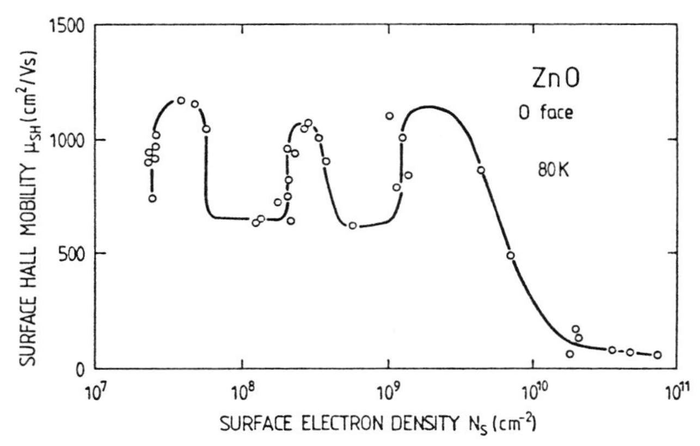

Fig.7.30. Surface Hall mobility versus electron density measured at 80 K on the cleaved polar (000$\bar{1}$) O face of ZnO. The surface electron density was generated by photolysis. Bulk mobility 1685 cm^2/Vs [7.30]

7.10 Magnetoconductance of a 2D Electron Gas: The Quantum Hall Effect

The application of a magnetic field causes the *dimensionality* of an electron gas to decrease; the 2D gas within an MOS inversion layer is further quantized. If a strong magnetic field B is applied normal to the surface of an MOS device (Fig.7.31b), the electrons in the inversion layer are forced to move in cyclotron orbits. Classically, the cyclotron frequency ω_c of an electron in such an orbit is calculated from the condition that the Lorentz force should equal the centrifugal force (Fig.7.31a). It is obtained as

$$\omega_c = eB/m_{\|}^{*} \tag{7.59}$$

where $m_{\|}^{*}$ is the principal effective mass parallel to the surface of the MOS device, i.e. normal to B. Without a magnetic field, the possible electronic levels of the 2D electron gas were given by the subband parabolas (7.51), where, parallel to the interface, the electrons can move freely with a $\mathbf{k}_{\|}$ being a good quantum number. Free movement parallel to the surface is no longer possible when $B \neq 0$. For the cyclotron orbits one obtains the so-called **Landau levels** as the energy eigenvalues within the magnetic field:

$$E_n^L = (n + \tfrac{1}{2})\hbar\omega_c\ , \quad n = 0, 1, 2, \ldots \ . \tag{7.60}$$

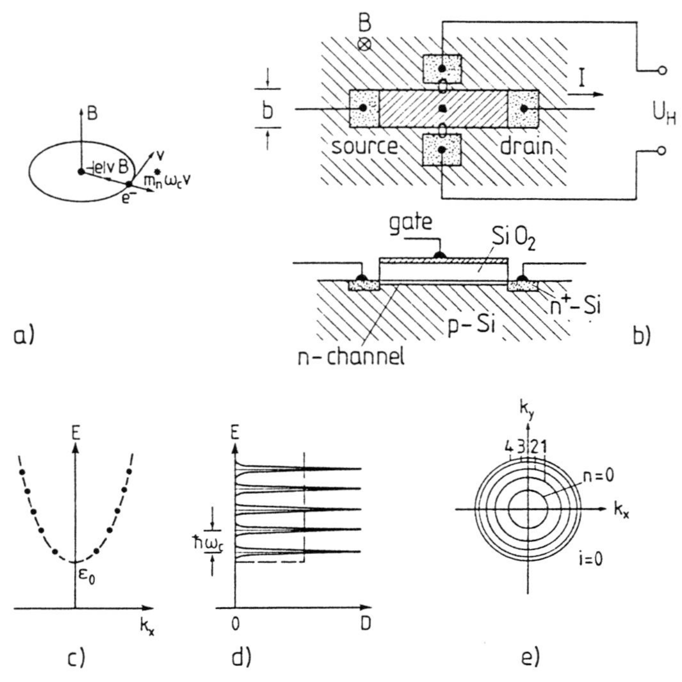

Fig. 7.31a-e. Quantization of a quasi 2D electron gas in a magnetic field B and its measurement using a Si MOS device. (**a**) Classical description of cyclotron orbit of an electron (velocity v) in a magnetic field B. The cyclotron frequency ω_C is determined by the balance between Lorentz force and centrifugal force. (**b**) MOS device with B field normal to the device connected for measuring the magneto-resistance and quantum Hall effect; (**c**) Subband parabola along k_x (wave vector parallel to the surface) with no magnetic field (*dashed line*). In a magnetic field $B \neq 0$ the continuous parabola splits into separate Landau levels $(n+1/2)\hbar\omega_C$ represented by the points. (**d**) Density of states for zero magnetic field (step function shown as dashed line) and with $B \neq 0$ (spikes, *full line*). (**e**) Representation of Landau levels in the k_x, k_y plane normal to the magnetic field

As is expected from the decomposition of an orbital movement into two linear oscillations, the quantum energies (7.60) of the Landau orbits are those of an oscillator. In a strong magnetic field one also has to take into account the fact that the spin degeneracy breaks down. The electron spins have two possible *orientations* in the B field. Instead of (7.51), the possible one-electron quantum energies for the 2D electron gas become

$$E_{i,n,s} = \epsilon_i + (n+1/2)\hbar\omega_c + sg\mu_B B \tag{7.61}$$

where ϵ_i is the energy of the i-th subband originating from "z-quantization" (Sect. 7.6). For a triangular potential well ϵ_i is proportional to $(i+3/4)^{2/3}$, see (7.57), g is the Landé g-factor, μ_B the Bohr magneton, and $s = \pm 1$ the spin quantum number. In multi-valley semiconductors the valley splitting ΔE_v has to be taken into account in (7.61). This can be done by a fourth term $v\Delta E_v$ with $v = \pm\frac{1}{2}$ as a valley quantum number. According to (7.61) the new quantization in the magnetic field in terms of Landau levels appears as a splitting of the continuous parabolic subbands into discrete levels (Fig.7.31c-d). For any particular subband (ϵ_0 in Fig.7.31c) the density of states being a step function (broken line in Fig.7.31d) for $B = 0$, is transformed at finite B into a series of δ-function-like spikes whose separation in energy is $\hbar\omega_c$ (Fig.7.31d). The states "condense" into these sharp Landau levels under the action of the magnetic field B. Because of state conservation the area under each spike, i.e. the degeneracy N_L of each Landau level (7.61) is obtained as

$$N_L = \hbar\omega_c D_0 , \tag{7.62}$$

where D_0 is the density of the subband at $B = 0$. In contrast to (7.53) the spin degeneracy has now been lifted by the B field and the density is obtained as

$$D_0 = \frac{m_{\parallel}^*}{2\pi\hbar^2} = \frac{(m_x m_y)^{1/2}}{2\pi\hbar^2} . \tag{7.63}$$

With (7.59), the degeneracy therefore follows as

$$N_L = eB/h . \tag{7.64}$$

If the Landau level is below the Fermi level, it is occupied by just N_L electrons (at sufficiently low temperature). A variation of the external magnetic field changes both the Landau level splitting $\hbar\omega_c$ (Fig.7.31d) via (7.59), and the degeneracy of each level, (7.64).

With increasing magnetic field the Landau levels shift to higher energies, cross the Fermi level E_F and are thereby emptied. It is easily seen that at low temperatures (sharp Fermi distribution) the system has its lowest free energy when a Landau level has just crossed the Fermi energy. With increasing B-field the free energy increases again to the point where the next Landau level crosses E_F and is emptied. Thus oscillations occur in the free energy as a function of the magnetic field and these can be detected in a variety of physical quantities. In the present context the electrical conductivity of the 2D electron gas is particularly interesting. Electrical conduction means that free carriers can acquire additional energy in an applied electric field and that they are scattered by impurities and phonons. In a strong magnetic field electrical conduction can be imaged as a "diffusion" of the centers of the cyclotron orbits in the direction of the electric field. An electron on a certain orbit may scatter into an arbitrary direction, but will then begin a new orbit. Electrons near the Fermi level E_F participate in these processes. When a Landau level crosses E_F, its occupancy changes and thus the density of available electrons near E_F changes. The corresponding oscillations observed in the electrical conductivity of an electron gas as a function of the externally applied magnetic field are called **Shubnikov-de Haas oscillations**. In a real system the Landau levels are, of course, broadened by imperfections, which decreases the oscillation amplitude to a certain extent. This oscillatory magnetoconductance, if observed for only one direction of the magnetic field, i.e. normal to the gate electrode in a MOS structure, gives evidence of the two-dimensionality of the conductance process. The strong anisotropy of the Shubnikov-de Haas effect is a general feature of 2D electron gases. It is also observed, for example in semiconductor heterostructures and superlattices (Sect. 8.6)

In an MOS structure (Fig. 7.31b) variation of the gate-voltage changes the width and depth of the inversion layer potential, i.e. the quantum-well width and depth for the 2D electron gas. The carrier concentration near E_F is changed, as is the spacing between the different subband energies ϵ_i. Thus a change of the gate voltage leads to shifts of the Landau levels, too, even with a fixed magnetic field B. Each time a Landau level crosses E_F under changing gate voltage, a maximum occurs in the electrical conductivity. Conductance oscillations are observed as a function of gate voltage at fixed magnetic field. Figure 7.32 shows experimental results obtained at 1.34 K and a magnetic field of 3.777 T on a circular MOSFET arrangement (Corbino disk). The oscillations are seen to be uniformly spaced as a function of gate voltage. Spin and valley splittings are not resolved in this example.

Another interesting phenomenon has been observed and analyzed by *von Klitzing* (Nobel prize 1985) [7.32] when he studied the magneto-conductance of a 2D electron gas in the n-channel of a Si MOS structure with electrodes in a Hall-effect configuration (Fig. 7.31b). Under the action of a magnetic field B directed normal to the gate electrode, a current I through the inversion layer causes a Hall voltage U_H normal to the current and to the B field direction. U_H is measured at zero current across a high resistance. As in the conventional Hall effect, the Lorentz force in the direction of the Hall voltage is then compensated by the Hall field U_H/b (b being the distance between Hall contacts, i.e. the gate width):

$$evB - eU_H/b = 0 \ . \tag{7.65}$$

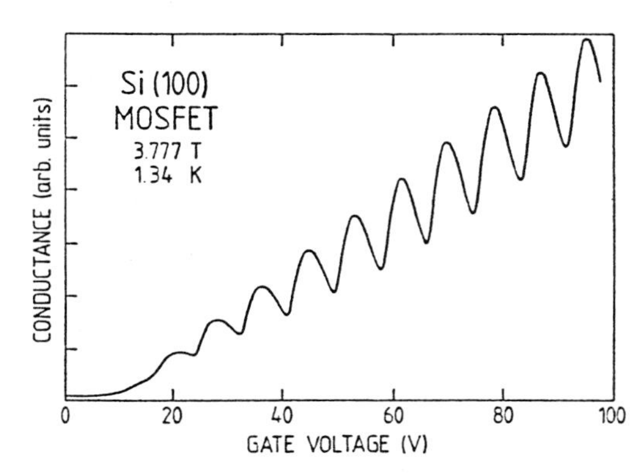

Fig. 7.32. Conductance oscillations measured on a (100) surface of an n-channel Si MOSFET with a circular plane or Corbino-disk geometry in the presence of a magnetic field of 3.777 T normal to the interfaces. The oscillations are seen to be uniformly spaced as a function of gate voltage. Spin and valley splittings are not resolved [7.31]

On the other hand, the current I can be expressed in terms of the 2D density of carriers N_s in the inversion layer

$$I = bN_s ev \ . \tag{7.66}$$

From (7.65, 66) we obtain

$$R_H = \frac{U_H}{I} = \frac{B}{N_s e} \ . \tag{7.67}$$

The quantity R_H, by dimension a resistance, is called the Hall resistance. In a strong magnetic field the electronic density of states now consists of a series of more or less sharp Landau levels with the density $N_L = eB/h$. When the 2D carrier density N_s is varied by means of the gate voltage, the Hall resistance R_H reaches the values

$$R_H = \frac{B}{\nu N_L e} = \frac{h}{e^2}\frac{1}{\nu} \ , \quad \nu = 1, 2, 3, \ldots \tag{7.68}$$

each time when the Fermi level has crossed the νth Landau level, such that this level is occupied. The unexpected behaviour, which is now called the **quantum Hall effect**, consits in the observation that each time when the condition (7.68) is satisfied by a variation of the 2D carrier density, a plateau occurs in the Hall resistance R_H. Figure 7.33 exhibits the results of Hall-resistance measurements on a Si MOS structure together with oscilla-

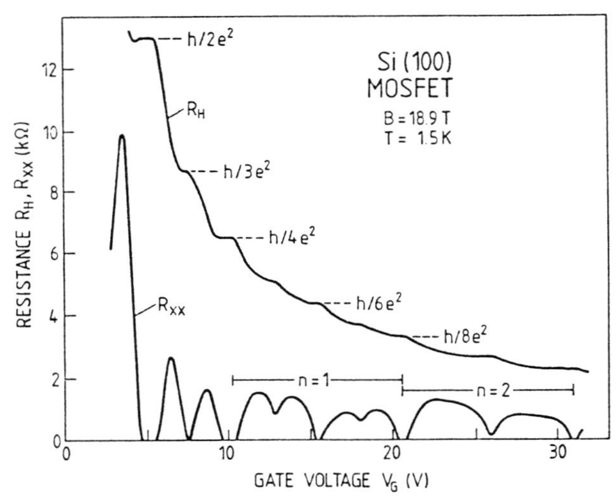

Fig. 7.33. Magnetoresistance R_{xx} ($\approx \rho_{xx}$) and Hall resistance R_H ($\approx \rho_{xy}$) measured on a Si MOSFET device (Fig. 7.31b) at T = 1.5 K in a constant magnetic field of 18.9 T (normal to the gate electrode) [7.33]

tions of the magnetoresistance R_H (analogous to Fig. 7.32). In the curve of R_{xx} versus gate voltage one also sees the spin splitting, not observed in Fig. 7.32. The Hall-resistance measurement demonstrates that R_H is a "staircase" function of the gate voltage. Measurements on a large number of different devices and from different laboratories have confirmed that the resistance values of the plateaus in R_H are independent of sample parameters and external quantities such as current, magnetic field or substrate bias. These experimental results form the basis for the assumption that the quantized Hall resistance depends only on the fundamental constant h/e^2. Correction terms are ruled out on an experimental basis by continuously decreasing errors in the measurements. One thus uses the quantum Hall effect to determine the fine-structure constant

$$\alpha = \frac{e^2}{h} \frac{\mu_0 c}{2} \approx \frac{1}{137} \tag{7.69}$$

to great accuracy. On the other hand, if α and therefore h/e^2 are known from other experiments, the quantum Hall effect can be used in conjunction with (7.68) as a resistance standard to determine the Ohm in units of h/e^2. But one should keep in mind that the present simple description on the basis of an ideal free electron gas does not justify this procedure theoretically.

The main theoretical problem consists of the explanation of the plateaus observed in R_H (Fig. 7.33). One possible interpretation is based on the assumption of localized states (the so-called **gap states**) in between the Landau levels. These localized states accept electrons when the Fermi level has crossed a Landau level but the electrons in these states are immobile and do not contribute to the current; the Hall resistance remains constant forming one of the plateaus. Only within the delocalized states at energies just around the Landau levels electrons are mobile and the Hall resistance changes from one plateau to the next one. There is another explanation for the occurance of the plateaus in the quantum Hall effect, which has been developed in connection with the observation of this phenomenon on III-V semiconductor heterostructures (Sect. 8.7). A final, consistent theoretical explanation of the phenomenon has not been found so far [7.34].

7.11 Two-Dimensional Plasmons

In strong narrow accumulation layers, where there is z-quantization perpendicular to the surface, the carriers are free in their movement parallel to the surface. To a good approximation, the situation can be described by the model of a 2D electron gas. A similar situation might also occur for thin but continuous metal overlayers of a few Ångstrom thickness on an insulator. In analogy to the 3D case, density fluctuations are also possible in the 2D system; the collective excitations are called 2D plasmons. However, these should not be confused with the 3D surface plasmons of a semi-infinite halfspace (Sect. 5.5). In contrast to these latter excitations, the charge is now spatially limited to within a sheet that is only a couple of Ångstroms thick.

In our model we assume a 2D charge distribution (density σ) in the xy plane ($z = 0$):

$$\rho = \sigma\delta(z) = n_s q \delta(z) \ . \tag{7.70}$$

The current density is likewise limited to this plane

$$\mathbf{j} = \mathbf{j}_x \delta(z) \ . \tag{7.71}$$

The potential ϕ of the 2D charge distribution must obey Poisson's equation

$$\nabla^2 \phi = \frac{-\sigma}{\epsilon\epsilon_0}\delta(z) \ , \tag{7.72}$$

or, after integration along z,

$$\left.\frac{\partial \phi}{\partial z}\right|_{z=+0} - \left.\frac{\partial \phi}{\partial z}\right|_{z=-0} = \frac{-\sigma}{\epsilon\epsilon_0} . \tag{7.73}$$

As in the case of the 3D surface plasmons of the semi-infinite halfspace, (7.72) is solved by a potential that decays exponentially in the z-direction and has plane-wave character along x:

$$\phi = \phi_0 \exp\left[i k_{\|} x - i\omega t - k_{\|} |z| \right] . \tag{7.74}$$

The current density j_x along x is related to the electric field

$$\mathscr{E}_x = \frac{-\partial \phi}{\partial x} \tag{7.75}$$

via the dynamic equation for charge transport within the plane. For simplicity, we neglect scattering and therefore ohmic damping, i.e. for an electron gas (q = -e)

$$m^* \dot{v}_x = - i m^* \omega v_x = q \mathscr{E}_x = e \frac{\partial \phi}{\partial x} . \tag{7.76}$$

For the velocity v_x we have assumed the same harmonic time dependence as in (7.74). From (7.76) follows that

$$j_x = - n_s e v_x = \frac{n_s e^2}{i\omega m^*} \frac{\partial \phi}{\partial x} . \tag{7.77}$$

Furthermore, the continuity equation

$$\frac{\partial \rho}{\partial t} + \nabla \mathbf{j} = 0 \tag{7.78}$$

must hold. With the representations (7.70, 71) we thus obtain from (7.77, 78):

$$- i\omega\sigma(z=0) + \frac{n_s e^2}{i\omega m^*} \left.\frac{\partial^2 \phi}{\partial x^2}\right|_{z=0} = 0 . \tag{7.79}$$

With (7.73, 74) this finally yields

$$2i\omega k_{\parallel}\epsilon\epsilon_0 \phi_0 \exp(ik_{\parallel}x - i\omega t) + \frac{n_s e^2}{i\omega m^*} k_{\parallel}^2 \phi_0 \exp(ik_{\parallel}x - i\omega t) = 0 \,. \quad (7.80)$$

Equation (7.74) is therefore a solution of the Poisson equation if we require

$$\omega^2 = \frac{1}{2}\frac{n_s e^2}{\epsilon\epsilon_0 m^*} k_{\parallel} \,. \quad (7.81)$$

This then is the dispersion relation for plasmons in a 2D free-electron gas without ohmic damping. The frequency vanishes with decreasing wave vector $k_{\parallel}$ as $(k_{\parallel})^{1/2}$, i.e. for small $k_{\parallel}$ or long wavelength the restoring force vanishes. This behavior can be understood qualitatively in the following way. For a 2D plasmon, the points of maximum charge lie on ideal parallel lines. With increasing line separation, i.e. increasing wavelength of the plasmon, the force between two lines vanishes logarithmically. This is not the case for a 3D plasmon: The charge maxima of a 3D plane wave lie on parallel sheets. With increasing sheet separation the field, and thus the force between the sheets, remains constant. Therefore, the frequency of a 3D plasmon stays finite for vanishing wave vectors.

Panel XII
Optical Surface Techniques

In optical spectroscopy [XII.1] light is irradiated onto a solid whose response is measured. The response might be an optical one, i.e. one records the reflected or transmitted light, or, in the Raman effect, inelastically scattered light is observed and analysed. In addition to these purely optical spectroscopies, other techniques like photovoltage or photoconductivity spectroscopy use light-induced electrical signals to gain information about the solid. In principle, photoemission spectroscopy in which photoemitted electrons are detected (Panel XI: Chap.6) also belongs to this class of methods. The latter techniques owe their surface sensitivity to the nature of the observed response, be it the absolute sensitivity with which an electrical signal as a photocurrent or work function (see also Panel XV: Chap.9) can be measured, or the escaped photoelectron which can only originate from a spatial region close to the surface (5 ÷ 50 Å depending on energy; Fig.4.1).

For optical spectroscopies in the strict sense, there is the problem of inherently low surface sensitivity. The probing depth of light in a solid, even in the spectral range of highest absorption, is on the order of 1000 ÷ 5000 Å. For an adsorbed monolayer of about 5 Å thickness, the contribution to the total optical signal, even in an optimized reflection experiment, amounts to only $\Delta R/R \approx 10^{-3}$ to 10^{-2}. Several approaches have thus been developed to exploit optical **reflection experiments with high precision** for the detection of excitations on solid surfaces or in thin overlayers. In principle, the surface sensitivity is always achieved by measuring difference signals which enhance the surface contribution with respect to that of the bulk or substrate. Reflectance spectroscopy measures spectral structures in the reflected light due to optical excitations. Depending on the particular spectral range, vibrational excitations such as normal-mode vibrations of adsorbed molecules or, in certain experimental geometries, surface and interface phonons are preferentially detected for photon energies $\hbar\omega$ below about 500 meV. On semiconductors, surface and interface plasmons can also be studied in this energy range. For energies higher than about 500 meV, i.e. in the near IR, visible and UV spectral ranges, both electronic interband transitions and collective plasmon excitations can be studied.

As an example of a high-precision difference reflectance experiment in the visible and UV range, we describe here the experiment of *Rubloff* et al. [XII.2] on well-defined W(100) surfaces. The change in reflectivity of the

W surface due to gas adsorption is measured by a set-up in which a rotating quartz-light pipe captures light alternately from the incident and the reflected light beams. An electronic gating circuit separates the output signal of the photomultiplier into two channels corresponding to the incident and the reflected beam. The signal corresponding to the incident beam is kept constant by a servo system which regulates the gain of the photomultiplier. This technique achieves a stability in R of the order $\Delta R/R \approx 10^{-5}$ with an absolute accuracy in R of about 10^{-2}. In this and other similar experiments, one always measures the reflectivity R of the clean surface and the reflectivity R′ of the adsorbate-covered surface; the $\Delta R/R$ spectrum, i.e. the relative reflectivity change, provides insight into changes in the electronic surface structure, due both to the adsorbate itself and to the topmost atomic layers of the substrate or space charge layers in the case of semiconductor interfaces.

To calculate such spectra theoretically, continuum-type models are usually applied, where a surface layer (adsorbate or topmost substrate atomic layer with electronic surface states) is described by a complex surface dielectric function $\epsilon_s(\omega) = \epsilon_s' - i\epsilon_s''$ on top of the substrate with its bulk dielectric function ϵ_b. For such a model the relative change of reflectivity for light polarized parallel and perpendicular to the plane of incidence is given [XII.3] by

$$\frac{\Delta R_{\parallel}}{R_{\parallel}} = \frac{8\pi d n_0 \cos\phi}{\lambda} \mathrm{Im}\left\{\frac{\epsilon_s - \epsilon_b}{\epsilon_0 - \epsilon_b} \, \frac{1 - (\epsilon_0/\epsilon_s\epsilon_b)(\epsilon_s + \epsilon_b)\sin^2\phi}{1 - (1/\epsilon_b)(\epsilon_0 + \epsilon_b)\sin^2\phi}\right\} , \quad \text{(XII.1)}$$

$$\frac{\Delta R_{\perp}}{R_{\perp}} = \frac{8\pi d n_0 \cos\phi}{\lambda} \mathrm{Im}\left\{\frac{\epsilon_s - \epsilon_b}{\epsilon_0 - \epsilon_b}\right\} . \quad \text{(XII.2)}$$

where ϕ is the angle of incidence, n_0 and ϵ_0 are the refractive index and the real dielectric constant of the surrounding medium, respectively ($n_0 = \epsilon_0 = 1$ for vacuum). Results for H_2 adsorption on W(100) are shown in Fig. XII.1. The measured change in reflectivity $\Delta R/R$ due to adsorption of H_2 (Fig.XII.1a) is analysed by means of (XII.1,2) in terms of changes of the surface dielectric function $\Delta\epsilon_s$ due to adsorption (Fig.XII.1b). The resulting spectral distribution of $\mathrm{Im}\{\Delta\epsilon_s\}$ is interpreted in terms of three different electronic transitions, namely three harmonic oscillators near 0.5, 2.5 and 5 eV. The latter two transitions, whose positive $\mathrm{Im}\{\Delta\epsilon_s\}$ implies an increase of the surface absorption with H_2 coverage must be due to surface state transitions characteristic of the adsorbed H_2. Because of the negative sign of $\mathrm{Im}\{\Delta\epsilon_s\}$ below 2 eV the low-energy structure is ascribed to a quenching of optical transitions within the topmost atomic layers of the W substrate

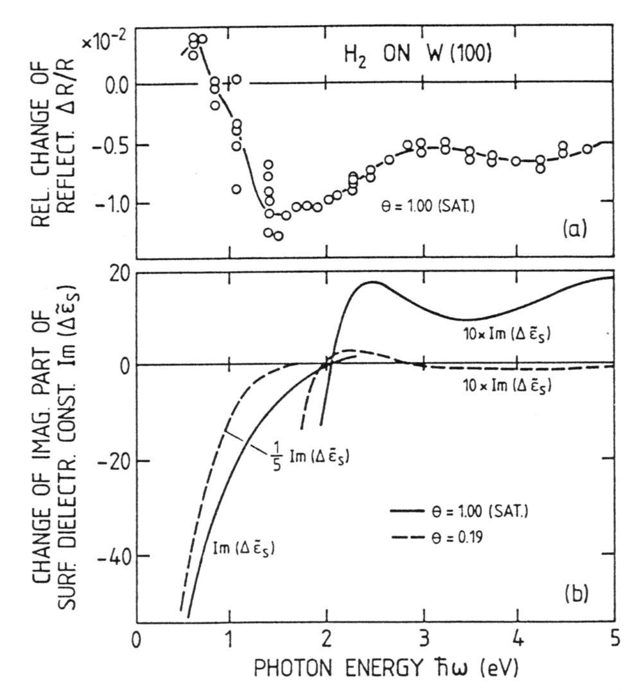

Fig. XII.1. (**a**) Spectral dependence of relative reflectance change due to saturation coverage of hydrogen, $\theta = 1$, taken to be two H atoms per W atom. The theoretical (solid) curve is obtained from an oscillator fit of $\tilde{\epsilon}_s$ to reproduce the $\Delta R/R$ data points. (**b**) Calculated change of the imaginary part of the surface dielectric constant $\mathrm{Im}\{\Delta\tilde{\epsilon}_s\}$ as caused by hydrogen coverages $\theta = 0.19$ and $\theta = 1$. Assumed thickness for the surface layer (scale factor): 5 Å ($\theta = 1$) and 1 Å ($\theta = 0.19$), respectively [XII.2]

due to H_2 adsorption. The interpretation in terms of quenching of transitions between electronic surface states of the clean W(100) surface due to H_2 adsorption is in agreement with results from other investigations.

Experimental set-ups based on the same principle are used for measuring reflectance difference spectra in the infrared spectral range for investigating adsorbate vibrations on metal surfaces. As an example, Fig. XII.2 shows a surface-sensitive double-beam IR reflection spectrometer for grazing incidence measurements on UHV prepared metal surfaces [XII.4]. Two symmetrically disposed monochromatic light beams pass through the UHV chamber and are focused onto a cooled PbSnTe detector. By means of a suitable chopper blade (frequency 13Hz) the two beams are pulsed with a phase difference of 180°. They pass the monochromator superimposed and one beam is reflected at the Pt surface at an angle of about 84° (grazing incidence). The intensity in the two channels can be equalized optically by means of a metallic grid (compensator in Fig. XII.2) to within $\Delta I/I \approx 10^{-4}$.

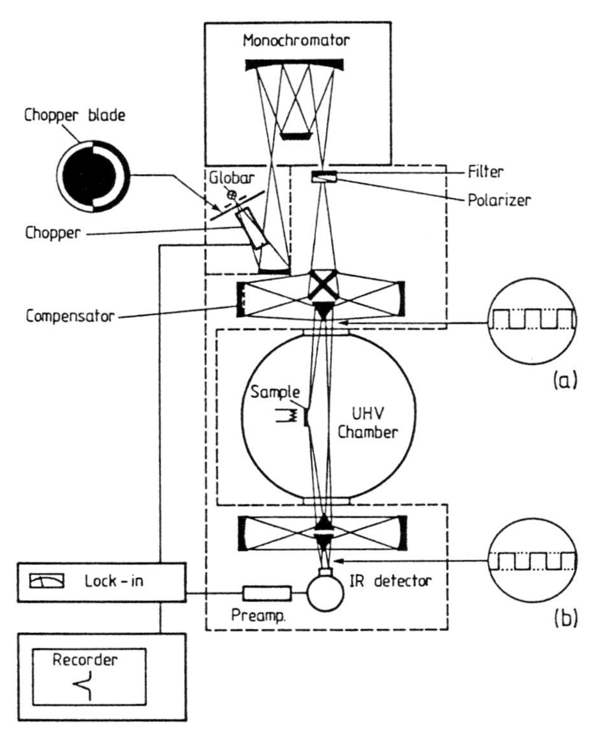

Fig. XII.2. Experimental set-up for a double beam IR grazing reflection experiment. The traces (*a*) and (*b*) show light intensity versus time curves of the two beams: compensated and unbalanced by an adsorbate vibration, respectively. The monochromator and the area enclosed by the dashed line can be flushed by dry air [XII.4]

Vibrational bands of an adsorbate on the sample surface cause an imbalance in the intensity of the two beams and are detected phase-sensitively. The whole optical path must be under vacuum conditions, or at least flushed by dry air or nitrogen in order to suppress background spectral bands due to atmospheric water absorption. Only light polarized parallel to the plane of incidence is used, since this polarization yields maximum surface sensitivity under grazing incidence conditions [XII.5]. Under these reflection conditions, Fresnel's formulae yield a maximum electric field strength at the metal surface and thus optimum coupling to vibrating dipoles of adsorbed molecules. For normal incidence, the electric field of the reflected light would be zero at the surface and thus there would be negligible coupling to adsorbed molecules (normal extension $2 \div 5$ Å) since the field reaches its maximum strength only at a distance of the order of micrometers ($\lambda/4$ of the light wave) above the surface.

Figure XII.3 shows as an example the two vibrational bands of CO molecules adsorbed on Pt(111) [XII.6]. The occurrence of two bands for the

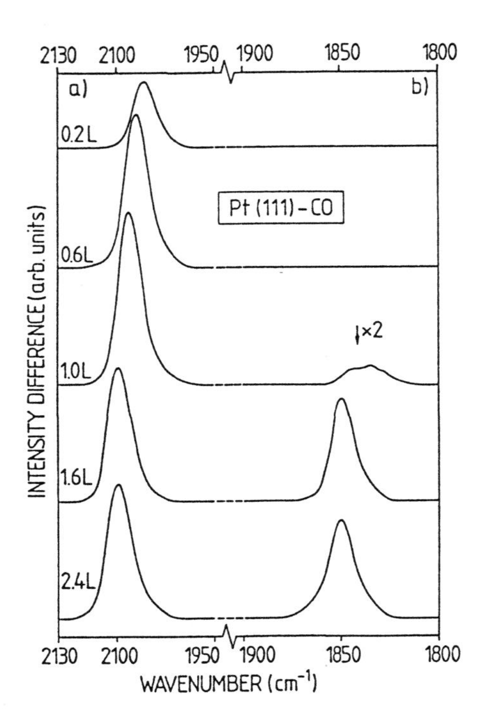

Fig. XII.3. Vibrational bands of CO adsorbed on Pt(111) surfaces as measured by the double beam IR reflection set-up of Fig. XII.2. At low exposures of less than 1 Langmuir (L) only the high energy band of CO at on-top sites is visible, whereas at higher exposures the band at around 1850 cm^{-1} appears due to CO molecules bonded at bridge sites between two Pt surface atoms [XII.6]

stretching vibration of the CO molecule indicates two different adsorption sites, the on-top position (CO molecule perpendicular to the surface on top of a Pt atom) associated with the high-energy vibration near 2100 cm^{-1} and the bridge site ($\approx 1870 cm^{-1}$) with CO molecules bonded to two surface Pt atoms [XII.7]. The latter configuration gives rise to lower vibrational frequencies since the adsorption bonding is thought to involve back donation of Pt d-orbital charge into the CO π^* antibonding orbitals, thereby weakening the intramolecular CO bonding and shifting the corresponding vibration to lower energies.

Other experimental approaches have successfully been employed to measure IR vibrational spectra of adsorbed species on metal surfaces. Since for grazing incidence light with a polarization normal to the metal surface is absorbed preferentially by adsorbed molecules, switching between normal and parallel polarization in a reflection experiment also allows a comparison between the reflectance of an adsorbate-covered and a *clean* surface (adsorbate not detected). This technique, which involves only a single light beam with a rotating polarizer, is used in connection with phase-sensitive lock-in detection; it also yields highly surface-sensitive adsorbate vibrational spectra [XII.8]

Since semiconductors are transparent for light of photon energy below the bandgap energy ($\hbar\omega < E_g$), surface excitations such as electronic surface state transitions and vibrations of adsorbed molecules, can be detected by **internal reflection** in which the light beam probes the optically absorbing surface region from inside the crystal. If, in the spectral range of the surface-layer absorption, the dielectric function of the bulk material ϵ_b is nearly constant, then the internal reflection spectrum resembles an ordinary absorption spectrum of the surface layer since $\Delta R/R$ is determined mainly by the absorption coefficient of the surface layer [XII.3]. The surface sensitivity of the method is enhanced by using a crystal whose shape allows multiple internal reflection. This is shown in Fig. XII.4 for the classic experiment by *Chiarotti* et al. [XII.9], in which dangling-bond surface state transitions on the clean cleaved Ge(111) surface were detected for the first time. The relative attenuation of the internally reflected light intensity was recorded for the clean cleaved Ge surface and for the same surface after oxygen adsorption. In the latter case the transmitted intenstiy was higher due to the removal of the surface state transitions. The corresponding spectral dependence of the surface state absorption [insert (a) of Fig. XII.4] indicates surface state transitions near 0.5 eV photon energy. The same technique of internal reflection spectroscopy has subsequently been applied with great success to the study of adsorbate vibrations on Si surfaces [XII.10]. In this case a Si wafer is shaped on opposite sides with slopes such that IR radiation can enter and leave at an angle for internal reflection inside the

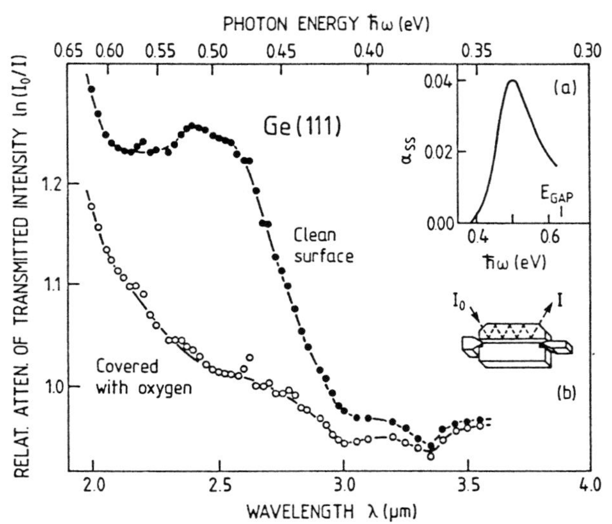

Fig. XII.4a,b. Natural logarithm of the intensity ratio I_0/I (I_0 incident, I transmitted) as a function of wavelength for the clean cleaved and oxygen-covered Ge(111) surface (ambient O_2 pressure $\approx 10^{-6}$ Torr). Insert (**a**): Surface absorption constant α_{ss} attributed to surface state transitions of the clean surface. Insert (**b**): Internal reflection element prepared by cleavage along (111) [XII.9]

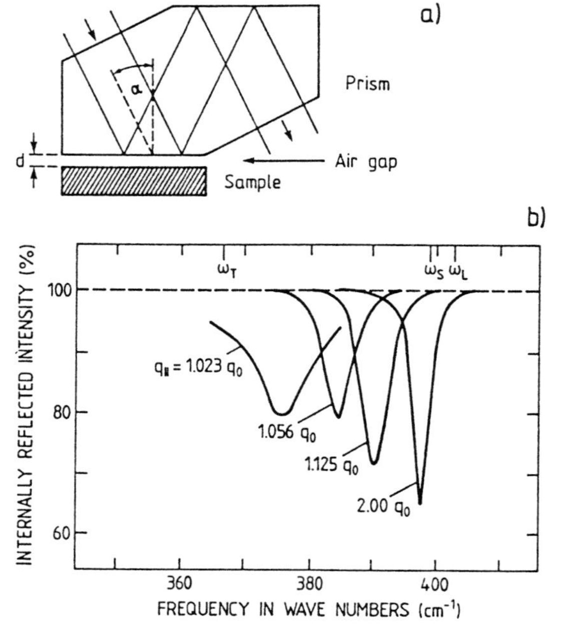

Fig. XII.5. (**a**) Experimental arrangement for the observation of frustrated total reflection. (**b**) Reflected intensity as a function of frequency (wave number), measured by means of an experimental arrangement as in (**a**) on a GaP sample. The parameter is the angle of incidence α. Together with the light frequency, the angle α determines the parallel component of the wavevector $q_{\parallel}$. The minima originate from excitation of surface phonon polaritons. For optimal observation the strength of the polariton coupling to the light wave must be adjusted by matching the air gap d to the inverse wave vector $q_{\parallel}^{-1}$ [XII.11]

sample [similar to insert (b) of Fig. XII.4]. The strongly absorbing vibrations of adsorbed molecules then give rise to sharp absorption bands in the transmitted IR intensity.

Surface polaritons (Chap. 5) cannot usually be detected by a normal absorption experiment since their dispersion curve $\omega(q_{\parallel})$ does not cross the light dispersion curve $\omega = ck$. However, using internal reflection in a so-called ***frustrated* total reflection experiment** [XII.11] (Fig. XII.5) by means of a prism in close proximity to the sample surface, surface phonons carrying an electromagnetic field outside the sample (surface polaritons) can be detected as dips in the internally reflected intensity (Fig. XII.5b). Light is totally reflected from the lower surface of the prism (made in this case of Si). The wave vector parallel to the surface is now $q_{\parallel} = (\omega/c)n\sin\alpha$ with $n = 3.42$. For a very small air gap, d, the exponentially decaying field outside the prism can be used to excite surface waves in the underlying semiconductor sample (in this case GaP). The excitation appears as a dip in the intensity reflected in the prism. Depending on the chosen angle α, i.e. on $q_{\|}$, the minima appear at different frequencies. The shift of the dip frequency with angle α allows one to calculate the surface polariton disperion curve $\omega(q_{\|})$. The same arrangement as in Fig. XII.5a can also be used to study adsorbate vibrations. Molecules adsorbed on the sample surface are also probed by the electromagnetic field travelling along the prism/sample surface air gap. Their vibrational absorption bands in the IR spectral range give rise to similar intensity minima in the transmitted light. Care has to be taken, of course, that only molecules adsorbed on the substrate are detected rather than species adsorbed on the opposite prism surface which leads the probing light beam.

For reasons of completeness **ellipsometric spectroscopy** also deserves mentioning (see also Chap. 3). Ellipsometry is essentially a reflectivity measurement in which one detects the change of polarization upon reflection rather than the intensity change [XII.12]. The high surface sensitivity of the technique is a result of the great accuracy with which angular changes of polarizer settings can be determined. In favorable cases, coverages of one-hundredth of a monolayer of an adsorbate can be detected. Since one measures two independent quantities, e.g., the ellipsometric angles Δ and ψ defined by the complex quantity

$$\rho = R_{\|}/R_{\perp} = \tan\psi e^{i\Delta} \ , \qquad \text{(XII.3)}$$

the two dielectric functions $\text{Re}\{\epsilon\}$ and $\text{Im}\{\epsilon\}$ of a semi-infinite half space can both be determined. Changes due to an adsorbate overlayer are measured as angular changes $\delta\Delta$ and $\delta\psi$ and are usually interpreted in terms of layer models with effective layer dielectric functions [XII.13].

Two techniques that have been successfully applied to study the electronic properties of semiconductor surfaces, e.g. electronic surface state transitions, are **Surface PhotoConductivity** (SPC) and **Surface PhotoVoltage** (SPV) **spectroscopy**. If electrons from occupied surface states below the Fermi level E_F are excited by light with energy below the bandgap energy E_g into the bulk conduction band (Process I in Fig. XII.6), they give rise to an increase in the density of the free electrons that contribute to electrical conductivity. Similarly, the excitation of electrons from the bulk valence band into empty surface states (Process II in Fig. XII.6) increases the density of free holes in the valence band. Both types of excitation can thus be measured in a photoconductivity experiment with light of the appropriate photon energy. Excitation of free carriers into the conduction or valence bands also changes the band bending and thus the work function $(E_{Vac} - E_F)$. A spectroscopy of work-function changes due to light irradiation can thus also yield similar information about excitation processes between electronic surface states and free carriers in bulk bands.

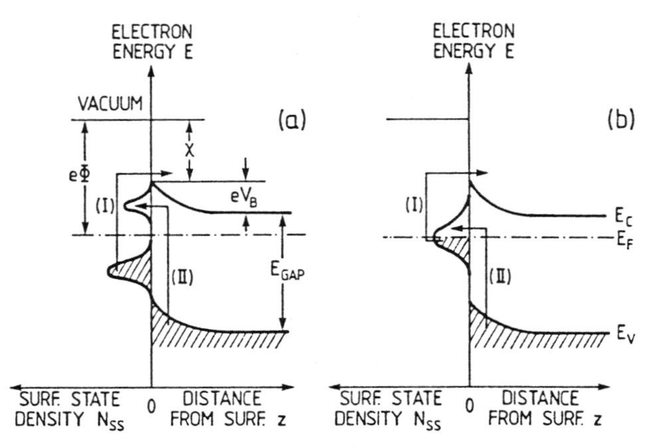

Fig. XII.6a,b. Surface state/bulk band transitions observable in SPC and SPV spectroscopy (I and II); E_C and E_V are the edges of conduction and valence band, E_F the Fermi energy, E_g the band gap energy, $e\Phi$ the work function, χ the electron affinity, and eV_B the band bending

Three different experimental methods for realizing SPV spectroscopy are currently in use (Fig. XII.7). The classic Kelvin method (see also Panel XV: Chap. 9) in which an opposite semi-transparent electrode vibrates and unmodulated illumination is used, allows an absolute determination of the contact potential difference between electrode and semiconductor surface

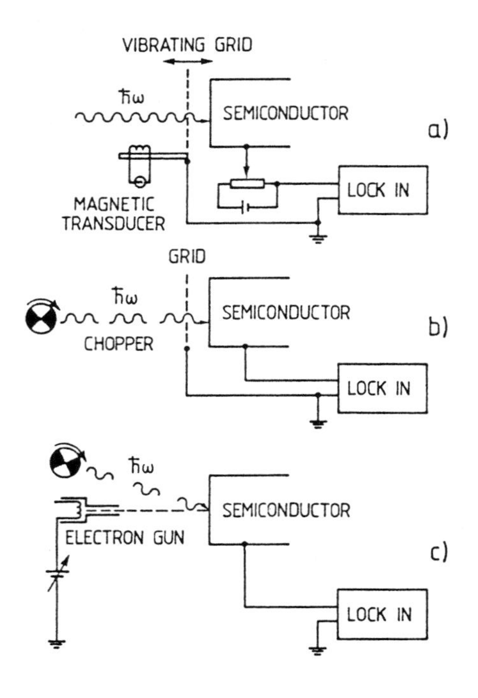

Fig. XII.7a-c. Schematic of an experimental set-up for SPV spectroscopy: (**a**) Kelvin probe with semi-transparent vibrating electrode (grid) and steady illumination; (**b**) chopped illumination modulates the contact potential difference between semiconductor surface and grid; (**c**) chopped illumination modulates the electron current between gun and semiconductor surface; a suitable chosen bias ensures maximum sensitivity near the threshold of the current-voltage characteristic by means of a compensation circuit. For the purpose of spectroscopy, the opposite electrode can be fixed and the light is AC modulated, giving rise to a modulated SPV signal which is recorded as a function of photon energy (Fig.XII.7b). Another interesting method [XII.14] uses an electron beam whose current is modulated by AC illumination of the surface (Fig.XII.7c). This method enables surface photovoltages on the order of 1 μV to be detected.

In order to understand the origin of SPC and SPV signals quantitatively, one has to keep in mind that the density of photocreated non-equilibrium carriers depends on both the generation and the recombination processes described by a mean lifetime τ of the free carriers [XII.15]. The generation rates of electrons and holes, $d(\Delta n)/dt$ and $d(\Delta p)/dt$, respectively, are proportional to the incident light intensity I and the absorption constant $\alpha(\hbar\omega)$, which, in the present case, carries the information about the surface state excitation as a function of photon energy $\hbar\omega$. The recombination rate of non-equilibrium carriers may be proportional to $\Delta n/\tau$ or to $(\Delta n)^2/\tau$. For surface state excitations with low intensity illumination, as considered here, the so-called **linear recombination** with $(\Delta n)/\tau$ and $(\Delta p)/\tau$ prevails. The measured AC photoconductivity Δg_{AC} is related in this case to the steady-state photoconductivity Δg_0 by

$$\Delta g_{AC} = \Delta g_0 \tanh(4\tau f)^{-1} , \qquad \text{(XII.4)}$$

where f is the modulation frequency. Equation (XII.4) allows one to determine the mean lifetime τ.

If carriers are excited out of surface states or bulk impurity states, the sign of the SPV signal depends both on the generation process (generally leading to band flattening because of a reduction of the space charge; see Chap.7) and on the recombination process. Since in many cases very little is known about the recombination mechanism (e.g., via surface or bulk traps), the analysis of experiments is often limited to spectral changes due to surface treatment. But this is sufficient to evaluate surface-state transitions. Even when the surface-state density is a relatively sharp band (Fig. XII.6), the SPV and SPC spectrum consists of a broad spectral shoulder due to the broad continuum of bulk conduction or valence-band states. The onset of this shoulder in the spectrum indicates the minimum transition energy and thus the energetic distance of the surface state band (or its onset) from the conduction or valence band edge (Fig.XII.6). Since excitations from and into bulk impurity states in the forbidden band of a semiconductor would show up in the same way as surface-state excitations, spectral changes due to a modification of the surface, e.g. by adsorption, are necessary to distinguish between surface and bulk excitations. However, such changes also have to be interpreted with great care, since even spectral structures arising

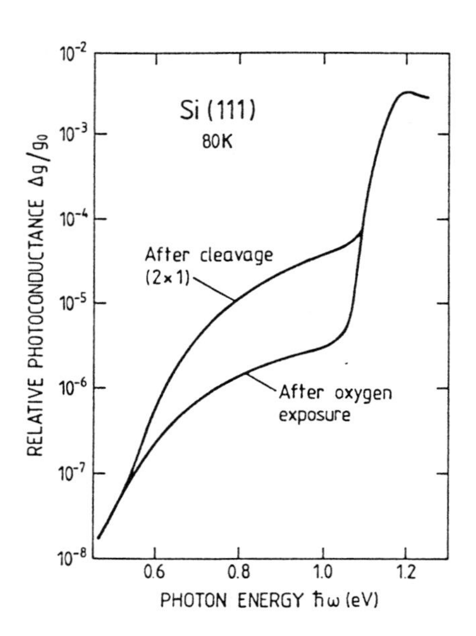

Fig. XII.8. Photoconductivity spectra of the cleaved Si(111) surface normalized to a photon flux density of $1 \cdot 10^{13}$ $cm^{-2} \cdot s^{-1}$. The spectra were measured at 80 K on the clean cleaved surface and after oxygen adsorption [XII.16]

from bulk excitations might undergo dramatic changes upon adsorption if the adsorbed species modifies the surface recombination process.

SPC spectroscopy has been used [XII.16] to determine the energetic position of the empty dangling-bond surface-state band (π^* band, Sect. 6.5.1) within the forbidden band of the clean, cleaved Si(111)-(2 × 1) surface. After cleavage in UHV the SPC spectrum at 80 K (Fig. XII.8) exhibits a steep threshold at around 1.1 eV photon energy due to bulk electron-hole pair production and a broad shoulder extending down to energies of around 0.5 eV. After exposure to oxygen this shoulder is strongly suppressed in intensity, whereas the spectrum above the band gap remains essentially unchanged. For energies lower than about 0.55 eV the two curves coincide. Since the measured lifetimes τ below the bandgap differ by less than 25% before and after oxygen adsorption (implying little change in recombination), the clean surface exhibits an additional generation process with an onset at about 0.55 eV, which is removed by oxygen adsorption. Since the Fermi level of Si(111)-(2 × 1) is pinned at 0.35 eV above the valence-band edge (Sect. 7.7), the SPC signal must be explained by excitation of electrons from valence-band states into empty surface states near midgap. Their energetic distance from the upper valence band edge thus amounts to about 0.55 eV. Based on a number of other experimental results (Sect. 6.5.1) the states are identified as the empty dangling-bond states of the π-bonded chain model of the Si(111)-(2 × 1) surface.

An example of an SPV spectrum measured by the modulated light-beam technique (Fig. XII.7b) is illustrated in Fig. XII.9 [XII.17]. On the

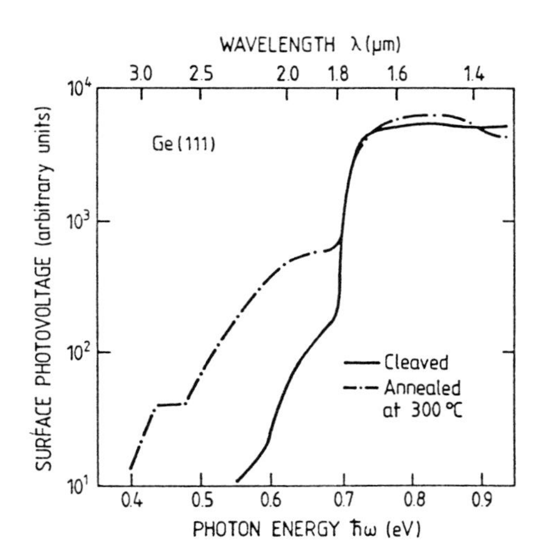

Fig. XII.9. Surface photovoltage spectra of the clean cleaved Ge(111) surface normalized to constant photon flux; light beam chopped at 13 Hz. The spectra are measured at room temperature, directly after cleavage and after a short anneal at 300 °C [XII.17]

clean, cleaved Ge(111)-(2 × 1) surface, besides the sharp onset of electron-hole excitation (≈0.7eV), one observes a slight shoulder with an onset near 0.55 eV. This shoulder is not reduced by oxygen adsorption and therefore allows no conclusions concerning intrinsic surface states on the Ge(111)-(2 × 1) surface. However, after annealing to temperatures above 200 °C in UHV, where the well-known surface phase transition from (2 × 1) to (8 × 8) occurs, the SPV spectrum changes significantly. A double shoulder with two onsets near 0.4 eV and 0.45 eV develops. Because of the Fermi level position (no more than 0.3eV above valence band edge), these shoulders must be due to electrons excited from the bulk valence band into empty surface states. The (8 × 8) superstructure on Ge(111) is thus accompanied by a double band of empty surface states in the upper half of the forbidden gap.

References

XII.1 H. Lüth: Appl. Phys. **8**, 1 (1975)
H. Kuzmany: *Festkörperspektroskopie* (Springer, Berlin, Heidelberg 1990)

XII.2 G.W. Rubloff, J. Anderson, M.A. Passler, P.J. Stiles: Phys. Rev. Lett. **32**, 667 (1973)

XII.3 J.D.E. McIntyre, D.A. Aspnes: Surf. Sci. **24**, 417 (1971)

XII.4 H.J. Krebs, H. Lüth: Proc. Int'l Conf. on Vibrations in Adsorbed layers, Jülich 1978; KFA Reports, Jül. Conf. **26**, 135 (1978)

XII.5 R.G. Greenler: J. Chem. Phys. **44**, 310 (1966); J. Chem. Phys. **50**, 1963 (1969)

XII.6 H. Moritz, H. Lüth: Vacuum (GB) **41**, 63 (1990)
H. Moritz: Wechselwirkungen im Koadsorbatsystem CO/Acetonotril auf Pt(111): Eine IRAS-Untersuchung. Diploma Thesis, RWTH Aachen (1989)

XII.7 H.J. Krebs, H. Lüth: Appl. Phys. **14**, 337 (1977)
XII.8 B.E. Hayden, A.M. Bradshaw: Surf. Sci. **125**, 787 (1983)
XII.9 G. Chiarotti, S. Nannarone, R. Pastore, P. Chiaradia: Phys. Rev. B **4**, 3398 (1971)
XII.10 Y.J. Chabal: Surf. Sci. Rep. **8**, 211 (1988)
XII.11 N. Marshall, B. Fischer: Phys. Rev. Lett. **28**, 811 (1972)
XII.12 R.M.A. Azzam, N.M. Bashara: *Ellipsometry and Polarized Light* (North-Holland, Amsterdam 1977)
XII.13 H. Lüth: J. Physique **38**, C5-115 (1977)
XII.14 F. Steinrisser, R.E. Hetrick: Rev. Sci. Instrum. **42**, 304 (1971)
XII.15 S.M. Ryvkin: *Photoelectric Effects in Semiconductors* (Consultants Bureau, New York 1964)
XII.16 W. Müller, W. Mönch: Phys. Rev. Lett. **27**, 250 (1971)
XII.17 M. Büchel, H. Lüth: Surf. Sci. **50**, 451 (1975)

Problems

Problem 7.1. A ZnO surface on a crystal with moderate n-type bulk doping, e.g. an electron bulk concentration of $n_b = 1.5 \cdot 10^{17}$ cm^{-3}, exhibits after treatment with atomic hydrogen a weak accumulation layer with a surface band bending $|eV_S|$ of 10 meV. Calculate the Debye length L_D, the thickness of the accumulation layer and the density of surface state charges which are produced by the hydrogen treatment. Are the responsible surface states acceptors or donors?

Problem 7.2. A clean, cleaved (110) surface of p-type GaAs (bulk hole concentration $p = 10^{17}$ cm^{-3} at 300K) exhibits a hole depletion layer with the Fermi level E_F at 300 meV above the valence band edge. How thick is the depletion layer? Calculate the space charge layer capacity.

Problem 7.3. A high-quality MOS (metal, SiO_2, Si) structure with a SiO_2 ($\epsilon = 3.9$) layer thickness of 0.2 μm is biased by an external voltage of 1 V. The negative bias is applied to the metal contact with respect to the Si substrate, which is n-doped with a room temperature bulk electron concentration of 10^{17} cm^{-3}. What surface-charge density is induced by the external bias and what is the spatial extension of the space-charge layer within the Si substrate.

Problem 7.4. An n-type semiconductor exhibits at its clean surface a moderately strong depletion layer due to acceptor-type surface states in the upper half of the forbidden gap. Plot the band scheme near the surface in the dark and with illumination by light with photon energies larger than the gap energy. Recombination of electron-hole pairs through surface traps is neglected. What processes are responsible for establishing a stationary non-equilibrium state under illumination?

8. Metal-Semiconductor Junctions, and Semiconductor Heterostructures

In comparison to the solid-vacuum interface, i.e. the clean, well-defined surface of a solid, other solid interfaces are of much more practical importance. The solid-liquid interface, for example, plays a major role in electrochemistry and biophysics. Studies of that particular interface have a long tradition in physical chemistry. A detailed treatment of solid-liquid interfaces is far beyond the scope of this text although certain general concepts, e.g. that of space-charge layers, are similar to those of the solid-vacuum and solid-solid interface.

The solid-solid interface is traditionally studied mostly within the framework of solid-state and interface physics. One reason is the enormous importance of metal-semiconductor junctions and semiconductor heterojunctions for device physics.

Metal-semiconductor junctions are found in every metal contact to a semiconductor device and the underlying physics is also the basis for rectifying devices. Semiconductor heterojunctions meanwhile are important in optoelectronic devices such as lasers, heterobipolar transistors, and field effect transistors produced from III-V semiconductor layer structures.

Recent years have witnessed a growing interest in metallic heterostructures due to possible applications in the fields of magnetism, memory devices and superconductivity. In the present chapter we shall concentrate on the electronic properties of semiconductor heterostructures and metal-semiconductor junctions, since most research work in the past has been devoted to these interfaces. The vibrational properties of solid interfaces, i.e. interface phonons and their dispersion, can be treated in a manner similar to that discussed in Chap.5 in connection with free surfaces and thin overlayers.

8.1 General Principles Governing the Electronic Structure of Solid-Solid Interfaces

As stressed in Chap.3., a solid-solid interface can be a complex quasi-2D system with unknown atomic geometry, interdiffusion and new chemical compound formation. Nevertheless, attempts have been made to establish

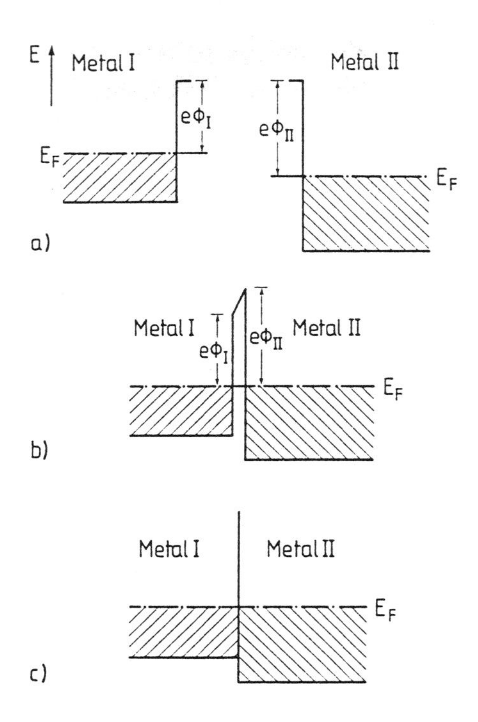

Fig. 8.1a-c. Schematic diagram of the formation of a metal-metal interface. (**a**) Simplified band schemes of the two metals I and II with work functions $e\phi_I$ and $e\phi_{II}$; the metals are assumed to be not in contact with each other; E_F is the Fermi energy. (**b**) Band schemes of the conduction bands at the interface for the two metals in contact. (**c**) Simplified representation of the metal-metal contact; the dipole layer at the interface is not shown

more or less general theoretical models to describe the electronic properties near such an interface. The success of these models depends, of course, on the degree of complexity. It is thus useful to consider first some ideal cases of interfaces with atomically sharp profiles and negligible intermixing of the two components. A simple case is the ideal interface between two metals I and II. The essential electronic properties are described by the bulk conduction bands, their band structure and the position of the Fermi levels, i.e. the work functions $e\phi_I$ and $e\phi_{II}$ (Fig. 8.1). If we bring the two metals into contact, thermal equilibrium requires that the chemical potential, i.e. the Fermi energy, is the same on both sides of the interface, i.e. the Fermi levels in the two metals have to be aligned. In order to establish this situation, electronic charge flows from metal I with the lower work function $e\phi_I$ into metal II with work function $e\phi_{II}$. The resulting space charge, positive in metal I (lack of electrons) and negative in metal II (additional free electrons), creates an interface dipole layer consisting of positive cores (metal I) and additional free electrons (metal II). The charge imbalance on each side of the interface is, of course, screened by the high density of conduction electrons in the metals. The size of the dipole layer can therefore easily be estimated by considering the screened Coulomb potential of a point charge [8.1]

$$\phi(r) = (Q/r)e^{-r/r_{TF}} , \quad (8.1)$$

where

$$r_{TF} \simeq 0.5(n/a_0{}^3)^{-1/6} \quad (8.2)$$

is the **Thomas-Fermi screening length**, a_0 the Bohr radius, and n the electron density. For Cu with $n = 8.5 \cdot 10^{22}$ cm^{-3}, r_{TF} is about 0.55 Å. The interface dipole layer therefore has a thickness of some Ångstroms. Two or three atomic layers further away from the interface, the electronic structure of metals I and II is essentially unperturbed.

Because of the small extent of this interface dipole layer, even though it is of atomic dimensions, a representation, as shown in Fig.8.1b, is adequate. The figure depicts the potential drop between the two metals along the interface dipole layer arising from the difference in work functions. One also encounters representations of the type shown in Fig.8.1c, in which the dipole layer is not plotted explicitly.

In this simple, idealized picture of a metal-metal interface a contact potential ΔV_c arises due to the difference in work function:

$$e\Delta V_c = e(\phi_{II} - \phi_I) . \quad (8.3)$$

Knowing the work functions for Ag (4.33eV), Cu (4.49eV), Au (4.83eV), Ni (4.96eV) and Al (4.29eV), one can readily estimate the contact potentials for the following metal contacts: Cu-Ag (0.16eV), Au-Ag (0.50eV), Ni-Ag (0.63eV), Al-Ag (−0.04eV). The difficulties involved in this simple approach are obvious: the work function as it is used here is a more or less theoretical quantity. The real work function is dependent on the particular type of surface orientation and is strongly affected by any rearrangement of the surface atoms, e.g. relaxations and reconstructions. The work function measured on a clean, well-defined surface under UHV conditions (Panel XV: Chap.9) depends on details of the given surface atomic structure. Within a real metal contact, even under ideal conditions (clean, well ordered, no interdiffusion, etc.), the atomic structure of the clean surface is rarely present; hetero-epitaxy of a layer of a second metal on top of a metal substrate usually changes relaxations and reconstructions. The work functions measured for the two metals separately therefore have no real meaning for the metal-metal interface. This problem affects all models that attempt to describe the electronic properties of a solid-solid interface in terms of simple work-function considerations.

We move on now to consider an ideal metal-semiconductor junction. As a first oversimplified approach one might again apply arguments based on the work function to give some initial insight into the problem. In ther-

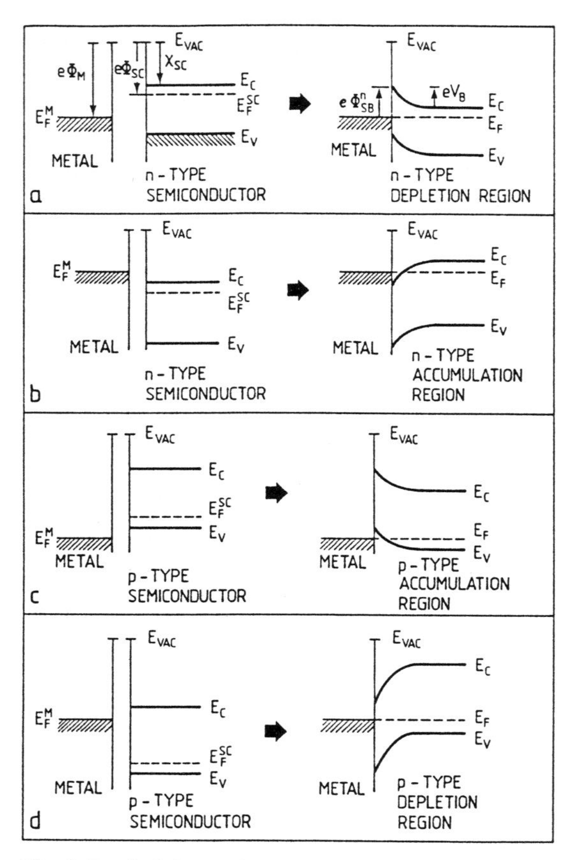

Fig. 8.2a-d. Schematic diagrams of band bending before and after metal-semiconductor contact: (**a**) High-work-function metal and n-type semiconductor, (**b**) low-work-function metal and n-type semiconductor, (**c**) high-work-function metal and p-type semiconductor, and (**d**) low-work-function metal and p-type semiconductor [8.2]

mal equilibrium the Fermi levels in the two materials must be aligned. Depending on the difference between the work function $e\phi_M$ of the metal and the electron affinity χ_{sc} of the semiconductor, different situations may arise, as shown in Fig. 8.2. When the two materials are brought into contact, the matching of the Fermi levels invariably causes charge to flow from one side to the other, and a dipole layer is built up at the interface. In the metal, the participating charge is screened within a few Ångstroms, as was discussed above; see (8.2). In a semiconductor, however, the free carrier concentration is orders of magnitude lower than in a metal and thus shielding is much less effective. Thus the space charge usually extends hundreds of Ångstroms into the crystal. In the semiconductor a normal space-charge layer, depletion or accumulation, is formed, as in the presence of surface states on a clean surface (Chap. 7). For example, for the high-work-function metal and n-type semiconductor depicted in Fig. 8.2a, electrons flow from the semiconductor to the metal after contact, depleting a characteristic surface region in the semiconductor of electrons. The upwards band bending in this depletion layer is correlated via Poisson's equation (7.9) with the positive space charge of ionized donors. In the present simple model this positive space charge is balanced on the metal side by a corresponding excess electronic charge extending only over an atomic distance. According to the calculations in Sect. 7.5, the total space charge is expected to be on the order of 10^{12} elementary charges per cm^2 or less. The maximum band bending eV_B at the interface is related to the potential barrier $e\phi_{SB}$ (Schottky barrier) which has to be overcome when an electron is excited from the metal into the conduction band of the semiconductor.

According to this simple picture (Fig. 8.2) a knowledge of the metal work function $e\phi_M$ and of the electron affinity χ_{sc} of the semiconductor would allow one to predict Schottky barrier heights. This simple approach in terms of work functions and electron affinities was first applied by Schottky to understand the rectifying action of metal-semiconductor junctions [8.3]. However, simple considerations based on the metal-metal contact already tell us that this work-function plus electron-affinity approach must fail in explaining the details of metal-semiconductor junctions.

Figure 8.3 shows measured Schottky barrier heights $e\phi_{SB}$ for metals with various work functions deposited on UHV-cleaved Si(111)-(2 × 1) sur-

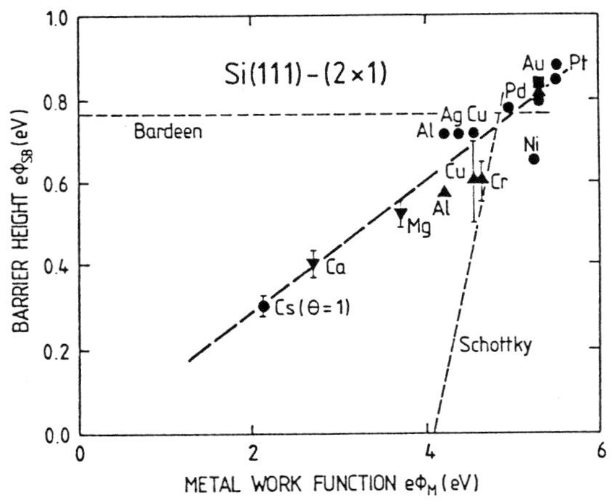

Fig. 8.3. Barrier heights $e\phi_{SB}$ of Si-Schottky contacts versus metal work function $e\phi_M$. The data were obtained by different researchers from measurements on the metal-covered Si(111)-(2 × 1) cleaved surface. For comparison the predictions of the Schottky model (no interface states) and the Bardeen model (high density of interface states) are given [8.4]

faces. According to the Schottky model one would expect a much greater variation of the barrier height with work function than is experimentally observed.

Apart from the fact that the work-function concept, as defined for a clean, adsorbate-free surface, cannot simply be transferred to a realistic solid-solid interface (see preceding discussion of metal-metal contact), there is a further problem which has not been taken into account in Fig.8.2. When the metal atoms come into close contact with the semiconductor surface, they will form chemical bonds whose strength will depend on the nature of the partners. The distribution of intrinsic surface states of the clean semiconductor surface will be changed. Additionally, charge will flow from one side to the other due to the formation of the bonds. This may be described by the formation of a dipole layer of atomic dimensions. The direction of the dipoles is determined by the difference between the work function $e\phi_M$ and the electron affinity of metal and substrate. Stronger interactions, such as the formation of alloys, interdiffusion, etc. are also possible, of course. In all cases one can expect the formation of new electronic interface states whose charge depends on their charging character (donors or acceptors) and on the position of the Fermi level. This situation is better described in a more extensive band scheme of the metal-semiconductor interface, which includes the interface region with its atomic dimensions (Fig.8.4). Using the metal work function $e\phi_M$ and the electron affinity χ_{sc} of the semiconductor (even though these quantities are not well defined within the junction) one might, at least in a model description, attribute an interface dipole Δ to the interface layer. The interface region ($\approx 1 \div 2$ Å) and the extent of the space charge layer (typically 500 Å) are, of course, not to scale in Fig.8.4. With the additional presence of interface states the total charge balance at the interface now includes the space charge in the semiconductor, the charge in the metal and the charge located in the interface

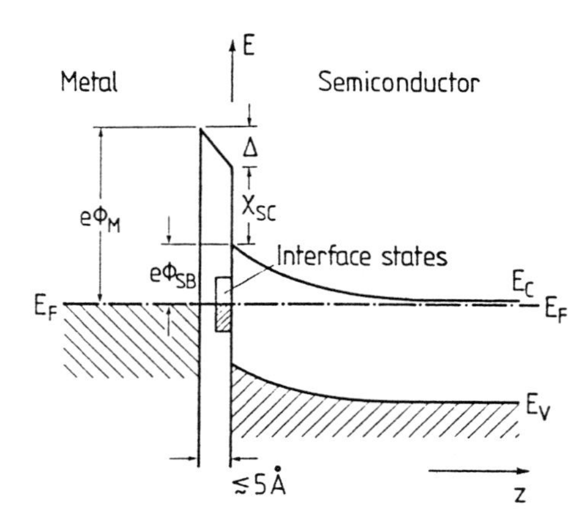

Fig.8.4. Band diagram of a metal-semiconductor junction in which the interface region (width ≈ 5 Å) is taken into account explicitly. The formation of new interface states of sufficient density pins the Fermi-level; $e\phi_M$ is the metal work function, χ_{sc} the electron affinity of semiconductor, $e\phi_{SB}$ the Schottky barrier height, and Δ the interface dipole energy

states. In thermal equilibrium no net current can flow through the junction. It is thus obvious that the band bending, i.e. the Schottky barrier height, depends sensitively on the nature of these interface states. The origin and character of these states therefore form the central topic of current research into Schottky-barrier formation.

The first step in this direction was made by *Bardeen* [8.5] in explaining the deviation of experimental data from the Schottky model (Fig. 8.3). Because of the lack of any knowledge about interface states he assumed that the surface states of the clean semiconductor surface persist under the metal overlayer and that they pin the Fermi level (Sect. 7.5). The work function of the deposited metal would thus have no effect on the position of the Fermi level at the interface. For metal coverages in the monolayer range which are unable to screen interface charge, surface-state densities as low as 10^{12} cm^{-2} could pin the Fermi level at a fixed position. The assumptions of the Bardeen model (Fig. 8.3) are, of course, incorrect, since the surface states of the clean surface are strongly affected by metal deposition.

The case of a semiconductor-semiconductor heterojunction (Fig. 8.5) is even more complex, since the two different length scales, that of *atomic-*

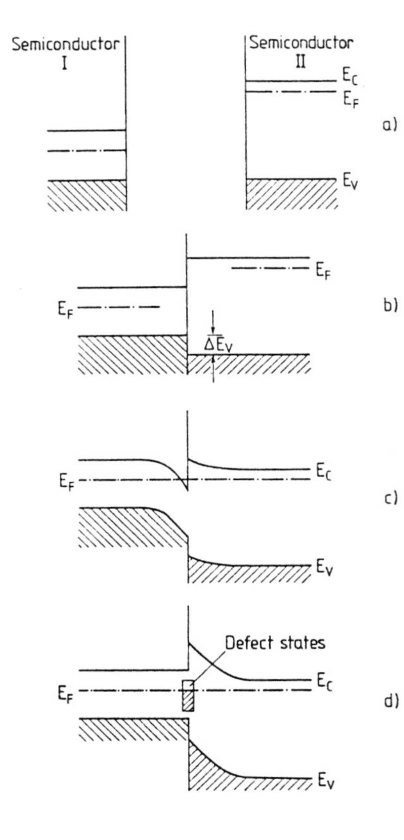

Fig. 8.5a-d. Formation of a semiconductor heterostructure shown qualitatively in its band-scheme representation. (**a**) The two semiconductors (I: moderately n-doped and II: highly n-doped) are separated in space. (**b**) The two semiconductors are brought into contact, but are not in thermal equilibrium (Fermi levels do not coincide). (**c**) The two semiconductors are in contact with thermal equilibrium established; ideal interface without any interface states. (**d**) Semiconductor heterostructure with a high density of (defect) interface states which pin the Fermi level at the interface near midgap

sized interface dipoles and that of quasi-macroscopic space charge effects (band bending) are both important in determining the final band scheme in thermodynamic equilibrium (Figs.8.5c, d). As an example we consider two semiconductors I and II with small and large band gaps, respectively (Fig. 8.5a). Semiconductor I is moderately n-doped, whereas semiconductor II is highly n-doped with the Fermi level slightly below the bulk conduction-band edge. When the two crystals come into contact – we assume for now an ideal contact with perfect lattice matching and no defects – two questions have to be answered: how do the two band schemes, in particular the forbidden bands, align with one another (Fig.8.5b) and what band bending is established, in order to match up the Fermi levels in thermal equilibrium (Fig. 8.5c)? The first question relates to the most important property of a semiconductor heterostructure, the band offset ΔE_V (Fig.8.5b), i.e. the relative positions of the two valence-band maxima (or equivalently of the two conduction-band minima: ΔE_C). The factors that determine this quantity are properties of the two intrinsic semiconductors. In the simplest, but highly questionable assumption (see discussion above), the relative positions of the two-band structures would follow from aligning the vacuum potentials of the two semiconductors. The band offset ΔE_C would then be determined as in the case of the Schottky model for the metal-semiconductur junction, by the difference in electron affinities χ_I and χ_{II}

$$\Delta E_C = \chi_I - \chi_{II} \ . \tag{8.4}$$

Because of the difficulties in attributing an electron affinity (determined for a free surface) to a solid-solid interface, other microscopic levels derived from the bulk band structure of the two semiconductors have been used as common levels for aligning. However, none of these attempts take into account that charged electronic states, even those of the ideal interface, might form a dipole layer of atomic dimensions. A reasonable procedure is thus to arrange the two band structures i.e. the band offset, such that a zero interface dipole results. The line-up in this case results from the electronic properties of a few atomic layers at the very interface, even though these properties are themselves determined essentially by the bulk band structures [8.6-9]. This promising approach to understanding band offsets in ideal semiconductor heterostructures is discussed in more detail in Sect.8.3.

Once the band offset ΔE_V (Fig.8.5b) has been determined theoretically, space-charge theory (Chap.7) is used to derive the actual band bending on each side of the heterostructure (Fig.8.5c). The particular bulk doping of the two semiconductors fixes the positions of the respective Fermi levels in relation to the conduction-band edges deep in the two materials. In thermal equilibrium the Fermi levels in the two materials have to be aligned.

As for a simple p-n junction, the heterostructure must be current free in thermal equilibrium. This determines the band bending in the two semiconductors. In the example of Fig. 8.5c the negative space charge in the strong accumulation layer in semiconductor I is balanced by the positive space charge of ionized donors in the depletion layer of semiconductor II. It must be emphasized that this effect of Fermi level alignment occurs on a scale completely different from that of the band structure alignment that gives rise to ΔE_V. The energy difference ΔE_V is maintained by charge compensation within one or two atomic layers and charge densities on the order of 10^{15} cm^{-2} (monolayer density), whereas the space-charge effects involve charge densities of about 10^{12} cm^{-2} elementary charges or less (Chap. 7), and this on a length scale of several hundred angstroms. This is the reason why the two alignment steps can usually be performed separately in the theory. This separation procedure might not be possible for extremely strong space charge layers whose spatial extent is comparable to that of the atomic interface dipoles.

Finally, it should be mentioned that interface defects such as impurity atoms, dislocations, uncompensated dangling bonds, etc. might give rise to electronic interface states in the forbidden band of the semiconductors. These might pin the Fermi level and contribute according to their charging character to the total charge balance. This, in turn, can significantly change the band bending (Fig. 8.5d) and thus conceivably also the band offset.

8.2 Metal-Induced Gap States (MIGS) at the Metal-Semiconductor Interface

Numerous experiments on a wide variety of metal-semiconductor systems prepared under UHV conditions indicate that the deposition of metal films produces interface states which determine the position of the Fermi level at the interface. Well-known examples are metal overlayers on cleaved GaAs (110) surfaces. A clean, well-cleaved GaAs(110) surface usually has flat bands, i.e. the forbidden band is free of surface states (Sect. 6.5). Deposition of metal atoms causes band bending in both p- and n-type materials at metal coverages far below one monolayer (ML) ($\lesssim 0.2 \div 0.5$ML) (Fig. 8.6). For n-type material an electron depletion layer is formed with the Fermi level E_F pinned at about 0.8 eV above the valence-band maximum E_V; on p-type crystals a hole depletion layer is built up with E_F between 0.5 and 0.6 eV above E_V (Fig. 8.7). A common characteristic of all III-V semiconductor surfaces studied so far is that a very low deposition (<0.5ML) is sufficient to establish a fixed Fermi level. This pinning of E_F is caused by interface

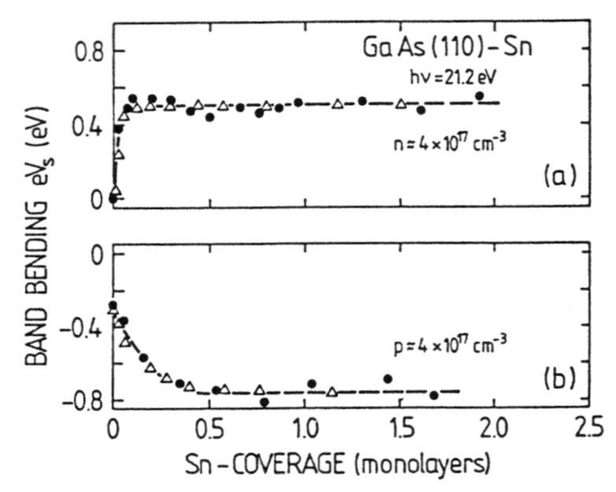

Fig. 8.6. Band bending eV_s versus Sn coverage [in monolayers referred to the surface atom density of (110) GaAs], measured by UV photoemission (HeI line, $h\nu = 21.2\,\mathrm{eV}$) (**a**) upwards bending on an n-type GaAs(110) surface (**b**) downwards bending on a p-type GaAs(110) surface [8.10]

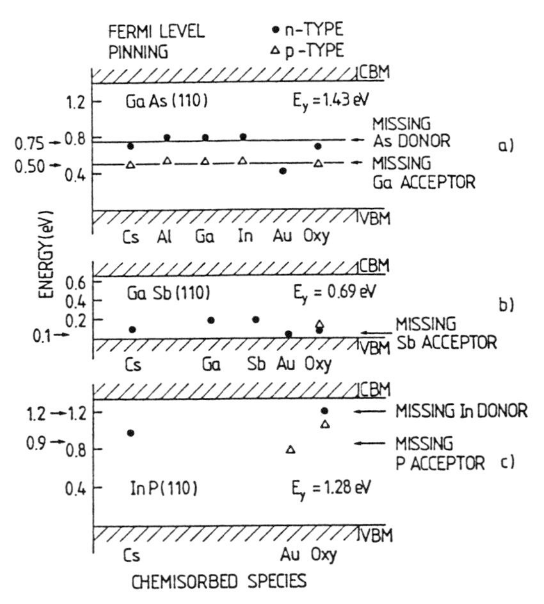

Fig. 8.7. Pinning positions within the band gaps of (**a**) GaAs, (**b**) GaSb, and (**c**) InP (as derived from photoemission (XPS) spectra) produced by monolayers or less of various adsorbates. Circles and triangles indicate energy positions for n- and p-type semiconductors, respectively, to an accuracy of $\pm\,0.1$ eV or better. Labels for various energy levels indicate their proposed electrical activity and chemical nature. CBM: conduction band minimunm, VBM: valence band maximum [8.11, 12]

(or surface) states with a minimum density of about 10^{12} cm^{-2} (Sect. 7.5). Further examples of such Metal-Induced Gap States (MIGS) are discussed in Sect. 8.4 in connection with other semiconductors.

What is the nature of these interface states? We will see that the situation is actually quite complex and that the states probably have a variety of different origins depending on the degree of complexity of the interface (Sect. 8.4). Nevertheless, some general conclusions can be drawn if one makes the assumption of an ideal metal-semiconductor interface. *Heine* [8.6] put forward the idea that metal (Bloch) wave functions tail into the semiconductor in the energy range in which the conduction band of the metal overlaps the forbidden band of the semiconductor. The situation is represented in detail in Fig. 8.8. The familiar band structure with real k-

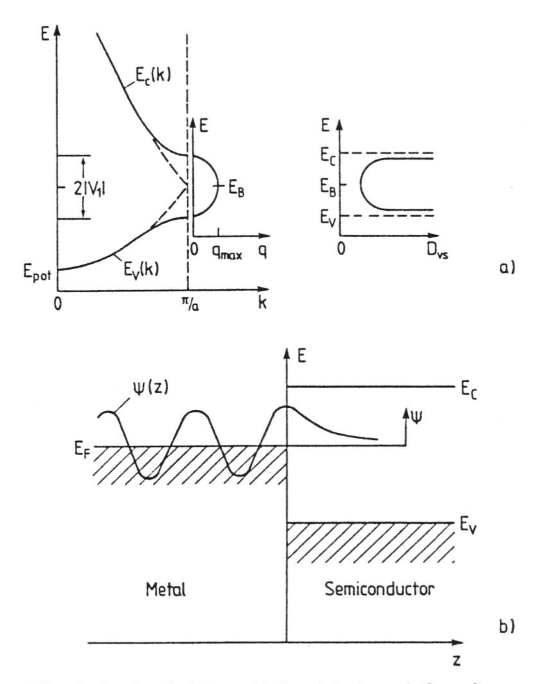

Fig. 8.8a,b. Origin of Metal-Induced Gap States (MIGS) at a metal-semiconductor interface. (**a**) Dispersion of a one-dimensional chain of atoms (model of semiconductor); the real band structure with $E_C(k)$ and $E_V(k)$ as conduction and valence bands, respectively, is obtained for an infinite chain. For a finite chain, exponentially decaying interface states with imaginary wave vectors $\kappa = -iq$ fill the gap of the semiconductor between E_C and E_V, and in the simplest case are symmetric about the so-called branching point energy E_B. The density of states D_{VS} of these states has singularities near E_C and E_V. (**b**) Qualitative representation of a metal Bloch state (near E_F), decaying into the semiconductor. The tail of the wave function arises because $\psi(z)$ cannot change abruptly to zero in the semiconductor, where no electronic states exist in the forbidden band

values is strictly applicable only to an infinite crystal (one-dimensional problem in Fig.8.8a). According to the arguments of Sect.6.1, the breakdown of periodicity at the interface introduces exponentially decaying interface states with imaginary wave vectors $\kappa = -iq$. In the simple, one-dimensional model of Fig.8.8a their dispersion curve E(q) "fills" the energy gap of the semiconductor symmetrically with respect to the band edges E_C and E_V. These gap states have a density of states with singularities near E_C and E_V (Fig.8.8a). The dispersion E(q) in Fig.8.8a indicates the range of theoretically possible states known sometimes as **Virtual Induced Gap States** (VIGS). Which of these states actually exist depends on the boundary conditions at the interface. In the case of a free surface (Sect.6.1), for example, the solutions to the Schrödinger equation in the adjacent vacuum have to be matched to the VIGS and one obtains only a single level for each $\mathbf{k}_{\|}$, namely the surface state $E(\mathbf{k}_{\|})$. In the present case of a metal-semiconductor interface (Fig.8.8b), the Bloch states of the metal conduction band have to be matched and a broad continuum of states results in the energy gap of the semiconductor. One might say that the continuum of metal states "leaks" into the VIGS which are themselves derived from the semiconductor band structure. A detailed calculation, as given in Sect.6.1, reveals that the VIGS are split off partly from the valence band and partly from the conduction band of the bulk. Each state in the gap is a mixture of valence- and conduction-band states. Accordingly they exhibit more donor or more acceptor character in the lower or upper half of the gap, respectively. There exists a neutrality level E_B (also called the **cross-over energy** or **branching point**) which separates the more donor-like VIGS from the more acceptor-like states. In the simple model of Fig.8.8a, E_B is obviously at midgap. Depending on the details of the 3D band structure of a real crystal, E_B is located somewhere near the middle of the semiconductor gap. The actual calculation of E_B, as performed, e.g., by *Tersoff* [8.13-16], is based on a summation over bulk states. The position of E_B is determined by the weights of conduction- and valence-band-derived densities of states. An approximate calculation is possible from a knowledge merely of the bulk band structure of the semiconductor.

The central importance of the branching point E_B for Schottky-barrier formation is obvious: if the Fermi level lies in the energetic range of the more donor-like VIGS, i.e. below E_B, ionized (empty) donors build up a large positive interface charge. On the other hand, if E_F is above E_B this leads to ionized (occupied) acceptor states and a negative interface charge. Energy considerations thus make it favorable for the Fermi level to be located close to the VIGS branching point E_B. Slight deviations of E_F from E_B are due to charge transfer within the chemical bonds of the interface. This can be taken into account by calculating the interface dipole Δ (Fig. 8.4) from the charge transfer δq using standard formulas (9.23). According to *Pauling* [8.17] (revised in [8.18]) the intrabond charge transfer between metal and semiconductor can be obtained from the empirical formula

$$\delta q = \frac{0.16}{eV} | X_M - X_{SC} | + \frac{0.035}{(eV)^2} | X_M - X_{SC} |^2 \qquad (8.5)$$

based on the electronegativities X_M and X_{SC} of the metal and the semiconductor, respectively. For Si and Ge, X_{SC} is 1.9 and 2.01, respectively, and for III-V and II-VI semiconductors X_{SC} should lie close to 2.0 ± 0.1, the average of the two elemental components. For most metals X_M is also close to 2. Thus, in a first approximation, δq and also the interface dipole Δ could be neglected. One can then estimate Schottky barrier heights in this model by identifying the Fermi level E_F at the interface with the branching point E_B of the VIGS. More elaborate treatments of course take into account the charge transfer [8.9]. They yield results which can clearly be observed experimentally.

The existence of interface states (VIGS) over the whole gap has also been derived from realistic, atomic calculations of metal-semiconductor interfaces. As an example, Fig. 8.9 shows results for the Si(111)-Al interface according to *Cohen* [8.19]. The interface is modeled by three slabs of Al (regions I-III) described by a jellium model and three slabs of Si (regions IV-VI) using the pseudopotential approach. For the different slab regions local state densities are calculated. Region I appears just like bulk Al, region VI just like bulk Si. Al states retain their bulk-like properties up to the very interface (region III). The strongest deviations from bulk properties are observed at the interface itself. In particular, the whole gap region of Si around E_F contains new interface states which decay rapidly into the Si (region IV). In general, the behavior shown in Fig. 8.9 is completely analogous to that discussed by *Heine* [8.6] for the leaking of Bloch states into VIGS (Fig. 8.8b): The bulk-like Al states decay into the Si whenever the semiconductor band gap overlaps the metallic band. Very similar results to Fig. 8.9 have been obtained by other researchers using different theoretical methods and other interface systems, e.g. metals on III-V [8.20, 21] and on II-VI semiconductors [8.21, 22]. The existence of MIGS or VIGS originating from decaying Bloch states of the metal conduction band is well established in all these cases.

Although these states are certainly important for Fermi-level pinning at the interface, they might not be the only factor. A common shortcoming of the calculations discussed above is the assumption of idealized atomic geometries at the interface. This assumption is justified only in a small minority of cases. There is nevertheless a large body of experimental data that supports the importance of MIGS (or VIGS) for a basic understanding of

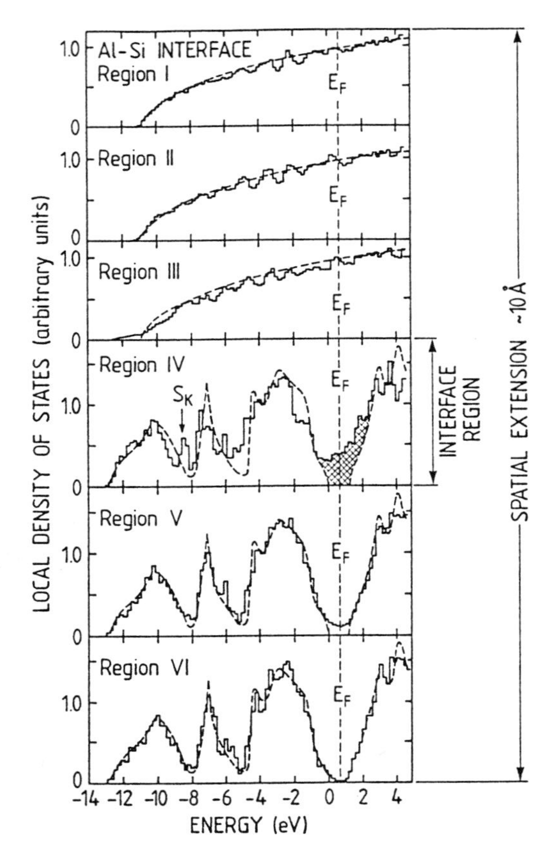

Fig. 8.9. Local density of states for the Al-Si(111) interface, as obtained from a slab model calculation; the Al is described in terms of a jellium model. The different regions (I-VI) describe a total spatial extent of about 10 Å symmetrically located about the Al-Si interface. For the true interface region (IV), metal-induced gap states (*shaded*) fill the forbidden band of the semiconductor [8.19]

Schottky barriers [8.4, 8, 9]. Table 8.1 presents the results of *Tersoff*'s calculations [8.14] of the branching-point energies E_B of VIGS for important elemental and III-V semiconductors using the complete bulk-band structures. The positions of E_B with respect to the upper valence-band edge compare quite well with the Fermi levels measured for the corresponding semiconductor-Au and Al interfaces. The deviations are typically in the order of $100 \div 150$ meV.

Figure 8.10 shows a compilation of measured Schottky barrier heights for Au on a number of semiconductors plotted versus the branching-point energies $(E_B - E_V)$ referred to the valence-band edge [8.4]. A strong correlation is found over a relatively wide energy range.

Tersoff has also suggested an approximate, semi-empirical method for obtaining the branching point energy E_B of the VIGS without performing complex mathematical sums over the bulk band structure in reciprocal space [8.15]. As was pointed out above, the branching point E_B falls near the center of the gap for the one-dimensional problem. In three dimensions the

Table 8.1. Semiconductor "midgap" energy E_B from [8.13-16] (*a*) and from [8.50] (*b*), and experimental Fermi-level positions E_F at metal (Au, Al)- semiconductor interfaces, relative to the valence-band maximum [8.13-16]]

	$E_B - E_V$ [eV]		$E_F(Au) - E_V$ [eV]	$E_F(Al) - E_V$ [eV]
	(*a*)	(*b*)		
Si	0.36	0.23	0.32	0.40
Ge	0.18	0.03	0.07	0.18
AlAs	1.05	0.92	0.96	
GaAs	0.50	0.55	0.52	0.62
InAs	0.50	0.62	0.47	
GaSb	0.07	0.06	0.07	
GaP	0.81	0.73	0.94	1.17
ZnSe	1.70	1.44	1.34	1.94
CdTe	0.85	0.73	0.68	
HgTe	0.34			

center of the gap has no fundamental meaning for the complex band structure of the VIGS; the point E_B, where conduction- and valence-band-derived states cross over, depends on a variety of wave-function contributions from different directions in the Brillouin zone. The main contributions to the sum, and those which most affect the position of E_B, come from regions where the local density of states is high. The band structure near the

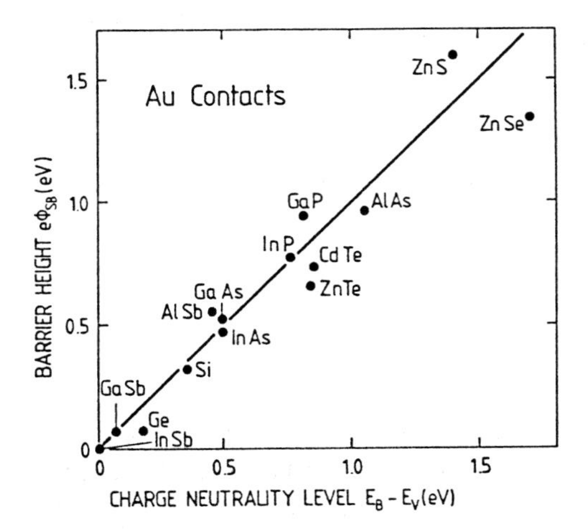

Fig. 8.10. Measured Schottky barrier heights for Au contacts on a number of semiconductor surfaces plotted versus the branching point energies (charge neutrality level, $E_B - E_V$) as determined theoretically by *Tersoff* [8.8]. After [8.4]

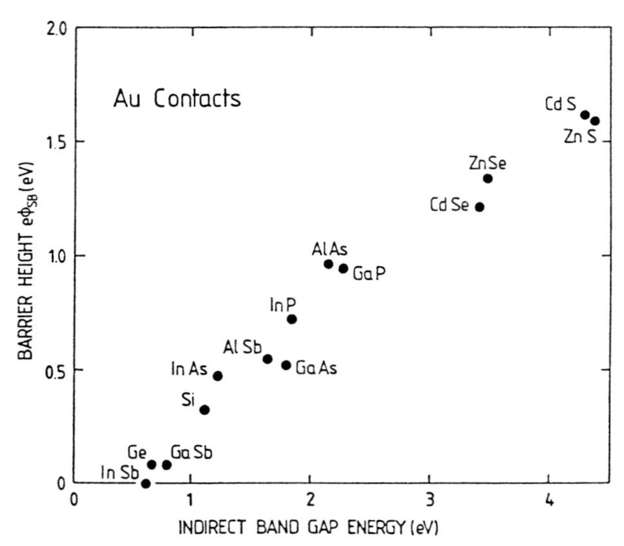

Fig. 8.11. Measured Schottky-barrier heights for Au contacts on a number of semiconductor surfaces plotted versus the indirect band-gap energy [8.4]

zone center (Γ) determines the direct gap in many III-V semiconductors, yet its energy bears little relation to the conduction band as a whole: Because of the usually very small effective electron mass at Γ, there is little k space associated with this minimum. On the other hand, minima associated with indirect gaps on or near the zone faces exhibit a large effective mass, they describe a large region of the conduction band in k space and are therefore much more characteristic of this band. Accordingly states from regions near the indirect minima have much more effect on the position of E_B than those from the minimum at Γ. Instead of estimating E_B by taking the middle of the direct gap at Γ, one should clearly refer to the indirect gap, regardless of the energy of the Γ minimum.

One indeed finds a strong correlation between Schottky-barrier heights measured on a number of Au-covered semiconductors and the corresponding indirect gap energy (Fig. 8.11). As a semi-empirical rule therefore, *Tersoff* [8.14-16] suggested that the neutrality level or branching point energy E_B be taken as

$$E_B = \frac{1}{2}\left(\bar{E}_V + E_C^i\right), \tag{8.6}$$

where E_C^i is the indirect conduction band minimum, not located at the Γ point; $\bar{E}_V$ is the maximum of the valence band at Γ, but because of the spin-orbit splitting Δ_{SO} at Γ, $\bar{E}_V$ must be an average value. Relative to the threefold degenerate band (without spin-orbit splitting) the twofold-degenerate valence band maximum is pushed up in energy by an amount $\Delta_{SO}/3$, while the split-off state is pushed down by $2\Delta_{SO}/3$. To take both contributions into account, it is thus necessary to consider not the actual valence band maximum E_V, but an effective maximum

$$\bar{E}_V = E_V - \frac{1}{3}\Delta_{SO} . \tag{8.7}$$

To calculate the Fermi level at the interface, i.e. the Schottky-barrier height $e\phi_{SB}$ (Fig.8.4), one might also take into account the difference δ_M between E_F and the branching point E_B due to the interface dipole Δ. Since δ_M describes the intrabond charge transfer, it will depend on the particular metal involved. For a given metal δ_M can be taken as a fit parameter and (8.6, 7) can be used to calculate the Schottky barrier heights for different semiconductors. With $E_g^i = E_C^i - \bar{E}_V$ as the minimum indirect gap energy the Schottky barrier height (Fig.8.4) follows as

$$e\phi_{SB} \simeq \frac{1}{2}\left(E_g^i - \frac{1}{3}\Delta_{SO}\right) + \delta_M . \tag{8.8}$$

Table 8.2 presents the relevant values for Au layers on a variety of semiconductors [8.14]. The branching points E_B and the Schottky-barrier heights $e\phi_{SB}$ (theory) are calculated according to (8.6-8). There is astonishingly good agreement with the experimental values $e\phi_{SB}$(exp). The parameter δ_M has been chosen to be $\delta_{Au} = -0.2$ eV so as to give the best overall fit to all Au data. Table 8.2 also demonstrates the success of the MIGS (or VIGS) model for describing Schottky barriers, at least for systems whose interface atomic structure is not too complex.

One must keep in mind, however, that the microscopic details of charge redistribution at the interface, atomic rearrangement or more complex behavior such as chemical reactions in deeper layers or defect formation, have not been included in the model so far. As will be shown in Sect. 8.4, these phenomena are important for certain interfaces and must be incorporated into the more refined models needed to describe more complex metal-semiconductor junctions.

One last point is worthy of mention. For many metal-semiconductor systems, e.g. metals on III-V semiconductor surfaces, Fermi-level pinning is observed even at very low coverages of less than a monolayer of metal ($\approx 10^{15}$ atoms/cm^2). This observation is by no means in contradiction to the MIGS model, although most of the discussion tacitly referred to thicker metal films. Submonolayer coverage means that no Bloch waves of bulk-like

Table 8.2. Semiconductor minimum indirect gap energies E_g^i, spin-orbit splittings Δ_{SO} and Schottky barrier heights $e\Phi_{SB}$ under gold contacts from theory and experiment [8.14]

	E_g^i [eV]	Δ_{SO} [eV]	Φ_{SB} (theory) [eV]	Φ_{SB} (expt) [eV]
Si	1.11	0.04	0.35	0.32
Ge	0.66	0.29	0.08	0.07
GaP	2.27	0.08	0.92	0.94
InP	1.87	0.11	0.70	0.77
AlAs	2.15	0.28	0.83	0.96
GaAs	1.81	0.34	0.65	0.52
InAs	1.21	0.39	0.34	0.47
AlSb	1.63	0.7	0.50	0.55
GaSb	0.80	0.75	0.07	0.07

metals can "leak" into the VIGS. But if the metal adsorbate atoms yield electronic states within the energy range of the semiconductor gap, as thin metal films obviously do, then these states can play the role of the Bloch states of a thick metal layer. A state density on the order of 10^{12} cm^{-2} is already sufficient to pin the Fermi level (Sect.7.5). This does not necessarily mean that the pinning position at submonolayer coverage coincides with that due to a thick metal film. Because of screening by the free metal electrons, a much higher density of interface states ($\gtrsim 10^{14}$cm^{-2}) is required for Fermi level pinning with a thick metal film than in the case of submonolayer coverage ($\approx 10^{12}$cm^{-2}); more details are given in Sect.8.4.

8.3 Virtual Induced Gap States (VIGS) at the Semiconductor Heterointerface

The VIGS concept [8.13-16] also contains the key to understanding band line-ups in ideal semiconductor heterostructures. As was already pointed out in connection with metal-semiconductor junctions (Sect.8.2), any electronic state in the gap of a semiconductor, including VIGS, is necessarily a mixture of valence- and conduction-band states. The closer the state is to the valence-band edge, the more valence character it has. Nevertheless it always includes a certain admixture of conduction-band wave function. On the other hand, gap states lying close to the bulk conduction-band edge are composed to a large extent of conduction-band wave functions. This is shown qualitatively in Fig.8.12a, where the conduction (broken line) and valence (full line) character of VIGS is plotted over the forbidden band of a model semiconductor with symmetric valence and conduction bands. The crossover point (also branching point or neutrality level) E_B occurs where the gap states have equal valence and conduction character, i.e. where a net donor-like behavior (lower part of gap) changes into a net acceptor-like behavior (upper part of gap). The occupation of a state in the gap leads to a local excess of electronic charge in proportion to its degree of conduction character. If the gap state is empty, there is a local positive charge (electron deficit) in proportion to the state's valence character. If the state lies near the bottom of the gap, filling it corresponds only to a slight negative charge, since little conduction character is involved. Leaving that state empty, however, results in a deficit of about one electronic charge (approximately one hole in the valence band). On the other hand, filling a state high in the gap gives a large negative charge, while leaving it empty gives a small local charge deficit. It is this behavior of VIGS that is responsible for controlling the Schottky barrier of an ideal metal-semiconductor junction (Sect.8.2). When the Fermi level E_F lies close to the branching point E_B of the VIGS, overall charge compensation occurs (Fig.8.12a). In the lower half of the gap the occupied states carry only tiny amounts of negative charge due to their weak conduction-band character; but an equally small amount of positive charge is related to the empty states in the upper part of the gap because of their small admixture of valence character. In particular, large interface charge is avoided, since the states with predominantly conduction character are empty and states with mainly valence character are occupied.

As is now easily seen from Figs.8.12b,c similar considerations are possible for an ideal semiconductor heterostructure consisting of two hypothetical semiconductors I and II with symmetrical valence and conduction bands but different energy gaps. Since semiconductor I has a smaller gap than semiconductor II, there are regions in the gap of II where the continuum of bulk valence- and conduction-band states of I leak into the gap of semiconductor II. Thus, in a limited energy range in the upper and lower parts of the gap of semiconductor II there exist continua of VIGS derived from the bands of both semiconductors I and II, as found in the whole gap in the case of a metal overlayer. Even when there are no VIGS at the Fermi level, a change in the band line-up can cause net excess charge (net dipoles) at the interface (Fig.8.12b,c). In the situation shown in Fig.8.12c, where the branching points E_B^I and E_B^{II} of the two semiconductors do not match, the negative charge in the VIGS below E_B^{II} exceeds (due to their weak conduction character) the tiny positive charge in the interface states in the upper

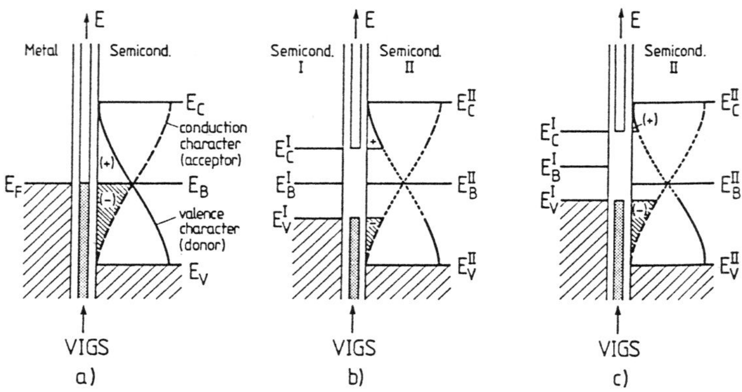

Fig. 8.12. Schematic explanation of the formation of a metal-semiconductor junction (**a**) and a semiconductor heterostructure (**b, c**) in terms of the Virtual Induced Gap States (VIGS) model. The band schemes of the metal and of the semiconductor are plotted with Fermi level E_F, conduction-band edge E_C and valence-band edge E_V. VIGS are plotted qualitatively within the interface (occupied states are shaded). Their charging character is also shown qualitatively within the forbidden band of the large gap semiconductor (II) (*dashed line*: acceptor- or conduction-band character; *full line*: donor- or valence-band character; *dotted line*: behavior in the energy range where VIGS are not present due to the totally forbidden band). (**a**) The position of the Fermi level with respect to the semiconductor band edges is determined by charge neutrality in the VIGS, i.e. E_F matches the branching point energy E_B where acceptor charging character equals donor character. (**b**) For a semiconductor heterostructure VIGS exist only in energy ranges in which one of the two semiconductors possesses conduction- or valence-band states. The VIGS are derived from wave functions of both semiconductors I and II. The matching of the two band schemes, i.e. the band off-sets, are determined by charge neutrality within the VIGS. Thus the two branching point energies E_B^I and E_B^{II} must coincide. (**c**) In contrast to (**b**) the branching point energies E_B^I and E_B^{II} are assumed not to coincide. Positive and negative charge in the VIGS is not balanced and the band scheme tends to adjust into a situation, as shown in (**b**)

half of the gap. This positive charge is, of course, due to the fact that these predominantly acceptor-type states nonetheless have a small amount of valence character, which carries positive charge if the state is not occupied. It is now obvious that both the positive and the negative interface charge residing in the VIGS is compensated when the branching points E_B^I and E_B^{II} in the two materials are aligned (Fig. 8.12b). For energetic reasons, the condition of zero interface dipole therefore requires alignment of the branching energies E_B. Tersoff's model describing the electronic properties of solid-solid interfaces is therefore general in so far as the energy level relevant for lining up the band structures of both metal-semiconductor junctions and semiconductor heterostructures is the branching point E_B of the VIGS. For an ideal semiconductor heterostructure the alignment of the

Table 8.3. Valence band offsets ΔE_V for various semiconductor interfaces derived theoretically from the branching point energies E_B in Table 8.1 (*a*) and from a similar theoretical approach [8.50] (*b*), together with experimental values from various sources. The listed heterostructures are to be understood as substrate materials (1st semiconductor) on which a layer, generally strained, is deposited (2nd semiconductor)

	Theory ΔE_V [eV]		Experiment ΔE_V [eV]
	(*a*)	(*b*)	
AlAs/GaAs	0.55	0.37	0.42 ± 0.08
InAs/GaSb	0.43	0.56	0.51
GaAs/InAs	0.00	0.18	0.17 ± 0.07
GaAs/Ge	0.32	0.51	0.42 ± 0.1
Si/Ge	0.18	0.12	0.17
CdTe/HgTe	0.51	0.67	0.35 ± 0.06

branching points E_B^I and E_B^{II} in the two semiconductors directly yields the valence-band offset (Fig.8.12b) as

$$\Delta E_V = (E_B^I - E_V^I) - (E_B^{II} - E_V^{II}) \, , \qquad (8.9a)$$

or if the branching point energies are referred to the valence band edges in each material ($\widetilde{E}_B = E_B - E_V$):

$$\Delta E_V = \widetilde{E}_B^I - \widetilde{E}_B^{II} \, . \qquad (8.9b)$$

Using the calculated branching point energies $\widetilde{E}_B$ of several semiconductors [8.13] in Table 8.1, expression (8.9b) readily yields the valence-band offsets for the heterostructures in Table 8.3. The comparison with experimental data must be made with some caution. Because of uncontrolled irreproducibilities in the epitaxial growth of those heterostructures (Sects. 2.4, 5), there are sometimes relatively large discrepancies between the data reported by different experimental groups. Also the theoretical values themselves are only reliable to within $0.1 \div 0.2$ eV because of the usual error limits in band structure calculations.

Another interesting consequence of the general connection between semiconductor heterostructures and Schottky-barrier heights within the VIGS model is seen by comparing Tables 8.1,3. Schottky-barrier heights measured on different semiconductors but with the same metal overlayer allow one to calculate valence band offsets. The measured Schottky-barrier heights $e\phi_{SB}$ have to be corrected for possible interface-dipole shifs δ_M (dependent on the metal, Sect. 8.2) to give the branching point energies E_B for each semiconductor. By then combining the two experimentally determined E_B values, one obtains the band discontinuity between the two semiconductors in the heterostructure according to (8.9b).

8.4 Structure- and Chemistry-Dependent Models of Interface States

The MIGS (VIGS) model of Schottky barriers is a general model that does not take into account any chemical reactions or structural changes occurring at the metal-semiconductor or semiconductor-semiconductor interface. It is therefore expected to be valid for largely ideal interfaces. However, film-growth studies (Chap. 3) show that for a number of metal-semiconductor systems quite complex behavior, including interdiffusion of substrate and film materials, is not uncommon. A clear dependence of the Schottky-barrier height on chemical reactivity has also been observed for many metal-semiconductor interfaces (Fig. 8.13) [8.23]. The experimentally determined barrier heights for many metals on ZnO, ZnS, Cds and GaP depend in a step-like way on the heats of interface chemical reaction ΔH_R, i.e., on the heats of formation of the most stable metal-anion bulk compounds. A steep transition between reactive ($\Delta H_R < 0$) and non-reactive ($\Delta H_R > 0$) junctions is observed.

Other models of interface states which emphasize the existence of interface reactions are therefore also of interest. The observation that a number of different metals and also adsorption of many gases including oxygen (Fig. 8.7) cause Fermi-level pinning at nearly the same energies on III-V semiconductor surfaces led *Wieder* [8.24], *Williams* et al. [8.25] and *Spicer* and his group [8.26] to assume that in all these cases defects at the interface are responsible for the pinning of the Fermi level. According to this **defect model** the deposition of metal atoms or atoms of a second semiconductor material leads to the formation of cation and/or anion vacancies (in GaAs, for example, V_{Ga} and V_{As}) or antisite defects, where a cation at the interface is replaced by an anion or vice versa (for example, As_{Ga} and Ga_{As} defect atoms). The energy required for defect formation is delivered to the surface by impinging adsorbate atoms. The defects destroy the charge neutrality in their vicinity, e.g., a Ga atom on a regular As site introduces a local negative charge deficiency and acts as an acceptor-type inter-

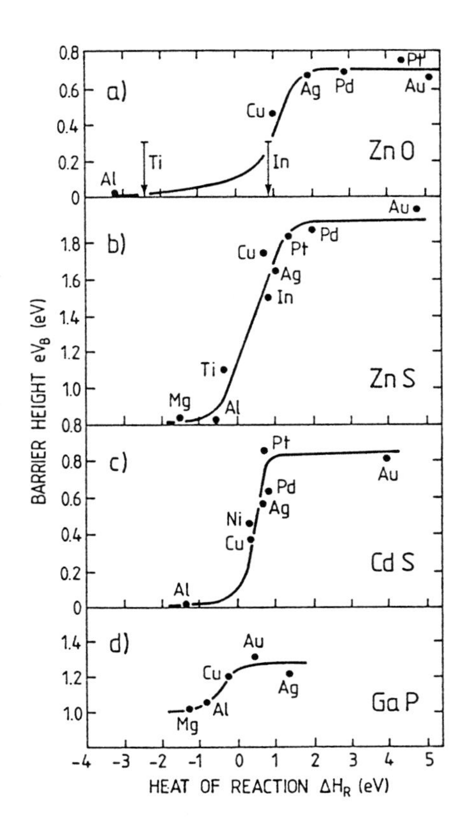

Fig.8.13. Schottky barrier heights correlated with the heat of reaction ΔH_R for metals on ZnO (**a**), ZnS (**b**), CdS (**c**) and GaP (**d**) [8.23]

face state. This Ga_{As} defect trap could conceivably capture one or two electrons. Conversely, one can show that an As_{Ga} antisite defect has donor character; it readily donates electrons into the conduction band. Detailed calculations of the corresponding energies of these defect states are now available. Despite the inherent inaccuracy of the calculated level energies, general trends are obtained (Fig.8.14) that are consistent with experimental findings. The defect levels shown in Fig.8.14 for GaAs surfaces might well explain the fact that the Fermi level is near midgap (Fig.8.7) for both n- and p-type material after the deposition of various metals (Cs, Al, Ga, etc.) and also after the adsorption of gases such as oxygen. For the exceptional case of InSb with its narrow band gap of about 180 meV most of the defect levels are found to be degenerate with the bulk conduction band states (Fig.8.14). The surface Fermi level on cleaved InSb(110) at Sn coverages (polycrystalline α-Sn) of more than 0.5 monolayers is experimentally determined to be 100 meV above the lower conduction-band edge (Fig. 8.15).

Another interesting case is that of the crystallographically well-matched GaAs/Ge interface (Fig.2.12). If Ge is deposited an a heated GaAs (110) substrate (T $\gtrsim$ 300°C), the surface mobility of the Ge atoms is high

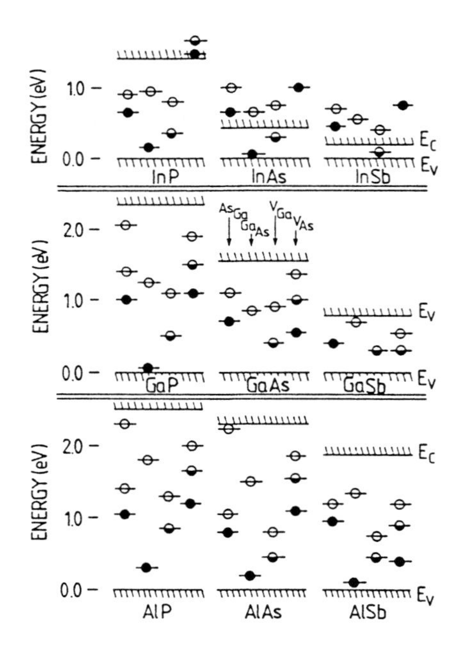

Fig. 8.14. Deep levels calculated for antisite defects and vacancies at (110) surfaces of III-V semiconductors. As indicated for GaAs, the levels are, from left to right, for anion on cation site, cation on anion site, cation vacancy and anion vacancy. The occupancy of the levels for the neutral charge state is shown; a black circle indicates that the level contains two electrons (spin up and down), a half circle one electron, and an open circle no electrons. (Charge-state splittings are neglected). For the In compound materials at the top, several resonances above the conduction band edge are also shown [8.27]

enough that a well-ordered epitaxial overlayer results. In this case no Fermi-level pinning is observed [8.29]. The band offset seems to be determined by VIGS (Sect. 8.3). On the other hand, when the GaAs substrate is at room temperature the deposited Ge layer is not well ordered and exhibits a high degree of imperfection at the interface. In this case one observes Fermi-

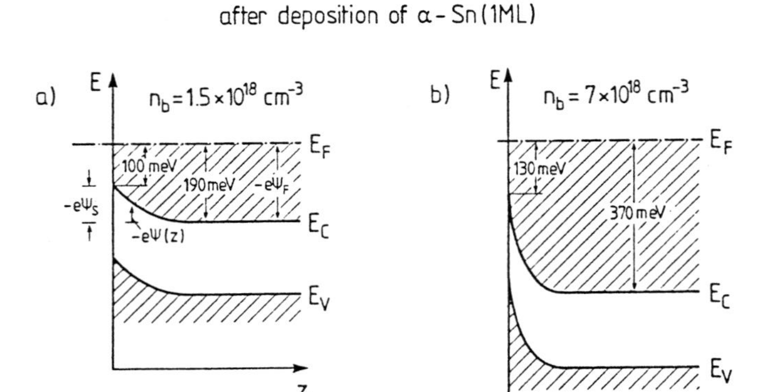

Fig. 8.15a,b. Band schemes of InSb(110) surfaces with different bulk doping n_b, on which under UHV conditions about one monolayer of amorphous α-Sn was deposited. z is the distance from the surface. Band bending and Fermi-level position of the surface were determined experimentally [8.28]

level pinning near midgap, exactly as would be expected on the basis of the defect model.

For thin metal overlayers there is a considerable body of experimental data to support the importance of defects in Schottky-barrier formation, in addition to the effects of VIGS. On the other hand, one encounters some theoretical difficulties if one tries to explain Fermi-level pinning under thick metal films solely on the basis of defect-induced states. For the sub-monolayer metal coverages on III-V surfaces needed to establish the major part of the Schottky barrier ($\theta \lesssim 0.5$ML) (Fig.8.6), an interface state density of less than 10^{12} cm^{-2} is required for Fermi-level pinning (Sect.7.5). Such a surface state density is consistent with the assumption of it being defect induced. However, for thick metal overlayers (>100 Å), the charge in the interface states is compensated not only by the space charge in the semiconductor (space-charge layer): Any interface charge located in defect states is also screened by the free electrons of the now bulk-like metal overlayer. The effect of these interface charges on the Schottky barrier height is therefore considerably diminished and their density would have to be much higher to establish a band bending of the order of 1 eV, as found for GaAs. To give a rough estimate of this density one has to compare with the shielding length in the semiconductor of a couple of hundred Ångstroms (the thickness of the space-charge layer). Using the simple formula for a parallel-plate capacitor, the areal charge density necessary to establish this potential, i.e. the band bending, would be about a hundred times higher in the presence of screening by a thick metal overlayer. Thus, under thick metal overlayers, a band bending of $0.5 \div 1$ eV, as observed for metals on GaAs, requires an interface state density of at least 10^{14} cm^{-2}. This rough estimate is also confirmed by more detailed calculations [8. 30, 31].

Electrical measurements (Panel XIII: Chap.8) on many metal/III-V semiconductor systems with thick metal overlayers show, however, that the Schottky barrier height ($0.6 \div 0.8$eV for GaAs) found far below monolayer coverage (Fig.8.6) are virtually unchanged under thick metal films. If one assumes that defect states alone are responsible for the barrier, then their density has to increase from less than 10^{12} cm^{-2} to at least 10^{14} cm^{-2} upon increasing the layer thickness. This is rather hard to explain and the tendency is to believe that defect-induced interface states are only partially responsible for the Schottky barrier under thick metal films.

Although the MIGS (VIGS) model yields a quite general explanation of the Schottky-barrier formation, it is clear that the chemical nature of the particular interface also has to be taken into account in a complete, detailed description of the electronic interface properties. The model of *Ludeke* and his group [8.32, 33] has been developed specifically for the case of *transition metal-III-V semiconductor interfaces*. The deposition of (d-band) transition metals such as Ti, V, Pd and Mn on III-V compound semicon-

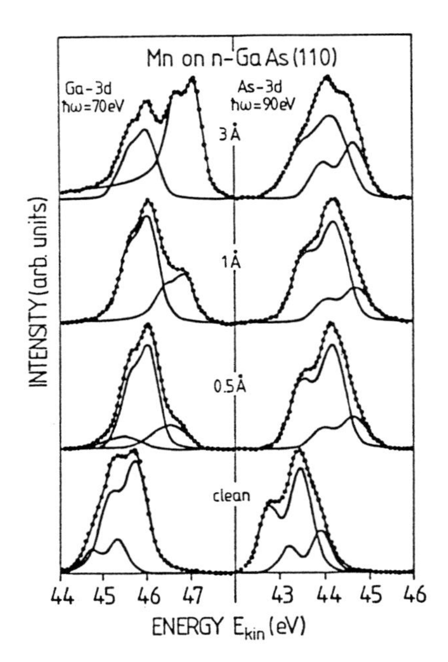

Fig. 8.16. Ga-3d and As-3d core level emission lines measured on clean n-GaAs(110) and after deposition of Mn films of various thicknesses (photon energies 70 eV and 90 eV). On the clean surface the spin-orbit split components $d_{3/2}$ and $d_{5/2}$ (at higher kinetic energy) are seen as bulk lines with high intensity (deconvoluted curves in solid line) and their surface counterparts with low intensity (deconvoluted) shifted to lower E_{kin} and higher E_{kin} for Ga and As, respectively. With increasing Mn coverage a shift is observed and satellite structures appear; the surface contributions disappear [8.34]

ductor surfaces, in particular GaAs, leads to strongly reactive interfaces, as is seen from the X-ray Photoemission Spectra (XPS) in Fig. 8.16. On the clean GaAs(110) surface the Ga(3d) and the As(3d) core-level photoemission spectra show the spin-orbit split doublet $d_{3/2}$, $d_{5/2}$ decomposed into a bulk- and surface-derived contribution. Due to the opposite directions of charge transfer in the Ga and As dangling bonds (Sect. 6.5), the surface peaks are shifted downwards and upwards in (kinetic) energy for Ga and As, respectively. The deposition of 0.5 Å of Mn causes an overall shift due to band bending (formation of a Schottky barrier) and the additional appearance of satellite structures at higher kinetic energies (lower binding energies). With increasing metal coverage the surface peaks decrease in intensity and finally disappear. For coverages higher than 1 Å the high-energy satellites become important. Their origin is assumed to be precipitates of elemental Ga, and probably As, too, in the metal overlayer. Thus XPS indicates strong interdiffusion of overlayer and substrate atoms for these transition metal-semiconductor interfaces.

UV Photoemission Spectroscopy (UPS) and inverse photoemission spectroscopy (Panel XI: Chap. 6) yield further information about interface states (Fig. 8.17). For Ti, V and Pd overlayers (each 0.2 Å thick) on GaAs (110), the fact that the Fermi level E_F is near midgap appears to be related to occupied and empty interface states whose maximum density is within the gap of the bulk GaAs states. An interpretation of these donor- and acceptor-like interface states (necessary to pin E_F on p-type and on n-type

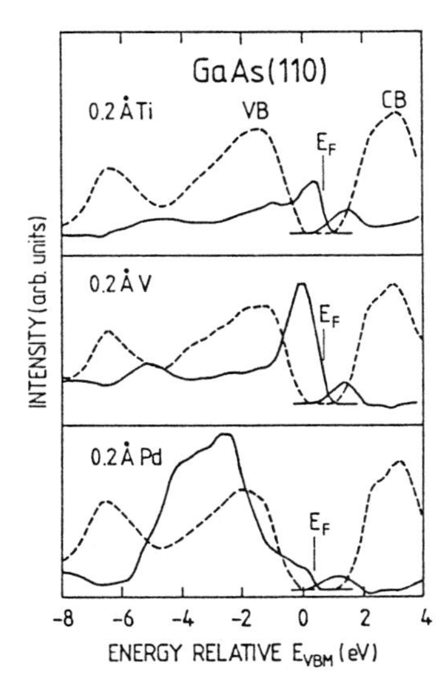

Fig. 8.17. Difference photoemission (at negative energies below E_F) and inverse photoemission curves (at positive energies above E_F) for GaAs(110) surfaces covered with 0.2 Å of Ti, V and Pd. The difference curves (*solid lines*) are obtained by substraction of the spectra for the clean surface from those for the metal-covered surfaces. The dashed curves show the experimental spectra for the clean surface [8.33]

material) can be given in terms of the d-electrons of the transition metal atoms in the GaAs host. Bonding (full) and antibonding (empty) d-states are probably involved. Independent of the details of the interpretation, transition metals on III-V semiconductor surfaces give rise to strongly reactive interfaces with new interface states which seem to be predominantly responsible for the Fermi-level pinning. In particular, the strong d-level chemical bonds of the metal component seem to be a determining factor in the Schottky-barrier formation.

Using the knowledge that interdiffusion, strong chemical reactions and the formation of new metallurgical compounds cannot be neglected for a profound understanding of many solid-solid interfaces *Freeouf* and *Woodall* [8.35, 36] developed another interesting scheme for predicting Schottky-barrier heights for such strongly reactive metal-semiconductor junctions. This could be called the **mixed interface Schottky model** or the **effective work-function model**. Basically the model represents the work-function electron-affinity matching scheme typically used as a starting point for discussing Schottky barrier heights (Sect. 8.1). In cases where a new interfacial compound is formed upon deposition of a metal, the assumption is that the dominant factor determining the Fermi level in the semiconductor substrate is the "chemistry" of this new compound. If one neglects the formation of surface defects, semiconductor reconstructions, etc., then it is the work function ϕ_{IC} of the new interfacial compound rather than that of the metal (overlayer) that, together with the electron affinity χ_{SC} of the semiconductor substrate, determines the Fermi level at the interface. The Schottky-barrier height ϕ_{SB} for an n-type semiconductor is then given by

$$e\phi_{SB} = e\phi_{IC} - \chi_{SC} . \tag{8.10}$$

In addition to neglecting other factors responsible for new interface states, the problem has now been reformulated in terms of the work function $e\phi_{IC}$ of the usually ill-defined interface compound. But providing at least its chemical composition is known, approximate values for ϕ_{IC} can be calculated or estimated and the model is then attractive because of its simplicity. There is a particular class of metal-semiconductor junctions for which the assumption that a new metallurgical compound is formed at the interface is certainly correct. Many transition metals and rare-earth elements (M) deposited onto Si surfaces form silicides of the composition M_2Si, MSi or MSi_2 (Sect.8.5). Depending on the deposition conditions, in particular the substrate temperature, more or less thick silicide interlayers or overlayers, sometimes even crystalline such as $CoSi_2$, $NiSi_2$, $FeSi_2$ and $CrSi_2$ are formed between the Si substrate and the metal overlayer or on the Si substrate, respectively. The applicability of the mixed interface Schottky model seems obvious. Accordingly *Freeouf* [8.37] modeled the effective silicide composition for a number of silicide/Si interfaces by a stoichiometry MSi_4. This assumes a graded interface, rather than a macroscopic layer of uniform composition. The silicide work function $\phi_{silicide} = \phi_{IC}$ is then approximated by the geometric mean of its components:

$$\phi_{silicide} \simeq (\phi_M \phi_{Si}^4)^{1/5} \tag{8.11}$$

and for χ_{Si} a value of 4.2 eV is taken. Based on these rough values a theoretical dependence of the Schottky-barrier height ϕ_{SB} is calculated as a function of the interface work function $\phi_{silicide}$ (Fig.8.18). The results indeed display the correct trend, as measured by various experimental techniques [I-V, C-V characteristics (Panel XIII: Chap.8), photo-response]. A much better agreement between theory and experiment can be achieved by simply shifting the theoretical curve, e.g., by changing the χ_{Si} value or assuming a different stoichiometry of the silicide compounds.

Similar arguments to these might also be applied to systems in which the interface region contains precipitates or a mixture of several metallurgical phases.

In conclusion, there are several model approaches to Schottky-barrier formation which emphasize the non-ideality of the metal/semiconductor

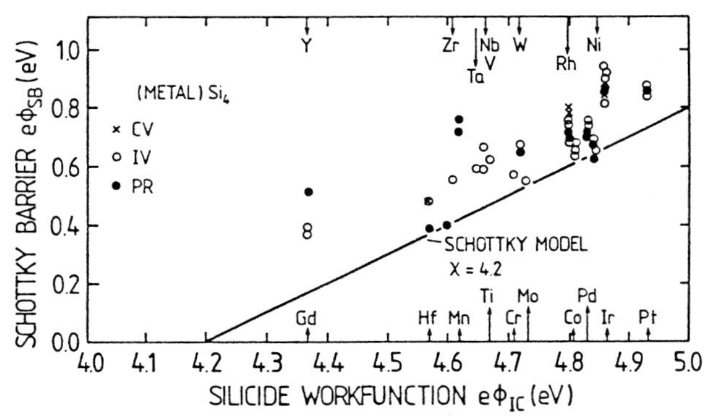

Fig. 8.18. Comparison between measured Schottky barrier heights of several metal silicides with effective work function derived assuming the composition of MSi_4. The silicide work function is calculated as $\phi_{silicide} = (\phi_M \phi_{Si}{}^4)^{1/5}$, where ϕ_M is taken from previous data [8.38], and ϕ_{Si} is assumed to be 4.76 eV. The experimental data used to obtain the barrier heights are C-V and I-V characteristics and photoresponse (PR) measurements [8.37]

junction. There are certainly cases for which the effects of defects, new interfacial compounds, etc. are important and it is then essential that these are included, in addition to the effect of MIGS. It is especially evident that a theoretical description of the electronic properties of a particular interface requires a detailed knowledge of its geometrical structure, degree of abruptness and metallurgical composition.

8.5 The Silicon-Silicide Interface

As noted in the previous chapter, there is a class of metal-semiconductor interfaces, the silicide forming systems, which are of great practical importance (in high stability contacts) for device technology and also of considerable fundamental interest. In combination with Mg and a number of transition and rare-earth metals, Si forms stable compounds of well-defined stoichiometry. With M representing the metal, the following compounds have so far been identified

M_2Si : Mg, Co, Ni, Pd, Pt:

MSi : Ti, Hf, Mn, Fe, Co, Rh, Ir, Ni, Pt;

MSi_2 : Ti, Zr, Hf, V, Nb, Ta, Cr, Mo, W, Mn, Fe, Co, Ni, Gd, Dy, Ho, Er.

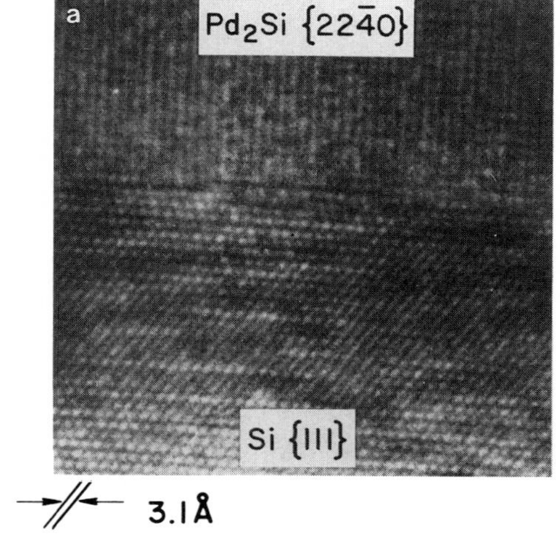

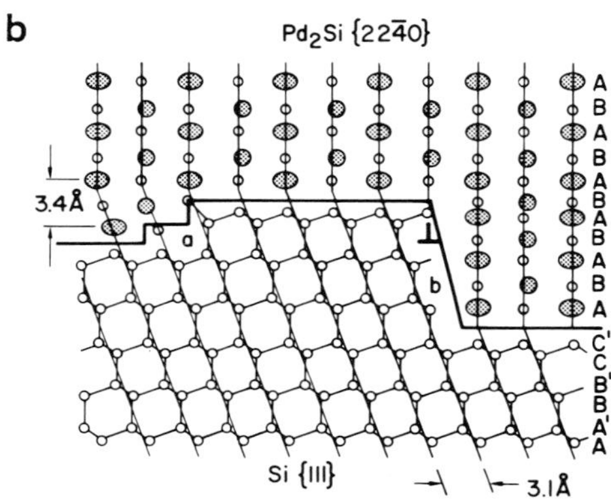

Fig. 8.19. (**a**) Cross-sectional transmission electron micrograph showing the lattice structure of an epitaxial Pd_2Si-Si(111) interface. (**b**) A schematic model of the interface. Image contrasts of lattice fringes in (**a**) are indicated by the fine lines; shown also are two atomic steps (**a**) and one misfit dislocation (**b**) [8.39]

Silicide films formed on a Si substrate can have a variety of different crystallographic structures: amorphous, polycrystalline or even epitaxial crystalline. The following silicides have been observed to grow as epitaxial films on Si with the crystal structures and at the growth temperatures indicated: PtSi (orthorhombic MnP, 300°C), Pd_2Si (hexagonal Fe_2P, 100 ÷ 700°C), $CoSi_2$ (cubic CaF_2, 550 ÷ 1000°C), $NiSi_2$ (cubic CaF_2, 750 ÷ 800°C). On Si(111) growth is possible in all cases; $CoSi_2$ and $NiSi_2$ also grow epitaxially on Si(100). Silicide epitaxy is usually performed in a UHV system by means of MBE (Sect. 2.4). Silicide formation occurs as a contact reaction at the metal-Si interface. The unreacted metal is deposited on the Si substrate. The sample is then annealed to promote interfacial reaction. This causes an initial layer of silicide to be formed at the metal-Si interface, representing the initial stage of the reaction. With further annealing more silicide is produced and a thicker interfacial silicide layer grows. For the growth process itself mass transport through the film is necessary, as is the breaking of Si-

Si bonds. The optimum growth parameters can vary widely for different silicides. With the refractory metal V, for example, no measurable reaction occurs at room temperature, while for the near-noble metal Pd there is a distinct silicide formation at 300 K involving the first 10 ÷ 15 Å of deposited metal.

Epitaxially grown silicide films can exhibit a quite perfect crystalline structure and a sharp interface to the Si substrate, albeit with steps and other microscopic imperfections. This is evident in the example of Pd_2Si on Si(111) (Fig.8.19).

Most silicides are highly conducting (metallic) materials with a considerable density of states at the Fermi level. The UPS data for the Pd-Si system (Fig.8.20) reveal a quasimetallic character with a density of states at E_F lower than that for metallic Pd(111), but considerably higher than that of Si(111). The formation of the silicide is seen to give rise to distinct new spectral features A, B, C, D, reflecting substantial changes in the character of the Pd and Si valence states. The silicide UPS spectrum with its intensity dominated by the Pd d-states indicates that the d-band becomes largely filled, with its main peak B shifted to approximately 2.75 eV below E_F. The states near E_F responsible for electronic conduction (feature A) are also composed to a large extent of Pd d-states.

The silicide with the highest known density of states at E_F, i.e. with the highest electrical conductivity, is crystalline $CoSi_2$. On the other hand, $FeSi_2$ (in a special β-modification) and $CrSi_2$, both crystalline materials, are known to have semiconducting character [8.40,41]. For thin β-$FeSi_2$ films

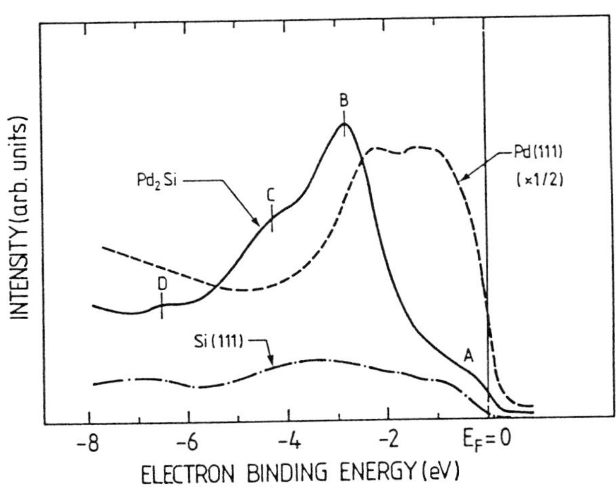

Fig.8.20. UV photoemission spectra (angle-integrated mode, photon energy 21.2eV) for clean Si(111), clean Pd(111) and bulk Pd_2Si. For Pd_2Si the main Pd(4d) states produce structures *B* and *C*, while Pd(4d)-Si(3p) bonding and antibonding states result in features *D* and *A*, respectively [8.39]

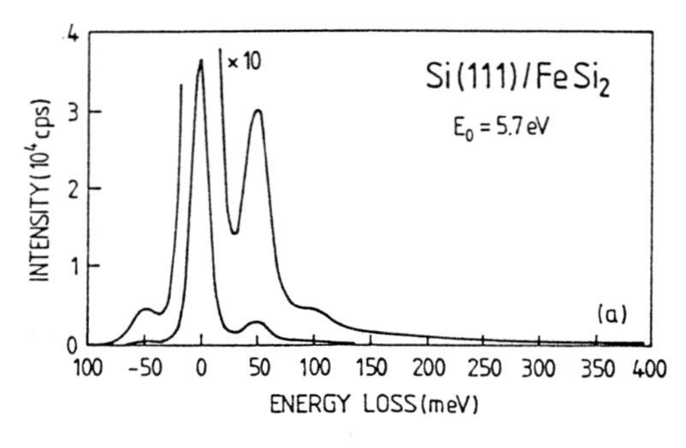

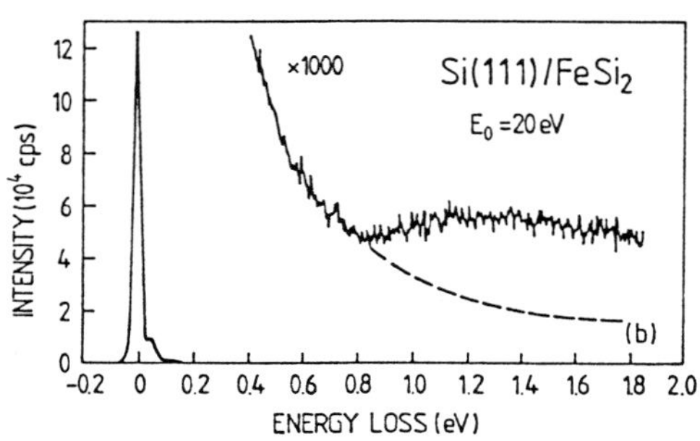

Fig. 8.21a,b. High-resolution electron energy loss spectra measured on a β-$FeSi_2$ film deposited by solid-phase reaction (deposition of Fe at 300 K and subsequent annealing to 900K) on Si(111). (**a**) Gain and loss peaks due to Fuchs-Kliewer phonons in the $FeSi_2$ film (excitation energy 50 meV). (**b**) Loss shoulder due to electronic interband transitions with an onset near 0.8 eV [8.41]

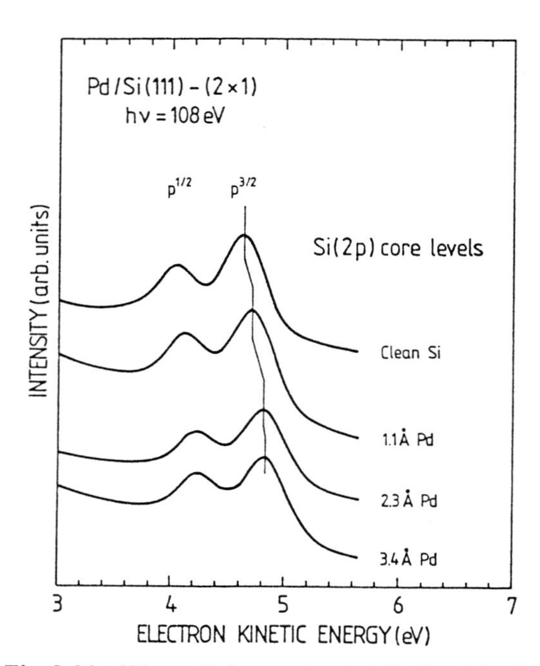

Fig. 8.22. Silicon (2p) core level shift for thin deposits of Pd on Si(111)-(2 × 1) as measured by synchrotron radiation photoemission at 108 eV photon energy. Substrate temperature during deposition was 25°C [8.42]

grown by a solid-phase reaction in UHV on Si(111) this is clearly seen from HREELS (Fig.8.21). In the low-loss energy region (Fig.8.21a), well-resolved gain and loss peaks are observed; these are due to excitation of the Fuchs-Kliewer surface phonons which give evidence of a semiconducting or insulating character of the silicide overlayer. In a metallic overalyer the long-range dipole fields of these surface phonons would be shielded by the free carriers and the collective excitations would not be observable. At higher loss energy, around 0.9 eV, the onset of the characteristic valence to conduction band transitions is observed as a shoulder in the spectrum (Fig.8.21b). The intensity of this loss shoulder is indicative of direct band-gap transitions in the semiconducting $FeSi_2$.

As an example of the formation of a silicide Schottky barrier at room temperature let us consider the Si(111)-(2 × 1) surface after deposition of a couple of Ångstroms of Pd (Fig.8.22). The photoemission spectra measured by synchrotron radiation ($h\nu$ = 108eV) reveal a gradual shift of the Si(2p) core level emission lines due to progressive band bending (Panel XV: Chap. 9). At a substrate temperature of 25°C during Pd deposition and for Pd coverages of up to 3.4 Å, one can assume complete reaction of the deposited Pd to Pd_2Si. The development of the Schottky barrier, i.e. the barrier height, as a function of increasing overlayer thickness is shown in Fig.8.23 for both the (2 × 1) and (7 × 7) Si(111) surfaces. The initial band bending is different for the two different reconstructions (Sect.7.7). But for both surfaces one observes nearly the same saturation value of about 0.75 eV for $e\phi_{SB}$. This value coincides with the barrier height found from bulk contact electrical measurements on thick films. In contrast to metal/III-V semiconductor systems (Fig.8.6) the final band bending value is attained more gradually for thicker overlayers.

Experiments on a number of silicide systems show that, in general, the barrier height ϕ_{SB} does not depend significantly on which silicide phase is

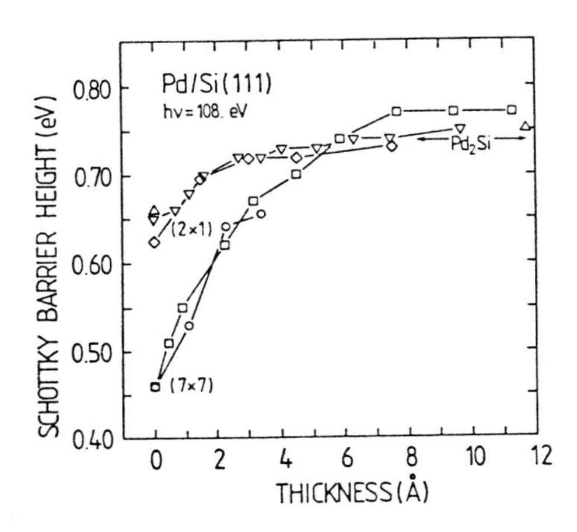

Fig.8.23. Schottky barrier height as a function of Pd coverage for the Si (111)-(2 × 1) and Si(111)-(7 × 7) surface as determined from shifts of the Si-2p core levels observed in photoemission (Fig.8.22). The saturation value of the barrier height coincides with the height determined from bulk contact electrical measurements (*arrows*) [8.43]

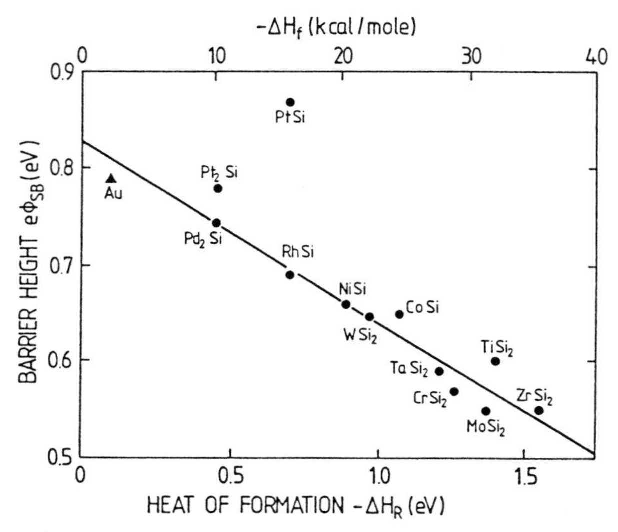

Fig.8.24. Schottky barrier heights $e\phi_{SB}$ of transition metal silicide-silicon interfaces plotted against the heats of formation ΔH_R of the transition-metal silicide [8.44]

present (e.g., Ni_2Si, NiSi, or $NiSi_2$). Furthermore, ϕ_{SB} is also relatively insensitive to the microstructure of the metallic overlayer (e.g., polycrystalline, textured, or epitaxial), and to the Si plane on which the contact is produced.

On the other hand, ϕ_{SB} is clearly dependent on the chemical reactivity of the two reaction partners. A plot of the heat of formation, ΔH_R, of transition metal silicides against the measured barrier height ϕ_{SB} shows good correlation (Fig.8.24), thus confirming the importance of the concept of chemical bonding for an understanding of Schottky barriers at silicon/ silicide interfaces. All the results so far obtained on these metal-Si interfaces suggest that the barrier height is determined at least partially by local chemical bonds between metal and Si atoms at the interface. However, other factors such as the role of interfacial defects, metal atom diffusion into the Si, etc. remain to be explored in detail.

It is also possible that one could apply the concept of MIGS, but this would require a better understanding of the intrinsic properties of the silicide. The same is true for the **mixed interface Schottky model** discussed in Sect.8.4. The problem there is that the microscopic details of the silicide interfaces are poorly understood and thus a reliable determination of the "effective" work function of the silicide interlayer is very difficult.

8.6 Some Applications of Metal-Semiconductor Junctions and Semiconductor Heterostructures

Although our understanding at the atomic level of Schottky-barrier heights and band discontinuities in semiconductor heterostructures is far from complete, metal-semiconductor junctions and heterostructures have found a wide variety of applications, in both fundamental science and device technology. Some important aspects will be discussed in the following sections.

8.6.1 Schottky Barriers

The band scheme of a metal-semiconductor junction (Figs. 8.2, 25) can be considered as that of half a p-n junction. Accordingly, one expects similar rectifying properties when a voltage is applied between metal and semiconductor. Furthermore, the depletion layer built up in an n-type semiconductor in contact with a metal electrode is the same as that induced by negatively charged surface states on a clean or gas-covered surface (Chap. 7). The band curvature, and thus also the barrier heigh ϕ_{SB}, are related via Poisson's equation (7.9) to the positive space charge due to ionized donors (density N_D) in the semiconductor. In the simple case of strong depletion (Schottky-type space-charge layer, Sect. 7.2) with band bending $|eV_s| \gg kT$, a rectangular space-charge density ρ (constant over the space-charge region d) is assumed and the double integration of Poisson's equation (7.12)

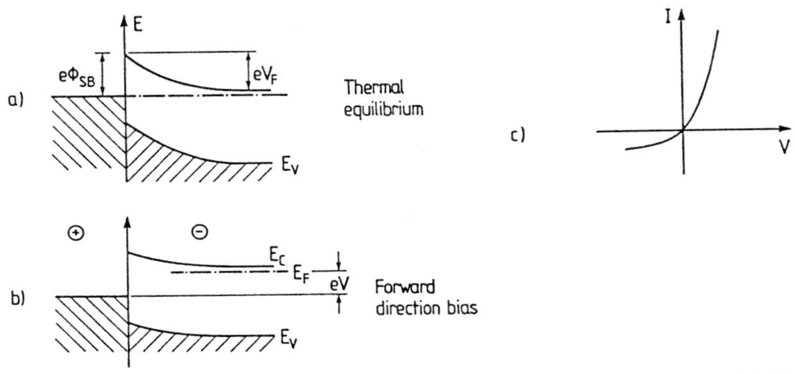

Fig. 8.25a-c. Electrical behavior of a metal semiconductor junction. **(a)** Simple band scheme of the metal-semiconductor junction in thermal equilibrium; $e\phi_{SB}$ Schottky barrier for a n-type semiconductor, eV_{if} maximum band bending at interface. **(b)** Band scheme for a junction biased in the forward direction; within the space charge region the Fermi energy E_F can no longer be defined (only quasi-Fermi levels). **(c)** Schematic current-voltage (I-V) characteristic of a metal-semiconductor junction

yields a quadratic potential dependence (7.14) with a maximum band bending V_{if} at the interface of

$$V_{if} = \frac{eN_D}{2\epsilon\epsilon_0} d^2 . \tag{8.12}$$

The thickness d of the depletion layer thus decreases with increasing doping level N_D as

$$d = \sqrt{\frac{2\epsilon\epsilon_0 V_{if}}{eN_D}} . \tag{8.13}$$

An externally applied voltage V causes a shift of the Fermi level in the bulk semiconductor with respect to that in the metal. The potential drop occurs across the space-charge layer (quasi-Fermi levels) and the interface layer of atomic dimensions (Fig. 8.4), whereas deep in the semiconductor and in the metal the respective Fermi levels remain distinct but constant because of the relatively high conductivity there. An external bias V therefore modifies the band bending ($V_{if}-V$) and the thickness of the depletion layer (8.13), becomes

$$d = \sqrt{\frac{2\epsilon\epsilon_0 (V_{if} - V)}{eN_D}} . \tag{8.14}$$

Depending on the strength of the applied potential, positive space charge and negative counter charge in the metal and in the interface states are more or less separated from one another. The total space charge Q_{sc} depends on the external bias according to

$$Q_{sc} = eN_D d = \left[2e\epsilon\epsilon_0 N_D (V_{if} - V) \right]^{1/2} ; \tag{8.15}$$

and by differentiating, one obtains the bias-dependent capacitance C_{sc} (per unit area) of the (Schottky) metal-semiconductor junction

$$C_{sc} = \left| \frac{dQ_{sc}}{dV} \right| = \sqrt{\frac{e\epsilon\epsilon_0 N_D}{2(V_{if}-V)}} = \frac{\epsilon\epsilon_0}{d} . \tag{8.16}$$

Taking into account the bias-dependent space charge (8.15) this is the simple formula for a parallel-plate capacitor. This property of a bias-dependent capacitance is used in a diode device called a **varactor** (variable reactor: device with bias-controlled reactance) whose essential component is a metal-semiconductor junction.

The rectifying action of a metal-semiconductor junction is easy to understand from the fact that electrons moving out of the metal into the semiconductor always have to overcome the energy barrier $e\phi_{SB}$, whereas electrons coming from the semiconductor "see" a barrier that is reduced from the thermal equilibrium value eV_{if} by the external potential energy eV (Fig. 8.25b).

Since electrons in the conduction band obey Boltzman statistics, the number that is able to overcome the barrier depends exponentially on its height. As for a p-n junction [8.1] the current-voltage characteristic thus follows as

$$I(V) = I_0(e^{eV/kT} - 1) \ . \tag{8.17}$$

The exact expression for the saturation current I_0 depends on the assumptions made about carrier transport (diffusion, thermionic emission etc. [8.45]). A more detailed derivation is given in Panel XIII (Chap. 8). It should be emphasized that a metal-semiconductor junction rectifier is a unipolar device unlike a p-n junction diode (bipolar device); only one type of carrier, electrons in an n-type metal-semiconductor junction, carry the current. In a bipolar device (p-n junction), both electrons and holes contribute to the current.

On many III-V semiconductors, in particular on GaAs, the Fermi level is always pinned near midgap under metal films. It is thus impossible to prepare a nonrectifying, ohmic contact to a metal. Instead, one inevitably obtains Schottky barriers with strongly nonlinear I-V characteristics. In semiconductor device technology one therefore uses a trick to prepare quasi-ohmic contacts by extremely high surface doping (Fig. 8.26). By subsequent ion implantation, or during epitaxy itself, an extremely highly n-doped (n^+) layer with donor concentrations N_D greater than $5 \cdot 10^{18}$ cm^{-3} is formed on top of the n-type substrate before the metal overlayer is deposited. Because the Fermi level is pinned near midgap, a considerable barrier exists but, due to the high doping level, the depletion layer becomes extremely thin, amounting to only $10 \div 20$ Å, see (8.13). This barrier can readily be penetrated by tunneling. For normal bias values, the exponential tunnel characteristic I(V) is approximately linear and the behavior of the junction is quasi ohmic.

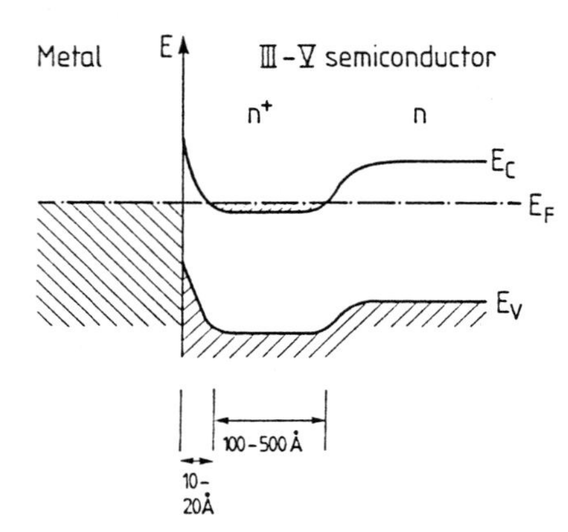

Fig.8.26. Band scheme of a quasi-ohmic metal - semiconductor contact. In an interface region of 100 ÷ 500 Å thickness the III-V semiconductor is highly n-doped (degenerate) such that the depletion space charge with a thickness of typically 20 Å is easily penetrated by tunneling.

8.6.2 Semiconductor Heterojunctions and Modulation Doping

The importance of semiconductor heterojunctions lies in the fact that they allow one to build into a semiconductor a variety of potential steps and even continuously varying potential profiles for the free electrons in the conduction band (Fig.8.27). This is achieved by the use of controlled epitaxy (MBE, MOMBE; Sects.2.4, 5). To introduce the necessary notation, Fig. 8.28 shows qualitatively an abrupt heterojunction between a narrow-gap n-type, and a wide-gap p-type semiconductor in thermal equilibrium. Near such a heterojunction Poisson's equation has to be used in its generalized form with a spatially varying $\epsilon(\mathbf{r})$.

$$\nabla\,[\epsilon_0\,\epsilon(\mathbf{r})\mathscr{E}(\mathbf{r})] = \rho(\mathbf{r})\;. \tag{8.18}$$

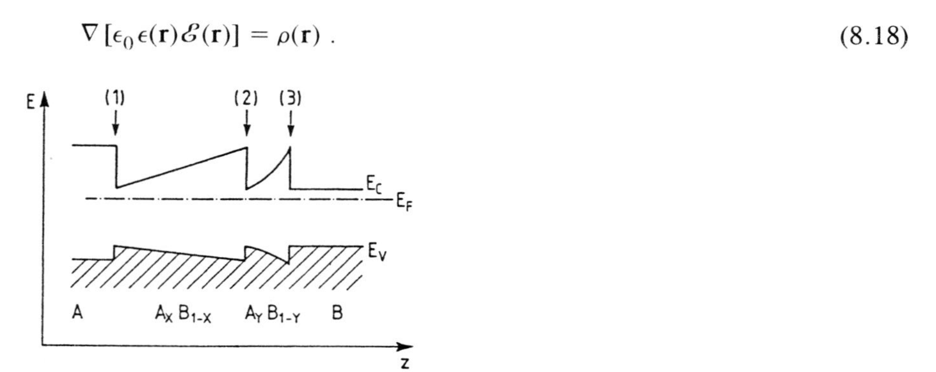

Fig.8.27. Schematic conduction-band edge E_C, valence-band edge E_V and Fermi level E_F versus coordinate z, for three semiconductor heterojunctions *1* - *3* between two n-type semiconductors A and B. Varying band gaps (graded profiles) are produced as a function of depth z by controlled modification of the AB composition during epitaxy (MBE, MOMBE, Sects.2.4, 5). A realistic example of such an artificially structured semiconductor layer system can be fabricated on the basis of GaAs/AlAs, i.e. by using the $Al_xGa_{1-x}As$ ternary alloy

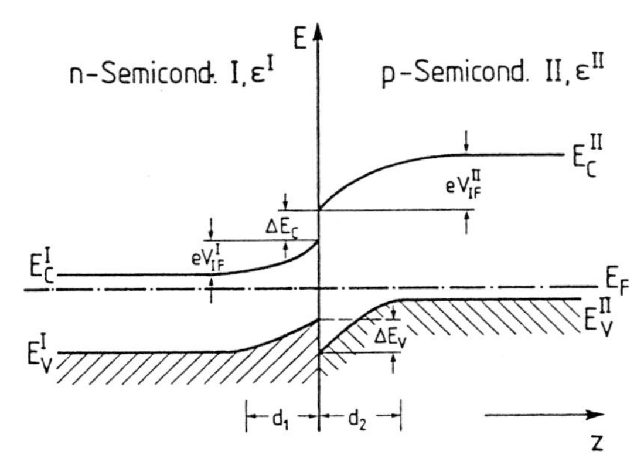

Fig. 8.28. Band scheme of a semiconductor heterojunction; an n-type semiconductor I with a small band gap (dielectric function ϵ^{I}) is in contact and thermal equilibrium with a p-type semiconductor II with a larger band gap. Such a junction is called a P-n junction. ΔE_C and ΔE_V are the conduction and valence band offsets, respectively; V_{if}^{I} and V_{if}^{II} are the built-in voltages on either side of the interface

For a layered (one-dimensional) structure this simplifies (with $\mathscr{E}\,||z$) to

$$\epsilon_0 \mathscr{E} \frac{d\epsilon}{dz} + \epsilon_0 \epsilon \frac{d\mathscr{E}}{dz} = \rho(z) \ . \tag{8.19}$$

Substituting for the electric field in terms of the potential V(z) and with N_D and N_A as donor and acceptor concentrations, respectively, this yields

$$\frac{d^2 V}{dz^2} = \frac{e}{\epsilon\epsilon_0}(N_D - n + p - N_A) - \frac{1}{\epsilon}\frac{d\epsilon}{dz}\frac{dv}{dz} \ . \tag{8.20}$$

For an abrupt heterojunction as in Fig. 8.26, the continuity of the electric displacement $\epsilon_0 \epsilon \mathscr{E}_{if}$ at the interface requires

$$\epsilon^{\mathrm{I}} \mathscr{E}_{if}^{\mathrm{I}} = \epsilon^{\mathrm{II}} \mathscr{E}_{if}^{\mathrm{II}} \ . \tag{8.21}$$

Poisson's equation (8.20) can now be solved with a step-like change of $\epsilon(z)$ from ϵ^{I} to ϵ^{II} across the interface. In each of the semiconductors one assumes Schottky-type depletion layers with $|V_{if}^{\mathrm{I}}| \gg kT$ and $|V_{if}^{\mathrm{II}}| \gg kT$. The total built-in potential at the interface V_{if} is equal to the sum of the partial built-in voltages, V_{if}^{I} and V_{if}^{II}, supported at equilibrium by semiconductors I and II (Fig.8.28):

$$V_{if} = V_{if}^{I} + V_{if}^{II} . \tag{8.22}$$

The addition of an applied bias causes a relative shift of the Fermi levels deep in materials I and II, thus producing the voltage drops V^{I} and V^{II} over the space charge layers in the two semiconductors:

$$V = V^{I} + V^{II} . \tag{8.23}$$

Using (8.21), one can solve Poisson's equation in a manner similar to that of Sect.7.2. For the thicknesses d_1 and d_2 of the depletion layers in the two semiconductors this yields

$$d_1 = \left[\frac{2N_A^{II} \epsilon^{I} \epsilon^{II} (V_{if} - V)}{eN_D^{I} (\epsilon^{I} N_D^{I} + \epsilon^{II} N_A^{II})} \right]^{1/2} , \tag{8.24a}$$

$$d_2 = \left[\frac{2N_D^{I} \epsilon^{I} \epsilon^{II} (V_{if} - V)}{eN_A^{II} (\epsilon^{I} N_D^{I} + \epsilon^{II} N_A^{II})} \right]^{1/2} , \tag{8.24b}$$

where N_D^{I} and N_A^{II} are the donor and acceptor concentrations in semiconductors I and II, respectively. The capacitance of the interface p-n junction is given by

$$C_{if} = \left[\frac{eN_D^{I} N_A^{II} \epsilon^{I} \epsilon^{II}}{2(\epsilon^{I} N_D^{I} + \epsilon^{II} N_A^{II})(V_{if} - V)} \right]^{1/2} . \tag{8.25}$$

The voltages V^{I} and V^{II} supported by each semiconductor under a bias V (8.23) are related to one another by

$$\frac{V_{if}^{I} - V^{I}}{V_{if}^{II} - V^{II}} = \frac{N_A^{II} \epsilon^{II}}{N_D^{I} \epsilon^{I}} . \tag{8.26}$$

Analogous formulas can be derived for heterojunctions with inverse doping and for equally doped n-n or p-p junctions [8.45].

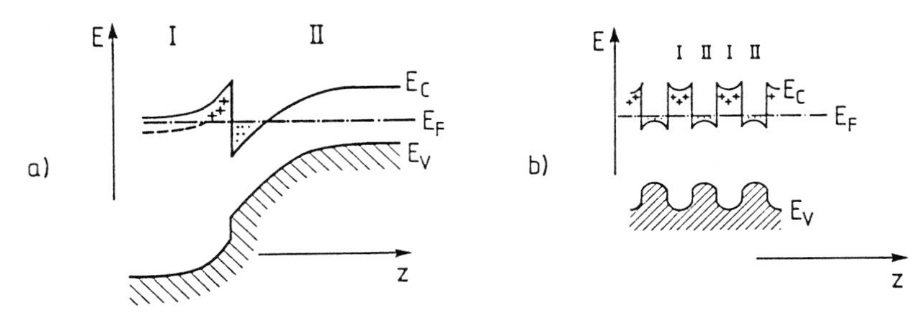

Fig. 8.29. (**a**) Modulation-doped heterojunction consisting of a highly n-doped wide gap semiconductor I and a lightly n-doped (or intrinsic) semiconductor II with smaller band gap; such a junction is called an N-n junction. (**b**) Band scheme of a modulation doped compositional superlattice. The layers of semiconductor I are highly n-doped, whereas semiconductor II is lightly n-doped or intrinsic

Particularly interesting new effects occur at semiconductor heterojunctions in which n-type doping is largely restricted to the semiconductor with the wider gap (Fig. 8.29). Semiconductor I in Fig. 8.29a with its large band gap is highly n-doped whereas material II with its smaller gap is intrinsic or very lightly n-doped. The Fermi levels in the bulk are fixed slightly below the conduction band edge (I) and near midgap (II). Furthermore, the band offset at the interface is fixed and is independent of doping. Total space charge neutrality therefore requires a strong depletion layer on side I which is balanced by a corresponding negative space charge in material II (accumulation). The free electrons originating from the ionized donors in material I accumulate in a narrow accumulation layer in semiconductor II. In this way the high concentration of free electrons is spatially separated from the ionized donors giving rise to impurity scattering.

In homogeneously doped material an increase in carrier density generally requires an increase in doping level which enhances impurity scattering. This, in turn, limits the mobility of the carriers. This disadvantage of the high doping levels necessary to yield high electron concentrations can be overcome by so-called **modulation doping** (Fig. 8.29) which separates the donor centers from the free electron gas. Electron mobilities in a modulation-doped $Al_xGa_{1-x}As/GaAs$ structure are shown in Fig. 8.30. If the electron mobility were determined by ionized Impurity Scattering (IS) and Polar Optical (PO) phonon scattering in the same way as in bulk GaAs, the mobility would be limited by these processes thus showing a decrease at low temperature (due to IS) and also at higher temperature (due to PO); a maximum would occur at intermediate temperatures. For an impurity concentration N_D of 10^{17} cm^{-3} this maximum of about $4 \cdot 10^3$ cm^2/Vs is reached near 150 K; for an N_D of $4 \cdot 10^{13}$ cm^{-3} a maximum mobility of $3 \cdot 10^5$ cm^2/Vs corresponds to a temperature of about 50 K. However, the experimental results for modulation-doped structures (shaded region) do not show any reduction

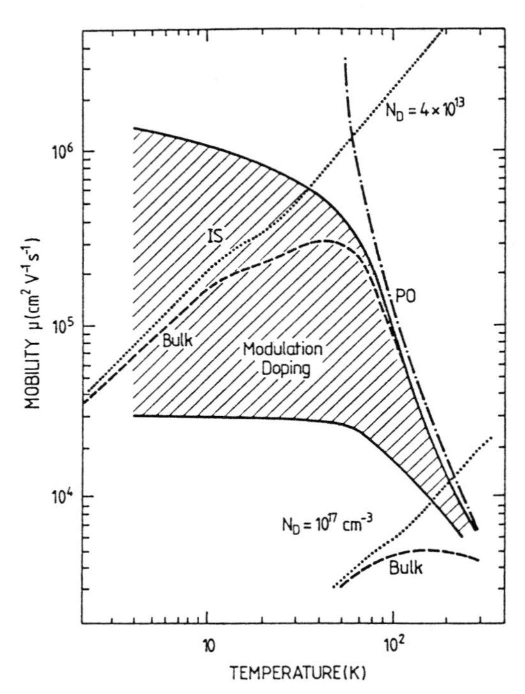

Fig. 8.30. The temperature dependence of the electron mobilities in bulk GaAs as compared with the electron mobilities observed in modulation doped AlGaAs/GaAs structures (*shaded area*); the variation in the observed mobilities depends on details of the quality of the interface and on the thickness of undoped AlGaAs spacer layers between the highly n-doped AlGaAs and the lightly doped GaAs layer. Theoretical mobilities due to Impurity Scattering (IS) and scattering from Polar Optical (PO) phonons are also displayed [8.46]

in mobility at low temperatures. The values obtained, as high as $2 \cdot 10^6$ cm^2/Vs, are much larger than the impurity-limited mobility. This effect of mobility enhancement in a modulation-doped heterostructure is even more pronounced when an undoped $Al_xGa_{1-x}As$ layer is placed between the highly doped AlGaAs region and the GaAs layer (Fig. 8.31). The thickness

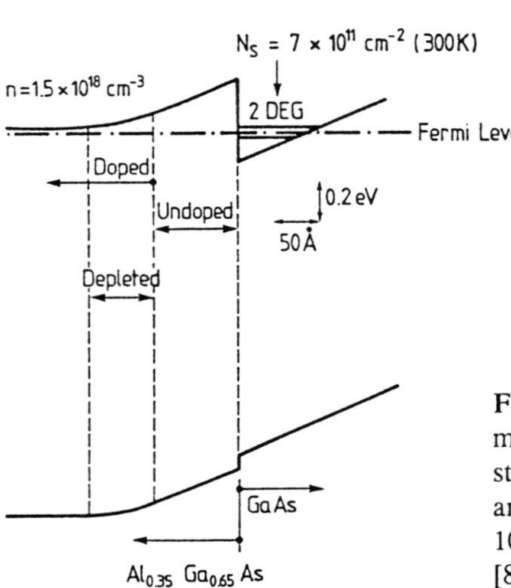

Fig. 8.31. Band diagram of a single interface modulation-doped $Al_xGa_{1-x}As$/GaAs heterostructure. The diagram is drawn to scale for an $Al_xGa_{1-x}As$ doping concentration of 1.5×10^{18} cm^{-3} and an AlAs mole fraction of 0.35 [8.46]

of this spacer layer is usually chosen to be on the order of 100 Å, such that the band scheme, drawn to scale for an $Al_{0.35}Ga_{0.65}As/GaAs$ interface with a donor concentration of $1.5 \cdot 10^{18}$ cm^{-3}, appears as in Fig.8.31. The conduction-band discontinuity is assumed in this case to be 0.3 eV. The free electrons in the GaAs accumulation layer are spatially confined in a region of about 70 Å in a triangular potential well, much the same as in a MOS structure (Sects.7.6, 9). Correspondingly the electronic wave functions are quantized along the z-direction normal to the AlGaAs/GaAs interface (Sect. 7.6). Free motion, i.e. Bloch character of the wave function, only occurs for coordinates parallel to the interface. The energy eigenvalues for electrons in the accumulation layer near the bottom of the conduction band thus lie on parabolas $E_i(\mathbf{k}_{\|})$ along wave-vector directions $\mathbf{k}_{\|}$ parallel to the interface (Fig.7.8b). In comparison to the situation for MOS structures described in Chap.7, however, the mobility of the electrons is now orders of magnitude higher (Fig.8.30). This makes the AlGaAs/GaAs heterostructure system, and comparable systems such as InGaAs/InP, extremely attractive for studying the properties of 2D electron gases (Sect.8.7), for example the quantum Hall effect (Sects.7.10 and 8.7).

Similar effects, in particular a strong mobility enhancement as in the accumulation layer at a semiconductor heterointerface (Fig.8.29a), are obtained in a superlattice consisting of an alternating series of semiconductor layers (I and II) grown epitaxially on one another (Fig.8.29b). Material I with the larger band gap (e.g., AlGaAs) is strongly n-doped, whereas material II with the smaller band gap is nearly intrinsic. As shown qualitatively in Fig.8.29b, the donors in semiconductor I have "given" their electrons to the energetically more favorable quantum wells in material II, thus giving rise to positive space charge in I (positive band curvature) and negative space charge in II (negative band curvature). For quantum-well sizes on the order of 100 Å and less, quantization occurs normal to the interfaces, whereas parallel to the layers one finds extremely high electron mobilities as in Fig.8.30. For a single quantum well, the energy eigenvalues are discrete parabola (subbands), as for the triangular potential (Fig.7.8), except that the quantized minimum energies ϵ_i are now spaced differently due to the different shape of the potential well (square versus triangular). If one now considers a multiple quantum-well structure as in Fig.8.29b, the wave functions in neighboring quantum wells can clearly overlap and a splitting of the single quantized levels is expected as in the tight-binding picture of the band structure of a periodic crystal. This splitting, of course, is only apparent for quantum wells that are sufficiently close to one another, i.e. for small repeat distances. Superlattices based on AlGaAs/GaAs exhibit discrete subbands ϵ_i for repeat distances exceeding about 100 Å (Fig.8.32). For smaller superlattice periods, a considerable broadening of the discrete energies ϵ_i into bands occurs.

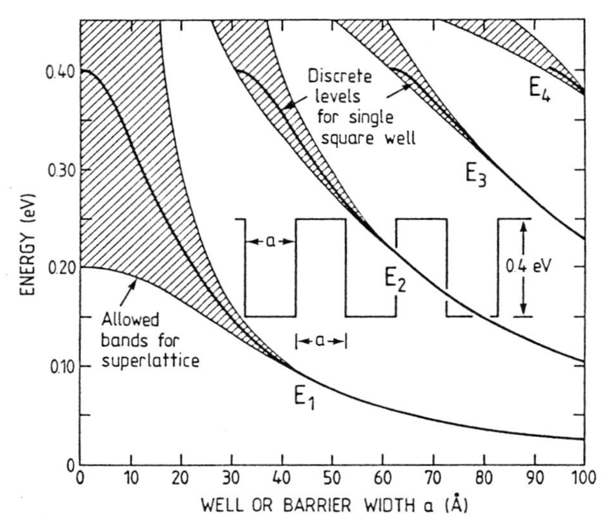

Fig. 8.32. Energy levels E_1, E_2, E_3, ... of electrons in the rectangular quantum wells of a compositional superlattice (inset shows the conduction band edge). The calculation was performed with an effective mass $m^* = 0.1\ m_0$. The solid curves are valid for single quantum wells of the corresponding width d_z. Quantum wells within a superlattice lead to wavefunction overlap when neighboring wells are sufficiently close, and the energy levels broaden into bands (*shaded areas*) [8.47]

8.6.3 The High Electron Mobility Transistor (HEMT)

One of the most interesting device applications of modulation-doped AlGaAs/GaAs heterostructures is the Modulation Doped Field Effect Transistor (**MODFET**), sometimes called the **TEGFET** (Two-Dimensional Electron Gas FET) or **HEMT** (High Electron Mobility Transistor). On a substrate of semi-insulating GaAs, a weakly doped or "undoped" GaAs layer is grown epitaxially (MBE, MOMBE; Sects. 2.4,5); then the wide-gap material $Al_xGa_{1-x}As$ is deposited, the first $20 \div 100$ Å undoped as a spacer layer, followed by heavily n^+-doped material, which supplies the free electrons for the 2D electron gas at the AlGaAs/GaAs interface (accumulation layer within the GaAs) (Fig. 8.33a). Source and drain contacts have to be prepared such that under the metal overlayers deep reaching zones of high n-type conductivity establish a good ohmic contact to the 2D electron gas at the interface. Sometimes n^+-GaAs layers are also deposited as contact areas. The preparation of these contact zones can be performed by ion implantation or by locally confined diffusion of Ge (after deposition of an AuGe alloy). In contrast, the gate electrode (diameter $\leq 0.5\,\mu$m) is prepared by deposition of metal onto a reset free area of the n-AlGaAs layer. In this

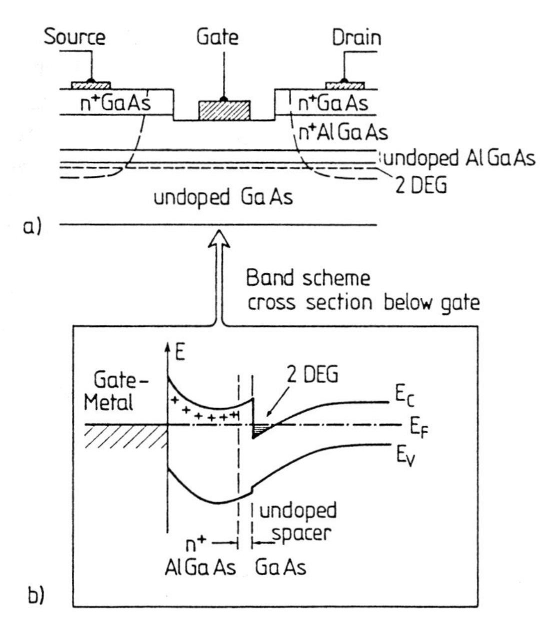

Fig. 8.33. (**a**) Schematic cross-sectional view of a High Electron Mobility Transistor (HEMT) consisting of a modulation doped $Al_xGa_{1-x}As/GaAs$ structure with a 2DEG at the interface, which carries the electron current between source and drain. Source and drain contacts are diffused in such that they reach the 2DEG plane. The gate metal (Al) is deposited after the channel has been recessed by chemical etching. (**b**) Band diagram along a coordinate normal to the 2DEG below the gate metal electrode. The gate electrode, which is used to control the conductivity channel of the 2DEG is electrically isolated from the 2DEG by the depletion layer below the Schottky metal-semiconductor junction

way a strong depletion layer (Schottky contact) is formed at the metal-semiconductor junction.

In thermal equilibrium, the band scheme along an intersection normal to the layers below the gate electrode appears as in Fig. 8.33b. The 2D electron gas with its high mobility carries the source-drain current. When a positive drain voltage is applied, the potential drop along the source-drain connection leads, of course, to a variation of the band scheme in Fig. 8.33b parallel to the AlGaAs/GaAs interface. Depending on the local potential, the accumulation layer is more or less emptied of electrons; the position of the Fermi level E_F with respect to the band edges varies along the current channel. Transistor action is possible since an additionally applied gate voltage shifts the Fermi level in the gate metal with respect to its value deep in the "undoped" GaAs layer (Fig. 8.33b). Because of the strong Schottky depletion layer just below the metal gate electrode (donors in the AlGaAs layer having been emptied; Sect. 8.2), most of the voltage drop occurs across this

AlGaAs layer, thus establishing a quasi-insulating barrier between gate electrode and 2D electron gas. The action of this Schottky barrier is similar to that of the SiO_2 layer in a MOSFET (Sect. 7.8). Depending on the gate voltage, the triangular potential well at the interface is raised or lowered in energy and the accumulation layer is emptied or filled. This changes the carrier density of the 2D electron gas and switches the source-drain current. Because of the high electron mobility of the 2D electron gas, extremely fast switching times can be achieved by this transistor.

To give a rough estimate of its performance, we consider a simplified model. With a gate of length L and width W the gate capacitance is given by

$$C_g = \epsilon\epsilon_0 LW/d_{AlGaAs} , \tag{8.27}$$

where ϵ and d_{AlGaAs} are the dielectric function and the thickness of the AlGaAs layer below the gate electrode. C_g determines via the external gate voltage V_g the charge induced in the 2D channel at the interface, i.e. in a linear approximation one has

$$C_g = \frac{dQ}{dV_g} \simeq \frac{en_s LW}{V_g - V_0} . \tag{8.28}$$

Here n_s is the 2D density of carriers and V_0 the built-in voltage at vanishing external gate bias. For normal performance the drain voltage is so high that almost all electrons move with saturation velocity v_s (on the order of 10^7 cm/s) independent of drain voltage, such that the drain-source current follows as

$$I_{DS} \simeq n_s Wev_s , \tag{8.29}$$

i.e., with (8.28) one obtains a nearly linear dependence on gate voltage in this simple model:

$$I_{DS} \simeq \frac{1}{L} C_g v_s (V_g - V_0) . \tag{8.30}$$

An important parameter describing the performance of a HEMT (or FET in general) is the transconductance

$$g_m = \left(\frac{\partial I_{DS}}{\partial V_g}\right)_{V_{DS}} \simeq C_g \frac{v_s}{L} = Wev_s \frac{dn_s}{dV_g} . \tag{8.31}$$

Transconductance and gate capacitance determine, according to (8.31), the transit time τ for an electron to pass under the gate

$$\frac{1}{\tau} = \frac{v_s}{L} = \frac{g_m}{C_g} . \tag{8.32}$$

For gate lengths on the order of 1 μm and saturation velocities v_s around 10^7 cm/s transit times of about 10 ps = 10^{-11} s are reached. This makes the HEMT extremely interesting for microwave applications.

8.7 Quantum Effects in 2D Electron Gases at Semiconductor Interfaces

In addition to their interesting applications in microelectronics (HEMT: Sect.8.6.3) 2D Electron Gases (2DEGs) prepared in modulation-doped semiconductor heterostructures also offer the possibility of studying interesting new quantum phenomena in which the wave nature of the electrons is strongly manifest. The possibility of observing such effects stems from the extremely high electron mobility at temperatures low enough that both inelastic phonon scattering (low T) and elastic impurity scattering (modulation doping) are suppressed. A prerequisite for the observation of quantum phenomena is the phase coherence of the electron wave function within spatial areas comparable in size with the dimensions of the system. Phase coherence means a constant electronic energy within that region, i.e. the absence of inelastic scattering processes which change the phase of the wave function. For sufficiently low temperature, the 2DEG at an AlGaAs/GaAs interface has an inelastic scattering time τ_{inel} that is longer than the elastic scattering time τ_{el} (elastic mean free path ℓ_{el}). The electron motion between phase breaking events is thus diffusive. The decisive length, the phase coherence length

$$\ell_\phi = (D\tau_{inel})^{1/2} \tag{8.33}$$

with D as the diffusion constant (determined by elastic scattering processes: $D = v_F \ell_{el}/2$ in a 2D system) can thus reach values of a couple of micrometers. Since lateral structuring by means of lithography is now able to produce structures with lateral dimensions in the range of 100 nm, so-called **ballistic transport** with suppressed inelastic scattering and phase coherence of the electron wave function can indeed be studied.

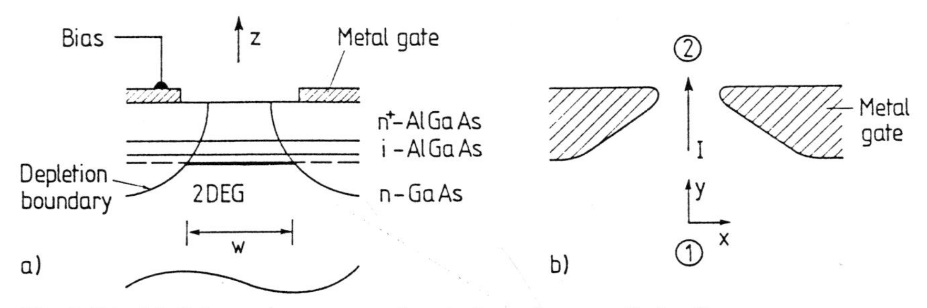

Fig. 8.34. (**a**) Schematic cross-sectional view of a so-called split-gate arrangement on top of a 2DEG in an AlGaAs/GaAs heterostructure. A suitable negative bias at the split metal gate electrode depletes the 2DEG below the metal gate and confines the 2DEG within a laterally restricted channel (quantum point contact: QPC). (**b**) Top view of the split metal gate electrode. The current I through the narrow channel within the 2DEG flows between contacts (1) and (2)

Particularly interesting phenomena can be observed when, in addition to the vertical confinement (z-quantization) within the narrow space-charge layer at the AlGaAs/GaAs interface (Sect. 8.6.3), a lateral confinement of the electronic wave functions is also achieved. In principle, this can be arranged by lateral constrictions in so-called **quantum point contacts** (*wires* and *dots*), or by an additional strong magnetic field perpendicular to the 2D electron gas giving rise to cyclotron orbit quantization as in the quantum Hall effect (Sect. 7.10).

Lateral restrictions for the generation of quantum dots or Quantum Point Contacts (QPC) in a 2DEG can be readily obtained by the deposition of a split metal (gate) contact on top of a modulation-doped semiconductor (AlGaAs/GaAs) heterostructure (Fig. 8.34). When this gate contact is suitably negatively biased, a depletion layer is formed below the contact, the

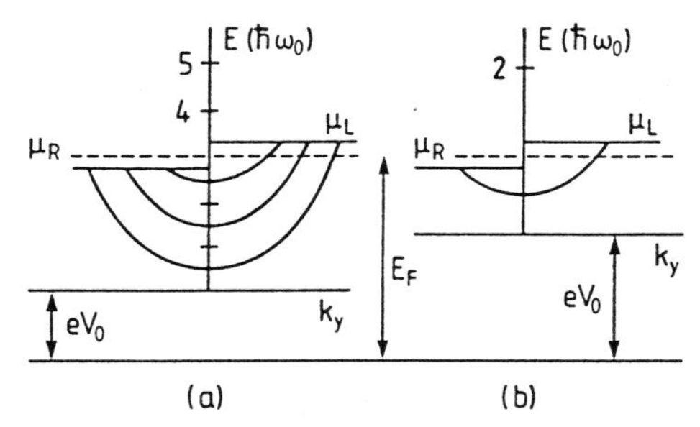

Fig. 8.35a,b. Occupied electron states in the 2DEG channel between the split gate arrangement (Fig. 8.34) at two different gate voltages, i.e. for two different widths of the QPC. In equilibrium the electron states are occupied up to the bulk Fermi energy E_F. When a current flows through the QPC the occupation is determined by the chemical potentials (Fermi energies) μ_1 and μ_2 at the contact areas *1* and *2*

band structure of the n-doped AlGaAs/GaAs heterostructure is raised in energy, and the triangular quantum well within the GaAs is emptied of electrons. Only within a spatial dimension of width w does the 2DEG continue to exist. There is now a spatial restriction within the QPC along z due to the limited spatial extension of the 2DEG along x because of the depletion zones left and right, and along y the point contact is defined by the dimensions of the metal gate in that direction. In a model description, the QPC is described as a channel with finite length, in which the electrons are laterally confined in the x-direction (Fig. 8.34b) by a parabolic potential (realistic approximation) of $0.5 m^* \omega_0^2 x^2$, m^* being the effective mass of the electrons, and ω_0 depends among other things on the width w and the doping; ω_0 can be calculated from space-charge theory. In such a parabolic confinement potential the quantized energy eigenvalues are those of a harmonic oscillator and one obtains the following dispersion relation for the electron states in the QPC:

$$E_n(k_y) = \left(n - \frac{1}{2}\right) \hbar\omega_0 + \frac{\hbar^2 k_y^2}{2m^*} + eV_0 , \tag{8.34}$$

where eV_0 is the electrostatic energy in the channel determined by the external bias at the gate, and k_y is the wave vector for motion along the channel. The spatial restriction of the 2DEG along x introduces different subbands (similar to a magnetic field B in Sect. 7.10), which are parabolic in k_y, the wave vector along the current (I) direction through the point contact. When an external voltage V is applied between the contacts (*1*) and (*2*) in order to induce a current through the contact (Fig. 8.34b) the chemical potentials (Fermi energies) μ_1 and μ_2 are different on the two sides and the relation

$$eV = \mu_1 - \mu_2 \tag{8.35}$$

holds. Depending on the external bias at the split gate electrode, a greater or lesser number of subbands are occupied within the QPC and the electrostatic potential V_0 will vary accordingly (Fig. 8.35). For the evaluation of the current I and the conductance I/V we assume that all electron states with positive velocity $v_y = 1/\hbar(dE_n/dk_y)$ are occupied up to μ_2 and that all electron states with negative v_y are occupied up to μ_1. Assuming also that no reflections occur at the QPC, the current through the contact derives from the difference in the chemical potentials as

$$I = \sum_{n=1}^{n_c} \int_{\mu_2}^{\mu_1} e\, D_n^{(1)}(E)\, v_n(E)\, dE , \tag{8.36}$$

where n is the number of the subband (n_c denotes the highest occupied subband), $D_n^{(1)}(E)$ the density of states in the one-dimensional subband n, and $v_n(E)$ the electron velocity in the nth subband. For the one-dimensional energy parabola $E = \hbar^2 k_y{}^2/2m^*$, the density of states (including spin degeneracy) is

$$D_n^{(1)} = \frac{1}{\pi}\left(\frac{dE_n}{dk_y}\right)^{-1} , \tag{8.37}$$

and the electron velocity in that subband follows as

$$v_n = \frac{1}{\hbar}\frac{dE_n}{dk_y} . \tag{8.38}$$

For the current through the QPC, (8.36), we thus obtain

$$I = \sum_{n=1}^{n_c} \frac{2e}{h}(\mu_1 - \mu_2) , \tag{8.39}$$

and for the conductance, with (8.35),

$$G_{QPC} = \frac{I}{V} = \sum_{n=1}^{n_c} \frac{2e^2}{h} . \tag{8.40}$$

According to (8.40) the conductance of such a QPC increases stepwise in jumps of the so-called **conductance quantum** ($2e^2/h$) up to a maximum value which is determined by n_c, the number of the highest occupied one-dimensional subband. Since the subbands result from quantization in the x-direction, the subband number is identical to the number of half wavelengths of the electronic wave functions within the slit width w. For a given electron energy, e.g. E_F, this number n varies with the width w of the QPC (extent of depletion below gate), i.e., the conductance of the QPC increases stepwise with decreasing negative gate voltage. This quantum effect in the QPC conductance has indeed been observed experimentally by *van Wees* et al. [8.48] on a QPC prepared in a 2DEG in a AlGaAs/GaAs heterostructure (Fig. 8.36).

For transport in the ballistic regime, where inelastic electron scattering is strongly suppressed, these QPCs can be used to generate *hot* electrons in

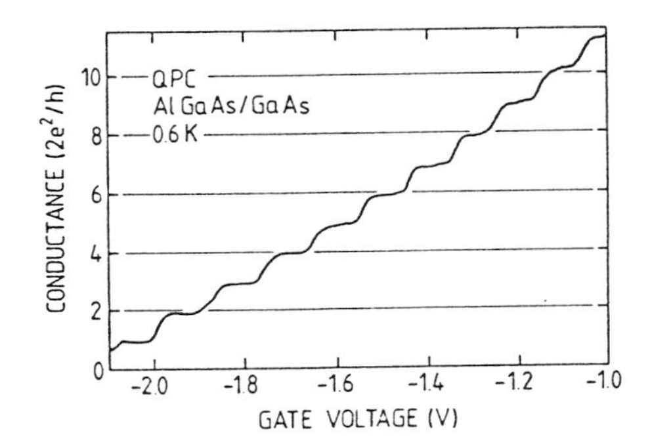

Fig. 8.36. Quantized conductance of a Quantum Point Contact (QPC) at 0.6 K prepared at a AlGaAs/GaAs interface (2DEG). The conductance was obtained from the measured resistance after subtraction of a constant series resistance of 400 Ω [8.48]

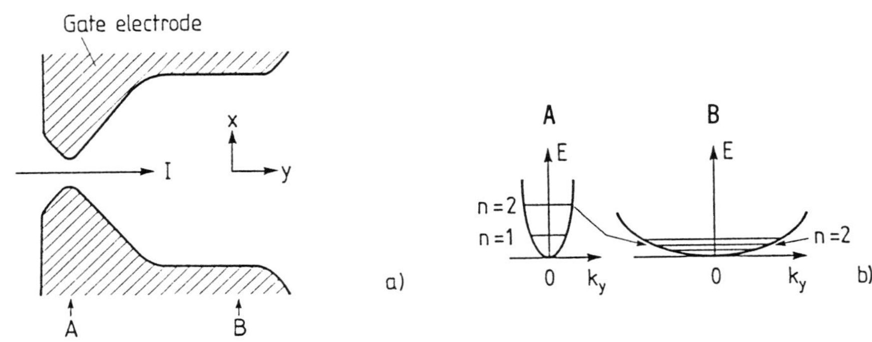

Fig. 8.37. (**a**) Top view of a split gate electrode arrangement for studying adiabatic, ballistic transport of electrons in a 2DEG. (**b**) Potential parabolas at points A and B within the 2DEG below the split gate electrodes. Also indicated are the quantized energy levels n = 1, 2, ...

a 2DEG. Gate electrodes shaped as in Fig. 8.37a and biased sufficiently negatively (Fig. 8.34a) provide a strong local constriction for the electron wave function at point *A* (QPC) whereas at point *B* the electron wave function can extend over a larger distance in the x-direction. Correspondingly the energy parabolas $E(k_y)$ are steep at A and flat at point *B*. The quantized subbands (numbered by n = 1,2,...) are thus quite close to each other at point B (Fig. 8.37b). For a ballistic current I in the y-direction within the 2DEG; electrons within the subband n = 2 change their potential energy considerably on moving from *A* to *B*, since the quantum number n is conserved in the case of ballistic transport. Furthermore, since the lack of inelastic scattering processes means conservation of total energy, the transport from *A* to *B* is connected with a gain of kinetic energy; the electrons leave the QPC (point *A*) to arrive as *hot* electrons in region *B*. This type of electron transport is called **adiabatic transport**.

A similar effect to that of lateral constrictions is induced by strong magnetic fields oriented perpendicular to the plane of the 2DEG. These can

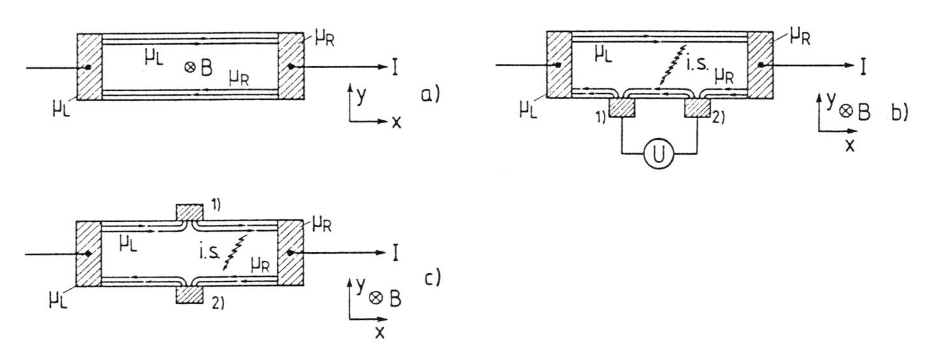

Fig. 8.38. (**a**) Ballistic electron flow in edge channels resulting from quantization in a strong magnetic field B normal to the 2DEG. Neglecting inelastic scattering between left (L) and right (R) edge channels the chemical potentials μ_L and μ_R are those of the ideal left and right contacts. (**b**) Edge channel conductance in a 2DEG in the presence of two additional contacts *1* and *2* for the measurement of Shubnikov-de Haas oscillations in a strong magnetic field B. Inelastic scattering (i.s.) between left and right channels induces a finite voltage drop between the contacts *1* and *2*. (**c**) Edge channel conductance in a 2DEG in the presence of two contacts (1) and (2) for measuring the quantum Hall effect

thus introduce a further quantization. This has already been discussed in connection with the 2DEG in narrow inversion layers under MOS structures (Sect. 7.10). In a model description we consider a simple geometry as in Fig. 8.38a with ideal contacts left (*L* source) and right (*R* drain or sink). The strong magnetic field B perpendicular to the 2DEG introduces Landau level quantization over the whole width (along y) of the current carrying stripe. The various sharp Landau levels (numbered n = 1,2,3,... in Fig. 8.39a) correspond classically to closed cyclotron orbits of the electrons between the two metal contacts. At the edges y_1 and y_2 closed cyclotron orbits are no longer possible; scattering of the electrons occurs at the boundaries (Fig. 8.39a) and this causes a stronger curvature of the electron wavefunctions in so-called **skipping orbits**. This, in turn, increases the energy eigenvalues and thus leads to an upwards bending of the Landau levels (n = 1,2,3,...) at the edges y_1 and y_2 (Fig. 8.39a). The Landau levels cross the Fermi energy E_F near the edges, leading to the formation of so-called **edge channels**, two (left and right) for each occupied Landau level, where electrons can move from one contact to the other (Fig. 8.38a). Electron transport can only occur within states near E_F. In strong magnetic fields as considered here, the skipping orbits do not allow backward scattering. Even scattering on impurities leads to transport in the forward direction only. Thus edge-channel transport does not exhibit electrical resistance, even in macroscopic samples. Therefore, the chemical potential μ_L in the left edge channels is that of the left contact, whereas on the backwards path (right) the chemical

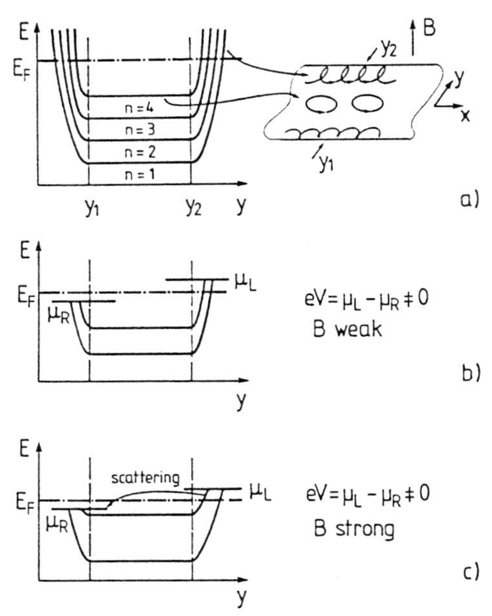

Fig. 8.39a-c. Explanation of edge channels in a 2DEG in the presence of a strong magnetic field B. (**a**) A strong magnetic field B normal to the plane of a 2DEG causes Landau level quantization ($n = 1, 2, 3, \ldots$) related to closed cyclotron orbits of the electrons. At the two borders y_1 and y_2, closed orbits are no longer possible; the Landau energy levels are shifted to higher energies thus forming so-called **edge channels** when they cross the Fermi energy E_F. (**b**) In the presence of a current (Fig. 8.38) the chemical potentials μ_R and μ_L of the right and left channels are different; their difference $\mu_L - \mu_R$ is directly related to the voltage drop across the contacts. (**c**) With increasing magnetic field B the splitting of the Landau levels increases and the uppermost occupied level approaches the Fermi level. Electronic states are available near E_F into which inelastic scattering can occur, thus bringing the left and right edge channels into contact

potential is that of the right contact μ_R (Fig. 8.38a). As in (8.36) the total ballistic current results from the difference in chemical potentials on the left and right, and each occupied edge channel contributes an amount

$$I_n = e v_n D_n^{(1)} (\mu_R - \mu_L) , \tag{8.41}$$

with $D_n{}^{(1)} = (2\pi)^{-1} (dE_n/dk_x)^{-1}$ as the density of states in the one-dimensional subband n. Since the electron velocity in that band is $v_n = \hbar^{-1} \times (dE_n/dk_x)$, each edge channel n transports a current of

$$I_n = \frac{e}{h} (\mu_R - \mu_L) . \tag{8.42}$$

Two contacts placed in series along the edge channels as in Fig.8.38b are used to measure Shubnikov de Haas oscillations (Sect.7.10). Since there is no voltage drop along the edge channels, zero resistance is measured as long as the occupied Landau levels between the edge channels are well separated from the Fermi energy E_F (Fig.8.39b). However, when the magnetic field B is increased (Fig.8.39c), the Landau level splitting is increased and the uppermost occupied level approaches E_F. Thus electronic states between the edge channels eventually reach the Fermi level and inelastic scattering between the left and right channels becomes possible. In this situation electrons are scattered from left to right (Fig.8.38b), transferring energy in the process and thus leading to a finite resistance ρ_{xx} between the two contacts in series. Each time a Landau level crosses the Fermi energy a maximum is observed in ρ_{xx} (Fig.8.40).

The same explanation holds for the quantum Hall effect. Voltage probes attached to either side of the 2DEG (Fig.8.38c) measure the electrochemical potential difference, i.e. the Hall resistance between the left and

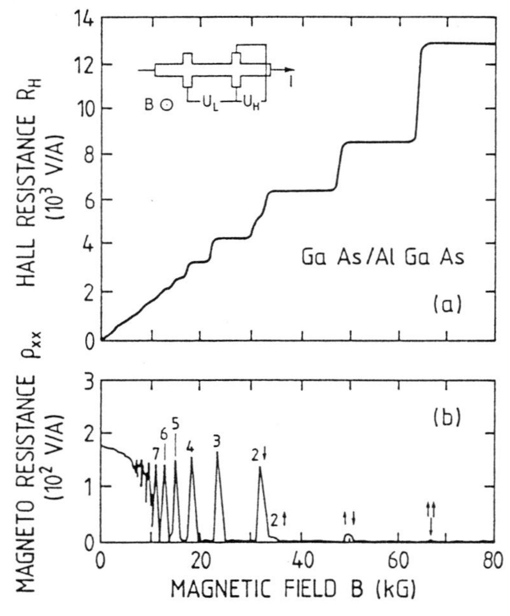

Fig.8.40. (**a**) Quantum Hall effect measured at 4 K on the 2DEG of a modulation-doped AlGaAs/GaAs heterostructure; the 2D electron density is $4 \cdot 10^{11}$ cm^{-2}, and the electron mobility $\mu = 8.6 \cdot 10^4$ cm^2/V·s. The Hall resistance $R_H = U_H/I$ is measured as a function of the magnetic field B as shown in the inset. (**b**) Shubnikov-de Haas oscillations of the magnetoresistance ρ_{xx} given by U_L/I as shown in the inset as a function of the magnetic field B. The number given at each maximum indicates the subband crossing the Fermi level, and the arrows give the corresponding spin orientations with respect to the B field [8.49]

the right edge channels. With (8.35) and (8.42) each edge channel contributes an amount of e^2/h to the Hall resistance. When a Landau level crosses E_F with increasing B field, the resistance increases by this amount. This is the elementary explanation for the stepwise change of the Hall resistance with increasing magnetic field in a quantum Hall effect measurement (Fig. 8.40a). It is worth mentioning that this explanation of the quantum Hall effect on the basis of edge channel conduction does not require localized states (due, to e.g., defects) in order to explain the formation of plateaus in the resistance as a function of magnetic field.

Panel XIII
Electrical Measurements of Schottky-Barrier Heights and Band Offsets

Schottky barriers at metal-semiconductor junctions and band-offsets between two semiconductors form internal potential barriers for carriers being transported through such interfaces. Electrical transport measurements can therefore give direct information about these barriers. The experiment requires thick metal or semiconductor layers on which electrical contacts can be prepared. A major advantage is that the measurements can also be performed ex-situ, the sandwich structure having been prepared under UHV conditions and then removed from the UHV chamber into the atmosphere.

The most direct approach is to measure the **current-voltage** (I-V) **characteristics**, from which the barrier height of a Schottky contact between a metal overlayer and a semiconductor substrate can be determined (Figs. 8.2, 4). Electrons contributing to the current through the Schottky contact can, in principle, tunnel through the barrier (if it is sufficiently thin) or more effectively overcome the barrier by means of their thermal kinetic energy (thermionic emission). For semiconductors with low electron mobilities the simple thermionic emission theory must be modified to take into account diffusion processes [XIII.1, 2]. In the simplest approximation the thermionic current density from the metal into the semiconductor is calculated for a direction z normal to the interface as

$$j_z^{m/s} = \int_{E_F + e\phi_B}^{\infty} e v_z D_C(E) f(E) dE \; , \tag{XIII.1}$$

where v_z is the velocity component along z of the free electrons in the semiconductor conduction band, $D_C(E)$ the conduction-band state density and f(E) the (Boltzmann) distribution function [XIII.3]. Since the energy E of an electron in the conduction band above E_C is kinetic energy, it follows that

$$E - E_C = \frac{1}{2} m^* v^2 \; , \quad dE = m^* v dv \; , \tag{XIII.2}$$

where m^* is the effective mass of the electrons. Using the square-root dependence of the density of states $[D_C \propto (E-E_C)^{1/2}]$ and

$$v^2 = v_x^2 + v_y^2 + v_z^2 , \quad 4\pi v^2 dv = dv_x dv_y dv_z , \tag{XIII.3}$$

we arrive at

$$j_z^{m/s} = 2e\left(\frac{m^*}{h}\right)^3 \exp\left(\frac{-e(E_C-E_F)}{kT}\right)\int_{v_{0z}}^{\infty} v_z \exp\left(\frac{-m^* v_z^2}{2kT}\right)dv_z$$

$$\times \int_{-\infty}^{\infty} \exp\left(\frac{-m^* v_x^2}{2kT}\right)dv_x \int_{-\infty}^{\infty} \exp\left(\frac{-m^* v_y^2}{2kT}\right)dv_y$$

$$= \left(\frac{4\pi e m^*}{h^3}\right)k^2 T^2 \exp\left(\frac{-e(E_C-E_F)}{kT}\right)\exp\left(\frac{-m^* v_{0z}^2}{2kT}\right) , \tag{XIII.4}$$

where v_{0z} is the minimum velocity in the z direction required to surmount the barrier. When an external voltage V is applied, v_{0z} is given by

$$\frac{1}{2} m^* v_{0z}^2 = e(V_{IF} - V) , \tag{XIII.5}$$

where eV_{IF} is the band bending at the interface (built-in potential at zero bias). Since the barrier height is given by

$$e\phi_B = eV_{IF} + E_C - E_F , \tag{XIII.6}$$

the thermionic current density is obtained as

$$j_z^{m/s} = A^* T^2 \exp\left(\frac{-e\Phi_B}{kT}\right)\exp\left(\frac{eV}{kT}\right) \tag{XIII.7}$$

with

$$A^* = 4\pi e m^* k^2 / h^3 \tag{XIII.8}$$

as the effective Richardson constant. For free electrons ($m^* = m_0$) the value of A^* is 120 $A/(cm^2 \cdot K^2)$.

The total current density through the interface consists of the component $j_z^{m/s}$ (XIII.7) from the metal into the semiconductor and the contribution $j_z^{s/m}$ from the semiconductor into the metal. The latter component is independent of external bias and equals the current density from the semiconductor into the metal (XIII.7) in thermal equilibrium ($V = 0$), i.e.

$$j_z^{s/m} = - A^* T^2 \exp(-e\phi_B/kT) . \quad \text{(XIII.9)}$$

The total thermionic current with external bias V thus follows as a sum of the contributions (XIII.7 and 9):

$$j_z = j_s [\exp(eV/kT) - 1] \quad \text{(XIII.10a)}$$

with

$$j_s = A^* T^2 \exp(-e\phi_B/kT) \quad \text{(XIII.10b)}$$

as the saturation current. An improved version of this theory, the so-called **thermionic emission-diffusion theory** also takes into account electron collisions within the depletion layer of the semiconductor [XIII.4]. It leads to the same expression (XIII.10a) for the current density through the interface, but the saturation current

$$j_s = A^{**} T^2 \exp(-e\phi_B/kT) \quad \text{(XIII.11a)}$$

now contains a generalized Richardson constant

$$A^{**} = f_P f_Q \frac{A^*}{1 + f_P f_Q v_R/v_D} , \quad \text{(XIII.11b)}$$

where f_P is the probability of electron emission over the potential maximum of the semiconductor into the metal without electron-optical-phonon backscattering; f_Q is the ratio of the total current flow including tunneling and quantum-mechanical reflection to the current flow obtained by omitting these effects. v_R and v_D are the recombination velocity and an effective diffusion velocity associated with thermionic emission.

From (XIII.10 and 11) it is obvious that the forward current density through a Schottky barrier, when extrapolated logarithmically to zero applied forward bias, has an intercept at the saturation current density j_s (XIII.11a). Thus the barrier height can be extracted from a plot of $\ln j_z$ versus applied forward voltage, as illustrated in Fig. XIII.1. For practical purposes, one often includes deviations from the ideal behavior described

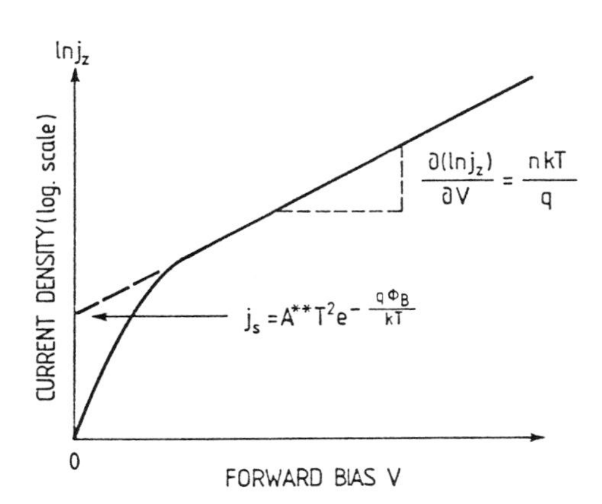

Fig.XIII.1. Forward current density plotted logarithmically as a function of applied voltage V for a metal-semiconductor contact (qualitative). The extrapolation to zero bias determines the saturation current density j_s, whereas the slope $\partial(\ln j_z)/\partial V$ allows one to determine the ideality factor n

by (XIII.10) in a so-called **ideality factor** n and writes in an heuristic way instead of (XIII.10a)

$$j_z = j_s[\exp(eV/nkT) - 1] . \qquad \text{(XIII.12)}$$

The ideality factor

$$n = \frac{2}{kT}\frac{\partial V}{\partial(\ln j_z)} = \left[1 + \frac{\partial \Delta\phi}{\partial V} + \frac{kT}{e}\frac{\partial(\ln A^{**})}{\partial V}\right]^{-1} \qquad \text{(XIII.13)}$$

contains a term $\partial\Delta\phi/\partial V$ which takes into account the voltage dependence of the barrier correction $\Delta\phi$ due to the combined effects of image force and applied field [XIII.5]. In addition, tunneling contributions to the current and recombination or trapping at interface states give rise to deviations of n from one. The ideality factor n can also be determined from the slope of the $\ln j_z$ versus V plot (Fig.XIII.1). Reliable values for the barrier height ϕ_B can only be obtained by the analysis described when the ideality factor n is close to one. Two experimental examples of I-V curves measured on W metal contacts on Si and on GaAs are shown in Fig.XIII.2.

In a similar way the measured I-V characteristic can also be used to determine conduction- or valence-band offsets in semiconductor heterostructures. For a single heterostructure the analysis is more complex [XIII.7], since the effective barrier for thermionic emission is dependent on both the band discontinuity and the band bending on either side of the interface (Fig.8.28). The determination of the band discontinuity ΔE_C or ΔE_V is more straightforward on a semiconductor double heterostructure, where a large band-gap material is sandwiched between two narrow bandgap layers. As an example Fig.XIII.3 presents experimental results [XIII.8] for a structure consisting of a layer of undoped AlGaAs (thickness between 50 and

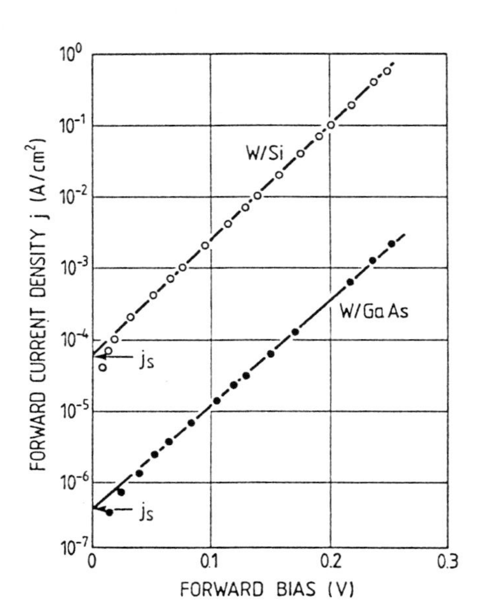

Fig. XIII.2. Measured forward current density as a function of applied voltage for a W/Si and a W/GaAs contact [XIII.6]

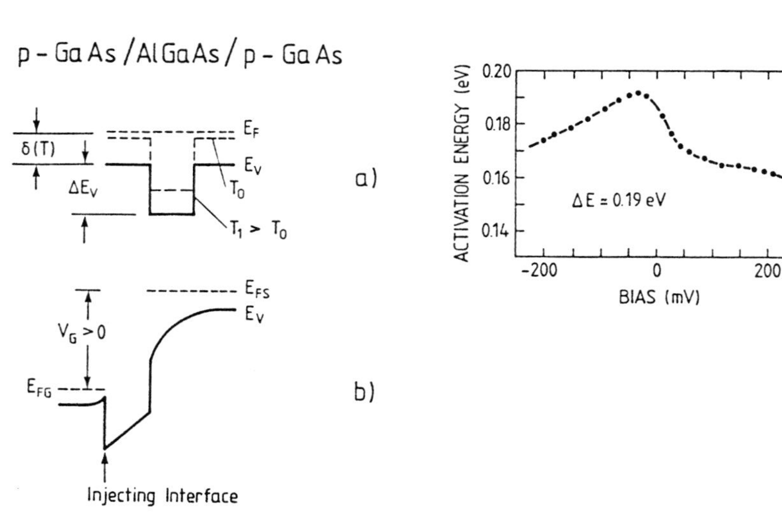

Fig. XIII.3a-c. Thermionic emission of holes over a p-GaAs/AlGaAs/ p-GaAs double heterostructure. (**a**) Qualitative plot of the valence band edge of the GaAs/AlGaAs/ GaAs double heterostructure for two temperatures T_0 and $T_1 > T_0$. (**b**) Qualitative band scheme of the lower energy barrier of (**a**) (T_1, full line) under applied bias; E_{FG} and E_{FS} are the quasi-Fermi levels. (**c**) Activation energy of the emission current over the energy barrier formed by the AlGaAs layer as a function of applied bias. The activation energy at zero voltage corresponds to the valence band discontinuity of a sample with a layer of $Al_{0.38}Ga_{0.62}As$ sandwiched between two GaAs layers [XIII.8]

100 Å) grown between two p-type GaAs layers forming a square potential barrier to the transport of holes (Fig. XIII.3a). Because of the layer thickness, tunneling can be neglected, and for sufficiently small bias (band bending and barrier distortion being negligible) the current density through the barrier at zero bias is given by (XIII.11a). The barrier height ϕ_B for hole transport is simply the valence-band discontinuity. A measurement of the thermal activation energy of the current density (current as a function of temperature) therefore yields ϕ_B. The results of such a procedure performed at various biases are shown in Fig. XIII.3c. The change in the activation energy with applied voltage is due to the effect of barrier deformation with bias. The valence-band offset of 0.19 eV is obtained from the data at zero bias for a AlGaAs layer containing 38% Al and 62% Ga.

The second experimental method for determining Schottky barrier heights and band offsets is the **Capacitance-Voltage** (C-V) **technique**. The space-charge layer at a Schottky contact or at a semiconductor heterointerface gives rise to a bias-dependent space-charge capacitance C_{sc}, i.e. the space-charge region has an effective complex impedance $R+(i\omega C_{sc})^{-1}$. Both the Ohmic part R and the capacitance C_{sc} can be determined by superimposing an AC voltage of frequency ω on a DC bias across the junction. Lock-in detection at frequency ω allows the separation of R and C_{sc}. According to (8.23) the capacitance of the depletion layer at a Schottky barrier is given by

$$C_{sc} = A\sqrt{\frac{e\epsilon\epsilon_0 N_D}{2(V_{IF}-V)}} \, , \qquad \text{(XIII.14)}$$

where A is the contact area, N_D the bulk donor concentration in the semiconductor, V_{IF} the band bending and V the external bias. Accordingly the band bending V_{IF} can be determined from a plot of $1/C_{sc}{}^2$ versus reverse bias V (Fig. XIII.4). A straight line with slope

$$\frac{d}{dv}C_{sc}{}^{-2} = \left(\frac{1}{2}\epsilon\epsilon_0 e N_D A^2\right)^{-1} \qquad \text{(XIII.15)}$$

is obtained and the intercept of this line with the abscissa (V scale) gives the band bending V_{IF} (Fig. XIII.4). For an n-type semiconductor the Schottky-barrier height is then given by (Fig. 8.2):

$$\phi_B = eV_{IF} + (E_C - E_F) - e\Delta\phi \, , \qquad \text{(XIII.16)}$$

Panel XIII

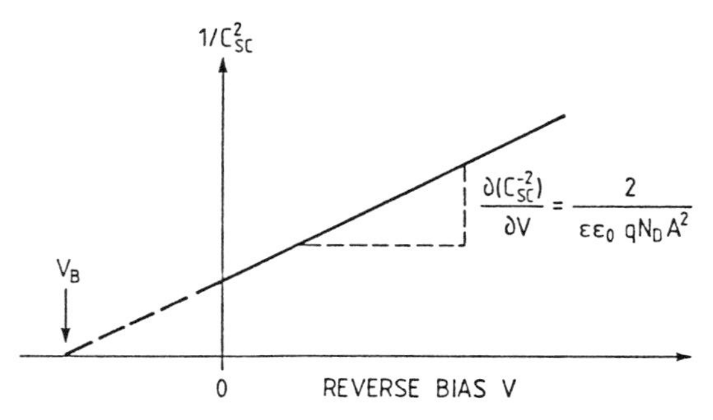

Fig.XIII.4. Capacitance-voltage (C-V) measurement on a metal-semiconductor junction. $1/C^2$ versus V plot for an ideal contact. The voltage intercept V_{IF} equals the band bending within the interface space-charge region

where E_C -E_F is the energetic separation of the conduction-band minimum and the Fermi level deep in the bulk of the semiconductor, and $\Delta\phi$ is the image force correction term [XIII.5]. $E_C - E_F$ is calculated directly from the bulk doping level using the effective mass of the conduction electrons. When applying the capacitance-voltage method to determine the Schottky-barrier height, one has to be aware of several possible sources of error: an insulating interface layer between the metal overlayer and the semiconductor might modify the total capacitance; furthermore, interface states not taken into account in the simple description (XIII.14) may be charged or emptied by a variation of the external bias.

Capacitance-voltage measurements can also be used to determine band discontinuities at semiconductor heterostructures. According to Fig.8.27 the total built-in potential (diffusion potential) at the interface is related to the conduction-band offset ΔE_C by

$$(E_V^{II} - E_F) - (E_C^{I} - E_F) = \Delta E_C + eV_{IF} \;. \qquad \text{(XIII.17)}$$

For the determination of ΔE one thus needs to know the total built-in potential V_{IF} as well as the bulk Fermi-level positions with respect to the conduction-band edges E_C^I and E_C^{II} in the two semiconductors. The latter values (E_C^I) and $(E_C^{II} - E_F)$ are given directly by the bulk doping levels, the effective masses and the temperature [XIII.5]. The total built-in potential V_{IF}, on the other hand, is obtained according to (8.23) as the intercept of a $1/C_{IF}{}^2$ versus V plot with the voltage axis (Fig.XIII.4). For other types of heterojunction the analysis may be more complex. Furthermore, severe errors can arise in the C-V method due to non-ohmic contacts, interfacial defects, bulk traps and non-uniform doping in the semiconductors I and II.

References

XIII.1 S.M. Sze: *Physics of Semiconductor Devices* (Wiley, New York 1981)

XIII.2 L.J. Brillson: The structure and properties of metal-semiconductor interfaces, Surf. Sci. Rep. **2**, 123 (1982)

XIII.3 H.A. Bethe: Theory of the boundary layer of crystal rectifiers, MIT Radiat. Lab. Rep. 43-12 (1942)

XIII.4 C.R. Crowell, S.M. Sze: Solid State Electron. **9**, 1035 (1966)

XIII.5 H. Ibach, H. Lüth: *Solid-State Physics – An Introduction to Principles of Material Science*, 2nd edn. (Springer, Berlin, Heidelberg 1996)

XIII.6 C.R. Crowell, J.C. Sarace, S.M. Sze: Trans. Met. Soc. AIME **233**, 478 (1965)

XIII.7 S.R. Forrest: Measurement of energy band offsets using capacitance and current measurement techniques in *Heterojunction Band Discontinuities: Physics and Device Applications*, ed. by F. Capasso, G. Margaritondo (Elsevier, Amsterdam 1987) p.311

XIII.8 J. Batey, S.L. Wright: J. Appl. Phys. **59**, 200 (1986)

Problems

Problem 8.1. Calculate the Fermi energy E_F and the Fermi wave vector $\mathbf{k}_F$ for a 2DEG at a modulation-doped AlGaAs/GaAs heterointerface. The 2D electron density at 4 K shall be $n_s = 5 \cdot 10^{11}\ cm^{-2}$. Plot the Fermi circle in the k_x,k_y-plane of the 2D reciprocal space and discuss the effect of quantum confinement in x-direction by means of a split gate arrangement (quantum point contact).

Problem 8.2. Calculate the density of states for a free electron gas, which is confined in 1, 2 and 3 directions (2DEG, quantum wire, quantum dot).

Problem 8.3. Estimate by means of Tersoff's model for the electronic structure of an ideal semiconductor interface the conduction and valence-band off-sets ΔE_C and ΔE_V of a Si/GaP heterojunction and plot the electronic band scheme near the interface.

a) For the case that Si and GaP are both moderately n-type doped.

b) For the case that Si is n-type and GaP is p-type doped.

Problem 8.4. Plot the band scheme (valence- and conduction-band edges E_V and E_C, respectively) as a function of space coordinate (z) for an AlAs/GaAs heterojunction,

a) when the heterostructure was grown by MBE with an ideal interface;and

b) when a high density of interface defects ($>10^{12}\ cm^{-2}$) was produced by electron irradiation at the AlAs/GaAs interface. Where in the forbidden band do you expect the highest density of interface states? The bulk doping in both semiconductors is assumed to be moderately n-type ($\approx 10^{17}\ cm^{-3}$).

Problem 8.5. Using the electronic band scheme of an Au/GaAs heterojunction, give an explanation why such a junction can be used as a Metal/Semiconductor (MS) photodetector for visible light. Make a suggestion for construction of a MS diode.

Problem 8.6. For a 2-Dimensional Electron Gas (2DEG) at a modulation-doped AlGas/GaAs heterojunction the 2D electron density at 1 K is $5 \cdot 10^{11}\ cm^{-2}$. By a split-gate point contact a ballistic electron beam is injected into the 2DEG with the average excess energy $\Delta = 50$ meV above the Fermi energy E_F. Discuss why, in contrast to a free electron gas in a metal, electron-electron scattering is the major scattering mechanism for the ballistic electrons at temperatures $T < 1$ K.

a) Consider phonon and defect scattering.

b) Consider electron-electron scattering by comparing Δ and E_F.

Problem 8.7. According to the Richardson-Dushman formula (see also Panel XIII) the work function of a metal surface can be determined from a measurement of the saturation current density $j_s = AT^2 \exp(-e\phi/kT)$ with $A = 4\pi mek^2/h^3 = 120\ A/K^2 \cdot cm^2$ upon thermionic emission of electrons from the metal into vacuum. Derive the Richardson-Dushman formula for j_s by assuming a metal at temperature T in thermal equilibrium with the vapor of emitted electrons above its surface. For both half spaces – metal and electron vapor – the electron density is given by

$$n = \int_0^\infty D(E)f(E)dE$$

with the same Fermi level E_F in the Fermi-distribution f(E).

Only, on the vacuum side the zero point of the energy scale is shifted by $E_{vac} = E_F + e\phi$ with respect to the metal. From a calculation of the electron densities in the vapor phase (n_v) and in the metal (n_M), and the assumption of equal electron current densities from and into the metal (dynamic equilibrium) derive j_s using classic kinetic gas theory.

Problem 8.8. Consider the interface between a superconductor (gap energy :2Δ) and a modulation doped semiconductor heterostructure with its 2DEG normal to the interface at a temperature $T \ll \Delta/k$. A hot electron in the 2DEG with a slight excess energy $\epsilon \ll \Delta$ above the Fermi energy E_F is emitted by a quantum point contact and moves towards the interface with a wave vector **k**. Since no single electron states are available on the superconductor side, the normal process is reflection of the electron at the interface. There is, however, a finite probability that the electron penetrates through the interface into the superconductor to form a Cooper pair;the second electron needed for the Cooper pair also originates from the 2DEG.

a) Show by considering the k vectors of the electron involved that the formation of the Cooper pair leads to the emission of a hole state with energy $E_F - \epsilon$ in the 2DEG. The hole moves backwards from the interface into the semiconductor with a wave vector $-\mathbf{k}$. This process is called **Andreev reflection**. Discuss the Andreev hole state in comparison with holes in the valence band of a semiconductor in thermal equilibrium.

b) By considering charge and effective mass of electron and hole states, show that Andreev reflection doubles the current of the initial primary hot electron.

9. Adsorption on Solid Surfaces

In previous chapters we have considered two types of interfaces, the solid/vacuum and the solid/solid interfaces. This last chapter is devoted to problems of the solid/gas interface. Some of the questions related to this interface have already been touched on in connection with film growth and the deposition of atoms and molecules to yield a second solid phase and thus a new solid/solid interface. In the present chapter we consider the interaction between a solid surface and foreign atoms in a more fundamental way.

At this point one might ask why the solid/liquid interface is not treated within the framework of the present book. The main reason is a methodological one: Most of the extremely powerful experimental techniques used to study solid/vacuum and solid/solid interfaces in UHV cannot be applied to solid/liquid interfaces, except in extreme cases of monolayer coverage. These, however, are identical with the "adsorption" systems treated in this chapter.

From a purely theoretical point of view, many features of the solid/liquid interface resemble those of the solid/gas interface. But the plaucity of experimental methods means that our understanding of the solid/liquid interface is less well developed. We thus concentrate here on adsorption processes, i.e. the interaction between a solid surface and atoms or molecules in the gas phase.

9.1 Physisorption

The adsorption of an atom or molecule on a solid surface involves the same basic forces that are known from the quantum-mechanical theory of chemical bonding. Now, however, one of the partners is a macroscopic medium with an "infinite" number of electrons, whose 2D surface is exposed to the other microscopic bonding partner, the atom or molecule. It turns out that many of the concepts of the theory of chemical bonding can be directly transferred to adsorption theory. In particular, there is a clear distinction between physisorption and chemisorption in the theory of adsorption. Generally speaking, physisorption is a process in which the electronic structure of the molecule or atom is hardly perturbed upon adsorption. The corresponding mechanism in molecular physics is **van der Waals bonding**. The attractive force is due to correlated charge fluctuations in the two bonding partners, i.e. between mutually induced dipole moments. In molecular physics, where these dipoles can be considered as "point" dipoles, the attractive potential is that between attracting dipoles.

In contrast, chemisorption is an adsorption process that resembles the formation of covalent or ionic bonds in molecular physics; the electronic structure of the bonding partners is strongly perturbed, new hybrid orbitals are formed and, as in the case of ionic bonding, there may be charge transfer from one partner to the other. In dissociative chemisorption one may even observe the formation of new molecules.

In spite of the similarity between adsorption and molecular bonding, certain features, such as the variation of forces with distance, might be different due to the different dimensionality of the two problems. Correspondingly one requires different models to describe bonding in molecular physics and in adsorption. This becomes quite clear when one considers a simple model of physisorption. In molecular physics the attractive potential of the van der Waals interaction between neutral molecules can be described by the interaction between mutually induced "point" dipoles. One dipole p_1 formed by momentarily occurring charge fluctuations induces an electric field $\mathcal{E} \propto p_1/r^3$ at the site of the other molecule at distance r. The induced dipole moment there is $p_2 \propto \alpha p_1/r^3$, where α is the polarizability of the molecule. The potential of this dipole p_2 in the field of the first dipole is proportional to $\mathcal{E}$ and to p_2; the attractive part of the van der Waals potential thus has an r^{-6} dependence.

In contrast, the physisorption of a non-reactive atom or molecule (e.g., He, Ne, CH_4) on a solid surface requires a different description [9.1, 2]. In Fig. 9.1 the physisorbed atom is modelled by an oscillator in which an electron executes simple harmonic motion (coordinate u) in one dimension. The atom is located outside the surface, which lies at a distance z from the positive nucleus. The van der Waals attraction between the solid and the atom arises from the time-dependent, non-retarded interaction of the valence

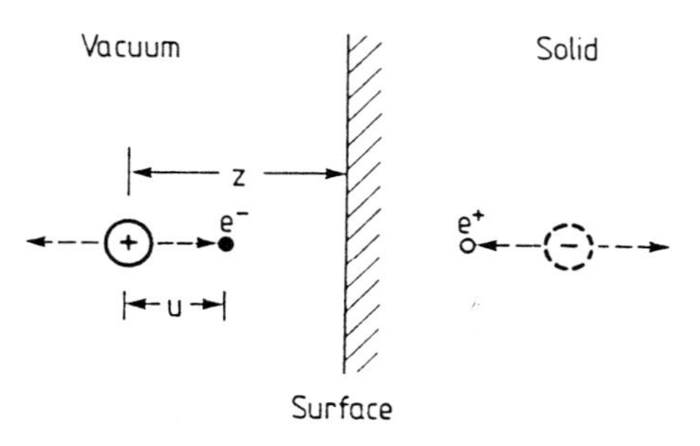

Fig. 9.1. Simple model of a physisorbed atom consisting of a positive ion and a valence electron e^-. The dynamics of the electron is described by a classical oscillation along a coordinate u normal to the solid surface. The attractive interaction with the solid is due to screening, i.e., it arises on forming image charges

electron and the nucleus (or core) with their images. The van der Waals interaction thus reduces to an image-charge attraction, describable in terms of the screening effects of the solid substrate. A point charge +e outside the surface of a semi-infinite medium with dielectric constant ϵ induces an image point charge.

$$q = \frac{1-\epsilon}{1+\epsilon} e \,, \tag{9.1}$$

positioned within the medium at the same distance from the surface. For a metal surface ($\epsilon \to \infty$, $q=-e$), the resulting potential energy between the real charge (distance z from surface) and its image is thus $V = -e^2/4\pi\epsilon_0 2z$. Setting $\tilde{q} = e^2/4\pi\epsilon_0$, the interaction energy between nucleus (core), electron and their images is thus obtained as

$$V(z) = -\frac{\tilde{q}^2}{2z} - \frac{\tilde{q}^2}{2(z-u)} + \frac{\tilde{q}^2}{(2z-u)} + \frac{\tilde{q}^2}{(2z-u)} \,. \tag{9.2}$$

The first term is the interaction of the nucleus (core) with its image, the second arises from the interaction of the electron with its image and the two repulsive terms are the interactions between the nucleus (core) and the electron image and vice versa. Expanding (9.2) in powers of u/z, one finds that terms with z^{-1} and z^{-2} cancel, and that the lowest order, non-vanishing term is

$$V(z) \simeq -\frac{\tilde{q}^2 u^2}{4z^3} \,. \tag{9.3}$$

The physisorption potential thus depends on the distance z between atom and surface as z^{-3}, in contrast to the r^{-6} dependence of molecular van der Waals bonding. Since the electron wave functions "leak out" of the surface of a metal (or solid in general), the image plane that serves as the reference for the z-coordinate in (9.3) is not identical with the surface itself, i.e. the plane defined by the coordinates of the nuclei of surface atoms. One therefore has to express the lowest-order physisorption potential as

$$V(z) \propto -(z - z_0)^{-3} \,, \tag{9.4}$$

with z_0 values on the order of half a lattice constant.

Calculations of more accurate physisorption potentials are possible using modern surface band structure and charge-density calculations. But they are tedious and require big computers. Some examples of calculated potential curves V(z) are given in Fig.9.2 for inert He atoms on Ag, Cu and Au surfaces. The repulsive part, not included in (9.1-3) is due to the repul-

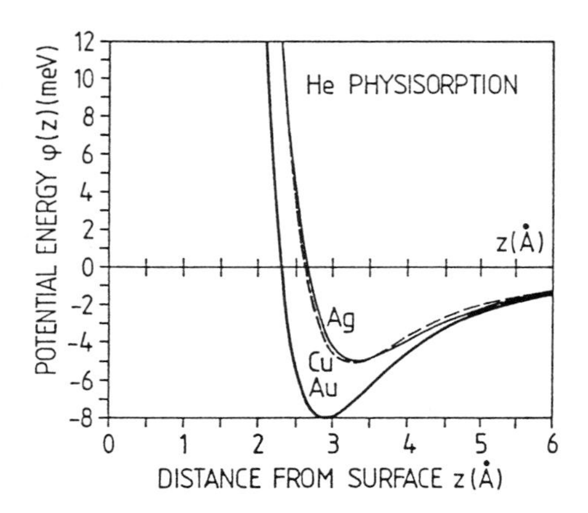

Fig.9.2. Calculated physisorption potentials $\phi(z)$ for He atoms outside Ag, Cu and Au surfaces. Each metal is described by a jellium model (homogeneously smeared-out charge), in which the properties of the particular metal are accounted for by different mean densities of positive background charge [9.3]

sion of overlapping electron shells, an effect that is included in the realistic calculations. The metal substrates in Fig.9.2 are described in terms of a jellium model with different mean densities of the smeared-out positive charge. An experimental method for investigating physisorption potentials is based on the analysis of scattering experiments (e.g., He atoms scattered from metal surfaces). The theoretical description of the experimentally determined scattering cross sections and angular distributions allows one to deduce certain features of the interaction potential between surface and scattered particle. Trial and error fits between the measured data and the curves calculated on the basis of assumed potentials yield a best-fit potential.

Physisorption potentials of the type shown in Fig.9.2 are characterized, in general, by a low binding energy (depth of the potential well) on the order of 10 to 100 meV, and by a relatively large equilibrium separation of $3 \div 10$ Å (distance between the potential minimum and the surface $z = 0$).

Physisorbed particles are therefore located at relatively large distances from the surface and are usually highly mobile in the plane parallel to the surface. As with the van der Waals interaction, the binding energy is quite low. Physisorption can only be observed when stronger chemisorption interactions are not present. In general, low temperatures are necessary to study physisorbed species, since at room temperature ($kT \approx 25$ meV) binding in a potential of the type shown in Fig.9.2 is not possible.

9.2 Chemisorption

Strong adsorbate bonding to a solid substrate must be understood in terms of a chemical reaction, similar to the case of molecular bonding. Covalent adsorption bonds obey essentially the same rules as do covalent bonds between atoms and molecules. The concept of orbital overlap is similarly important, and, for a qualitative approach at least, the same theoretical methods can be used as in the theory of chemical bonding.

In order to demonstrate the general principles underlying chemisorption bonding, let us consider a fairly simple adsorption system (Fig.9.3), namely a transition metal with an energetically sharp, partially-filled d-band, and a molecule with a partially-filled molecular orbital M. When the molecule approaches the metal surface, one expects covalent bonding between the partially filled orbitals of the two partners, i.e., the orbital overlap between M and d should lead to chemisorption with rehybridisation and the formation of new Md orbitals. In a simplified model description we represent the metal d-band by a single energy level (partially filled) and neglect interactions with the s- and p-states of the metal and with molecular orbitals other than M. An approximate wavefunction for the adsorbate-metal system may then be formulated as

$$\psi = a\psi_1(M^-, d^+) + b\psi_2(M^+, d^-) , \qquad (9.5)$$

where $\psi_1(M^-, d^+)$ and $\psi_2(M^+, d^-)$ represent so-called **charge-transfer states**. $\psi_1(M^-, d^+)$ describes a state in which an electron is transferred from the metal states into the molecular orbital M, whereas $\psi_2(M^+, d^-)$ refers to the reverse situation in which the molecule has donated an electron from its orbital M into the empty part of the metal d-band. The calculation of the

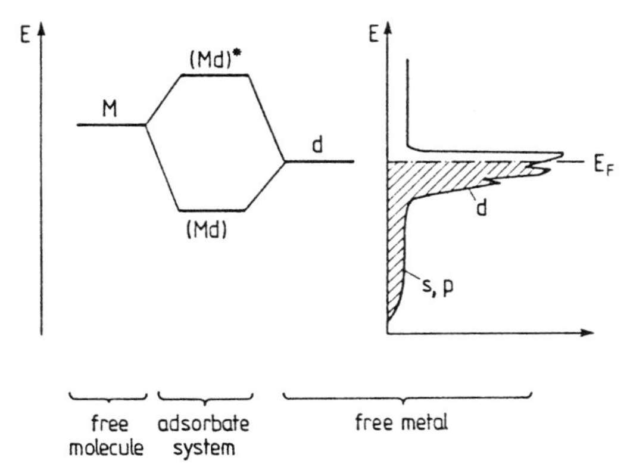

Fig.9.3. Simple model of covalent chemisorption bonding between a molecule (partially filled molecule orbital M) and a transition metal with partially filled d bands. Bonding with s and p metal states is neglected. In the adsorbate system, bonding (Md) and antibonding (Md)* states are formed

new chemisorption energy levels is performed by minimizing the energy functional

$$\widetilde{E} = \frac{\langle\psi|\mathscr{H}|\psi\rangle}{\langle\psi|\psi\rangle} . \tag{9.6}$$

where $\mathscr{H}$ is the total Hamiltonian (molecule plus metal substrate) and ψ as the trial charge-transfer wavefunction (9.5). We define $S = \langle\psi_1|\psi_2\rangle$ as the overlap integral between the two "ionic" charge-transfer states and $H_1 = \langle\psi_1|\mathscr{H}|\psi_1\rangle$ and $H_2 = \langle\psi_2|\mathscr{H}|\psi_2\rangle$ as the total energies of the states in which an electron is transferred from the metal to the molecule and vice versa. With $H_{12} = H_{21} = \langle\psi_2|\mathscr{H}|\psi_1\rangle$ as the interaction energy between the two "ionic" charge-transfer states (9.6) becomes

$$\widetilde{E}(a^2 + b^2 + 2abS) = (a^2 H_1 + b^2 H_1 + 2abH_{12}) . \tag{9.6a}$$

The wave functions ψ_1 and ψ_2 are assumed to be normalized. Minimization of $\widetilde{E}$ requires

$$\frac{\partial\widetilde{E}}{\partial a} = 0 \quad \text{and} \quad \frac{\partial\widetilde{E}}{\partial b} = 0 . \tag{9.7}$$

This yields the secular equations

$$a(\widetilde{E} - H_1) + b(S\widetilde{E} - H_{12}) = 0 , \tag{9.8a}$$

$$a(S\widetilde{E} - H_{12}) + b(\widetilde{E} - H_2) = 0 , \tag{9.8b}$$

whose solutions are given by the vanishing of the determinant

$$\begin{vmatrix} \widetilde{E} - H_1 & S\widetilde{E} - H_{12} \\ S\widetilde{E} - H_{12} & \widetilde{E} - H_2 \end{vmatrix} = 0 . \tag{9.9}$$

Two energy eigenvalues are obtained from (9.9):

$$\widetilde{E}_{\pm} = \frac{1}{2}\,\frac{H_1 + H_2 - 2SH_{12}}{1 - S^2}$$

$$\pm \sqrt{\frac{H_{12} - H_1 H_2}{1 - S^2} + \frac{1}{4}\left(\frac{H_1 + H_2 - 2SH_{12}}{1 - S^2}\right)^2} . \tag{9.10}$$

To demonstrate the qualitative behavior, we assume weak overlap between ψ_1 and ψ_2 and neglect second order terms in S and H_{12} ($S^2, H_{12}{}^2, SH_{12}$). In this linear approximation (9.10) yields

$$\tilde{E}_{\pm} = \frac{H_1 + H_2}{2} \pm \sqrt{\frac{H_1{}^2 + H_2{}^2}{2} + H_{12}} \; . \tag{9.11}$$

Compared to the average ionic energy $(H_1 + H_2)/2$, Eq.(9.11) yields two values $\tilde{E}_+$ and $\tilde{E}_-$, which, for positive H_{12}, are respectively higher and lower in energy. They belong to the (Md) chemical bond (Fig.9.3) and the corresponding antibonding orbital $(Md)^*$. The decrease of the total energy (9.10, 11) in $\tilde{E}_-$ favors a chemisorption bond in which electrons are transferred back and forth between adsorbate and substrate. For a more accurate description of the chemisorption bond the ansatz (9.5) for the total wave function is too simple. Better approximations take into account the wave function ψ_0 (M, met) of the no-bond state [separated substrate (met) and molecule (M), no charge transfer] and charge transfer to all unoccupied metal Bloch states ($k > k_F$, k_F : Fermi wave vector) and from all filled metal states ($k < k_F$) into the molecular orbital M; i.e., instead of (9.5) one uses a trial wave function [9.4]

$$\psi = N\psi_0(M, met) + \sum_{k < k_F} a_k \psi_k (M^-, met^+) + \sum_{k > k_F} b_k \psi_k (M^+, met^-) \tag{9.12}$$

to minimize the energy functional (9.6)

As in the orbital theory of molecular bonding, the concept of **frontier orbitals** is also useful in a description of chemisorption bonds. The strongest interaction with the adsorbing molecule occurs for an overlap between occupied and unoccupied orbitals, i.e. by electron transfer into the **Lowest Unoccupied Molecular Orbital (LUMO)** and by electron donation from the **Highest Occupied Molecular Orbital (HOMO)** into an empty substrate state. In the simplified picture of Fig.9.3 LUMO and HOMO are identical, since the highest energy molecular state M carrying a valence electron is assumed to be partially occupied. Carrying out the minimization procedure of (9.6) by means of (9.12) leads to the equation [9.4]

$$\tilde{E} - E_0 = \sum_{k<k_F} \frac{|U_{Lk}|^2}{E_k - E_{LUMO}} + \sum_{k>k_F} \frac{|U_{Hk}|^2}{E_{HOMO} - E_k} \tag{9.13}$$

for the total energy difference between the bonding situation ($\tilde{E}$) and the non-bonding state (E_0), where the molecule and the substrate are not in contact. E_k are the energies of the unperturbed metal Bloch states (or possibly surface states involved in the bonding); E_{LUMO} and E_{HOMO} are the unperturbed molecular orbital energies, whilst U_{Lk} and U_{Hk} are interaction matrix elements between the metal orbital k and the LUMO ($k<k_F$) and the HOMO ($k>k_F$), respectively.

More sophisticated theoretical approaches to chemical bonding on solid surfaces also exist, but these are far beyond the scope of this book. Particular emphasis to the local nature of a chemisorption bond is provided by cluster models, which are very useful in applying the methods of quantum chemistry to chemisorption bonding. In these calculations the solid surface is modelled by a finite number of substrate atoms ($3 \div 20$) and the chemisorption bond is described as a chemical bond between this cluster of substrate atoms and the particular chemisorbed atom or molecule. Since the cluster is bonded back to the whole (semi-infinite) solid substrate, it is inert to "backward adsorption". This property is sometimes taken into account by saturating all dangling bonds (apart from the chemisorption bond) with hydrogen atoms.

Chemisorption potentials $\phi(z)$ as a function of the distance z between adsorbate atoms or molecules and the surface are generally characterized by a short equilibrium separation z_0 of $1 \div 3$ Å (Fig.9.4a) and a relatively high binding energy E_B on the order of a couple of eV. Chemisorption is accompanied by a rearrangement of the electronic orbitals, i.e. of the electronic shell of the adsorbate atom or molecule; the shape of the adsorbate is thus changed due to its new chemical bonds to the substrate.

In the case of chemisorption of molecules, this rearrangement of the electronic shell can lead to dissociation and formation of new adsorbate species (Fig.9.4b). This so-called **dissociative adsorption** occurs, for instance, for hydrogen molecules on many transition metal surfaces at room temperature. When the clean metal surface is exposed to molecular hydrogen, H_2, rapid adsorption occurs, accompanied by dissociation of the mole-

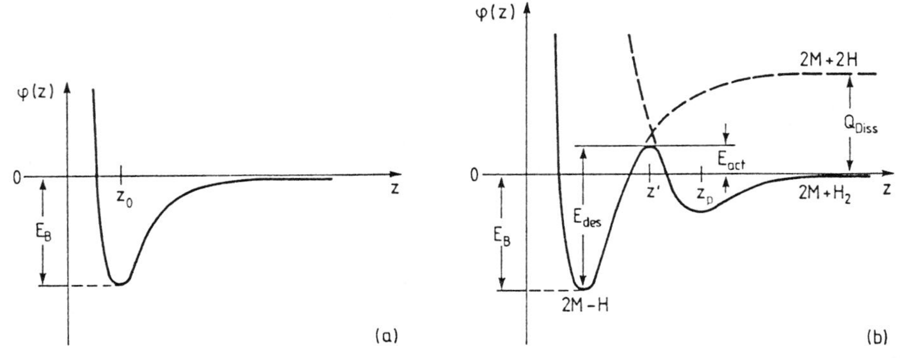

Fig.9.4. (**a**) Qualitative shape of a chemisorption potential ϕ as a function of the distance z of the adsorbed atom or molecule from the solid surface. The equilibrium distance z_0 is on the order of $1 \div 3$ Å and the binding energy E_B on the order of an electron volt. (**b**) Combination of a chemisorption and a physisorption potential shown qualitatively for the example of dissociative hydrogen (H_2) bonding on a metal (M) surface; Q_{Diss} is the dissociation energy of H_2 in the gas phase, E_B is the binding energy in the chemisorption state 2M-H, E_{act} the activation energy for adsorption of H_2, E_{des} the activation energy for desorption of 2H

cules into the atomic species H, which is bonded to the surface. The potential diagram for a hydrogen molecule approaching the surface along a coordinate z (normal to surface) can be described qualitatively as a combination of the potential for physisorption of molecular H_2 and chemisorption of atomic H (Fig.9.4b). A hydrogen molecule approaching the surface from a large distance z "sees" a potential, which leads into a physisorption state with equilibrium distance z_p (potential minimum). Closer approach to the surface would cause a rapid increase of potential energy due to overlap between the molecule's electronic shell and the metal states. Hydrogen atoms, however, can be bonded on the surface in a chemisorption state with much higher binding energy E_B and a smaller equilibrium distance. The corresponding potential curve for two H atoms differs from that of molecular H_2 at large distances z by exactly the dissociation energy Q_{dis}. This is the energy that must be supplied to dissociate H_2 into 2H in the gas phase.

According to Fig.9.4b the two potential curves for H_2 and 2H (each time referred to a complete system: two metal atoms (2M) plus hydrogen) intersect at a distance z'. A hydrogen molecule with enough kinetic energy to overcome the activation barrier E_{act} thus prefers to follow the chemisorption potential curve; near z' it is dissociated into two H atoms, which chemisorb on the surface forming two M-H bonds with an adsorption energy E_B. Near z' the electronic structure of the adsorbing particle is completely changed. The molecular orbitals of H_2 transform into atomic orbitals of H. From Fig.9.4 it is evident that chemisorption of molecular hydrogen into its

Table 9.1. Metal ion-binding energies [eV] on different single crystal tungsten surfaces, as determined from experiment [9.5, 6]

	Adsorbate			
Substrate	Na	K	Pt	Re
W {100}		2.28	5.0	9.3
W {110}	2.46	2.05	5.5	10.15
W {111}	2.45	2.02		

atomic adsorption state requires a minimum kinetic energy of E_{act}, the activation energy for chemisorption. Since this activation barrier is lower than the dissociation energy Q_{Diss} in the gas phase, dissociation is favored by adsorption on the metal. The decrease of the activation barrier by the presence of a solid surface for dissociation is a feature of catalytic decomposition. From Fig.9.4 one can also deduce that desorption of chemisorbed atomic H from the metal surface requires a minimum energy E_{des}, the desorption energy. The desorbing H atoms recombine near z′ to form molecular H_2, which is detected in the gas phase. For this process of activated adsorption the characteristic energies, E_B (chemisorption energy), E_{des} (desorption energy) and E_{act} (activation energy for chemisorption) are related to one another by

$$E_{des} = E_B + E_{act} \ . \tag{9.14}$$

Since particles adsorbed in the chemisorption potential minimum always have a certain finite energy, even at zero temperature, E_B, as shown in Fig.9.4b, must be corrected for this small zero-point energy. Some experimental values of chemisorption binding energies are given in Table 9.1 for metal atoms adsorbed on single crystal surfaces of tungsten. The effect of the d-electrons on the bonding strength for Pt and Re is particularly evident.

9.3 Work-Function Changes Induced by Adsorbates

Adsorbed atoms and molecules generally have a significant influence on the electronic structure of a surface: They rearrange the electronic charge within the chemical bond and may also add elementary dipoles if the adsorbed molecule has its own static dipole moment. It is thus necessary to consider the work function of a solid surface in more detail, in particular in the presence of an adsorbed species.

In previous chapters (Chaps.6 and 8) the work function $e\phi$ was introduced in an intuitive way as the energy difference between Fermi level E_F and the vacuum energy E_{vac}. The precise definition of $e\phi$ is based on a gedanken experiment in which an electron is removed from inside the bulk crystal and transferred through the surface to a region outside, but not too far away from the surface. The distance of the electron from the crystal face should be so large that the image force can be neglected (typically 10^{-4} cm $= 10^4$ Å), but it should be small compared with the distance from any other face of the crystal with a different work function. Otherwise it is not possible to discriminate between work functions of different crystal faces. In this definition the work function is the energy difference between two states of the whole crystal. As in a photoemission experiment (Sect.6.3), the initial state is the ground state of a neutral crystal containing N electrons with energy E_N. In the final state one electron is removed to the outside, where it has only electrostatic energy described by the vacuum level E_{vac}. The crystal with the remaining N−1 electrons is assumed to be in its new ground state with energy E_{N-1}. We thus obtain for the work function at zero temperature

$$e\phi = E_{N-1} + E_{vac} - E_N \ . \tag{9.15}$$

For finite temperatures this process is described as a thermodynamic change of state. The difference E_N-E_{N-1} has to be replaced by the derivative of the free energy F with respect to the electron number (T = const, V = const). This derivative $(\partial F/\partial N)_{T,V}$ is the electrochemical potential of the electrons (or the Fermi energy E_F at finite temperature). A rigorous expression for the work function is thus

$$e\phi = E_{vac} - \mu = E_{vac} - E_F \ . \tag{9.16}$$

Even on a clean, well-defined surface in UHV, the microscopic interpretation of $e\phi$ might contain several contributions. On a metal surface a major contribution is due to the fact that the electron density "leaks out" from the relatively rigid framework of positive ion cores (Fig.3.7a). This gives rise

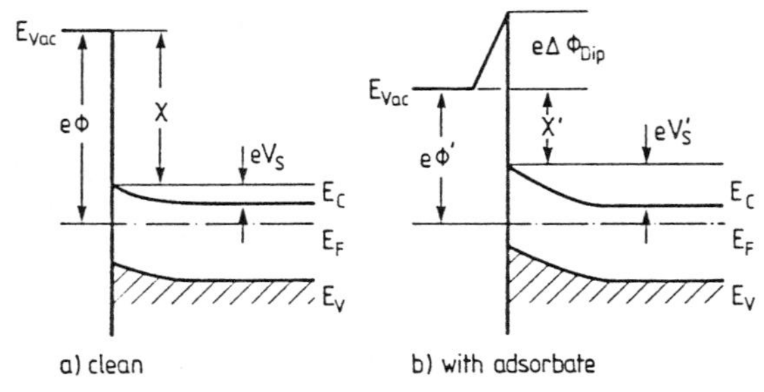

Fig.9.5. Qualitative electronic-band diagrams of clean and adsorbate-covered semiconductor surfaces; (**a**) clean surface with work function $e\phi$, electron affinity χ, band bending at the surface eV_S, conduction band and valence band edges E_C and E_V (**b**) chemisorption bonding of an adsorbate generally changes the band bending into $eV_S{}'$; charge transfer within the chemisorption bond induces dipoles within the surface and thus changes the work function and the electron affinity into $e\phi'$ and χ', respectively (dipole contribution $\Delta\phi_{Dip}$)

to a dipole layer at the surface which the emitted electron must pass through. Similar effects occur at steps, which thus also modify the work function of a clean surface (Fig.3.7b).

In the case of strong chemisorption, charge is shifted from the substrate to the adsorbed atom or molecule, or vice versa, thus giving rise to additional dipoles whose field acts on emitted electrons. This effect is described by a change of the work function $e\Delta\phi$ due to adsorption. Even in the case of physisorption, image charges just below the surface are created (Fig.9.1) by screening. The resulting dipole moments give rise to work function changes. For semiconductors, one has the additional effect of band bending (Chap.7), which also contributes to the total work function change (Fig.9.5). For a semiconductor it is convenient to describe the total work function by means of three terms:

$$e\phi = \chi + eV_s + (E_C - E_F)_{bulk} ,$$

where χ is the electron affinity. The effect of dipoles (due to the adsorbed atoms or molecules) $e\Delta\phi_{Dip}$ is assumed to change the electron affinity from χ to χ', and there is an additional band bending change ΔV_s. Thus the total work function change $e\Delta\phi$ of a semiconductor due to adsorption is obtained as

$$e\Delta\phi = \Delta\chi + e\Delta V_s = e\Delta\phi_{Dip} + e\Delta V_s . \qquad (9.17)$$

The two contributions can be determined separately in a photoemission experiment (Panel XV: Chap.9).

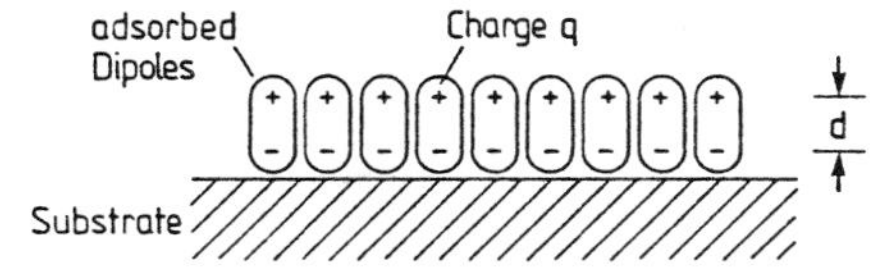

Fig. 9.6. Schematic representation of a well-ordered monolayer of highly-polar molecules with molecular dipole moment qd

The dipole contribution $e\Delta\phi_{Dip}$ of a monolayer of adsorbate is often calculated on the basis of simplifying assumptions about the nature and magnitude of the dipoles. In a simple model (Fig. 9.6) one can describe the dipole-induced work function change in terms of emitted electrons crossing a parallel plate capacitor (plate separation d), which carries a total charge density $n_{Dip}\,q$, n_{Dip} being the surface density of adsorbed dipoles. The corresponding work function change is

$$e\Delta\phi = -q\mathcal{E}d\,, \tag{9.18}$$

where

$$\mathcal{E} = \frac{n_{Dip}\,q}{\epsilon_0} \tag{9.19}$$

is the electric field within the dipole layer (between the capacitor plates). With $p = qd$ as the dipole moment of the adsorbed particle, one has the simple relation

$$e\Delta\phi = \frac{-e}{\epsilon_0} n_{Dip}\,p\,. \tag{9.20}$$

In a more rigorous treatment one has to take into account that the electric field at the site of a particular dipole is modified by all the surrounding dipoles [9.7]. A depolarization effect occurs such that in (9.18) an effective field $\mathcal{E}_{eff}$ has to be used

$$\mathcal{E}_{eff} = \mathcal{E} - f_{dep}\,\mathcal{E}_{eff}\,, \tag{9.21}$$

where f_{dep}, a so-called **depolarization factor**, takes into account the field due to dipoles in the vicinity. According to *Topping* [9.8], for a square array of uniformly arranged dipoles, this depolarization factor is obtained as

$$f_{dep} \simeq \frac{9\alpha n_{Dip}^{3/2}}{4\pi\epsilon_0}\,, \tag{9.22}$$

where α is the polarizibility of the adsorbed particles (or adsorbate/substrate complexes). From (9.18-22) one thus obtains the dipole-induced work function change as

$$e\Delta\phi = -\frac{e}{\epsilon_0} p n_{Dip} \left[1 + \frac{9\alpha n_{Dip}^{3/2}}{4\pi\epsilon_0}\right]^{-1}\,. \tag{9.23}$$

Apart from simple cases such as strong ionic chemisorption, it is difficult to apply (9.23) to real experiments, since neither the dipole moment p nor the polarizability α of the adsorbed particle is well known.

On the other hand, the measurement of work-function changes upon adsorption often yields interesting information about different adsorbed species. Figure 9.7 shows as an example work-function changes $e\Delta\phi$ measured by UPS for a Cu(110) surface which was exposed to H_2O at about 90 K [9.9]. After an initial decrease by about 0.9 eV the work function increases in several steps with increasing temperature. Each step indicates a new adsorbed species which also gives rise to different LEED patterns. The identification of the different species as *physisorbed* H_2O, strongly chemisorbed "H_2O", OH and atomic oxygen (O) was made on the basis of the observed photoemission spectra.

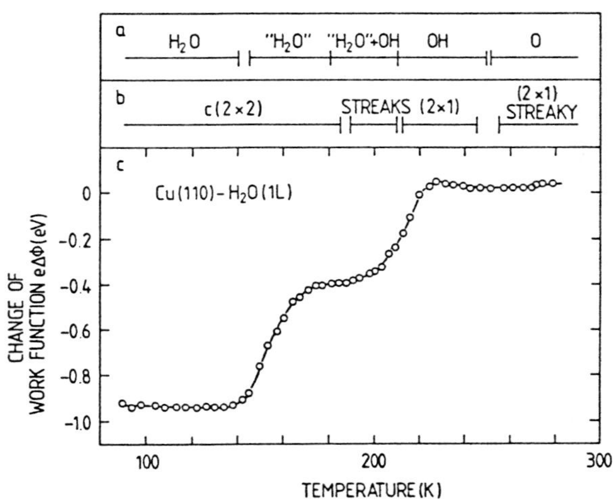

Fig. 9.7. Work-function change $e\Delta\phi$ of a water-covered Cu(110) surface as a function of annealing temperature; the surface was initially exposed to 1L = 10^{-6} Torr·s of water (**c**); (**a**) adsorbed species as identified by UPS measurement (**b**) corresponding superstructures as observed by LEED [9.9]

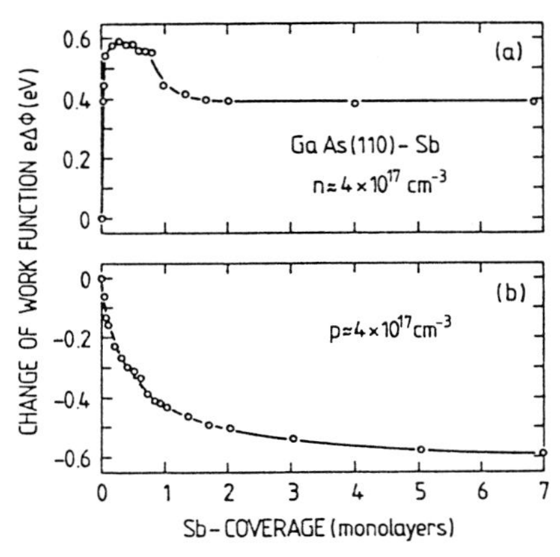

Fig. 9.8a,b. Work-function changes for GaAs(110) surfaces as a function of Sb coverage. (**a**) for n-type material with an electron concentration n $\simeq 4 \cdot 10^{17}$ cm^{-3} with nearly-flat bands prior to deposition of Sb (**b**) for p-type material with a hole concentration p $\simeq 4 \cdot 10^{17}$ cm^{-3} and an initial band bending of -0.3 eV before deposition of Sb [9.10]

In Fig. 9.8 work-function changes measured by UPS on cleaved GaAs (110) are shown as a function of Sb coverage [9.10]. For p- and n-type GaAs completely different curves are obtained, since the band bending contributions $e\Delta V_s$ (9.17) are different. On n-type material the bands are bent upwards due to Sb deposition whereas on p-type material downwards band bending changes are induced. Since the dipole contribution to the work function is related to the microscopic properties of the chemisorption bond, it is assumed to be equal for both dopings. From a more detailed analysis of the data it is found to decrease monotonically up to a coverage of about one monolayer of Sb (Fig. 9.9). This dipole contribution $e\Delta\phi_{\mathrm{Dip}}$ is responsible for the step-like behavior of the total work-function change $e\Delta\phi$ near one monolayer coverage on n-type GaAs (Fig. 9.8a)

A particularly interesting system with respect to changes in the work function is the adsorption of cesium both on metal and semiconductor surfaces. Upon adsorption, Cs atoms donate an electron to the substrate and chemisorb as positive Cs^+ with strong ionic chemisorption bonds. A dipole layer with negative charge in the substrate and Cs^+ on the surface is formed. An emitted electron is accelerated within this dipole field on its way from the crystal into the vacuum. The work function is thus considerably decreased by the adsorption of Cs. The effect is demonstrated in Fig. 9.10 for Cs adsorption on different surfaces of W.

On GaAs surfaces the work function or, more specifically, the dipole contribution (i.e., the electron affinity), is reduced by Cs adsorption to such an extent that the vacuum level resides below the conduction band minimum

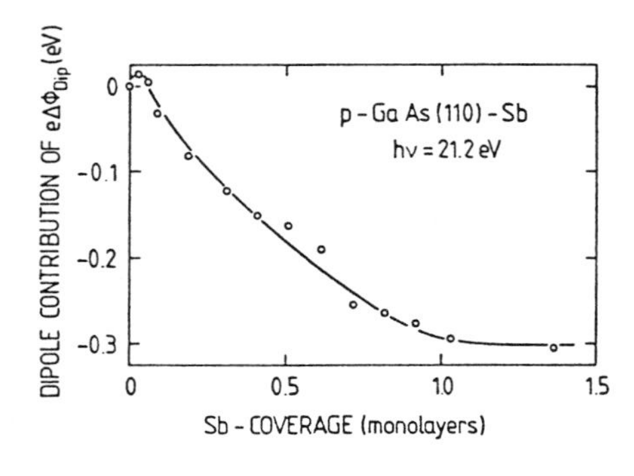

Fig.9.9. Dipole contribution $e\Delta\phi_{Dip}$ to the work-function change for a GaAs (110) surface due to adsorption of Sb. The data are obtained from UPS measurements on p-doped GaAs surfaces, which exhibited a saturated initial band bending due to irradiation with HeII photons before deposition of Sb [9.10]

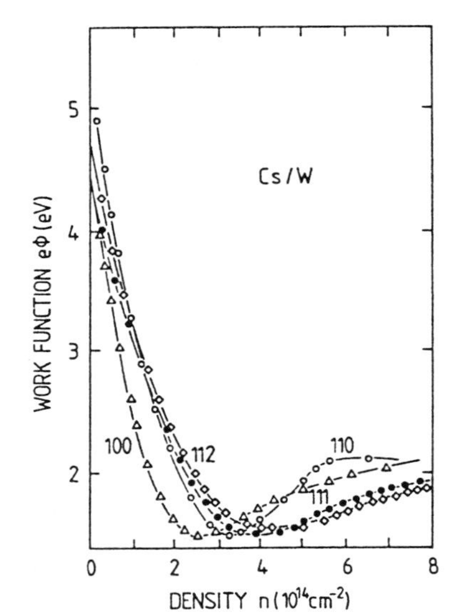

Fig.9.10. Work function of several tungsten surfaces as a function of coverage with Cs atoms [9.11]

when the additional downwards band bending on p-type material is taken into account (Fig.9.11). The effect is even more pronounced if oxygen is coadsorbed with the Cs atoms. Therefore, any electron which is pumped into the conduction band spills out into the vacuum without having to surmount any energy barrier. GaAs substrates covered with Cs are thus used as high flux sources in electron photoemission. In addition, relativistic effects (spin-orbit splitting) cause the electron states at the top of the valence band to be of such a symmetry that excitation to the conduction band minimum produces free electrons which are highly spin polarized. The additional effect of Cs deposition then allows the fabrication of an effective source of spin polarized electrons.

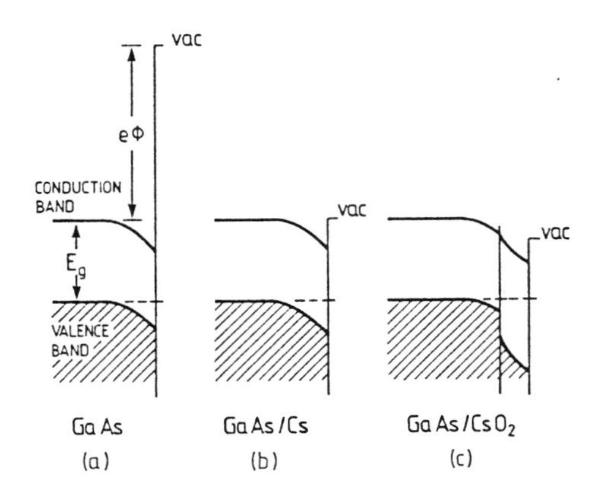

Fig. 9.11a-c. Qualitative band diagram for Cs adsorption on a p-type GaAs surface; the dipole contributions are shown in terms of a changed electron affinity. (**a**) clean GaAs surface (**b**) after deposition of Cs (**c**) after coadsorption of Cs and oxygen [9.12]

9.4 Two-Dimensional Phase Transitions in Adsorbate Layers

Figure 9.12 shows the frequencies $\omega_{0\|}$ and $\omega_{0\perp}$ of oxygen atoms vibrating parallel and normal to a Ni(100) surface. Atomic oxygen forms a chemisorbed overlayer on Ni(100), which gives rise to a c(2×2) superstructure in LEED. The frequencies $\omega_{0\|}$ and $\omega_{0\perp}$ have been measured by inelastic scattering of low energy electrons (HREELS) under various scattering angles, i.e. with angular resolution [9.13]. Thus non-negligible wave vector transfers $q_{\|}$ have been obtained and a strong dispersion at least of the $\omega_{0\perp}$ vibra-

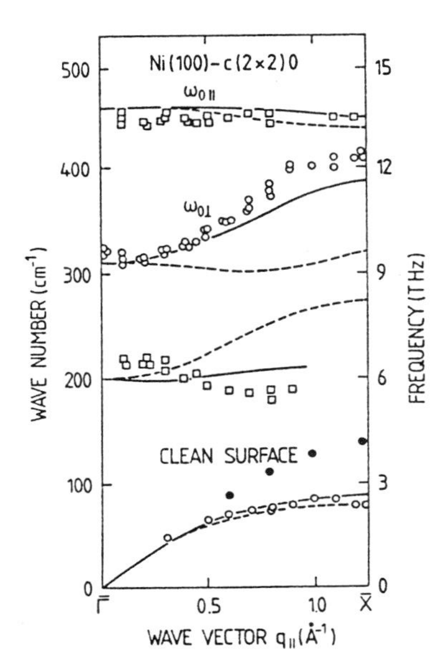

Fig. 9.12. Surface phonon dispersion curves measured on an oxygen covered Ni(100) surface with c(2× 2) LEED pattern. Full dark circles describe surface phonons of the clean Ni(100) surface. $\omega_{0\|}$ and $\omega_{0\perp}$ are vibrational modes of the oxygen atoms with displacements predominantly parallel and normal to the sample surface, respectively [9.13]

tion is measured. This dispersion clearly shows – in analogy to the 3D solid – that there is a strong mutual interaction between the atoms forming the ordered 2D array or 2D lattice. Mutual interactions between adsorbed molecules and atoms can also be deduced from the changes in vibration frequency measured (from HREELS or IRS) as a function of coverage. For low coverages, far below a monolayer, interactions within the layer itself cannot be significant. At high coverage in the monolayer range these interactions are important and in analogy to the 3D case one can consider the ordered array of oxygen atoms in Fig.9.12 as a 2D crystal.

For coverages below a monolayer, two different situations can occur (Fig.9.13). Case (a), where the adsorbed atoms or molecules are adsorbed in a random and dilute way, can be described in terms of a 2D lattice gas. In Fig.9.13b the adsorbate layer grows in islands which already possess the internal order of the completed monolayer. This situation can be described as the growth of 2D crystallites. Adsorbate islands with dense packing but with no long-range internal order are described as 2D liquid droplets. Variations of temperature can cause case (b) to change into case (a). This is a 2D phase transition on the surface, in which a 2D crystal or liquid "evaporates" to form a 2D gas.

The picture of 2D phases (gas, liquid, solid) of adsorbates appears quite natural where the *vertical* interaction between substrate and adsorbed atoms or molecules is small compared with the "lateral" interaction within the adsorbed layer itself. This case, however, is exceptional: Even for physisorbed layers of inert gas atoms, the van der Waals forces to the substrate can be comparable in strength to those between the adsorbate atoms. In situations of strong chemisorption, the substrate-adsorbate interaction is often much stronger than the lateral interaction. Nevertheless, at elevated temperature, the lateral mobility of the adsorbed species can be considerable and, if the 2D phase is dense enough at higher coverages, the lateral forces cannot be neglected. A description of strongly chemisorbed adsorbates in terms of 2D phases is also useful to describe lateral order changes, i.e., 2D phase

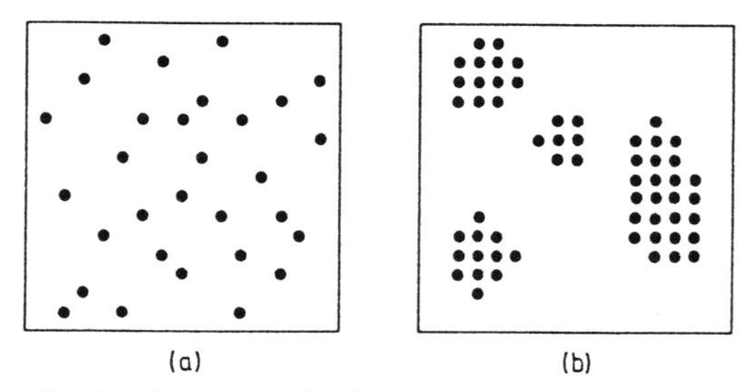

Fig.9.13a,b. Qualitative representation of 2D adsorbate phases; dots represent adsorbed atoms or molecules. **(a)** random, dilute phase of a 2D gas **(b)** adsorbate islands with 2D internal order, i.e. 2D crystallites

transitions. The *vertical* forces to the substrate, of course, influence the critical parameters (T_c etc.) of such chemisorbed layers. But, in a phenomenological description, these *vertical* forces are not considered explicitly; they are regarded merely as being responsible for establishing the 2D system, i.e., they maintain the adsorbed atoms or molecules in a single plane above the surface.

The relative strength of lateral and vertical interactions also determines whether an ordered adsorbate overlayer (2D crystal) is in or out of registry with the substrate surface periodicity. Strong *vertical* forces impose registry with the surface.

What is the physical nature of the lateral interaction between adsorbed atoms or molecules? Several sources of such interactions might be considered. The most readily identified are:

(i) The **van der Waals attraction**, which is due to correlated charge fluctuations (Sect.9.1) and is not characteristic for any particular adsorbed atom or molecule. Van der Waals attraction is the only important force for physisorbed inert gas atoms at low temperature. For most other systems, stronger interactions are superimposed and these dominate the net interaction.

(ii) **Dipole forces** may be related to permanent dipole moments of adsorbed molecules (e.g., H_2O, NH_3, etc.) or to the permanent dipoles formed by the adsorption bond due to a charge transfer between substrate and adsorbed atom or molecule. The interaction between parallel dipoles is, of course, repulsive.

(iii) **Orbital overlap** between neighboring atoms or molecules in a densely packed adsorbate layer also leads to a repulsive interaction. The behavior of CO at high coverages on transition metals is most probably dominated by this type of interaction.

(iv) **Substrate-mediated interactions** can have two origins. A strongly chemisorbed atom or molecule modifies the substrate electronic structure in its vicinity due to the chemisorption bond. A depletion or accumulation of charge in the substrate over distances of a few Ångstroms can increase or decrease the interaction strength with a second adsorbate particle and may therefore cause an indirect interaction between the adsorbate atoms or molecules.

A similar interaction can be mediated through the elastic properties of the substrate. A strongly chemisorbed atom might attract substrate atoms in its neighborhood due to the strong charge rearrangement. There is thus a local contraction of the lattice which must be compensated by an expansion at more distant points. Suppose a second adsorbate atom also attracts the substrate atoms. This second adsorbate has to do more work in order to contract the lattice at adsorption sites with lattice expansion than at those that are already contracted. Thus, depending on their separation, a repulsive or attractive interaction can arise between the adsorbate atoms. The indirect substrate-mediated lateral interactions are usually weaker than the direct dipole-dipole interactions.

As in the case of 3D solids, information about the interaction forces can be deduced from a study of phase transitions and critical parameters such as critical temperature T_c, pressure p_c, and density n_c. Via the equation of state, these parameters determine the conditions for phase transitions. A whole variety of equations of state are possible for a 2D system [9.14], but we restrict ourselves here to the simplest type of equation, the van der Waals equation. For 3D systems this reads

$$(p + a n^2)(1 - nb) = nkT . \tag{9.24}$$

This is the simplest and most plausible semi-empirical formula for the description of a non-ideal fluid in the liquid-vapor regime. n is the volume density of particles ($n = N/V$), and b is the volume of a particle, i.e. the volume which is excluded for another particle. For interparticle potentials that are not of the hard-core type, a is given approximately by $4\pi R^3/3$, where R is the separation at which strong repulsion between particles becomes important. The constant b takes into account the (two-body) interparticle potential $\phi(r)$ between two atoms or molecules at distance r. Within the approximations involved in deriving the van der Waals equation [9.14], a is given in terms of the interparticle potential $\phi(r)$ by

$$\frac{a}{V} = -\frac{N}{2}\int_{2R}^{\infty} \phi(r)\,\frac{N}{V} 4\pi r^2 dr , \tag{9.25}$$

where N is the number of particles and V the corresponding volume. The parameter a is essentially the total potential energy due to attractive interactions with other particles in the neighborhood of the one considered.

The simplest equation of state that allows a rough description of the phase transition 2D gas $\leftrightarrows$ 2D liquid is obtained be rewriting (9.24, 25) in two dimensions. The thickness of the quasi-2D system, i.e., of the adsorbed layer, is d ($= 1 \div 3$ Å). With θ as the area density of particles (number per cm^2) one has

$$nd = \theta ; \tag{9.26}$$

and with f_p as the minimum area that has to be attributed to one particle, it follows that

$$b = d f_p \ . \tag{9.27}$$

For the 2D problem one has to introduce a so-called **splay pressure**, π, which is defined as the force acting on a line element, i.e., for the present system π is related to the usual pressure p by

$$\pi = dp \ . \tag{9.28}$$

Using (9.26-28) one obtains from (9.24) the van der Waals equation for a 2D adsorption system:

$$\left(\pi + \frac{a\theta^2}{d}\right)\left(\frac{1}{\theta} - f_p\right) = kT \ . \tag{9.29}$$

Instead of using the absolute area density θ one often refers to the relative coverage θ_r which is related to the coverage θ_0 of a completed monolayer by $\theta_r = \theta/\theta_0$.

The van der Waals isotherms $\pi(\theta, T=\text{const.})$ describing a 2D liquid-gas system (Fig.9.14) are essentially identical to those of a normal 3D liquid-gas system. For temperatures below a critical temperature T_c there is a certain surface-density range between A and B, where the liquid and the gaseous adsorption phases coexist on the surface. For θ^{-1} above A only the

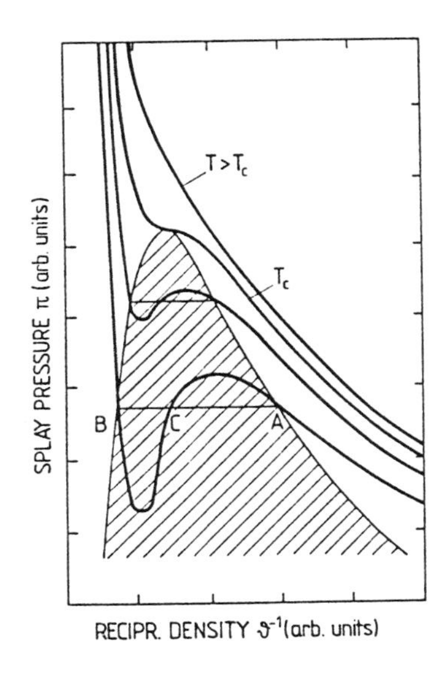

Fig.9.14. Qualitative 2D liquid-gas phase diagram (isotherms) for an adsorbate system forming dense liquid-like and dilute gas-like structures on the surface. Above the critical temperature T_c only the 2D gas phase exists, below T_c gaseous and liquid phases coexist for reciprocal densities θ^{-1} between A and B

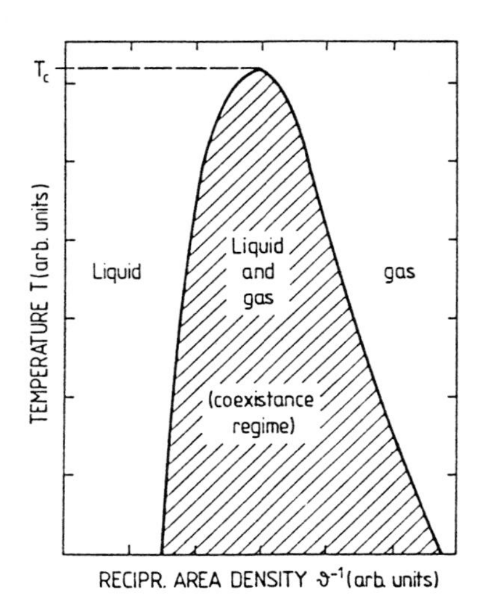

Fig.9.15. Qualitative 2D phase diagram for an adsorbate that exists both as a 2D liquid and as a 2D gas on the surface. The coexistence curve with enclosed coexistence regime (*shadowed*) is plotted in the plane of substrate temperature T versus reciprocal area density θ^{-1} of the adsorbed particles. T_c is the critical temperature above which a distinction between liquid and gas is no longer possible

less dense gaseous phase of the adsorbate is present, whereas for θ^{-1} below B the adsorbate is in its more dense liquid phase. For temperatures higher than the critical temperature there is no distinction between the liquid and the gas phase. The thermal kinetic energy is so high that "condensation" cannot occur; the splay pressure π is correlated monotonically with the surface density θ. This property of the gas-liquid system is also obvious from a temperature versus density (or T versus θ^{-1}) plot (Fig.9.15). If one plots the densities θ (or θ^{-1}) at which the transitions gas-coexistence phase (point A in Fig.9.14) and coexistence phase-liquid (point B in Fig.9.14) occur, as a function of temperature, a phase diagram as shown in Fig.9.15 is obtained. Below T_c liquid and gas phases are separated by the coexistence regime. For temperatures higher than T_c the transition from gas to liquid is continuous and there is no real phase separation between the two.

For the simple van der Waals equation of state (9.29), the critical parameters, temperature T_c, density θ_c, and splay pressure π_c are obtained by finding that solution of the cubic equation (9.29) for which all three roots θ_c coincide. For this particular θ_c, (9.29) represented as

$$\theta^3 - \frac{1}{f_p}\theta^2 + \left(\frac{\pi d}{a} + \frac{d}{af_p}kT\right)\theta - \frac{\pi d}{af_p} = 0 \qquad (9.30)$$

must become

$$(\theta-\theta_c)^3 = \theta^3 - 3\theta_c\theta^2 + 3\theta_c^2\theta - \theta_c^3 = 0\ . \qquad (9.31)$$

by comparison of the θ^n coefficients one obtains

$$3\theta_c = 1/f_p \ , \tag{9.32}$$

$$3\theta_c^2 = \left[\frac{\pi_c d}{a} + \frac{d}{af_p} kT_c\right] , \tag{9.33}$$

$$\theta_c^3 = \frac{\pi d}{af_p} \ . \tag{9.34}$$

This yields the following critical parameters

$$\theta_c = \frac{1}{3f_p} \ , \tag{9.35}$$

$$\pi_c = \frac{a}{27df_p^2} \ , \tag{9.36}$$

$$T_c = \frac{8}{27} \frac{a}{kdf_p} \ . \tag{9.37}$$

An experimental evaluation of the critical parameters, temperature T_c, splay pressure π_c, and critical density θ_c, thus gives direct insight into the interesting interaction parameters f_p (minimum particle area) and the total interaction energy a between one particle and its neighbors (9.25)

Simple 2D phase diagrams, as shown qualitatively in Fig.9.15, are indeed found for certain adsorption systems. Atomic hydrogen (H) on Ni (111) can form an ordered, *crystalline* phase with a (2×2) superstructure seen in LEED [9.15]. But the occurrence of this structure is critically dependent on the coverage and the substrate temperature. A phase diagram $T_c(\theta)$ can be determined experimentally by LEED studies (Fig.9.16), and this indeed shows the qualitative features of a simple van der Waals phase diagram (Fig.9.15).

For other systems, e.g. oxygen on Ni(111), more complex phase diagrams have been observed experimentally (Fig.9.17). In addition to the random gas and liquid states, several 2D crystalline states exist. These exhibit various characteristic superstructures, such as $p(2 \times 2)$ or $(\sqrt{3} \times \sqrt{3})$ R30°.

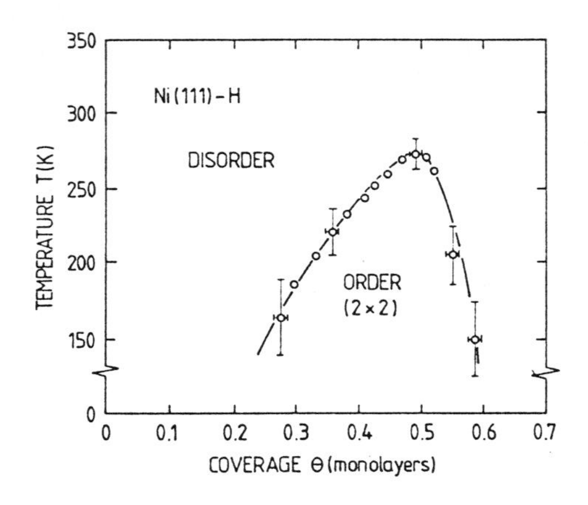

Fig. 9.16. Experimental 2D phase diagram for atomic hydrogen adsorbed on Ni(111). Depending on the coverage θ and substrate temperature T, one may observe an ordered adsorbate phase with (2×2) superstructure or a disordered phase [9.15]

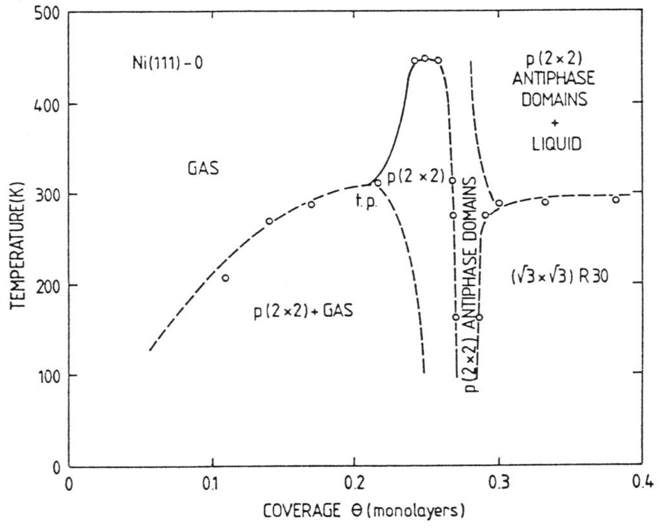

Fig. 9.17. Experimental phase diagram (o) and a postulated phase diagram for the oxygen on Ni(111) system; continuous phase boundaries are shown as *solid lines*, first order phase boundaries as *dotted lines*, t.p. is the tricritical point [9.16]

9.5 Adsorption Kinetics

So far we have considered the microscopic details of single adsorbed molecules or atoms and the structure and properties of the adsorbed phase. In order to analyse a real adsorption experiment, a more phenomenological framework is also needed to describe measured quantities such as adsorption rate and degree of coverage. These quantities depend, of course, on the details of the particular adsorption process, but their relationship to the microscopic picture is often complex and not well understood. Nevertheless, the so-called **kinetic description** on a more phenomenological level can yield first important information, which, in combination with more refined spectroscopic data, can lead to a deeper understanding of adsorption interactions.

Adsorption kinetics is a thermodynamic approach describing the interplay between the adsorbed species and the ambient gas phase; adsorption and desorption are the two processes which determine the macroscopic coverage on a solid surface exposed to a gas. The adsorption rate depends on the number of particles striking the surface per second and on the so-called **sticking coefficient**, which is the probability that an impinging particle actually sticks to the substrate. According to kinetic gas theory (Sect.2.1) the rate at which particles impinge on a surface (per unit area and time) is given by

$$\frac{dN}{dt} = \frac{p}{\sqrt{2\pi mkT}} . \tag{9.38}$$

The adsorption rate, i.e. the number of adsorbing particles per unit time and surface area is then obtained as

$$u = S\frac{dN}{dt} = S\frac{p}{\sqrt{2\pi mkT}} , \tag{9.39}$$

where m is the mass of the impinging particle.

Since the coverage θ (number of adsorbed particles per unit area) is given by

$$\theta = \int u dt = \int S\frac{dN}{dt} dt , \tag{9.40}$$

one can determine the sticking coefficient S from a measurement of the coverage θ (e.g., by AES) as a function of dosage (Fig.9.18).

According to (9.38,40) one has

$$S = \sqrt{2\pi mkT}\,\frac{u}{p} = \sqrt{2\pi mkT}\,\frac{1}{p}\frac{d\theta}{dt} . \tag{9.41}$$

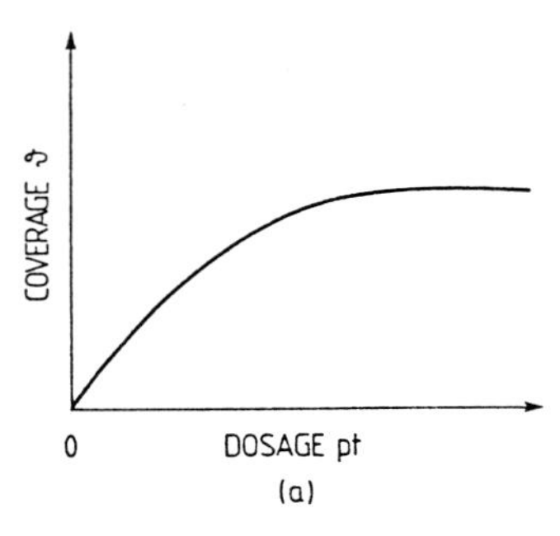

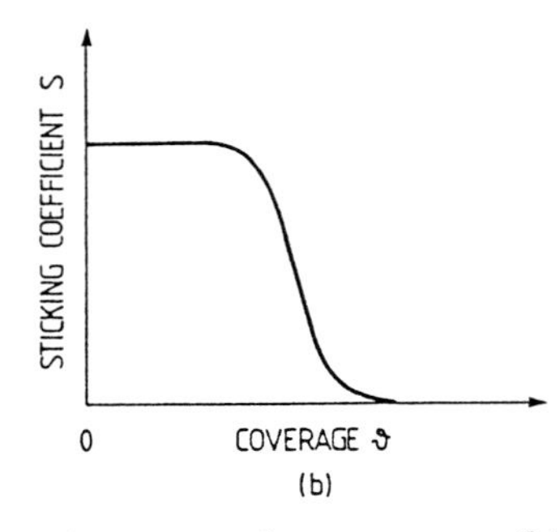

Fig.9.18. (**a**) Qualitative dependence of coverage θ on exposure (**b**) corresponding dependence of the sticking coefficient S on coverage θ

Typical coverage versus dosage (pressure·time) dependences, as shown in Fig.9.18a, yield, after differentiation, a sticking coefficient $S(\theta)$ as in Fig. 9.18b. The general shape of the $S(\theta)$ dependence is easy to understand on the basis that the first chemisorbed molecules or atoms bond to "free" valence orbitals (dangling bonds) of the surface and thus decrease its reactivity to further bonding as more and more sites become occupied. The sticking coefficient S implicitly reflects the details of the microscopic adsorption process. Several important factors affect the quantity S:

(i) In many cases (e.g., see Fig.9.4b) an activation barrier E_{act} has to be overcome before chemisorption can occur. Only atoms or molecules whose impact energy exceeds E_{act} can stick to the surface. In this case of activated adsorption the sticking coefficient must contain a Boltzman term $\exp(-E_{act}/kT)$.

(ii) In order for an impinging atom or molecule to be chemisorbed, its electronic orbitals must have a particular orientation with respect to the dangling-bond orbitals of the surface (steric factor). Besides the orientation of the molecules, their mobility on the surface and the site of impact are also important. The adsorption potential varies locally along the surface due to the atomic structure of the substrate.

(iii) During adsorption, an incident atom or molecule must transfer at least part of its remaining kinetic energy to the substrate, otherwise it will be desorbed again after approximately one vibrational period. Excitations of the substrate, such as surface phonons and plasmons are thus also involved in the adsorption kinetics.

(iv) Adsorption sites must of course be available to an impinging atom or molecule. The more sites are occupied, the fewer particles can be adsorbed. For particles adsorbed in a precursor (intermediate) state, the diffusion path to a final sticking site becomes longer; this enhances the probability of desorption and decreases the sticking probability.

A convenient description of the sticking coefficient for activated adsorption, taking into account the above-described phenomena, is thus

$$S(\theta) = \sigma f(\theta) \exp(-E_{act}/kT) , \tag{9.42}$$

where σ, the so-called **condensation coefficient** contains the effects of molecular orientation (steric factor), energy transfer to the surface, etc. $f(\theta)$ is the occupation factor, which describes the probability of finding an adsorption site. For non-dissociative adsorption (mobile or immobile adsorption) a site is occupied or unoccupied and $f(\theta)$ is simply

$$f(\theta) = 1 - \theta , \tag{9.43}$$

where θ is now the relative coverage, i.e., the ratio between occupied sites and the maximum number of available sites in the first completed adsorbed layer. For dissociative adsorption, where the impinging molecule dissociates into e.g. two adsorbed radicals, the second radical must find an empty site directly neighboring the first radical, at least for an immobile adsorbate. With z as the maximum number of adjacent sites for the second radical, the number of available sites is

$$f_2(\theta) = \frac{z}{z - \theta}(1 - \theta) . \tag{9.44}$$

For adsorption of the whole molecule, i.e. two immobile radicals one obtains

$$f(\theta) = f_1(\theta) f_2(\theta) = \frac{z}{z - \theta}(1 - \theta)^2 . \tag{9.45}$$

For low coverages ($\theta \ll 1$) one has

$$f(\theta) \simeq (1 - \theta)^2 . \tag{9.46}$$

This expression is, of course, also valid for dissociative adsorption of mobile complexes, since, for low enough coverage, sufficient sites are free that there is no real restriction due to prior occupation of neighboring sites. The condensation coefficient σ depends on the various states in which the adsorbed molecule, the free gas phase molecule and the adsorbant surface can exist. A detailed statistical theory [9.17] for the calculation of σ describes the adsorption process as a transition from the initial state of the free surface plus free molecule (S+M) via an excited transition state $(SM)^*$

into the final adsorption state (SM). In the transition state $(SM)^*$ the system is in a state of excitation, its total energy includes the activation energy which has to be supplied before adsorption occurs. Adsorption may thus be considered as the decay of the transition state into the adsorption state. Rate theory then yields the result that σ is given essentially by the ratio

$$\sigma \propto Z_{(SM)^*}/Z_M Z_S , \tag{9.47}$$

where Z are the partition functions of the excited transition complex $(SM)^*$, of the free molecule (M) and the free surface (S).

These partition functions are sums over the various possible states, e.g. for a molecule with energy eigenvalues ϵ_i (degree of degeneracy g_i) one has

$$Z_M = \sum_i g_i \exp(-\epsilon_i/kT) . \tag{9.48}$$

The calculation of σ for a realistic system clearly requires a detailed knowledge of the reaction path and of the quantum-mechanical properties of the various constituents. Table 9.2 gives some charactgeristic values for simple diatomic gases in both mobile and immobile adsorbate layers. σ is dependent mainly on the degree of freedom of the adsorbed molecule.

Experimentally, one sometimes finds an exponential dependence of the sticking coefficient S on the coverage θ:

$$S \propto \exp(-\alpha\theta/kT) \tag{9.49}$$

(Elovich equation). This dependence is easily understood according to (9.42), if the activation energy E_{act} is assumed to depend on coverage as $E_{act} = E_0 + \alpha\theta$.

Table 9.2. Some characteristic condensation coefficients σ for diatomic molecules adsorbed in mobile and immobile configurations [9.17]

Adsobate	Immobile adsorbate	Mobile adsorbate loss of rotation	Mobile adsorbate no loss of roation
H_2	$3\cdot10^{-2} \div 0.2$	0.52	1
O_2, N_2	$10^{-4} \div 3\cdot10^{-2}$	0.12	1
CO_2	$7\cdot10^{-5} \div 0.02$	0.1	1

The **desorption process** is described phenomenologically by a desorption rate v, i.e. the number of desorbing particles per unit time and surface area. For desorption to occur, an adsorbate particle must acquire enough energy to surmount the desorption barrier $E_{des} = E_B + E_{act}$, which comprises the binding energy E_B and the activation energy for adsorption (Fig.9.4b). The desorption rate v is thus proportional to the exponential term $\exp(-E_{des}/kT)$, but as for adsorption, the number of adsorbed particles also enters via an occupation factor $\bar{f}(\theta)$, as do the detailed steric and mobility factors by means of a desorption coefficient $\bar{\sigma}(\theta)$. The quantities $\bar{f}(\theta)$ and $\bar{\sigma}(\theta)$ describe a process which is inverse to adsorption; accordingly they are complementary to f and σ, i.e. inversely dependent on coverage and on the partition functions of adsorbate, substrate and transition complex. In the simplest case of desorption of one atom from a single site, one has

$$\bar{f}(\theta) = \theta . \tag{9.50}$$

In the case of a molecular process where the desorbing molecule originates from two radicals at different sites, there is an approximate dependence

$$\bar{f}(\theta) \simeq \theta^2 . \tag{9.51}$$

As a whole, the desorption process is described by a desorption rate

$$v = \bar{\sigma}(\theta)\bar{f}(\theta)\exp(-E_{des}/kT) . \tag{9.52}$$

Thermal equilibrium between the gas phase and the solid surface is characterized by equal adsorption and desorption rates. At a constant temperature, there thus exists an equilibrium adsorbate coverage $\theta(p,T)$, which is described by the so-called **adsorption isotherm**. To calculate this, one equates the adsorption and desorption rates u and v:

$$u = v . \tag{9.53a}$$

With the simple assumptions of (9.42 and 52) one thus obtains

$$u = \sigma(\theta)f(\theta)e^{-E_{act}/kT}\frac{p}{\sqrt{2\pi mkT}} = \bar{\sigma}(\theta)\bar{f}(\theta)e^{-E_{des}/kT} = v , \tag{9.53b}$$

or, with $E_{des} = E_B + E_{act}$,

$$p = \frac{\bar{\sigma}}{\sigma}\sqrt{2\pi mkT}\, e^{-E_B/kT}\frac{\bar{f}(\theta)}{f(\theta)} = \frac{1}{A}\frac{\bar{f}(\theta)}{f(\theta)} . \tag{9.54}$$

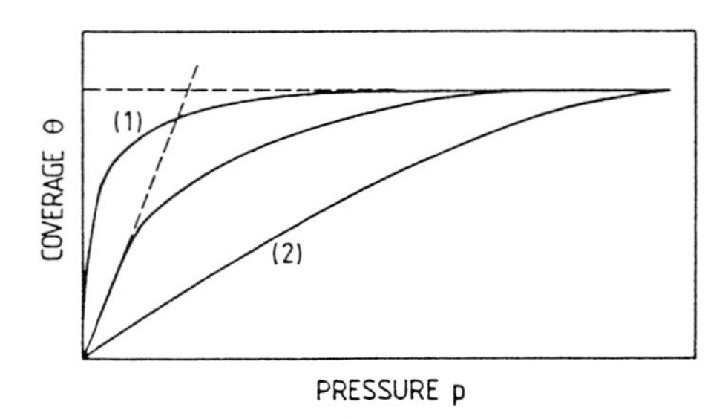

Fig. 9.19. Qualitative shapes of Langmuir-type isotherms of coverage versus pressure $\theta(p)$. Curve *1* describes the case of strong adsorption with large adsorption energy, curve *2* represents the case of weak adsorption. Between these extremes there is a gradual transition from type *1* to *2*

This is the general form of the so-called **Langmuir isotherm**. For the special case of non-dissociative adsorption (9.43), where $f(\theta) = 1-\theta$ and $\bar{f}(\theta) = \theta$, one obtains the simple form

$$p(\theta) = \frac{\theta}{A(1-\theta)}, \quad \text{or} \quad \theta(p) = \frac{Ap}{1+Ap}, \tag{9.55}$$

where A is a constant at a fixed temperature, see (9.54). A measurement of the equilibrium adsorbate coverage θ as a function of ambient pressure p therefore allows a determination of the constant A, which in turn yields, according to (9.54), the chemisorption (or binding) energy E_B, provided the condensation and desorption coefficients σ and $\bar{\sigma}$ are known. Figure 9.19 shows the qualitative dependence of coverage θ on pressure p as expected from the Langmuir isotherm (9.55b). For low pressures, the curves can be approximated by a linear relationship whose slope ($\simeq A$) increases exponentially with adsorption energy E_B, i.e. with the strength of the adsorption process.

For many realistic adsorption systems the Langmuir isotherm fails to correctly describe the dependence of coverage on pressure in thermal equilibrium. In particular, the neglect of multilayer adsorption is unrealistic. Much better agreement was achieved by a theory of Brunauer, Emmett and Teller (BET isotherm), in which multilayer adsorption was also taken into account. Each adsorbed particle in the first layer serves as a site for adsorption into the second layer, and each particle in the second layer serves as a site for adsorption into the third layer and so forth. In even more refined approaches, activation energy, adsorption energy, etc., are assumed to be layer dependent. In this way more parameters enter into the theory, but a variation of these parameters allows an accurate description of a variety of experimentally observed isotherms (Fig. 9.20).

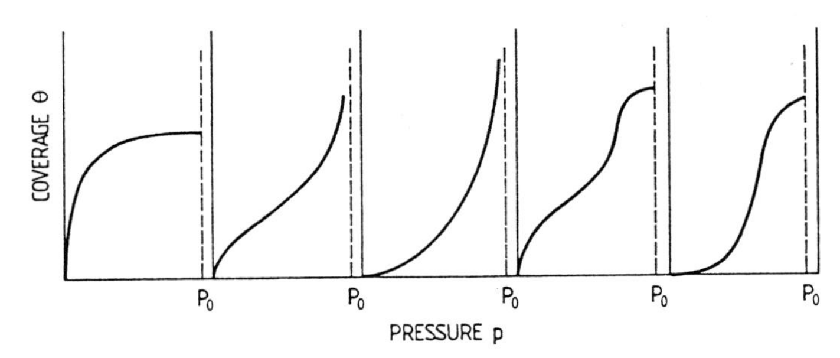

Fig. 9.20. Various possible physical adsorption isotherms

Panel XIV
Desorption Techniques

Much essential information about adsorption processes and surface chemical reactions is derived from desorption experiments. The entire class of desorption techniques has the common feature that a clean surface under UHV conditions is exposed to a well-defined gas atmosphere or a molecular beam. Subsequent desorption of the resultant adsorbate is performed by thermal annealing of the surface or by irradiating with light or energetic particles. The desorbing species can be analysed mass-spectroscopically or the particle beam can be optically imaged on a screen to yield information about possible anisotropy in the angular distribution of the desorbing atoms or molecules.

The simplest technique which gives useful information, particularly about simple adsorption systems, is the so-called **Thermal Desorption Spectroscopy** (TDS), where thermal annealing of the adsorbate-covered surface gives rise to desorption [XIV.1]. A straightforward measurement of the pressure increase in the UHV chamber as a function of sample temperature yields interesting information about the desorption energy, etc. The mathematical description of the desorption process is based on the pumping equation (I.2). The desorbing particles are pumped away (pumping speed $\tilde{S}$) but give rise to a temporary pressure increase in the UHV vessel. With v as the desorption rate, particle conservation thus yields

$$vA = \frac{-Ad\theta}{dt} = \frac{V_v}{kT}\left[\frac{dp}{dt} + \tilde{S}\frac{p}{V_v}\right], \qquad (XIV.1)$$

where θ is the relative coverage of the sample surface (area A), V_v the volume of the UHV chamber and p the pressure (background subtracted).

In the limit of negligible pumping speed the rate of the pressure increase would reflect the desorption rate ($d\theta/dt \propto dp/dt$). On the other hand, with modern pumping equipment $\tilde{S}$ is extremely high (for cryopumps $\tilde{S}$ can reach values of 10000 ℓ/s and (XIV.1) can be approximated by

$$v = \frac{-d\theta}{dt} \propto p, \qquad (XIV.2)$$

such that monitoring the pressure directly yields interesting information about the desorption rate. The desorption rate might be described as in (9.52) by means of

$$v = \frac{-d\theta}{dt} = \bar{\sigma}\bar{f}(\theta)\exp(-E_{des}/kT) , \tag{XIV.3}$$

with E_{des} as the desorption energy. In the simplest experimental set-up the temperature T of the sample is controlled by a computer program such that it changes linearly with time t (Fig. XIV.1a)

$$T = T_0 + \beta t \quad (\beta > 0) . \tag{XIV.4}$$

The pressure rise as a function of temperature T is then given by

$$p \propto \frac{-d\theta}{dT} = \frac{\bar{\sigma}}{\beta}\theta^n e^{-E_{des}/kT} , \tag{XIV.5}$$

where for the general case of a desorption process of order n, the occupation factor $\bar{f}(\theta)$ is assumed as θ^n (Sect. 9.5). For monomolecular and bimolecular desorption n equals 1 and 2, respectively. The measured pressure as a function of sample temperature (Fig. XIV.1b) reaches a maximum at a characteristic temperature T_p and decreases again when the surface coverage decreases by desorption. The pressure rise is determined by the exponential term in (XIV.5), whereas the decrease of p ($\propto \theta^n$) also depends on the order of the desorption process. The temperature of the maximum of p(T) is determined by

$$\frac{-d^2\theta}{dT^2} = \frac{d}{dT}(\bar{\sigma}\theta^n e^{-E_{des}/kT}) = 0 . \tag{XIV.6}$$

Inserting the expression for $d\theta/dT$ (XIV.5), one obtains for an nth-order desorption process

$$\ln\left[T_p{}^2 \frac{1}{\beta}\theta^{n-1}(T_p)\right] = \frac{E_{des}}{kT_p} + \ln\left(\frac{E_{des}}{n\bar{\sigma}k}\right) , \tag{XIV.7}$$

or for a simple monomolecular process

$$\ln\left(T_p{}^2/\beta\right) = \frac{E_{des}}{kT_p} + \ln\left(\frac{E_{des}}{\bar{\sigma}k}\right) . \tag{XIV.8}$$

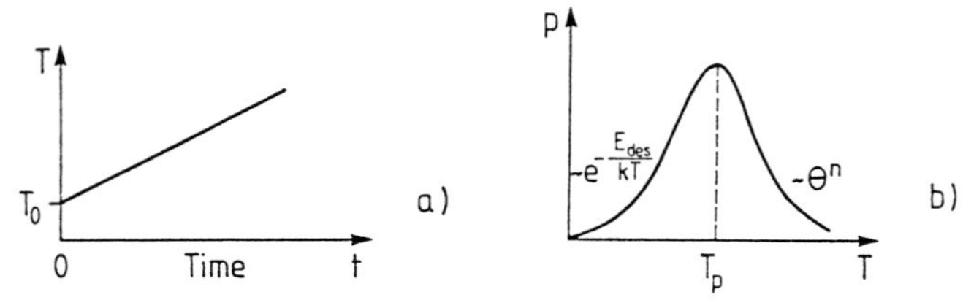

Fig. XIV.1a,b. Qualitative description of a Thermal Desorption Spectroscopy (TDS) experiment. (**a**) The sample temperature T is increased linearly with time t, starting from an initial value T_0. (**b**) Due to desorption the pressure in the UHV vessel increases and decreases again with increasing sample temperature. The initial increase is mainly determined by the desorption barrier E_{des}, whereas the pressure drop gives information about the order n of the desorption process

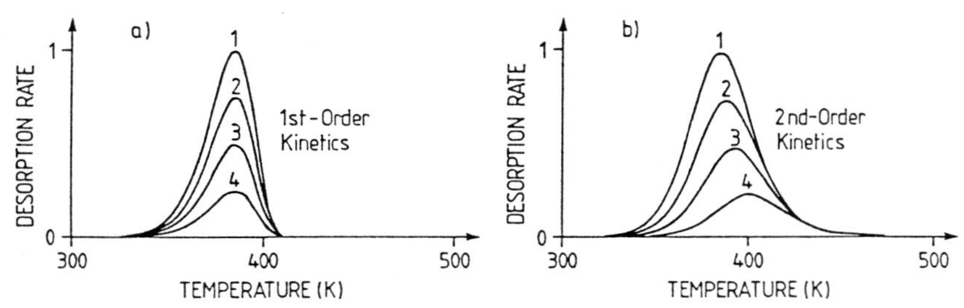

Fig. XIV.2a,b. Thermal desorption spectra, i.e. desorption rate (dimensionless) versus sample temperature T. For the calculation a desorption energy E_{des} of 25 kcal/mole and different fractional surface coverages have been assumed: *(1)* $\theta = 1.0$, *(2)* $\theta = 0.75$, *(3)* $\theta = 0.5$, *(4)* $\theta = 0.25$. The desorption process has been assumed to involve (**a**) first-order kinetics, and (**b**) second-order kinetics.

With reasonable assumptions about the steric factor $\bar{\sigma}$ (XIV.7, 8) are used to determine the desorption energy E_{des} by recording the p versus T dependence (Fig. XIV.1b).

From the mathematical form of the pressure (or desorption rate) versus temperature curve, it is obvious that only for a monomolecular process is the temperature of the maximum T_p independent of θ and thus also of θ_0, the initial coverage (Fig. XIV.2a). A shift of the desorption peak with an initial coverage variation indicates a desorption process of higher order as in Fig. XIV.2b [XIV.2].

Further information about the order of the process is obtained from the shape of the desorption curve [XIV.2]. Second-order curves are symmetrical with respect to T_p, whereas first-order desorption causes less symmetric bands (Fig. XIV.3). Another experimental example of second-order desorption in TDS is given in Fig. XIV.4, where the desorption of N_2 from Fe(110) surfaces is monitored [XIV.3]. The description in terms of a sec-

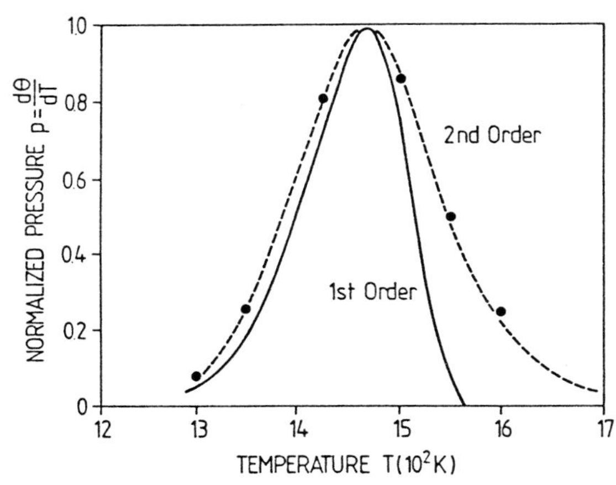

Fig. XIV.3. Normalized desorption rate as a function of temperature for a first-order (E_{des} = 91.5kcal/mole) and a second-order (E_{des} = 87.5kcal/mole) reaction, calculated for a linear temperature sweep. The experimental data (*dark circles*) are obtained from desorption experiments on the so-called β-phase of N_2 on W, adsorbed at 300 K [XIV.1]

ond-order process yields the information that nitrogen is adsorbed dissociatively and the desorption energy E_{des} is estimated to be roughly 7 eV per atom.

In another type of desorption experiment the surface bearing the adsorbate is irradiated and it is the incident energy that gives rise to desorption. Depending on the particular type of radiation one distinguishes several techniques:

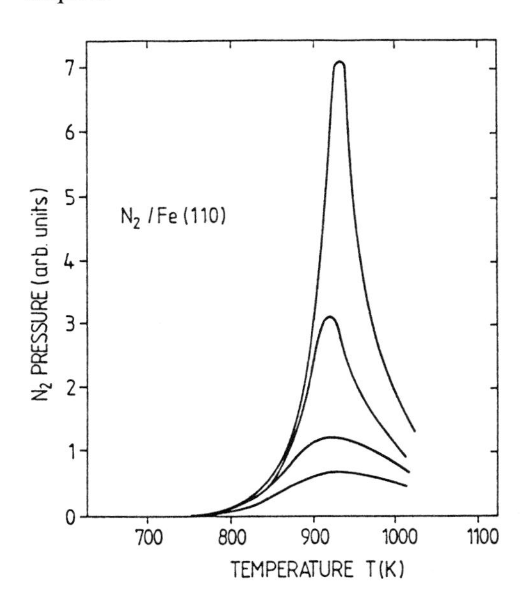

Fig. XIV.4. Thermal Desorption Spectra (TDS) of nitrogen (N_2) desorbed from a Fe(110) surface. The mathematical description in terms of a second-order process gives a desorption energy E_{des} of 7 eV per atom [XIV.3]

• In **Ion Impact Desorption** (IID) ions of typically 100 eV primary energy, e.g. Ar ions, are accelerated onto the sample and adsorbate particles are desorbed by direct momentum transfer. Mass-spectroscopic detection generally reveals only the chemical nature of the adsorbate.

• In **Field Desorption** (FD) high electric fields ($\approx 10^8$ V/cm) are applied by a counter-electrode, e.g. in a field-ion microscope, and the desorbing particles can be made visible on a fluorescent screen. Local adsorption geometry is sometimes studied but energetic questions are rarely tackled by this method.

• In **Photodesorption** (PD) experiments light of sufficient photon energy ($3 \div 10$ eV) is used to excite electrons from the adsorbate bond into antibonding orbitals. This disrupts the adsorption bond and leads to desorption. PD is usually accompanied by heat transfer and the effect is sometimes difficult to distinguish from thermal desorption.

• Of considerable importance in adsorption studies is **Electron Stimulated Desorption** (ESD). In this technique electrons with primary energies up to about 100 eV are incident on the adsorbate covered surface; the desorbing products are either detected mass-spectroscopically or one uses a multichannel plate array backed by a fluorescent screen to obtain a spatial image of the desorption direction of the removed particles. This technique of visualizing the angular distribution of the desorbing atoms or molecules is called ESDIAD (Electron Stimulated Desorption of Ion Angular Distributions). Figure XIV.5 shows a typical experimental set-up, which allows both mass-spectroscopic detection (ESD) and the collection of ESDIAD data [XIV.4]. Mass spectroscopy yields information about the chemical nature of the desorbing species, whereas detection of the angular distribution gives insight into the local geometrical arrangement of the adsorbed complex. Inversion of the bias at the Multichannel Plate (MCP) array enables the detection of electrons, i.e. the optical display of a diffraction pattern. This allows the simultaneous observation of the LEED pattern (Panel VIII: Chap. 4) of the adsorption system.

The theoretical description of ESD processes is based on two limiting cases. In the classical model of *Menzel* and *Gomer* [XIV.5] intramolecular excitations (indicated by the shaded area in Fig.XIV.6) within the adsorbed molecule lead to non-bonding and antibonding neutral or ionic final states. At the crossover points of the corresponding potential curves (Fig.XIV.6) the adsorbed molecule can change into the new state and might desorb. In the case of ion desorption an electron can be captured by tunneling from the solid to a desorbing particle.

This type of excitation, however, does not explain ESD results which have been observed on $TiO_2(001)$ surfaces on which tiny amounts of hydrogen were adsorbed (Fig.XIV.7) [XIV, 6, 7]. Under irradiation with quasi-

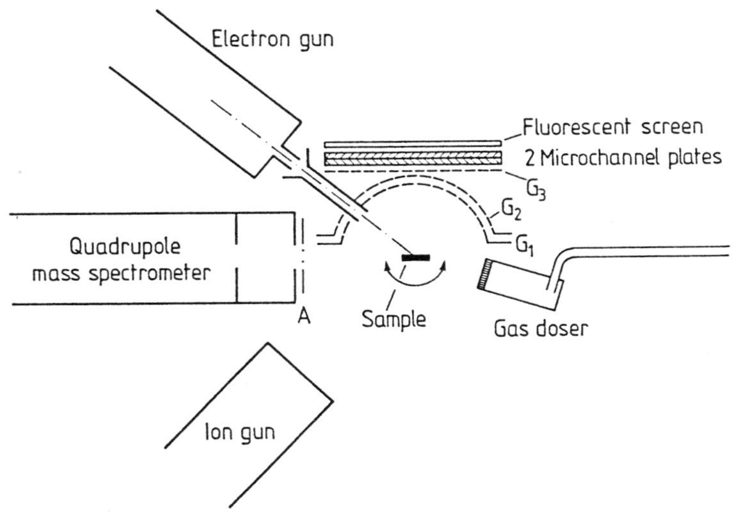

Fig. XIV.5. Schematic picture of an Electron Stimulated Desorption (ESD) and ESDIAD apparatus. The sample S can be rotated about an axis normal to the plane of the drawing. ESD ions are mass-analyzed in the quadrupole mass spectrometer, and ESDIAD patterns are displayed using the grid MicroChannel-Plate (MCP) plus fluorescent screen array. The radius of curvature of G_1 is 2 cm, and the active area of each MCP has a diameter of 4 cm. For most ESDIAD measurements typical potentials are $G_1 = G_2 = 0V$, $G_3 = -70$ V, MCP entrance: -700 V, MCP midpoint: 0V, MCP exit: $+700$ V, fluorescent screen: $+3800$ V. Electron gun filament potential $V_f = -100$ V, crystal potential $V_B = 0$ to $+100$ V. Electron energy $E_e = e(|V_f| + |V_B|)$ [XIV.4]

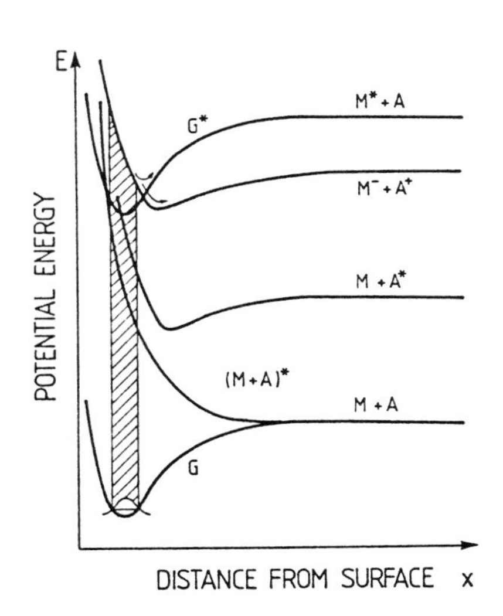

Fig. XIV.6. Potential energy diagrams for an adsorbate system. G: adsorbed ground state; $M^- + A^+$: ionic state; $(M+A)^*$: antibonding state; $M + A^*$: excited state of the adsorbate; $M^* + A$: adsorbate ground state with excitation energy in the metal (vertically shifted replica of G). The vibrational distribution in G and resulting ESD ion energy distribution are indicated [XIV.2]

Panel XIV

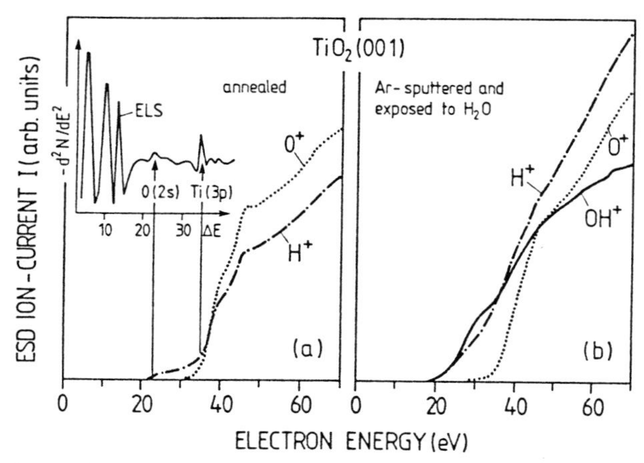

Fig. XIV.7a,b. Ion yields (O^+, H^+, OH^+) from Electron-Stimulated Desorption (ESD) measurements on: (**a**) a clean annealed TiO_2(001) surface (the H^+ yield probably results from slight contamination); (**b**) an Ar-sputtered TiO_2(001) surface after exposure to H_2O. For comparison, part (**a**) contains as an insert a second-derivative Electron Energy Loss Spectrum (EELS) of the same annealed surface. Its loss scale (ΔE) is identical with the scale of the ESD primary energy. Transitions from the O(2s) and Ti(3p) levels to the vacuum level are indicated [XIV.6]

monoenergetic electrons of varying primary energy, thresholds in the desorption flux of H^+ and O^+ ions are observed near 21 eV and 34 eV, respectively. As is seen from additional double differentiated EELS measurements (inset), these energies correspond to the O(2s) and Ti(3p) core-level excitations. The interpretation of these desorption experiments on highly ionic materials involves the formation of core-level holes in the O(2s) and Ti(3p) orbitals and subsequent interatomic transitions between the O and Ti core levels. In detail, primary electrons create a Ti(3p) core hole, and owing to an interatomic Auger process an O(2p) electron decays into the Ti(3p) state, with the emission of a second or third O(2p) electron to dissipate the energy released in the decay. This fast process is responsible for the relatively large charge transfer in the transformation of the O^{2-} lattice ion into O^+, which is observed in ESD. The H^+ desorption spectrum exhibiting two thresholds (Fig. XIV.7) may be interpreted in terms of two types of hydrogen, that bonded to O and that bonded to Ti surface atoms. Energy-dependent ESD measurements can thus give detailed information about atomic-scale features of the electron-induced desorption process.

Experimental examples of the use of ESDIAD in the determination of an adsorption geometry are given in Fig. XIV.8. From TDS it is known that H_2O and NH_3 adsorb molecularly on the Ru(001) surface at 90 K. Irradiation by electrons with primary energies below 100 eV produces angularly-resolved desorption patterns of H^+ ions, as shown in Fig. XIV.8b-d [XIV.4].

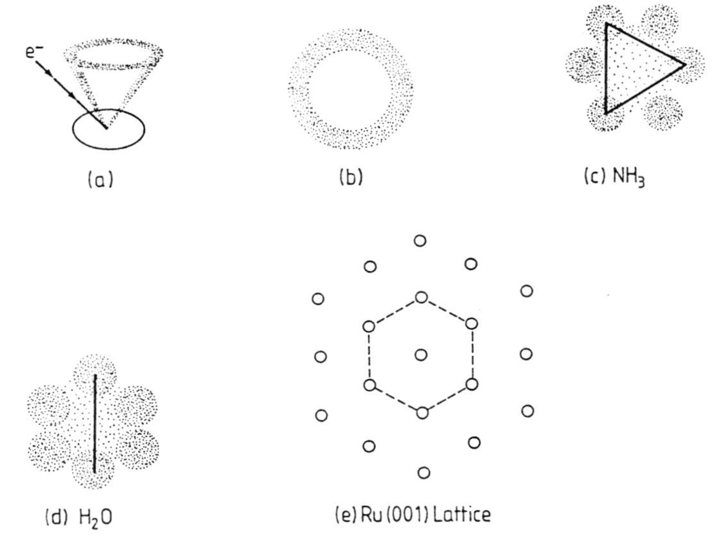

Fig. XIV.8a-e. Schematic ESDIAD patterns for H_2O and NH_3 adsorbed on Ru(001) at 90 K [XIV.4]. (**a**) Formation of hollow cone of H^+ ions from adsorbed NH_3 and H_2O (at low coverages). (**b**) "Halo" H^+ pattern characteristic of low coverages of NH_3 and H_2O. (**c**) Hexagonal H^+ pattern characteristic of intermediate NH_3 coverages ($0.5 \leq \theta < 1$). (**d**) Hexagonal H^+ pattern chracteristic of intermediate H_2O coverages ($0.2 \leq \theta < 1$). (**e**) Ru(001) substrate with respect to the ESDIAD patterns above

For low coverages a halo-type pattern is observed. Assuming that the desorbing H^+ ions leave the surface along the direction of the intramolecular chemical bond, a low-coverage bonding geometry is derived in which the H_2O and NH_3 molecules are bonded with their O and N atoms closest to the surface. The orientation with respect to the 2D lattice planes of the surface is irregular or statistical. At increased coverages a hexagonal symmetry becomes visible in the ESDIAD patterns, which demonstrates that the molecular orientation is now in registry with the underlying substrate (Fig. XIV.8c-e). The molecules have lost one degree of freedom, the free rotation around an axis normal to the surface. With certain assumptions about the microscopic desorption process, ESDIAD thus allows detailed conclusions concerning the local adsorption geometry.

References

XIV.1 F.M. Lord, J.S. Kittelberger: Surf. Sci. **43**, 173 (1974)

XIV.2 D. Menzel: In *Interactions on Metal Surfaces*, ed. by R. Gomer, Topics Appl. Phys. Vol.4 (Springer, Berlin, Heidelberg, 1974) p.124

XIV.3 F. Bozso, G. Ertl, M. Weiss: J. Catalysis **50**, 519 (1977)

XIV.4 T.E. Madey, J.T. Yates: Proc. 7th Int'l Vac. Congr. and Int'l Conf. Solid Surfaces, Wien (1977)
XIV.5 D. Menzel, R. Gomer: J. Chem. Phys. **41**, 3311 (1964)
XIV.6 M.L. Knotek: Surface Sci. **91**, L17 (1980)
XIV.7 M.L. Knotek, P.J. Feibelman: Phys. Rev. Lett. **40**, 964 (1978)

Panel XV
Kelvin-Probe and Photoemission Measurements for the Study of Work-Function Changes and Semiconductor Interfaces

The adsorption of atoms or molecules on a solid surface, i.e. the first steps of the formation of a solid/solid interface, is generally associated with a change of work function (Sect.9.3), and on semiconductors also with a change in band bending (because of the formation of new interface states). These effects can be studied in situ both by photoemission spectroscopy (UPS and XPS; Sect.6.3) and by Kelvin-probe measurements. The latter technique, in particular, is useful for work-function measurements on metal surfaces, where space-charge layer effects are negligible (spatial extension of some Ångstroms).

Kelvin probes for the determination of work-function changes consist of an electrode (usually point-like) which can be positioned in front of the surface being studied (Fig.XV.1a). This counterelectrode is driven electromagnetically by a solenoid or by piezoceramics such that it vibrates with frequency ω against the sample surface. Sample and vibrating electrode are connected electrically through an ammeter (A) and a battery which allows a variable biasing (U_{comp}).

The principle of the work-function measurement becomes clear if we consider that for two solids [sample *S* and probe *P*] in electrical contact the electrochemical potentials, i.e. the Fermi energies E_F^S and E_F^P, are equal in thermal equilibrium ($E_F^S = E_F^P$). Since in general the work function ($e\phi = E_{vac} - E_F$ is different for the sample surface and for the Kelvin probe, a so-called **contact potential** U^{SP} is built up between the sample and the probe.

Since

$$E_F^S = E_{vac}^S - e\phi^S = E_{vac}^P - e\phi^P = E_F^P \, , \tag{XV.1}$$

the contact potential is obtained as

$$U^{SP} = -\frac{1}{e}(E_{vac}^S - E_{vac}^P) = -(\phi^S - \phi^P) \, . \tag{XV.2}$$

Measurement of this contact potential therefore determines the difference in work function between the Kelvin probe and the sample surface. In an experimental set-up as in Fig.XV.1 the voltage between sample and probe is

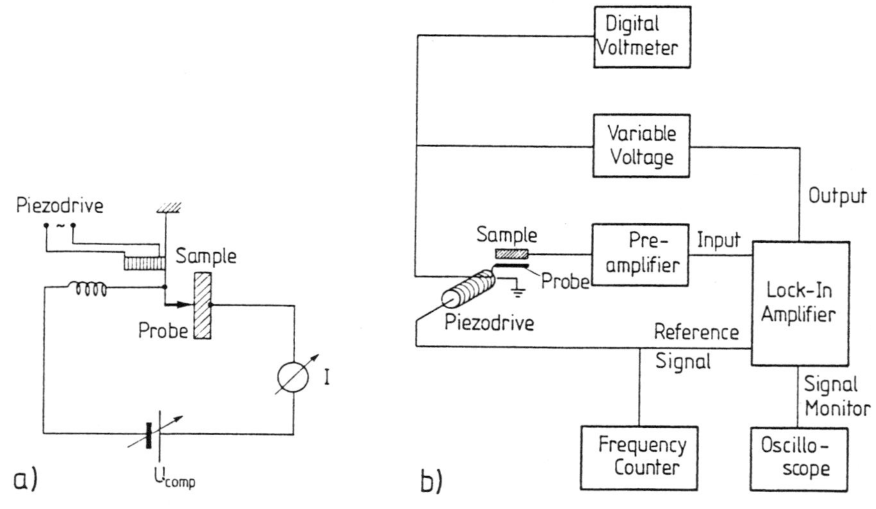

Fig. XV.1. (**a**) Principle of a Kelvin probe measurement; the compensation voltage U_{comp} compensates the AC current driven by the vibrating probe. (**b**) Schematic circuit for the Kelvin probe measurement

$$U = -(\phi^S - \phi^P) + U_{comp} \, . \tag{XV.3}$$

The capacitor formed by sample and probe thus carries a charge (C being the capacitance)

$$Q = C[-(\phi^S - \phi^P) + U_{comp}] \, , \tag{XV.4}$$

and the vibration of the probe electrode (frequency ω) gives rise to an oscillating current

$$I = \frac{dQ}{dt} = \frac{dC}{dt}[-(\phi^S - \phi^P) + U_{comp}] \, . \tag{XV.5}$$

By means of the compensation voltage U_{comp} the oscillating current I is compensated to zero and the particular value of U_{comp} yields the difference in work function

$$U_{comp} = \phi^S - \phi^P \, . \tag{XV.6}$$

For practical purposes Kelvin-probe measurements are usually performed by automatic compensating circuits, for example, of the type shown in Fig. XV.1b. The AC current between sample and vibrating probe is amplified and detected phase-sensitively by a lock-in amplifier. The reference signal from the lock-in is also used to control the frequency of the AC voltage supplying the piezodrive. The DC output of the lock-in, which is proportional to its AC input amplitude, controls a variable voltage source which compensates the contact potential between sample surface and probe. The compensating voltage U_{comp} is read out by a digital voltmeter and gives directly the required contact potential difference (XV.6).

If the work function $e\phi^p$ of the reference electrode (probe) is known, the work function of the sample surface is determined. The method is readily applicable if the sample surface can be covered by an adsorbate without affecting the probe surface. This can be achieved, e.g., if the Kelvin probe can be removed during evaporation onto the sample surface (with geometrically well-defined beam). Difficulties arise when adsorption from an ambient is studied and the probe surface is exposed to the same gas atmosphere. The accuracy of Kelvin probe measurements is quite high. Relative changes in ϕ^S can be determined within error limits of about 10 meV. Absolute measurements, of course, depend on a knowledge of the work function of the probe. Absolute measurements are sometimes performed by means of comparison with well-defined surfaces for which the work function is known from other measurements. Two measurements are then needed, one on the known surface and one on the sample under study. The necessary exchange of the two samples in front of the Kelvin probe decreases the accuracy of the measurement considerably.

For semiconductor surfaces a knowledge of the work-function change $e\Delta\phi = e\phi' - e\phi$ due to adsorption does not give direct insight into atomic properties. The work-function change contains contributions due to band-bending changes and in addition a surface dipole contribution which may be described as a change of the electron affinity χ. These two contributions can be determined separately in a photoemission experiment with UV light (UPS) or X-ray excitation (XPS). According to Fig. XV.2 an adsorption process giving rise to extrinsic Surface States (SS) in the gap and thus an upwards band bending (depletion layer) causes the work function $e\phi$ to change into

$$e\phi' = e\phi + eV_S + e\Delta\phi_{Dip} \, , \tag{XV.7}$$

where V_s is the band bending (change in Fig. XV.2); $e\Delta\phi_{Dip}$ is a dipole contribution arising from the elemental dipoles of the adsorbed molecules or atoms (Δ) and may also include a change of the electron affinity $\Delta\tilde{\chi}$ due to a surface reconstruction during adsorption

$$e\Delta\phi_{Dip} = \Delta + \Delta\tilde{\chi} \, . \tag{XV.8}$$

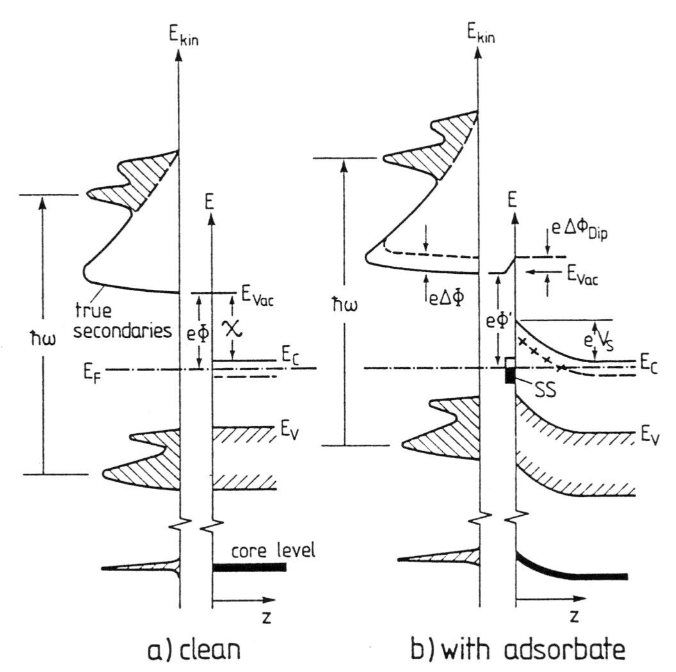

Fig. XV.2a,b. Explanation of adsorbate-induced changes in the photoemission spectrum of a semiconductor: (**a**) Photoemission process on the clean surface of a semiconductor. Photons of energy $\hbar\omega$ excite electrons from the valence band (upper edge E_V) into empty states, from where they leave the crystal and are detected with a kinetic energy above the vacuum energy E_{vac}. True secondary electrons arise from multiple scattering events within the crystal. (**b**) An adsorbate induces extrinsic Surface States (SS) in the bulk band gap and an upwards band bending eV_S. This changes the work function from $e\phi$ to $e\phi'$. Simultaneously with the shift in valence and conduction band states, core level states also shift upwards in energy at the surface

The distinction between Δ and $\Delta\tilde{\chi}$ is rather arbitrary and can be avoided by using a single change of the electron affinity $\Delta\chi$, such that the change in the work function due to adsorption is given as in (9.17).

Figure XV.2 is a schematic drawing of a photoemission experiment (UPS) in which electrons are emitted from occupied valence band states (of density given by shaded area) by irradiation with photons of energy $\hbar\omega$ and detected with a kinetic energy E_{kin}. The detected spectral distribution (also shaded) thus resembles the density of occupied states, but is superimposed on a background of true secondaries, which have undergone several inelastic processes on their way from the point of excitation to the solid surface. Since the probing depth in such an experiment is only a couple of Ångstroms (Sect. 6.3), small in comparison with the thickness of the space charge layer, the measured electron distribution yields information about the electronic band structure at the very surface. In the experiment the Fermi energy E_F is the general reference point (chemical potential of electrically connected sample and analyser) and all measured energies are related to this. E_F is usually determined at the end of the different measurements by evaporating a metal film onto the sample surface and determining the high energy onset of emission which, on a metal surface, is given by E_F.

As is evident from Fig.XV.2 a change of work function $e\Delta\phi = e\phi' - e\phi$ is directly detected as a change of the energy width of the entire spectral distribution of the emitted electrons. The exact position of the low energy flank of the true secondaries with respect to the experimentally determined position of E_F gives the absolute values of the work functions $e\phi$ with and without adsorbate present. In principle, both the work-function and the band-bending changes due to adsorption can be separately determined. The high-energy flank of the distribution of emitted electrons corresponds to the upper valence-band maximum (for normal emission). A shift of this flank due to adsorption indicates a shift of the valence-band edge with respect to the Fermi level and thus a band-bending change ΔV_S (or V_S for initially flat bands as in Fig.XV.2). This is only true if the adsorption process produces no new surface states in the gap. Such extrinsic gap states would modify the spectral distribution at the high-energy flank and the determination of any shift of the onset would be impossible. A second possibility for determining a band-bending change is via a characteristic emission band that is clearly recognized as due to bulk states (rather than surface states, Sect.6.3.3). If this spectral band does not change its shape significantly upon adsorption (due to new surface states in the neighborhood), its position can be determined with sufficient accuracy on the clean surface and after adsorption. An observed shift gives direct information about the band bending change (Fig.XV.2). For example, the adsorption of metallic Sn on GaAs (110) surfaces has been studied by UPS in order to gain information about the Schottky-barrier formation (Fig.XV.3). The emission band near 4.7 eV binding energy (marked by an arrow) shifts to lower binding energy (i.e., towards the Fermi level E_F) with increasing Sn coverage as does the emission onset corresponding to the upper valence band edge. This indicates an upwards band-bending change, i.e., the formation of a depletion layer on n-type material. A quantitative evaluation of the band bending is not possible from the shift of the emission onset since new metal-induced surface states cause a strong deformation of the spectrum. This is clearly seen from the plot in Fig.XV.4. The information about the band bending change must be extracted from the energetic position of the bulk emission band (filled circles).

A similar procedure for investigating band-bending changes upon adsorption can also be performed by XPS on core-level emission bands, since

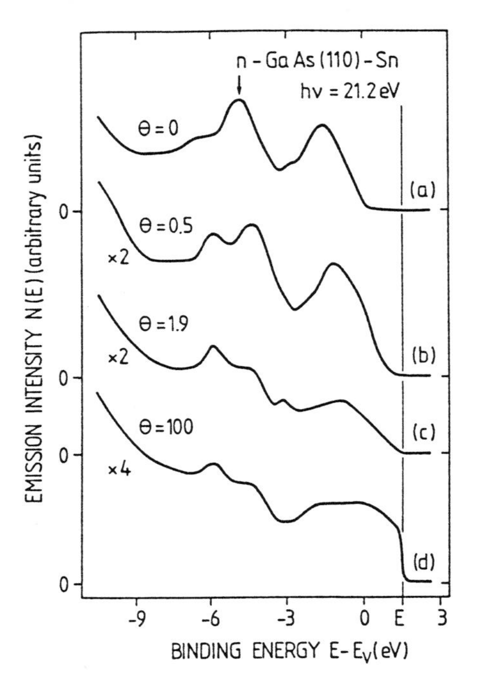

Fig.XV.3. UPS electron energy distribution curves of a clean (*a*) and Sn covered (*b-d*) n-GaAs(110) surface taken with He I radiation ($h\nu = 21.2$eV). The coverage θ is given in monolayers (1 ML contains $8.85 \cdot 10^{14}$ atoms/cm^2). The binding energy is defined relative to the energetic position of the valence band maximum at the clean surface. The arrow in spectrum (*a*) shows an emission band originating from bulk electronic states; its shift with coverage reflects the band bending change [XV.1]

these shift in the same way as valence states (Fig.XV.2). Standard XPS equipment, however, does not usually offer sufficient energy resolution, and so optical monochromators are needed. Furthermore, severe problems in the analysis of the data can occur when chemical bonding shifts (Sect. 6.3) are superimposed on the band-bending shifts.

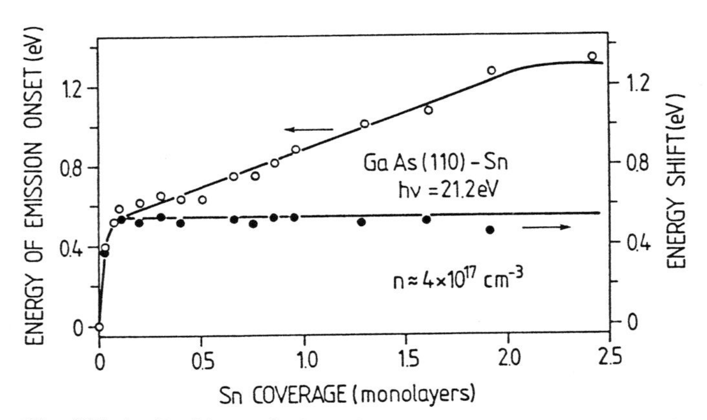

Fig.XV.4. Position of the high-energy emission onset of the UPS spectra of Fig.XV.3, relative to the position of the bulk valence-band maximum (*open circles; left-hand ordinate*) and energy shift of the GaAs valence-band emission peak (arrow in Fig.XV.3) at 4.7 eV below VBM (*full circles; right-hand ordinate*) versus Sn coverage for n-GaAs [XV.1]

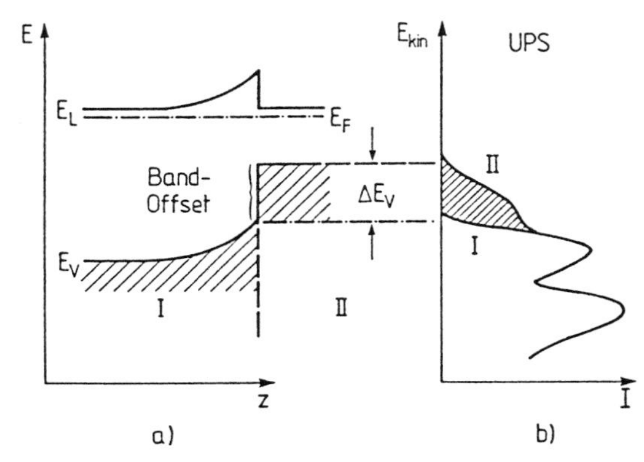

Fig. XV.5a,b. Qualitative explanation of the determination of semiconductor valence band offsets ΔE_V by means of UV photoemission. On the clean surface of semiconductor I thin epitaxial layers of semiconductor II are grown (**a**) and the UPS spectra are measured in situ. (**b**) The shoulder II represents the valence band emission onset of semiconductor II on top of the valence band emission onset I of semiconductor I

Photoemission spectroscopy, in particular UPS, also yields the most direct way to measure band offsets (discontinuities) (Sect. 8.1) in situ between different epitaxy steps. The principle of the method is explained in Fig. XV.5. On the same energy scale one plots the band structure of an uncompleted semiconductor heterostructure (*a*) together with the corresponding (kinetic) energy distribution of the emitted electrons (*b*). In an ideal case the clean surface spectrum of semiconductor I shows an emission onset which indicates the energetic position of its upper valence band maximum. After growing one or two monolayers of semiconductor II, a new shoulder appears whose emission onset characterizes the upper valence-band edge of semiconductor II. The difference between the two thresholds is simply the valence-band discontinuity. It is clear that this technique only works when the valence-band edge of semiconductor II occurs at energies higher than that of semiconductor I.

Furthermore, the measurement can only be made on epilayers of thicknesses up to a couple of Ångstroms (information depth of UPS); but this is usually sufficient to allow the complete development of the band structure of semiconductor II. The method is illustrated in Fig. XV.6 for Ge overlayers on a ZnSe substrate [XV.2]. The energy of the new emission onset due to the Ge overlayer is insensitive to the crystallographic order of the Ge. An amorphous film exhibits the same band discontinuity as two other Ge overlayers which have been annealed and are crystalline.

In general, it should be emphasized that photoemission techniques give quite direct information about band-bending changes, work function and band discontinuities; but the accuracy in determining the absolute values is

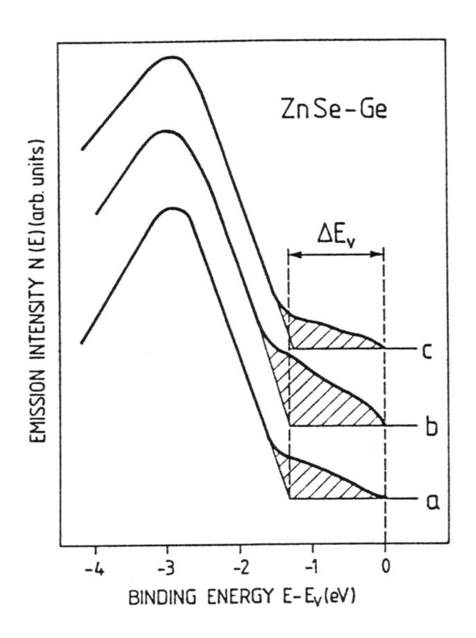

Fig.XV.6. ZnSe-Ge photoemission spectra showing the valence band offset ΔE_V. These spectra were taken on unannealed, amorphous Ge overlayers (curve *a*) and on two different annealed Ge overlayers, exhibiting good LEED patterns (curves *b* and *c*). The spectra suggest that the order or disorder of the overlayer is not an important factor in ΔE_V for this particular system [XV.2]

typically between 20 and 100 meV. This is usually inadequte for narrow gap semiconductors such as InSb or InAs, in particular for explaining the electronic properties of interfaces.

References

XV.1 M. Mattern-Klosson, H. Lüth: Surf. Sci. **162**, 610 (1985)

XV.2 G. Margaritondo, C. Quaresima, F. Patella, F. Sette, C. Capasso, A. Savoia, P. Perfetti: J. Vac. Sci. Technol. A **2**, 508 (1984)

Panel XV

Problems

Problem 9.1. A cesium ion (Cs^+) has an ionic radius of 3 Å. Calculate the approximate surface dipole moment for a (Cs^+) ion adsorbed on a tungsten (W) surface and discuss the result with respect to the observed work-function changes for Cs-adsorption on a W surface (Fig. 9.10).

Problem 9.2. Calculate the time at which 10% of the adsorption sites of a (100) W surface are occupied by nitrogen molecules, when the surface is exposed to an N_2 pressure of $2.67 \cdot 10^{-7}$ Pa at a temperature of 298 K. At this temperature the sticking probability is 0.55. The surface density of the adsorption sites amounts to $1 \cdot 10^{15}$ cm^{-2}.

Problem 9.3. Desorption studies of oxygen on tungsten (W) show that equal amounts of the gas are desorbed within 27 min at 1856 K, within 2 min at 1987 K and within 0.3 min at 2070 K. What is the activation energy for desorption of oxygen from W? What is the time needed for desorption of the same amount of oxygen at the temperatures 298 K and 3000 K?

Problem 9.4. At 300 K, gas molecules have a sticking coefficient of $S = 0.1$ on a freshly prepared, clean semiconductor surface. The adsorption is thermally activated with a sticking coefficient $S \propto \exp(-E_{act}/kT)$ and an activation energy per molecule of $E_{act} = 0.1$ eV. How high are the adsorbate coverages after a one hour exposure at 300 K and at 70 K, respectively? Are the adsorbate coverages detectable by Auger Electron Spectrosopy (AES)?

References

Chapter 1

1.1 H. Ibach (ed.): *Electron Spectroscopy for Surface Analysis*, Topics Curr. Phys., Vol.4 (Springer, Berlin, Heidelberg 1977)

For further reading see, for example, the monographs:

Desjonqueres M.C., D. Spanjaard: *Concepts in Surface Physics*, 2nd edn. (Springer, Berlin, Heidelberg 1996)

Henzler M., W. Göpel: *Oberflächenphysik des Festkörpers* (Teubner, Stuttgart 1991)

Lannoo M., P. Friedel: *Atomic and Electronic Structure of Surfaces*, Springer Ser. Surf. Sci., Vol.16 (Springer, Berlin, Heidelberg 1991)

Mönch W.: *Semiconductor Surfaces and Interfaces*, 2nd edn., Springer Ser. Surf. Sci., Vol.26 (Springer, Berlin, Heidelberg 1995)

O'Connor D.J., B.A. Sexton, R.St.C. Smart (eds.): *Surface Analysis Methods in Materials Science*, Springer Ser. Surf. Sci., Vol.23 (Springer, Berlin, Heidelberg 1992)

Yu P.Y., M. Cardona: *Fundamentals of Semiconductors* (Springer, Berlin, Heidelberg 1996)

Zangwill A.: *Physics at Surfaces* (Cambridge Univ. Press, Cambridge 1988)

Or the conference proceeding:

Annual proceedings of the "European Conferences on Surface Science (ECOSS)" published in Surf. Sci. (North-Holland, Amsterdam)

Cardona M., J. Giraldo (eds.): *Thin Films and Small Particles* (World Scientific, Singapore 1989)

Castro G.R., M. Cardona (eds.): *Lectures on Surface Science* (Springer, Berlin, Heidelberg 1987)

Ponce F.A., M. Cardona (eds.): *Surface Science*, Springer Proc. Phys., Vol.62 (Springer, Berlin, Heidelberg 1991)

Vanselow R., R. Howe (eds.): *Chemistry and Physics of Solid Surfaces*, Vols.I-III (CRC Press, Boca Raton, FL 1976, 1979, 1982); Vols.IV-VIII: Springer Ser. Chem. Phys., Vol.20 and 35, and Springer Ser. Surf. Sci., Vols.5, 10 and 22 (Springer, Berlin, Heidelberg 1982, 1984, 1986, 1988, and 1990)

Wette F.W. de (ed): *Solvay Conference on Surface Science*, Springer Ser. Surf. Sci., Vol.14 (Springer, Berlin, Heidelberg 1988)

Chapter 2

2.1 M.A. Van Hove, W.H. Weinberg, C.-M. Chan: *Low-Energy Electron Diffraction*, Springer Ser. Surf. Sci., Vol.6 (Springer, Berlin, Heidelberg 1986)

2.2 W. Espe: Über Aufdampfung von dünnen Schichten im Hochvakuum in *Ergeb-*

nisse der Hochvakuumtechnik und Physik dünner Schichten, herausg. von M. Auwärtner (Wissenshaftliche Verlagsgesellschaft, Stuttgart 1957) p.67
2.3 A.A. Chernov: *Modern Crystallography III*, Springer Ser. Solid-State Sci., Vol. 36 (Springer, Berlin, Heidelberg 1984)
2.4 L.L. Chang, K. Ploog (eds): *Molecular Beam Epitaxy and Heterostructures*, NATO ASI Series, No.87 (Nijhoff, Dordrecht 1985)
2.5 E.H.C. Parker (ed.): *The Technology and Physics of Molecular Beam Epitaxy* (Plenum, New York 1985)
2.6 M.A. Herman, H. Sitter: *Molecular Beam Epitaxy*, 2nd edn., Springer Ser. Mat. Sci., Vol.7 (Springer, Berlin, Heidelberg 1996)
2.7 H. Künzel, G.H. Döhler, K. Ploog: Appl. Phys. A **27**, 1-10 (1982)
2.8 F. Capasso (ed.): *Physics of Quantum Electron Devices*, Springer Ser. Electron. Photon., Vol.28 (Springer, Berlin, Heidelberg 1990)
2.9 J.H. Neave, B.A. Joyce, P.J. Dobson, N. Norton: Appl. Phys. A **31**, 1 (1983)
2.10 J.R. Arthur: J. Appl. Phys. **39**, 4032 (1968)
2.11 E. Kasper, H.J. Herzog, H. Dämbkes, Th. Ricker: Growth mode and interface structure of MBE grown SiGe structures. *Two-Dimensional Systems: Physics and New Devices*, ed. by G. Bauer, F. Kuchar, H. Heinrich, Springer Ser. Solid-State Sci., Vol.67 (Springer, Berlin, Heidelberg 1986) p.52
E. Kasper: Silicon germanium - heterostructures on silicon substrates. *Festkörperprobleme*, ed. by P. Grosse, **27**, 265 (Vieweg, Braunschweig 1987)
2.12 H. Ibach, H. Lüth: *Solid-State Physics*, 2nd edn. (Springer, Berlin, Heidelberg 1996)
2.13 M.G. Craford: Recent developments in LED technology. IEEE Trans. ED-**24**, 935 (1977)
2.14 R.F.C. Farrow, D.S. Robertson, G.M. Williams, A.G. Cullis, G.R. Jones, I.M. Young, P.N.J. Dennis: J. Cryst. Growth **54**, 507 (1981)
M. Mattern, H. Lüth: Surf. Sci. **126**, 502 (1983)
2.15 F. Arnaud d'Avitaya, S. Delage, E. Rosencher: Surf. Sci. **168**, 483 (1986)
2.16 J.R. Waldrop, R.W. Grant: Phys. Lett. **34**, 630 (1979)
2.17 E. Veuhoff, W. Pletschen, P. Balk, H. Lüth: J. Cryst. Growth **55**, 30 (1981)
2.18 H. Lüth: Metalorganic molecular beam epitaxy (MOMBE). Proc. ESSDERC, (Cambridge, GB 1986), Inst. Phys. Conf. Ser. **82**, 135 (1987)
2.19 M.B. Panish, H. Temkin: *Gas Source Molecular Beam Epitaxy*, Springer Ser. Mater. Sci., Vol.26 (Springer, Berlin, Heidelberg 1993)
2.20 M.B. Panish, S. Sumski: J. Appl. Phys. **55**, 3571 (1984)
2.21 O. Kayser, H. Heinecke, A. Brauers, H. Lüth, P. Balk: Chemitronics **3**, 90 (1988)
G.B. Stringfellow: *Organometallic Vapor-Phase Epitaxy: Theory and Practice* (Academic, Boston 1989) p.23
2.22 M. Weyers, N. Pütz, H. Heinecke, M. Heyen, H. Lüth, P. Balk: J. Electron. Mater. **15**, 57 (1986)
2.23 K. Werner, H. Heinecke, M. Weyers, H. Lüth, P. Balk: J. Cryst. Growth **81**, 281 (1987)
2.24 R. Kaplan: J. Vac. Sci. Technol. A **1**, 551 (1983)
2.25 G.L. Price: Collected Papers of the 2nd Int'l Conf. on Molecular Beam Epitaxy and Related Clean Surface Techniques (Jpn. Soc. Appl. Phys., Tokyo 1982) p.259
2.26 H. Lüth: Surf. Sci. **299/300**, 867 (1994)

Chapter 3

3.1 J.W. Gibbs: Reprinted in J.W. Gibbs: *The Scientific Papers* (Dover, New York 1961) Vol.1
3.2 E.A. Guggenheim: *Thermodynamics* (North-Holland, Amsterdam 1959)
3.3 G. Wulff: Z. Kristallogr. Mineral. **34**, 449 (1901)
3.4 J.M. Blakely, M. Eizenberg: Morphology and composition of crystal surfaces. *The Chemical Physics of Solid Surfaces and Heterogeneous Catalysis*, ed. by D.A. King, D.F. Woodruff (Elsevier, Amsterdam 1981) Vol.1
3.5 C.B. Duke, R.J. Meyer, A. Paton, P. Mark, A. Kahn, E. So, J.L. Yeh: J. Vac. Sci. Technol. **16**, 1252 (1979)
3.6 K.C. Pandey: Phys. Rev. Lett. **49**, 223 (1982)
3.7 B.K. Vainshtein: *Fundamentals of Crystallography*, 2nd edn., Modern Crystallography, Vol.1 (Springer, Berlin, Heidelberg 1996)
M.A. Van Hove, W.H. Weinberg, C.-M. Chan: *Low-Energy Electron Diffraction*, Springer Ser. Surf. Sci., Vol.6 (Springer, Berlin, Heidelberg 1986)
W. Ludwig, C. Falter: *Symmetries in Physics*, 2nd edn., Springer Ser. Solid-State Sci., Vol.64 (Springer, Berlin, Heidelberg 1996)
3.8 E.A. Wood: J. Appl. Phys. **35**, 1306 (1964)
3.9 P. Chaudhari, J.W. Matthew (eds.): Grain Boundaries and Interfaces. Surf. Sci. **31** (1972)
3.10 E. Kasper: Silicon germanium - heterostructures on silicon substrates. *Festkörperprobleme* **27**, 265 (Vieweg, Braunschweig 1987)
3.11 A.A. Chernov: *Modern Crystallography III, Crystal Growth*, Springer Ser. Solid-State Sci., Vol.36 (Springer, Berlin, Heidelberg 1984)
3.12 J.A. Venables, G.D.T. Spiller, M. Hanbücken: Rep. Progr. Phys. **47**, 399 (1984)
3.13 J.A. Venables, G.D.T. Spiller: In *Surface Mobility on Solid Surfaces*, ed. by Thien Bink (Plenum, New York 1981) p.339
3.14 M.A. Herman, H. Sitter: *Molecular Beam Epitaxy*, 2nd edn., Springer Ser. Mat. Sci., Vol.7 (Springer, Berlin, Heidelberg 1996)
3.15 E. Bauer: Z. Kristallog. **110**, 372 (1958)
E. Bauer, H. Popper: Thin Solid Films **12**, 167 (1972)
3.16 S. Sugano: *Microcluster Physics*, Springer Ser. Mat. Sci., Vol.20 (Springer, Berlin, Heidelberg 1991)
3.17 T.E. Gallon: Surf. Sci. **17**, 486 (1969)
3.18 M.P. Seah: Surf. Sci. **32**, 703 (1972)
3.19 M. Mattern, H. Lüth: Surf. Sci. **126**, 502 (1983)
3.20 N. Bündgens, H. Lüth, M. Mattern-Klosson, A. Spitzer, A. Tulke: Surf. Sci. **160**, 46 (1985)
3.21 L. Reimer: *Scanning Electron Microscopy*, Springer Ser. Opt. Sci., Vol.45 (Springer, Berlin, Heidelberg 1985)
3.22 H.-J. Güntherod, R. Wiesendanger (eds.): *Scanning Tunneling Microscopy I*, 2nd edn., Springer Ser. Surf. Sci., Vol.20 (Springer, Berlin, Heidelberg 1994)
3.23 Priv. communication by H. Niehus (IGV, Research Center Jülich, 1990)
3.24 L. Reimer: *Transmission Electron Microscopy*, 3rd edn., Springer Ser. Opt. Sci., Vol.36 (Springer, Berlin, Heidelberg 1993)
3.25 Priv. commun. by D. Gerthsen (IFF, Research Center Jülich, 1990)
3.26 Priv. commun. by A. Förster and D. Gerthsen (ISI and IFF, Research Center Jülich, 1990)

3.27 M. Cardona (ed.): *Light Scattering in Solids I*, 2nd edn., Topics Appl. Phys., Vol.8 (Springer, Berlin. Heidelberg 1983)
3.28 W. Pletschen: Dissertation, RWTH Aachen (1985)
3.29 R.M.A. Azzam, N.M. Bashara: *Ellipsometry and Polarized Light* (North-Holland, Amsterdam 1977)
3.30 A. Tulke: Elektronenspektroskopische Untersuchung von Sb, As und P Schichten auf III-V Halbleiteroberflächen, Dissertation, RWTH Aachen (1988)

Chapter 4

4.1 G. Ertl, J. Küppers: *Low Energy Electrons and Surface Chemistry*, 2nd edn. (VHC, Weinheim 1985)
4.2 M.P. Seah, W.A. Dench: Surf. Interf. Anal. I (1979). Compilation of experimental data determined with various electron energies for a large variety of materials
4.3 M.A. Van Hove, W.H. Weinberg, C.-M. Chan: *Low-Energy Electron Diffraction*, Springer Ser. Surf. Sci., Vol.6 (Springer, Berlin, Heidelberg 1986)
4.4 K. Christmann, G. Ertl, O. Schober: Surf. Sci. **40**, 61 (1973)
4.5 G. Ertl: In *Molecular Processes on Solid Surfaces*, ed. by E. Dranglis, R.D. Gretz, R.I. Jaffee (McGraw-Hill, New York 1969) p.147
4.6 J.B. Pendry: *Low Energy Electron Diffraction* (Academic, New York 1974)
4.7 G. Capart: Surf. Sci. **13**, 361 (1969)
4.8 E.G. McRae: J. Chem. Phys. **45**, 3258 (1966)
4.9 R. Feder (ed.): *Polarized Electrons in Surface Physics* (World Scientific, Singapur 1985)
4.10 H. Ibach, D.L. Mills: *Electron Energy Loss Spectroscopy and Surface Vibrations* (Academic, New York 1982)
4.11 E. Fermi: Phys. Rev. **57**, 485 (1940)
4.12 J. Hubbard: Proc. Phys. Soc. (London) A **68**, 976 (1955)
H. Fröhlich, H. Pelzer: Proc. Phys. Soc. (London) A **68**, 525 (1955)
4.13 H. Ibach, H. Lüth: *Solid-State Physics*, 2nd edn. (Springer, Berlin, Heidelberg 1996)
4.14 H. Lüth: Surf. Sci. **126**, 126 (1983)
4.15 R. Matz: Reine und gasbedeckte GaAs(110) Spaltflächen in HREELS, Dissertation, RWTH Aachen (1982)
4.16 P. Grosse: *Freie Elektronen in Festkörpern* (Springer, Berlin, Heidelberg 1979)
4.17 Ph. Lambin, J.-P. Vigneron, A.A. Lucas: Solid State Commun. **54**, 257 (1985)
4.18 A. Ritz, H. Lüth: Phys. Rev. B **32**, 6596 (1985)
4.19 N. Bündgens: Elektronenspektroskopische Untersuchungen an Sn-Schichten auf III-V Halbleiteroberflächen, Diploma Thesis, RWTH Aachen (1984)
M. Mattern, H. Lüth: Surf. Sci. **126**, 502 (1983)
4.20 A. Spitzer, H. Lüth: Phys. Rev. B **30**, 3098 (1984)
4.21 S. Lehwald, J.M. Szeftel, H. Ibach, T.S. Rahman, D.L. Mills: Phys. Rev. Lett. **50**, 518 (1983)
4.22 R.F. Willis: Surf. Sci. **89**, 457 (1979)
4.23 H. Lüth, R. Matz: Phys. Rev. Lett. **46**, 1952 (1981)
4.24 R. Matz, H. Lüth: Surf. Sci. **117**, 362 (1982)
4.25 L.C. Feldman, J.W. Mayer: *Fundamentals of Surface and Thin Film Analysis* (North-Holland, New York 1986)

4.26 J.T. McKinney, M. Leys: 8th Nat'l Conf. on Electron Probe Analysis, New Orleans, LA (1973)
4.27 J.F. van der Veen: Ion beam crystallography of surfaces and interfaces. Surf. Sci. Rpts. **5**, 199 (1985)
4.28 L.C. Feldman, J.W. Mayer, S.T. Picraux: *Materials Analysis by Ion Channeling* (Academic, New York 1982)
4.29 Priv. communication by S. Mantl, ISI, Research Center Jülich (1990)
4.30 J. Haskell, E. Rimini, J.W. Mayer: J. Appl. Phys. **43**, 3425 (1972)
4.31 J.U. Anderson, O. Andreason, J.A. Davis, E. Uqgerhoj: Rad. Eff. **7**, 25 (1971)
4.32 R.M. Tromp: The structure of silicon surfaces, Dissertation, University of Amsterdam (1982)

Chapter 5

5.1 M. Born, R. Oppenheimer: Ann. Phys. (Leipzig) **84**, 457 (1927)
5.2 G. Benedek: Surface Lattice Dynamics. *Dynamic Aspects of Surface Physics*, LV III Corso (Editrice Compositori, Bologna 1974) p.605
A.A. Maradudin, R.F. Wallis, L. Dobrzinski: *Handbook of Surface and Interfaces III, Surface Phonons and Polaritons* (Garland STPM Press, New York 1980)
5.3 L.M. Brekhovskikh, O.A. Godin: *Acoustics of Layered Media I,II*, Springer Ser. Wave Phen., Vols.5,10 (Springer, Berlin, Heidelberg 1990, 1992)
5.4 L.M. Brekhovskikh, V. Goncharov: *Mechanics of Continua and Wave Dynamics*, 2nd edn., Springer Ser. Wave Phen., Vol.1 (Springer, Berlin, Heidelberg 1994)
5.5 E.A. Ash, E.G.S. Paige (eds.): *Rayleigh-Wave Theory and Application*, Springer Ser. Wave Phen., Vol.2 (Springer, Berlin, Heidelberg 1985)
5.6 L.D. Landau, E.M. Lifshitz: *Theory of Elasticity VII, Course of Theoretical Physics* (Pergamon, London 1959) p.105
5.7 J. Lindhard: Kgl. Danske Videnskab Selskab Mat.-Fys. Medd. **28**, No.8, 1 (1954)
5.8 A. Stahl: Surf. Sci. **134**, 297 (1983)
5.9 R. Matz, H. Lüth: Phys. Rev. Lett. **46**, 500 (1981)
5.10 M. Hass, B.W. Henvis: J. Phys. Chem. Solids **23**, 1099 (1962)
5.11 H. Ibach, D.L. Mills: *Electron Energy Loss Spectroscopy and Surface Vibrations* (Academic, New York 1982)
5.12 H. Lüth: Vacuum (GB) **38**, 223 (1988)
5.13 F.W. De Wette, G.P. Alldredge, T.S. Chen, R.E. Allen: Phonons. Proc. Int'l Conf. (Rennes), ed. by M.A. Nusimovici (Flamarion, Rennes 1971) p.395
R.E. Allen, G.P. Alldredge, F.W. De Wette: Phys. Rev. Lett. **24**, 301 (1970)
5.14 G. Benedek: Surface collective excitations. Proc. NATO ASI on Collective Excitations in Solids (Erice 1981); ed. by B.D. Bartols (Plenum, New York 1982)
5.15 G. Brusdeylins, R.B. Doak, J.P. Toennies: Phys. Rev. Lett. **46**, 437 (1981)
5.16 S. Lehwald, J.M. Szeftel, H. Ibach, T.S. Rahman, D.L. Mills: Phys. Rev. Lett. **50**, 518 (1983)
5.17 R.E. Allen, G.P. Alldredge, F.W. De Wette: Phys. Rev. B **4**, 1661 (1971)

Chapter 6

6.1 H.-J. Güntherodt, R. Wiesendanger (eds.): *Scanning Tunneling Microscopy I*, 2nd edn., Springer Ser. Surf. Sci., Vol.20 (Springer, Berlin, Heidelberg 1994)
R. Wiesendanger, H.-J. Güntherodt (eds.): *Scanning Tunneling Microscopy II, III*, 2nd edn., Springer Ser. Surf. Sci., Vols.28,29 (Springer, Berlin, Heidelberg 1995, 1996)
R. Wiesendanger: *Scanning Probe Microscopy and Spectroscopy* (Cambridge Univ. Press, Cambridge, GB 1994)
6.2 H. Ibach, H. Lüth: *Solid-State Physics*, 2nd edn. (Springer, Berlin, Heidelberg 1996)
6.3 W. Shockley: Phys. Rev. **56**, 317 (1939)
6.4 I. Tamm: Phys. Z. Soviet Union **1**, 733 (1932)
6.5 M. Cardona, L.Ley (eds.): *Photoemission in Solids I,II*, Topics Appl. Phys., Vols.26,27 (Springer, Berlin, Heidelberg 1978,1979)
S. Hüfner: *Photoelectron Spectroscopy*, 2nd edn., Springer Ser. Solid-State Sci., Vol.82 (Springer, Berlin, Heidelberg 1996)
6.6 A. Einstein: Ann. Phys. **17**, 132 (1905)
6.7 I. Adawi: Phys. Rev. **134**, A 788 (1964)
J.G. Endriz: Phys. Rev. B**7**, 3464 (1973)
6.8 B. Feuerbacher, R.F. Willis: J. Phys. C **9**, 169 (1976), and references therein
6.9 H. Puff: Phys. Status Solidi **1**, 636 (1961)
C.N. Berglung, W.E. Spicer: Phys. Rev. **136**, A1030 (1964)
6.10 A. Spitzer, H. Lüth: Phys. Rev. B **30**, 3098 (1984)
6.11 P. Heimann, H. Miosga, N. Neddermeyer: Solid Stat. Commun. **29**, 463 (1979)
6.12 K. Siegbahn, C. Nordling, A. Fahlmann, R. Nordberg, K. Hamrin, J. Hedman, G. Johansson, T. Bergmark, S.E. Karlsson, I. Lindgren, B. Lindberg: ESCA-atomic, molecular and solid state structure, studied by means of electron spectroscopy. *Nova Acta Regiae Societatis Scientiarum Upsaliensis*, Ser. IV, **20**, 21 (Almquist and Wirsel, Uppsala 1967)
6.13 G.V. Hansson, S.A. Flodström: Phys. Rev. B **18**, 1562 (1978)
6.14 A. Zangwill: *Physics at Surfaces* (Cambridge Univ. Press, Cambridge 1988) p.80
6.15 E. Caruthers, L. Kleinman, G.P. Alldredge: Phys. Rev. B **8**, 4570 (1973)
6.16 G.V. Hansson, S.A. Flodström: Phys. Rev. B **18**, 1572 (1978)
6.17 P. Heimann, H. Neddermeyer, H.F. Roloff: J. Phys. C **10**, L17 (1977)
P. Heimann, J. Hermanson, H. Miosga, H. Neddermeyer: Surf. Sci. **85**, 263 (1979)
6.18 P. Heimann, J. Hermanson, H. Miosga, H. Neddermeyer: Phys. Rev. B **20**, 3050 (1979)
6.19 J.G. Gay, J.R. Smith, F.J. Arlinghaus: Phys. Rev. Lett. **42**, 332 (1979)
6.20 M. Posternak, H. Krakauer, A.J. Freeman, D.D. Koelling: Phys. Rev. B **21**, 5601 (1980)
6.21 J.C. Campuzano, D.A. King, C. Somerton, J.E. Inglesfield: Phys. Rev. Lett. **45**, 1649 (1980)
6.22 M.I. Holmes, T. Gustafsson: Phys. Rev. Lett. **47**, 443 (1981)
6.23 L.F. Mattheis, D.R. Hamann: Phys. Rev. B **29**, 5372 (1984)
6.24 J.A. Appelbaum, D.R. Hamann: Solid State Commun. **27**, 881 (1978)
6.25 M. Mehta, C.S. Fadley: Phys. Rev. B **20**, 2280 (1979)

6.26 D.E. Eastman, F.J. Himpsel, J.F. van der Veen: J. Vac. Sci. Techn. **20**, 609 (1982)
6.27 A. Goldmann, V. Dose, G. Borstel: Phys. Rev. B **32**, 1971 (1985)
6.28 V. Dose, W. Altmann, A. Goldmann, U. Kolac, J. Rogozik: Phys. Rev. Lett. **52**, 1919 (1984)
6.29 F.G. Allen, G.W. Gobeli: Phys. Rev. **127**, 150 (1962)
6.30 G. Chiarotti, S. Nannarone, R. Pastore, P. Chiaradia: Phys. Rev. B **4**, 3398 (1971)
6.31 H. Lüth: Appl. Phys. **8**, 1 (1975)
6.32 D.E. Eastman, W.D. Grobman: Phys. Rev. Lett. **28**, 16 (1972)
6.33 I. Ivanov, A. Mazur, J. Pollmann: Surf. Sci. **92**, 365 (1980)
6.34 D. Haneman: Phys. Rev. **121**, 1093 (1961)
6.35 J.A. Appelbaum, D.R. Hamann: Phys. Rev. B **12**, 1410 (1975)
K.C. Pandey, J.C. Phillips: Phys. Rev. Lett. **34**, 2298 (1975)
6.36 K.C. Pandey: Phys. Rev. Lett. **47**, 1913 (1981)
6.37 J.E. Northrup, M.L. Cohen: J. Vac. Sci. Techn. **21**, 333 (1982)
6.38 R.I.G. Uhrberg, G.V. Hansson, J.M. Nicholls, S.A. Flodström: Phys. Rev. Lett. **48**, 1032 (1982)
6.39 K.C. Pandey: Phys. Rev. Lett. **49**, 223 (1982)
6.40 R. Matz, H. Lüth, A. Ritz: Solid State Commun. **46**, 343 (1983)
6.41 P. Chiaradia, A. Cricenti, S. Selci, G. Chiarotti: Phys. Rev. Lett. **52**, 1145 (1984)
6.42 J.E. Northrup, M.L. Cohen: Phys. Rev. B **27**, 6553 (1983)
6.43 J.M. Nicholls, P. Martensson, G.V. Hansson: Phys. Rev. Lett. **54**, 2363 (1985)
6.44 K. Takayanagi, Y. Tanishiro, M. Takahashi, S. Takahashi: Surf. Sci. **164**, 367 (1985)
6.45 D.E. Eastman, F.J. Himpsel, J.A. Knapp, K.C. Pandey: Physics of Semiconductors, ed. by B.L.H. Wilson (Inst. of Physics, Bristol 1978) p.1059
D.E. Eastman, F.J. Himpsel, F.J. van der Veen: Solid State Commun. **35**, 345 (1980)
6.46 G.V. Hansson, R.I.G. Uhrberg, S.A. Flodström: J. Vac. Sci. Techol. **16**, 1287 (1979)
6.47 P. Martensson, W.X. Ni, G.V. Hansson, J.M. Nicholls, B. Reihl: Phys. Rev. B **36**, 5974 (1987)
6.48 F. Houzay, G.M. Guichar, R. Pinchaux, Y. Petroff: J. Vac. Sci. Technol. **18**, 860 (1981)
6.49 T. Yokotsuka, S. Kono, S. Suzuki, T. Sagawa: Solid State Commun. **46**, 401 (1983)
6.50 J.M. Layet, J.Y. Hoarau, H. Lüth, J. Derrien: Phys. Rev. B **30**, 7355 (1984)
6.51 J.A. Appelbaum, G.A. Baraff, D.R. Hamann: Phys. Rev. B **14**, 588 (1976)
6.52 F.J. Himpsel, D.E. Eastman: J. Vac. Sci. Technol. **16**, 1297 (1979)
6.53 D.J. Chadi: Phys. Rev. Lett: **43**, 43 (1979)
6.54 J. Pollmann: On the electronic structure of semiconductor surfaces, interfaces and defects at surfaces and interfaces. *Festkörperprobleme* ed. by J. Treusch (Vieweg, Braunschweig 1980) p.117
I. Ivanov, A. Mazur, J. Pollmann: Surf. Sci. **92**, 365 (1980)
6.55 J.R. Chelikowsky, M.L. Cohen: Phys. Rev. B **20**, 4150 (1979)
6.56 A. Huijser, J. van Laar: Surf. Sci. **52**, 202 (1975)
6.57 R.P. Beres, R.E. Allen, J.D. Dow: Solid State Commun. **45**, 13 (1983)

6.58 G.P. Williams, R.J. Smith, G.J. Lapeyre: J. Vac. Sci. Techol. **15**, 1249 (1978)
6.59 A. Huijser, J. van Laar, T.L. van Rooy: Phys. Lett. **65A**, 337 (1978)
6.60 V. Dose, H.-J. Gossmann, D. Straub: Phys. Rev. Lett. **47**, 608 (1981); ibid. Surf. Sci. **117**, 387 (1982)
6.61 P.K. Larsen, J.D. van der Veen: J. Phys. C **15**, L 431 (1982)
6.62 S.J. Lee, J.D. Joannopoulos: J. Vac. Sci. Technol. **17**, 987 (1980)
6.63 C.B. Duke: Chem. Rev. **96**, 1237 (1996)
6.64 R. Dorn, H. Lüth, M. Büchel: Phys. Rev. B **16**, 4675 (1977)
6.65 G.V. Hansson, R.I.G. Uhrberg: Photoelectron spectroscopy of surface states on semiconductor surfaces. Surf. Sci. Rep. **9**, 197 (1988)

Chapter 7

7.1 H. Ibach, H. Lüth: *Solid-State Physics*, 2nd edn. (Springer, Berlin, Heidelberg 1996)
7.2 A. Many, Y. Goldstein, N.B. Grover: *Semiconductor Surfaces* (North Holland, Amsterdam 1965)
7.3 H. Lüth, M. Büchel, R. Dorn, M. Liehr, R. Matz: Phys. Rev. B **15**, 865 (1977)
7.4 E. Veuhoff, C.D. Kohl: J. Phys. C: Solid State Phys. **14**, 2395 (1981)
7.5 T. Ando, A.B. Fowler, F. Stern: Electronic properties of two-dimensional systems. *Rev. Mod. Phys.* **54**, 437 (AIP, New York 1982)
7.6 F.J. Allen, G.W. Gobeli: Phys. Rev. **127**, 152 (1962)
7.7 K.C. Pandey: Phys. Rev. Lett. **47**, 1913 (1981)
7.8 M. Henzler: Phys. Status Solidi **19**, 833 (1967)
W. Mönch: Phys. Status Solidi **40**, 257 (1970)
7.9 J. von Wienskowski, W. Mönch: Phys. Status Solidi B **45**, 583 (1971)
G.W. Gobeli, F.G. Allen: Surf. Sci. **2**, 402 (1964)
7.10 F. Himpsel, D.E. Eastman: J. Vac. Sci. Technol. **16**, 1287 (1979)
W. Mönch, P. Koke, S. Krüger: J. Vac. Sci. Technol. **19**, 313 (1981)
7.11 Private communication by H. Wagner (ISI, Research Center Jülich, 1988)
7.12 J.M. Nicholls, B. Reihl: Phys. Rev. B **36**, 8071 (1987)
F.J. Himpsel, D.E. Eastman, P. Heimann, B. Reihl, C.W. White, D.M. Zehner: Phys. Rev. B **24**, 1120 (1981)
P. Mårtensson, W. Ni, G. Hansson, J.M. Nicholls, B. Reihl: Phys. Rev. B **36**, 5974 (1987)
7.13 P. Balk (ed.): The Si-SiO_2 System, in *Materials Science Monographs* **32** (Elsevier, Amsterdam 1988)
7.14 H. Ibach, H.D. Bruchmann, H. Wagner: Appl. Phys. A **29**, 113 (1982)
7.15 M.H. White, J.R. Cricchi: *Characterization of Thin Oxide MNOS Memory Transistors*, IEEE Trans. ED-**19**, 1280 (1972)
7.16 F.J. Grunthaner, P.J. Grunthaner, R.P. Vasquez, B.F. Lewis, J. Maserjian, A. Madhukar: J. Vac. Sci. Technol. **16**, 1443 (1979)
7.17 F.J. Grunthaner, B.F. Lewis, J. Maserjian: J. Vac. Sci. Techol. **20**, 747 (1982)
7.18 S.P. Svensson, J. Kanski, T.G. Andersson, P.-O. Nilsson: J. Vac. Sci. Technol. B **2**, 235 (1984)
7.19 A. Förster, H. Lüth: Surf. Sci. **189/190**, 307 (1987)
7.20 K. Smit, L. Koenders, W. Mönch: J. Vac. Sci. Technol. B **7**, 888 (1989)
7.21 H. Moormann, D. Kohl, G. Heiland: Surf. Sci. **80**, 261 (1979)

7.22 G. Heiland, H. Lüth: Adsorption on oxides, in *The Chemical Physics of Solid Surfaces and Heterogeneous Catalysis*, Vol.3, ed. by D.A. King, D.P. Woodruff (Elsevier, Amsterdam 1984) p.137
7.23 G. Heiland, P. Kunstmann: Surf. Sci. **13**, 72 (1969)
7.24 D. Kahng, M.M. Atalla: *Silicon-Silicon Dioxide Field Induced Surface Devices*, IRE Solid State Device Res. Conf., Carnegie Institute of Technology, Pittsburgh, Pa., 1960
D. Kahng: *A Historical Perspective on the Development of MOS Transistors and Related Devices*, IEEE Trans. ED-**23**, 655 (1976)
7.25 H.C. Pao, C.T. Sah: "Effects of diffusion current on characteristics of metal-oxide (insulator) semiconductor transistors (MOST)", Solid State Electron. **9**, 927 (1966)
7.26 S.M. Sze: *Physics of Semiconductor Devices*, 2nd edn. (Wiley, New York 1981) p.431
7.27 A. Kamgar, P. Kneschaurek, G. Dorda, J.F. Koch: Phys. Rev. Lett. **32**, 1251 (1974)
7.28 D.C. Tsui, G. Kaminsky, P.H. Schmidt: Phys. Rev. B **9**, 3524 (1974)
7.29 A. Hartstein, A.B. Fowler, M. Albert: Surf. Sci. **98**, 181 (1980)
7.30 C.-D. Kohl, G. Heiland: Surf. Sci. **63**, 96 (1977)
7.31 A.B. Fowler, F.F. Fang, W.E. Howard, P.J. Stiles: Phys. Rev. Lett. **16**, 901 (1966)
7.32 K. von Klitzing, G. Dorda, M. Pepper: Phys. Rev. B **28**, 4886 (1983)
7.33 K. von Klitzing: The fine structure constant α, a contribution of semiconductor physics to the determination of α. *Festkörperprobleme XXI* (Advances in Solid State Physics), ed. by J. Treusch (Vieweg, Braunschweig 1981) p.1
7.34 T. Chakraborty, P. Pietiläinen: *The Fractional Quantum Hall Effect*, 2nd ed., Springer Ser. Solid-State Sci., Vol.85 (Springer, Berlin, Heidelberg 1995)

Chapter 8

8.1 H. Ibach, H. Lüth: *Solid-State Physics*, 2nd edn. (Springer, Berlin, Heidelberg 1996)
8.2 L.J. Brillson: The structure and properties of metal-semiconductor interfaces. Surf. Sci. Rep. **2**, 123 (1982)
8.3 W. Schottky: Z. Physik **113**, 367 (1939)
8.4 W. Mönch: In *Festkörperprobleme*, ed. by P. Grosse (Vieweg, Braunschweig, 1986) Vol.26
8.5 J. Bardeen: Phys. Rev. **71**, 717 (1947)
8.6 V. Heine: Phys. Rev. A **138**, 1689 (1965)
8.7 C. Tejedor, F. Flores, E. Louis: J. Phys. C **10**, 2163 (1977)
8.8 J. Tersoff: Phys. Rev. Lett. **30**, 4874 (1984)
8.9 W. Mönch: Rep. Prog. Phys. **53**, 221 (1990)
8.10 M. Mattern-Klosson, H. Lüth: Surf. Sci. **162**, 610 (1985)
8.11 W.E. Spicer, I. Lindau, P. Skeath, C.Y. Su: J. Vac. Sci. Technol. **17**, 1019 (1980)
8.12 W.E. Spicer, R. Cao, K. Miyano, T. Kendelewicz, I. Lindau, E. Weber, Z. Liliental-Weber, N. Newman: Appl. Surf. Sci. **41/42**, 1 (1989)
8.13 J. Tersoff: Phys. Rev. **30**, 4874 (1984)

8.14 J. Tersoff: Phys. Rev. B **32**, 6968 (1985)
8.15 J. Tersoff: Surf. Sci. **168**, 275 (1986)
8.16 J. Tersoff: Phys. Rev. Lett. **56**, 2755 (1986)
8.17 L. Pauling: *The Nature of the Chemical Bond* (Cornell Univ. Press, Ithaca 1960)
8.18 N.B. Hanney, C.P. Smith: J. Am. Chem. Soc. **68**, 171 (1946)
8.19 M.L. Cohen: Adv. Electron. Electron Phys. **51**, 1 (1980)
M.L. Cohen, J.R. Chelikowsky: *Electronic Structure and Optical Properties of Semiconductors*, 2nd edn., Springer Ser. Solid-State Sci., Vol.75 (Springer, Berlin, Heidelberg 1989)
8.20 J.R. Chelikowsky, S.G. Louie, M.L. Cohen: Solid State Commun. **20**, 641 (1976)
8.21 S.G. Louie, J.R. Chelikowsky, M.L. Cohen: Phys. Rev. B **15**, 2154 (1977)
8.22 S.G. Louie, J.R. Chelikowsky, M.L. Cohen: J. Vac. Sci. Technol. **13**, 790 (1976)
8.23 L.J. Brillson: Phys. Rev. Lett. **40**, 260 (1978)
8.24 H.H. Wieder: J. Vac. Sci. Technol. **15**, 1498 (1978)
8.25 R.H. Williams, R.R. Varma, V. Montgomery: J. Vac. Sci. Technol. **16**, 1418 (1979)
8.26 W.E. Spicer, P.W. Chye, P.R. Skeath, I. Lindau: J. Vac. Sci. Technol. **16**, 1422 (1979)
8.27 R.A. Allen, O.F. Sankey, J.D. Dow: Surf. Sci. **168**, 376 (1986)
8.28 A. Förster, H. Lüth: Surf. Sci. **189**, 190 (1987)
8.29 H. Brugger, F. Schäffler, G. Abstreiter: Phys. Rev. Lett. **52**, 141 (1984)
8.30 A. Zur, T.C. McGill, D.L. Smith: Phys. Rev. B **28**, 2060 (1983)
8.31 C.B. Duke, C. Mailhiot: J. Vac. Sci. Technol. B **3**, 1170 (1985)
8.32 R. Ludeke, G. Landgren: Phys. Rev. B **33**, 5526 (1986)
8.33 R. Ludeke, D. Straub, F.J. Himpsel, G. Landgren: J. Vac. Sci. Technol. A **4**, 874 (1986);
8.34 D.E. Eastman, T.C. Chiang, P. Heimann, F.J. Himpsel: Phys. Rev. Lett. **45**, 656 (1980)
8.35 J.L. Freeouf, J.M. Woodall: Appl. Phys. Lett. **39**, 727 (1981)
8.36 J.L. Freeouf, J.M. Woodall: Surf. Sci. **168**, 518 (1986)
8.37 J.L. Freeouf: Surf. Sci. **132**, 233 (1983)
8.38 H.B. Michaelson: J. Appl. Phys. **48**, 4729 (1977)
8.39 P.S. Ho, G.W. Rubloff: Thin Solid Films **89**, 433 (1982)
8.40 M.C. Bost, J.E. Mahan: J. Appl. Phys. **58**, 2696 (1985); ibid. **64**, 2034 (1988); ibid. **63**, 839 (1988);
A. Borghesi, A Piaggi, A. Franchini, G. Guizetti, F. Nava, G. Santoro: Europhys. Lett. **11**, 61 (1990)
8.41 A. Rizzi, H. Moritz, H. Lüth: J. Vac. Sci. Technol. A **9**, 912 (1991)
8.42 R. Purtell, J.G. Clabes, G.W. Rubloff, P.S. Ho, B. Reihl, F.J. Himpsel: J. Vac. Sci. Technol. **21**, 615 (1982)
8.43 G.W. Rubloff: Surf. Sci. **132**, 268 (1983)
8.44 J.M. Andrews, J.C. Phillips: Phys. Rev. Lett. **35**, 56 (1975)
8.45 S.M. Sze: *Physics of Semiconductor Devices*, 2nd ed. (Wiley, New York 1981)
8.46 H. Morkoc: "Modulation doped $Al_xGa_{1-x}As$/GaAs Heterostructures" in *The Technology and Physics of Molecular Beam Epitaxy* ed. by E.H.C. Parker (Plenum, New York 1985) p.185

8.47 G. Bastard: *Wave Mechanics Applied to Semiconductor Heterostructures* (Les Edition de Physique, Paris 1988);
L. Esaki: "Compositional Superlattices" in *The Technology and Physics of Molecular Beam Epitaxy* ed. by E.H.C. Parker (Plenum, New York 1985) p.185
8.48 B.J. van Wees, H. van Houten, C.W.J. Beenakker, J.W. Williamson, L.P. Kouwenhoven, D. van der Marel, C.T. Foxon: Phys. Rev. Lett. **60**, 848 (1988)
8.49 M.A. Paalonen, D.C. Tsui, A.C. Gossard: Phys. Rev. B **25**, 5566 (1982)
8.50 M. Cardona, N. Christensen: Phys. Rev. B **35**, 6182 (1987)

Chapter 9

9.1 J.N. Israelachvili, D. Tabor: Van der Waals Forces: Theory and Experiment, Prog. Surf. Membrane Sci. **7**, 1 (1973)
9.2 J.N. Israelachvili: Quart. Rev. Biophys. **6**, 341 (1974)
9.3 E. Zaremba, W. Kohn: Phys. Rev. B **15**, 1769 (1977)
9.4 T.B. Grimley: Theory of Chemisorption, *The Chemical Physics of Solid Surfaces and Heterogeneous Catalysis*, Vol.2, ed. by D.A. King, D.P. Woodruff (Elsevier, Amsterdam 1983) p.333
9.5 E.W. Plummer, T.N. Rhodin: J. Chem. Phys. **49**, 3479 (1968)
9.6 E. Bauer: In *The Chemical Physics of Solid Surfaces and Heterogeneous Catalysis*, Vol.3, ed. by D.A.King, D.P. Woodruff (Elsevier, Amsterdam 1984) p.1
9.7 E.P. Gyftopoulos, J.D. Levine: J. Appl. Phys. **33**, 67 (1962)
9.8 J. Topping: Proc. R. Soc. London A **114**, 67 (1927)
9.9 A. Spitzer, H. Lüth: Surf. Sci. **120**, 376 (1982)
9.10 M. Mattern-Klosson: Photoemissionsspektroskopie zur Untersuchung der Schottky-Barrieren von Sn und Sb auf GaAs(110). Dissertation, Aachen Technical University (RWTH)
M. Mattern-Klosson, H. Lüth: Solid State Commun. **56**, 1001 (1985)
9.11 L.A. Bol'shov, A.P. Napartovich, A.G. Naumovets, A.G. Fedorus: Uspekhi Fiz. Nauk **122**, 125 (1977) [English transl.: Sov. Phys. - Uspekhi **20**, 432 (1977)]
9.12 D.T. Pierce, F. Meier: Phys. Rev. B**13**, 5484 (1977)
9.13 T.S. Rahman, D.L. Mills, J.E. Black, J.M. Szeftel, S. Lehwald, H. Ibach: Phys. Rev. B**30**, 589 (1984)
9.14 R.H. Fowler, E.A. Guggenheim: *Statistical Thermodynamics* (Cambridge Univ Press, Cambridge 1949)
9.15 R.J. Behm, K. Christmann, G. Ertl: Solid State Commun. **25**, 763 (1978)
9.16 A.R. Kortan, R.L. Park: Phys. Rev. B **23**, 6340 (1981)
9.17 J.M. Thomas, W.J. Thomas: *Introduction to the Principles of Heterogeneous Catalysis* (Academic, New York 1967);
D. Hayword, B. Trapnell: *Chemisorption* (Butterworths, London 1964)

Subject Index